Physics and Mathematics of Strings

Vadim Knizhnik (20 February 1962 – 15 December 1987)

Memorial Volume for Vadim Knizhnik

Physics and Mathematics of Strings

Editors

L. Brink

D. Friedan

A. M. Polyakov

World Scientific
Singapore • New Jersey • London • Hong Kong

Published by

World Scientific Publishing Co. Pte. Ltd.
5 Toh Tuck Link, Singapore 596224
USA office: 27 Warren Street, Suite 401-402, Hackensack, NJ 07601
UK office: 57 Shelton Street, Covent Garden, London WC2H 9HE

British Library Cataloguing-in-Publication Data
A catalogue record for this book is available from the British Library.

PHYSICS AND MATHEMATICS OF STRINGS
Memorial Volume for Vadim Knizhnik

ISBN-13 978-9971-5-0980-4
ISBN-10 9971-5-0980-6
ISBN-13 978-9971-5-0981-1 (pbk)
ISBN-10 9971-5-0981-4 (pbk)

FOREWORD

Dima Knizhnik appeared in our community like a meteor. In the fall of '83 he approached me and asked me to help him with some problem. He was at his last year at the Moscow Institute for Physics and Technology. At that time Weigman and I found the Bethe–*Ansatz* solution to the Wess–Zumino–Novikov–Witten model. Although we got some anomalous dimensions, the solution was clumsy and I was planning to get by studying "current blocks", analogous to the conformal blocks of Virasoro algebra. So I asked Dima to analyze these "current blocks," thinking to put them together to form minimal models of current algebra *etc.* So, I thought that if he was really good, then in half a year he will (with my help) compute the spectrum of anomalous dimensions.

Instead, Dima appeared with the answer for this spectrum in two weeks. He did not follow my conservative suggestions; instead, he invented Sugawara energy momentum tensors and solved the problem. My first reaction was very subdued — I thought he was wrong and told him so. But Dima was a very self-confident man — and, also, he was right. So in a couple of days I gave up. He proceeded in a collaboration with Zamolodchicov to investigate the problem and finally they transformed a brilliant guess into a beautiful and complete theory.

After that incident, I always believed in his guesses. That was the way he worked — he got the right answers out of nowhere. After that followed a period of finding proofs. He was incredibly fast both in solving problems and in learning new physics and mathematics. It has been very obvious to everybody that in the close future he was going to become a superstar. His death is a tragedy for the entire physics community, and it has certainly delayed developments of physics.

After his death I received his notes. The last entry was just a day before. He was fascinated by the connection of gauge fields and gravity in two dimensions; he was obviously moving ahead very fast, working very hard and was very happy (judging by his exclamation marks after an especially nice formula).

A. M. Polyakov

There are some things that never die...

When the message about Dima Knizhnik's death came from Sasha Polyakov one gloomy day the winter before last, it was almost a repetition of a situation I had encountered before. One day in 1980 when I came to my office there was a telex on my desk saying that my oldtime friend Joël Scherk was dead. In both cases I had seen the person shortly before his death. In October 1987 some of us had been invited to Japan and for Dima this was the first time in his life that he was able to go abroad. I think it is fair to say that he took the Japanese physicists by a storm. With his appearance as a young Russian bear he delivered several lectures and there did not seem to be any impedance in his scientific work. He computed the measure in the two-loop graph by methods for which parts only existed in his brain and he was versatile in almost any subject connected to the physics and mathematics of strings. I knew Dima from visits to Moscow and was fully aware of his capacity but for someone seeing him for the first time he must have made an unforgettable impression. The mere fact that he was only twenty-five years old and looked physically very strong made the impression still stronger that here was a man who would be a leader in physics for many years to come. Two months later he was dead, one of these incomprehensible acts of the universe that we so intensely want to understand.

Joël Scherk also started out as a young genius. When he was twenty-five he was already a leading figure in dual models as string theory was called at that time. He had renormalized the one-loop graph, he had introduced the zero-slope limit, which became instrumental for convincing an older generation that the dual amplitudes were generalizations of field theory amplitudes, and he had had a finger in many of the other developments too. Also he was a scientist for which there was no impedance. He used to come to his office around ten o'clock. He then took up his pad and wrote continuously except for a lunch break up to five o'clock when he put down his pad in his desk and went home. In the evenings he often studied Chinese history or some similar subject very remote from physics. There never seemed to be any friction in his pen. Joël, however suffered from a difficult case of diabetes which he had contracted in the beginning of his career. The disease influenced him quite a lot. It was only the scientific mind that was completely intact. In this sense his death at the age of thirty-three was not so surprising even though it was a great shock to his friends and to the scientific community. Also that time I had met him a month earlier when we had shared a hotel room at a meeting in Erice. He did not look well and when I returned home I wrote a letter to John Schwarz expressing my worry for Joël and that we must do something. However, we never got the time to do anything.

Both deaths have been serious blows to the developments of physics. However, even so life goes on and the progress of science, if slowed down is irrevocable. We should though not forget neither Joël nor Dima and we shall remember them as leaders of their generations who did pioneering contributions and who would certainly have done even greater achievements had they had the luck of living longer.

This book is a contribution to the memory of Dima Knizhnik written by physicists who met him and whom he respected during his very short scientific career. It is my personal reflection that we should also bear Joël in remembrance at this occasion. I am sure that all the contributors agree with me here. Both Joël and Dima are dead but the memories of them are very much alive.

Göteborg in June, 1989

Lars Brink

FOREWORD

Daniel Friedan

I met Dima Knizhnik only three times – first in Moscow in the spring of 1983, again in Moscow at the Landau-Nordita meeting in June, 1984 and, finally, at the Yukawa Symposium in Kyoto in October, 1987. Because of the political situation in those years, direct contacts were infrequent between physicists working in the Soviet Union and physicists working in the U.S. or even Europe.

On several occasions Dima and I were thinking independently about the same subjects at the same times, usually from somewhat different points of view. It might have been interesting if we had had more freedom to interact. Beginning in 1984, Steve Shenker and I made Dima a standing offer of a position, but he was not able even to consider visiting until just before his death.

Consequently, I was forced to know Dima mostly from his published work. It is clear that he was in a state of explosive intellectual growth throughout his short career. It is impossible to imagine where time might have led him.

Theoretical physics is, by and large, an improvisational ensemble work, although the psychology of physicists seems to require that the official history be written otherwise. When such a creative voice as Dima Knizhnik is lost, we are all irreversibly diminished.

CONTENTS

GAUGE TRANSFORMATIONS AND DIFFEOMORPHISMS *

A. M. Polyakov†

Center for Theoretical Physics
Laboratory for Nuclear Science
and Department of Physics
Massachusetts Institute of Technology
Cambridge, Massachusetts 02139 U.S.A.

* This work is supported in part by funds provided by the U. S. Department of Energy (D.O.E.) under contract #DE-AC02-76ER03069.

† Permanent address:

I. INTRODUCTION

If we still harbor hope for understanding string dynamics, it is absolutely unavoidable to learn more about quantum gravity in two dimensions, representing internal fluctuations of the string's world sheet. Two years ago, I found a rather unexpected gauge symmetry (in the $SL(2, R)$ group) which is present in this problem. After that, Knizhnik, Zamolodchicov and I computed a set of fractal dimensions of random surface which appeared to be connected with the weights of representations of $SL(2, R)$ current algebra. It has been very gratifying to find complete agreement with results on discrete lattice models of random surfaces, developed by Midgal, Kazakov *et al.*[1]

However, the geometrical origin of $SL(2, R)$ remained unclear. In this paper I shall try to gain some understanding of this question. Namely, it will be demonstrated how to get diffeomorphisms from the restricted $SL(2, R)$ gauge transformations. The mechanism resembles well-known phenomenon — transformation of isotopic spin into ordinary spin in the field of magnetic monopoles. This happens because the field of the monopole "solders" isotopic space and ordinary space. In our case we will take a certain background gauge field, which performs the same task and show that residual gauge transformations in this background are precisely diffeomorphism. After that, connection between the Wess–Zumino–Novikov–Witten action (WZNW) and its gravitational analogue will come automatically. Another consequence will be a "composition formula" for the gravitational action.

The same kind of relations exist for other groups and other gauge restrictions. We shall show that there are two ways of restricting a gauge in the $SL(3)$ case.

One leads to W-algebra (which might have been expected from Drinfeld–Sokolov[2] and Belavin's works, while the other surprisingly gives some new algebra with two spin-3/2 generators, perhaps connected to $N = 2$ superconformal algebra.

The main stimulus for finding this gauge representation lies in the hope of mastering the case of $c > 1$ in quantum gravity. It is possible that in this strong gravity "region description" in terms of metric tensor breaks down and gauge fields should become fundamental variables (just as the π-meson variables in the σ-model are not good when the symmetry is restored). If so, we encounter one of the most exciting situations in physics.

II. DIFFEOMORPHISMS OUT OF GAUGE TRANSFORMATIONS

Let us consider two-dimensional gauge fields $A^a_\pm$, where $a = \pm 0$ are isotopic indices for $SL(2,R)$ algebra, and we describe ordinary space by light-cone coordinates $x^\pm$. One of the standard ways to fix a gauge is to set $A^a_- = 0$. After that, the remaining gauge freedom consists of functions depending on x^+ only. Let us now do something different. Let us fix a gauge only partially by imposing two conditions instead of three:

$$A^-_- = 1 \ ; \quad A^0_- = 0 \ ; \quad A^+_- \equiv T_{--} \quad \text{– not fixed} \ . \tag{2.1}$$

It is obvious that the remaining gauge freedom contains one independent function of x and x^-. *What comes as a geometrical surprise* is that these residual gauge transformations act on T_{--} by Virasoro group. Let us check this. The general form of the gauge transformation is given by:

$$\delta A_-^+ = \partial_- \epsilon^+ + A_-^+ \epsilon^0 - A^0 \epsilon^+$$
$$\delta A_0^0 = \partial_- \epsilon^0 + A_-^+ \epsilon^- - A_-^- \epsilon^+ \tag{2.2}$$
$$\delta A_-^- = \partial_- \epsilon^- + A_-^0 \epsilon^- - A_-^- \epsilon^0$$

conditions of preserving (2.1) are:

$$\delta A_-^- = 0 \ , \qquad \delta A_0^0 = 0 \ . \tag{2.3}$$

Substituting (2.3) and (2.1) into (2.2) we obtain

$$\epsilon^0 = \partial_- \epsilon^-$$
$$\epsilon^+ = \partial_- \epsilon^0 + T_{--} \epsilon^- = \left(\partial_-^2 + T_{--}\right) \epsilon^- \tag{2.4}$$
$$\delta T_{--} = \partial_-^3 \epsilon^- + \epsilon^- \partial_- T_{--} + 2(\partial_- \epsilon^-) T_{--} \ .$$

Here we recognize the standard action of Virasoro algebra on energy momentum tensors. What has happened? The main point is that by choosing a "background field" $A_-^- = 1$ we have soldered isotopic space and x space. That means that all fields in such a background acquire extra orbital spin equal to their isotopic spin (this phenomenon is well-studied in the case of magnetic monopoles). Because of that, the field A_-^+, which originally had spin-1 and isotopic spin-1 will have orbital spin-2. We denoted it by T_{--} and indeed proved the standard Virasoro transformation law.

It is known that the Virasoro group is not a subgroup of $\widehat{SL}(2,R)$. That means that any attempt to constrain parameters of $\widehat{SL}(2,R)$ is not supposed to lead to Virasoro. In our case, the key point is that the constraint (2.4) is field-dependent.

III. CONSEQUENCES FOR WZNW ACTIONS

In this section we will show that the geometrical observation of the previous section has dynamical consequences. Namely, it implies connection between Wess–Zumino–Novikov–Witten action (WZNW) for the $SL(2,R)$ group and gravitational action, introduced in Ref. [3]. Let us consider the gauge invariant functional

$$S(A_+, A_-) \sim \log \mathcal{D}\mathrm{et}(\not{\partial} + \not{A}) \tag{3.1}$$

for Dirac operators in an $SL(2,R)$ gauge field. It is known that it has a form:[4]

$$S(A_+, A_-) = \Gamma_+(A_+) + \Gamma_-(A_-) = \mathrm{Tr} \int A_+ A_- \tag{3.2}$$

with $\Gamma_\pm$ being the WZNW action:

$$\Gamma_\pm(A_\pm) \sim \int \left[\mathrm{Tr}\left(\Omega^{(\pm)-1} d\Omega^{(\pm)}\right)^2 \pm \mathrm{Tr}\left(\Omega^{(\pm)-1} d\Omega^{(\mp)}\right)^3 \right] \tag{3.3}$$

where we introduced Ω by:

$$\partial_\pm \Omega_+^{(\pm)} \Omega^{(\pm)-1} = A_\pm \quad . \tag{3.4}$$

Now, we are going to exploit gauge restriction (2.1) in (3.2). We expect to obtain an action which possesses symmetry $x^- \Longrightarrow f(x^-, x^+)$ and must be connected with the "gravitational WZNW" of Ref. [3]. Let us denote $A_+^- \equiv h_{++}$ and, as before $A_-^+ \equiv T_{--}$, and begin with the term $\Gamma_-(A_-)$ in (3.2). We naturally introduce:

$$\begin{aligned} S_-(T_{--}) &= \Gamma_-\left(A_-^+ = T_{--};\; A_-^0 = 0;\; A_-^- = 1\right) \\ &= \int \left\{ \mathrm{Tr}\,(\Omega^{-1}\Omega)^2 - \mathrm{Tr}\,(\Omega^{-1} d\Omega)^3 \right\} \end{aligned}$$

where:

$$\partial_- \Omega = \begin{pmatrix} 0 & T_{--} \\ 1 & 0 \end{pmatrix} \Omega \ . \tag{3.5}$$

Let us compute the variation of S_-:

$$\begin{aligned} \delta S_-(T_{--}) &= \int \frac{\delta \Gamma_-}{\delta A_-^+} \delta A_-^+ \\ \delta \Gamma_- &= \int T_2(J_+ \delta A_-) \\ &= \int T_2(J_= \nabla_- \epsilon) = \int (A_- \partial_+ \epsilon) \\ &= \int (\partial_+ A_-^+) \cdot \epsilon^- = \int \epsilon^- (\partial_+ T_{--}) \ . \end{aligned} \tag{3.6}$$

If we introduce now a field g_{++} by the relation:

$$\partial_-^3 g_{++} + g_{++} \partial_- T_{--} + 2(\partial_- g_{++}) T_{--} = \partial_+ T_{--} \quad m$$

we obtain after integration by parts in (3.6):

$$\delta S_-(T_{--}) = \int (g_{++} \delta T_{--}) \tag{3.8}$$

and hence g_{++}, satisfying anomaly equation (3.7), is a component of the metric tensor. If we now introduce a Legendre transform for the action $S_-(T_{--})$:

$$S_+(h_{++}) \underset{(\mathrm{def})\{T_{--}\}}{=} \min \left(S_-(T_{--}) - \int T_{--} h_{++} \right) \tag{3.9}$$

then, according to standard properties of this transformation we get:

$$\delta S_+(h_{++}) = \int \theta_{--} \delta h_{++} \tag{3.10}$$

where the energy-momentum tensor θ_{--} satisfies the same equation as (3.7):

$$\partial_+ \theta_{--} - h_{++} \partial_- \theta_{--} - 2(\partial_- h_{++}) \theta_{--} = \partial_-^3 h_{++} \tag{3.11}$$

This is just the equation, which in Ref. [3] determined "gravitational WZNW."

It is obvious from these variational formulas that the action

$$S(h_{++}, T_{--}) = S_+(h_{++} + S_-(T_{--}) - \int T_{--} h_{++} \tag{3.12}$$

is invariant under Virasoro symmetry:

$$\begin{aligned} \delta T_{--} &= \partial_-^3 \epsilon^- + \epsilon^- \partial_+ T_{--} + 2(\partial_+ \epsilon^-) T_- \\ \delta h_{++} &= \partial_+ \partial^- + \epsilon^- \partial_- h_{++} - (\partial_+ \epsilon^-) h \end{aligned} \tag{3.13}$$

This result can be checked explicitly, but also can be understood as a residual gauge symmetry of (3.2), when A_- is restricted by (2.1). By substituting this expression into (3.2) we also become able to identify:

$$\begin{aligned} S_+(h_{++}) &= \min_{\{A_+^+, A_+^0\}} \left\{ \Gamma_+(A_+^+, A_+^0, h_{++}) - \int A_+^+ \right\} \quad . \\ S_+(T_{--}) &= \Gamma_-(A_-^+ = T_{--}; A_-^0, A_-^- = 1) \end{aligned} \tag{3.14}$$

The first of these formulas can be understood as follows. According to Ref. [3], correlation functions of h_{++} are expressed through that of $SL(2, R)$ currents. That means that the effective action for h_{++} is given by:

$$e^{-S_+ h_{++}} = \int \mathcal{D}B \, e^{-\Gamma_+(B)} \times \delta\left(h_{++} - \left(B_+^- - 2B_+^0 x^- + B_+^+ (x^-)^2\right)\right) \quad . \tag{3.15}$$

(For the moment, we ignore the central charges and their renormalizations. Formally, all I say is true in the classical limit only, and will be corrected (trivially) later.) The expression inside the δ-function looks as something gauge-rotated, as indeed it is. It is easy to check that gauge matrix:

$$\Omega = \begin{pmatrix} 1 & x^- \\ 0 & 1 \end{pmatrix} \tag{3.16}$$

which transforms B into A:

$$A = \Omega^{-1} B \Omega \ , \tag{3.17}$$

transforms (3.15) into the expression

$$e^{-S_+(h_{++})} = \int \mathcal{D}\mathcal{A}\, e^{-\left(\Gamma_+(A) \mp \int A_+^+\right)} \times \delta(h_{++} - A_+^-) \tag{3.18}$$

which in the classical limit gives (3.14) again.

Finally, let us present the solutions of (3.7) and (3.11) in the explicit form (they have been given in Ref. [3]).

If we choose

$$\partial_+ f = h_{++} \partial_- f \tag{3.19}$$

then θ_{--}, satisfying (3.11), is just the Schwarz derivative

$$\theta_{--} = \mathcal{D}_- f = \frac{\partial_-^3 f}{\partial_- f} - \frac{3}{2} \frac{(\partial_-^2 f)^2}{(\partial_- f)^2} \ , \tag{3.20}$$

and, analogously, for T_{--} and g_{++}.

Let us recapitulate. We have two actions, related by the Legendre transform, $S_+(h_{++})$ and $S_-(T_{--})$. The first action, S_+, according to (3.15), performs geometrical quantization of the $SL(2,R)$ group. By that we mean that correlation functions, computed with this action are expressed through that of $SL(2,R)$ current algebra. In the same sense, the dual action $S_-(T_{--})$ serves for geometrical quantization of the Virasoro algebra. In some sense it can be said that the Virasoro algebra appears as a Fourier transform of $SL(2,R)$ and vice versa.

From the gauge invariance (3.13) of the action (3.12) we can readily derive the analogue of the composition formula. It stems from the fact that the transformation

(3.13) is an infinitesimal version of the following finite transformations. Let us define new fields to replace T_{--} and h_{++}:

$$\partial_+ F = h_{++}\partial_- F$$
$$T_{--} = \mathcal{D}_- G \ . \tag{3.20}$$

Then the finite version of (3.13) has the form:

$$F(x^+,x^-) \Longrightarrow F\left(x^+, f(x^+,x^-)\right)$$
$$G(x^+,x^-) \Longrightarrow G\left(x^+, (x^+,x^-)\right) \tag{3.21}$$
$$f(x^+,x^-) \approx x^- + \epsilon^-(x^+,x^-) \ .$$

We come to the conclusion that the functional

$$W(F,G) = S_+(F) + S_-(G) - \int \left(\frac{\partial_+ F}{\partial_- F}\right) \mathcal{D}_- G \tag{3.22}$$

is invariant under (3.21):

$$W(F,G) = W(F \circ f, g \circ f) \tag{3.23}$$

(with $F \circ f \equiv F\left(x^+, f(x^+, x^-(\right)$. This invariance means that

$$W(F,G) = S_+(F \circ G^{-1}) = S_-(G \circ F^{-1}) \tag{3.24}$$

In particular,

$$S_+(F) = S_-(F^{-1}) \cdot \tag{3.25}$$

Here by $G^{-1}(x^+,x^-)$ we denoted the inverse function

$$G\left(x^+, G^{-1}(x^+,x^-)\right) = x^- \ . \tag{3.26}$$

Finally we get the useful composition formula

$$S_+(F \circ G^{-1}) = S_-(G \circ F^{-1}) = S_+(F) + S_-(G) - \int \frac{\partial_+ F}{\partial_- F} \mathcal{D}_- G \tag{3.27}$$

IV. THE CASE OF SL(3)

It is tempting to exploit the idea of "soldering" for other gauge groups. In this section I shall analyze the case of $SL(3)$. Whether this case has any practical interest is not clear, but it extends our understanding of possible symmetries.

In $SL(3)$ we have eight gauge fields $\mathcal{A} = (A^{\pm 0}, B^{\pm}, C^{\pm}, \mathcal{D}^0)$. Here $A^{\pm 0}$ form $SL(2)$ subgroups and have quantum numbers of π-mesons, B and C correspond to K-mesons and $\mathcal{D}^0$ to η-mesons. Commutation relations for $SL(3)$ are most easily expressed using "bosonic quarks" u, d, s with commutators:

$$[u, \bar{u}] = [d, \bar{d}] = [s, \bar{s}] = 1 \tag{4.1}$$

and with relations:

$$\begin{aligned} &A^+ = \bar{d}u && B^+ = \bar{s}u && C^+ = \bar{d}s \\ &A^- = \bar{u}d && B^- = \bar{u}s && C^- = \bar{s}d \\ &A^0 = \tfrac{1}{2}(\bar{d}d - \bar{u}u) && D^0 = \tfrac{1}{2}(\bar{d}d + \bar{u}u - 2\bar{s}s) \end{aligned} \tag{4.3}$$

with these relations one easily obtains the explicit form of the gauge transformation law

$$\delta\mathcal{A} = \partial\epsilon + [\mathcal{A}, \epsilon] \tag{4.3}$$

with

$$\epsilon = (\epsilon^{\pm 0}, \xi^{\pm}, \eta^{\pm}, \omega^0) \tag{4.4}$$

which generalizes (2.2). (I skip these long formulas.)

Now comes the question of possible solderings. One option would be to choose

$$\begin{aligned} &B_-^- = C_-^- = 1 \ ; \quad A_-^- = 0 \ ; \quad A_-^0 = 0 \\ &T = C_-^+ = B_-^+ \ ; \quad \mathcal{D}_-^0 = 0 \ . \quad W = A_-^+ \end{aligned} \tag{4.5}$$

With this restriction we see that we can identify the isotopic spin 1/2 (for the B fields) with orbital spin-1. In this case the field $A_-^+ = W$ will have an extra spin-2 (since its isospin is 1) and total spin = 3. The field T will have spin-2. As a result, soldering of degree 2 (extra spin = 2× isotopic spin) leads to the Zamolodchicov W-algebra.

The connection of $SL(3)$ and W-algebra was anticipated before Ref. [5] and we shall not discuss it anymore. The only general point to be added is that from the geometrical point of view, quadratic relations of Zamolodchicov means the following. We can compute the commutator of our transformations

$$[\delta_{\epsilon_1}\delta_{\epsilon_2}]\,\mathcal{A} = \delta_{\epsilon_3}\mathcal{A} \tag{4.6}$$

In Lie algebra cases we have a certain composition law

$$\epsilon_3 = \epsilon_3(\epsilon_1, \epsilon_2) \tag{4.7}$$

($e.g.$ $\epsilon + 3 = \epsilon_1'\epsilon_2 - \epsilon_2'\epsilon_1$ for Virasoro algebra).

In cases of W-algebra, (4.7) is replaced by

$$\epsilon_3 = \epsilon_3(\epsilon_1, \epsilon_2, \text{ fields}) \tag{4.8}$$

and, also, transformation laws of the fields is non-linear. Explicit formulas for the W-algebra can be easily obtained from (4.3) and (4.5).

Now let us consider soldering of degree 1. In this case we impose the following constraints:

$$\begin{aligned} &A_-^- = 1 \;\; ; \quad B_-^- = 0 \;\; ; \quad C_-^- = 0 \;\; ; \quad A_0^0 = 0 \\ &A_-^+ = T \;\; ; \quad B_-^+ = B \;\; ; \quad C_-^+ = C \;\; ; \quad \mathcal{D}_-^0 = \mathcal{D} \;\; . \end{aligned} \tag{4.9}$$

Now, we have the following quantum numbers: T has spin-2; B and C have spin-3/2 $(1_{\text{orb}} + (1/2)_{\text{iso}})$ and $\mathcal{D}$ has spin-1. We find the spin content of the $N = 2$ super-Virasoro algebra in this case. Fields B and C are bosonic, but we know that in two dimensions they can be equivalent to fermionic fields. Let us present transformation laws for this case. From the conditions

$$\delta A^- = \delta C^- = \delta B^- = \delta A^0 = 0$$

one finds relations:

$$\begin{cases} \epsilon^0 = \partial\epsilon^- \\ \epsilon^+ = \left(-\frac{1}{2}\partial^2 + T\right)\epsilon^- - \frac{1}{2}(B\xi^- + C\eta^-) \\ \eta^+ = \left(\partial - \frac{3}{2}\mathcal{D}^0\right)\xi^- + C\epsilon^- \equiv \nabla\xi^- + C\epsilon^- \\ \xi^+ = -\left(\partial_+\frac{3}{2}\mathcal{D}^0\right)\eta^- + B\epsilon^- \end{cases} \tag{4.10}$$

Transformation laws for the fields are

$$\begin{aligned} \delta T &= -\frac{1}{2}\partial^3\epsilon + \epsilon\partial T + 2(\partial\epsilon)T - \frac{1}{2}\partial(B\xi c\eta) + B\nabla\xi + C\nabla\eta \\ \delta B &= -\nabla^2\eta + \nabla(B\epsilon) - \frac{3}{2}\omega^0 B\eta T + \frac{1}{2}(\partial\epsilon)B \\ \delta C &= \nabla^2\xi + \nabla(C\epsilon) + \frac{3}{2}\omega^0 C \\ &\quad - \xi T + \frac{1}{2}(\partial\epsilon)C \\ \delta\mathcal{D} &= \partial\omega^0 + B\xi - C\eta \ . \end{aligned} \tag{4.11}$$

The question of whether this extension of the Virasoro algebra is equivalent to $N = 2$ super-Virasoro algebra is not quite clear, and must be further investigated. Here I shall outline the scheme of such an investigation.

The first step should be to replace the transformation laws by the operator products, by the use of the relations:

$$\delta\phi = \left[\oint (T\epsilon + B\xi + C\eta)\, dx^{-}, \phi\right] \tag{4.12}$$

(where ϕ stands for T, B, C). From (4.11) and (4.12) one extracts the operator algebra. Schematically (modulo coefficients) it looks as the following:

$$\begin{aligned}
&T(z)T(0) = x^{-4} + z^{-2}T) -_{+} z^{-1}T'(0) + \ldots \\
&\mathcal{D}(z)\mathcal{D}(0) = z^{-2} \\
&T(z)B(0) = x^{-2}B + z^{-1}(\partial_z + \mathcal{D}B) \\
&T(z)C(0) = z^{-2}C + z^{-1}(\partial_z - \mathcal{D}C) \\
&T(z)\mathcal{D}(0) = 0 \\
&J(z)B(0) = z^{-1}B \\
&J(z)C(0) = z^{-1}C \\
&B(z)C(0) = z^{-3} + z^{-2}\mathcal{D} + z^{-1}(T + \mathcal{D}^2 + \partial_z\mathcal{D}) \ .
\end{aligned} \tag{4.13}$$

My conjecture is that it is possible to redefine fields (in particular to replace $T \to T + \mathcal{D}^2$ so that (4.13) will become $N = 2$ SUSY. However, I have not proved that.

ACKNOWLEDGEMENTS

Many of the ideas of this paper were discussed with Dima Knizhnic. In particular, he found the first of the relations (3.14) and helped me to find the exact form of the composition formula (3.27). He also attracted my attention to the papers by Drinfeld and Sokolov[2] and by Belavin[5] with which the subject of this paper is connected (although a precise relationship is not clear to me).

Recently, papers by Alekseev and Shatashvili[6] and by Bershadsky and Ooguiri[7] have appeared, which partly overlap this article. Nevertheless, I hope that it is still worthwhile to publish these results.

REFERENCES

1. V. Kazakov and A. Midgal, Niels Born Institute preprint NBI–HE–88/28 and references therein.

2. V. Brienfeld and V. Sokolov, *Journal of Soviet Mathematics* **30**, 1975 (1984).

3. A. Polyakov, *Mod. Phys. Lett* **A2**, 893 (1987).

4. A. Polyakov and P. Wiegman, *Phys. Lett.* **141B**, 233 (1984).

5. A. Belavin, in *Proceedings of the Yakawa Memorial Symposium* N. Kawamoto and T. Kago, eds. (Nishinomiya, Japan, 1987).

6. A. Alekseyev and S. Shatashvily, LOMI preprint (1988).

7. M. Bershadsky and M. Ooguri, Institute for Advanced Study preprint IASSNS–HEP–89/09.

TOPICS IN CONFORMAL FIELD THEORY

L. ALVAREZ-GAUMÉ, AND G.SIERRA†
Theory Division, CERN, CH-1211 Geneva 23, Switzerland

C.GOMEZ*
Département de Physique Théorique, Université de Genève, Genève, Switzerland

†Address after October 1st: Instituto de Estructura de la Materia, CSIC, Serrano 119, Madrid, Spain.

*Permanent address: Departmento de Física, Universidad de Salamanca, Salamanca, Spain.

TABLE OF CONTENTS

In Memory of V. Knizhnik

1. Introduction

There has been au enormous amount of activity recently in Conformal field theory (for a review of standard material and references, see Refs. 1 and 2) and string Theory (for details and references see Refs. 3,4 and 5). In these notes we have selected a number of Topics partly because they do not appear in most reviews on the subject, and also because they contain some of the most recent research trends. We will begin with a general introduction in Sec. 1 which reviews some of the aspects of the seminal paper by Belavin, Polyakov and Zamolodchikov[6] that will be used in later sections. We have also collected here a number of properties of the braid group with n-strands [7] and its relation with monodromy and duality properties of Conformal Field Theories (CFT). The first classic examples of CFT appear in Sec. 3. There we study the minimal series and the unitary series[6,8], we present in detail the Coulomb gas representation of the correlation functions[9], null vectors, the Kac table of degenerate representations and the fusion rules. We derive the Rocha-Caridi[10] formula for the Virasoro characters of the minimal models. In Sec. 4 we study in some detail the Wess-Zumino-Witten (WZW) models[11,12,13], the representations of the relevant Kac-Moody algebras and the possible modular invariants[14] combination of characters for the SU(2) level k characters. Sec. 5 is dedicated to review the properties of the fusion algebra and formulate the Verlinde conjecture [15] and its proof by G. Moore and N. Seiberg [16,17,18]. We also present the rationality contraints in the conformal dimensions and central term of the Virasoro algebra from the assumption that the number of primary fields in the theory is finite[19,20]. Theories whose Hilbort space consists only of a finite number of representations are known as Rational Conformal Field Theories (RCFT) and they were introduced in Ref. 21 in the "Universal Moduli Space" formulation of CFT.

In Sec. 6 we concentrate on the duality properties of RCFT and motivate the use of quantum groups as the natural structure to characterize and classify RCFT[22,23,24,25]. The second part of this section gives a quick overview of the theory of Quantum Groups[26,27,28,29,30] and their representations specially when the deformation parameters are roots of unity[31,32,25]. In Sec. 7 we apply the techniques of quantum groups to derive the duality properties of WZW models, Rational Gaussian Theories and one of the more interesting results in the connection between RCFT and quantum groups which computes using only properties of the co-multiplication the modular transformation matrix S_{ij} determining the behavior of the characters under the modular transformation on the torus $\tau \to -1/\tau$[25]. This also allows us to obtain a proof of the Verlinde conjecture in this context and to connect with Witten's formulation of CFT and Knot and Link invariants in terms of Topological Field Theories in three dimensions[33]. In Sec. 8 we quickly review those aspects of integrable lattice models[34,35,36] related to Conformal Field Theories and knot invariants. This section is included for completeness and also in preparation for Sec. 9 where we detail the connection between knot invariants, Conformal Field Theory and 3D Chern Simons gauge theories. We first present some of the basic notions in Knot Theory[37,38] and connect with integrable models and Quantum Groups through Turaev's[39] construction of Knot invariants in terms of extended Yang-Baxter systems. Next following Jones original work[40,41] we show now the fusion rules of RCFT and the associated matrix S generates a collection of Knot invariants. This will clarify what are the few ingredients of a RCFT needed to construct Knot polynomials regardless of how one wishes to study these theories in detail. In the last subsection in Sec. 9 we present the rudiments of Witten's construction of Knot invariants in arbitrary three manifolds[33].

2. General Properties of a Conformal Field Theory

2.1 *Conformal Invariance and Conformal Ward Identities*

In this section we briefly review the basic properties of CFT in two dimension[6] (see also Ref. 1). For an early application of conformal invariance in field theory, see Ref. 42. To avoid infrared problems we consider field theories on a cylinder $S^1\times \mathbb{R}$, described by coordinates $(\sigma,\tau), 0\le\sigma\le 2\pi, -\infty<\tau<+\infty$. The light cone coordinates are defined as $x^{\pm}=\tau\pm\sigma$. Since the line element in two-dimensional Minkowski space has the form $ds^2=dx^+dx^-$, the group of transformations leaving the light cone invariant is infinite dimensional, and it is given by arbitrary reparametrizations of x^+ and x^- : $x^+=f(x^+), x^-=g(x^-)$. In the analysis of CFT it is convenient to rotate to Euclidean space and use radial quantization. Under Euclidean rotation:

$$\begin{aligned}\tau+\sigma &\to -i(\tau+i\sigma) = -iw\\ \tau-\sigma &\to -i(\tau-i\sigma) = -i\bar{w}\end{aligned}$$

mapping now the cylinder to the plane.

$$z=e^{w} \tag{2.1}$$

we find that the $\tau=0$ "surface" becomes the unit circle on the z-plane, $\tau=-\infty$ is mapped to $z=0$, and $\tau=+\infty$ to $z=+\infty$. The infinitesimal conformal transformations are generated by the vectors

$$l_n=z^{n+1}\frac{d}{dz}\quad \bar{l}_n=\bar{z}^{n+1}\frac{d}{d\bar{z}}\ . \tag{2.2}$$

A CFT is characterized by its scale invariace. As in any local field theory, scale invariance implies full conformal invariance. Scale invariance is equivalent to the vanishing of the trace of the energy-momentum tensor. In complex coordinates this means that $T_{z\bar{z}}=0$. Hence there are only two independent components of the energy-momentum tensor $T_{zz}, T_{\bar{z}\bar{z}}$ and their conservation law becomes:

$$\begin{aligned}\partial^z T_{zz}&=0 \qquad \partial_{\bar{z}}T_{zz}=0\\ \partial^{\bar{z}} T_{\bar{z}\bar{z}}&=0 \qquad \partial_{z}T_{\bar{z}\bar{z}}=0\end{aligned} \tag{2.3}$$

implying that $T_{zz}\equiv T(z)$ $(T_{\bar{z}\bar{z}}\equiv\bar{T}(\bar{z}))$ is a holomorphic (antiholomorphic) function of $z(\bar{z})$. The generators of infinitesimal conformal transformations are:

$$L_n=\oint_0\frac{dz}{2\pi i}z^{n+1}T(z) \tag{2.4}$$

and similarly for $\bar{L}_n$. The contour circles the origin only once. Its shape is irrelevant as a consequence of (2.3) and Cauchy's theorem. Using (2.4) we can write:

$$T(z) = \sum_{n\in\mathbb{Z}} L_n z^{-n-2} \quad \bar{T}(\bar{z}) = \sum_{\nu\in\mathbb{Z}} \bar{L}_n \bar{z}^{-n-2} . \tag{2.5}$$

Out of all the fields in the CFT we can distinguish some behaving like tensors under conformal transformations. These are called primary fields and they behave as $(h, \bar{h})$ tensors, i.e.

$$\phi_{h,\bar{h}}(z,\bar{z})dz^h d\bar{z}^h \tag{2.6}$$

is invariant under conformal transformations. If we use two different coordinates systems z, z', then $\phi_{h,\bar{h}}$ transforms according to:

$$\phi'_{h,\bar{h}}(z',\bar{z}') = \phi_{h,\bar{h}}(z,\bar{z})(\frac{dz'}{dz})^{-h}(\frac{d\bar{z}'}{d\bar{z}})^{-\bar{h}} \tag{2.7}$$

locality implies as we will see that $h - \bar{h}$ be an integer. Using (2.7) we can easily relate the mode expansions on the cylinder and on the plane. If we take for example $\bar{h} = 0$ and ϕ_n a holomorphic field, the standard Fourier expansion on the cylinder is:

$$\phi_h(w) = \sum_{n\in\mathbb{Z}} \phi_n e^{-nw} . \tag{2.8}$$

Using (2.1) and (2.7) we obtain the expansion on the plane:

$$\phi_h(z) = \sum_{n\in\mathbb{Z}} \phi_n z^{-n-h} \tag{2.9}$$

since the energy-momentum tensor has dimension 2, $h = 2$, and this also explains the expansion (2.5). The infinitesimal form of (2.7) is:

$$\begin{aligned} \delta_\epsilon \phi_{h,\bar{h}}\, , &= (\epsilon(z)\frac{d}{dz} + h\epsilon'(z))\phi_{h,\bar{h}} = [T(z), \phi_{h,\bar{h}}(z)] \\ T(\epsilon) &\equiv \oint_0 \epsilon(z)T(z) \end{aligned} \tag{2.10}$$

(from now on the contour integral symbol includes $dz/2\pi i$). Equation (2.10) can be expressed more conveniently in terms of the operator product expansion (OPE) between $T(z)$ and $\phi(z,\bar{z})$. Since $T(z)$ is analytic, the commutator in (2.10) is equivalent to a contour integral of $\epsilon(w)T(w)$ around $\phi(z)$:

$$[T(\epsilon), \phi_{h,\bar{h}}(z)] = \oint_z \epsilon(w)T(w)\phi_{h,\bar{h}}(z,\bar{z}) \tag{2.11}$$

since the contour around z can be chosen as small as we wish, the only contributions to (2.11) come from the singularities in the OPE. From (2.10) and (2.11) we obtain:

$$T(z)\phi(w) = \frac{h}{(z-w)^2}\phi(w) + \frac{1}{z-w}\partial_w\phi(w) + \text{regular} \ . \tag{2.12}$$

This is the operator form of the conformal ward identities. By the analyticity properties of $T(z)$ the correlation function

$$\langle T(z)\phi_1(z_1)\ldots\phi_N(z_N)\rangle \tag{2.13}$$

can be evaluated using (2.12). As a function of $z, T(z)$ is a meromorphic quadratic differential ($h = 2$) on the sphere. The only singularities appear at the points $z_1, \ldots, z_N$ and they are determined by the OPE (2.12). Hence,

$$\langle T(z)\phi_1(z_1)\ldots\phi_N(z_N)\rangle = \sum_{j=1}^{N}\left(\frac{h_j}{(z-z_j)^2} + \frac{1}{z-z_j}\frac{\partial}{\partial z_j}\right) \langle\phi_1(z_1)\ldots\phi_N(z_N)\rangle \tag{2.14}$$

The "in" vacuum $|0\rangle$ can be represented by the insertion of the unit operator at the origin of the z-plane. If we consider the holomorphic vectors without poles at $z = 0, z^{n+1}d/dz, n \geq -1$, and since the unit operator is a scalar, we conclude that:

$$L_n|0\rangle = 0 \qquad n \geq -1 \tag{2.15a}$$

as we will see later this ground state is invariant under the action of the Möbius group SL_2 (C) of conformal isometries of the sphere. Taking Hermitian conjugates, $|0\rangle^+ = \langle 0|, L_n^+ = L_{-n}$, therefore the out vacuum $\langle 0|$ is invariant under the infinitesimal conformal transformations regular at $z = \infty$:

$$\langle 0|L_{-n} = 0 \qquad n \geq -1 \ . \tag{2.15b}$$

The primary states are defined by:

$$|h, \bar{h}\rangle = \lim_{z,\bar{z}\to 0} \phi_{h,\bar{h}}(z, \bar{z})|0\rangle \ . \tag{2.16}$$

Writing (2.10) in components:

$$[L_n, \phi_n(z)] = \left(z^{n+1}\frac{d}{dz} + (n+1)z^n h\right)\phi_h(z) \tag{2.17}$$

we find that:

$$\begin{aligned} L_n|h,\bar{h}\rangle &= 0 \quad n>0 \\ L_0|h,\bar{h}\rangle &= h|h,\bar{h}\rangle \end{aligned} \tag{2.18}$$

hence every primary field generates a highest weight representation of the conformal algebra $\{L_n\}$. The states of the representation $V_{h,\bar{h}}$ are constructed by acting on the primary state $|h,\bar{h}\rangle$ with arbitrary polynomials in $\{L_{-n}, \bar{L}_{-m}, n,m \geq 1\}$. Such representations are known as Verma modules. In the operator language the states $L_{-n_1} \dots L_{-n_N}|h,\bar{h} >$ are represented as:

$$\oint_{C_1} z_1^{-n_1+1}T(z_1) \oint_{C_2} z_2^{-n_2+1}T(z_2) \dots \oint_{C_N} z_N^{-n_N+1}T(z_N)\phi_{h,\bar{h}}(z) \tag{2.19}$$

where $C_1 \supset C_2 \supset \dots \supset C_N \supset \{z\}$. The three generators $L_{\pm 1}, L_0$ are associated to infinitesimal Möbius transformations. They are obtained from the vector fields $1, z, z^2$. These are the conformal killing vectors on the sphere. The integrated form of the infinitesimal transformations is the group of fractional linear transformations $SL_2(\mathbb{C})$:

$$z' = \frac{az+b}{cz+d}, a,b,c,d \in \mathbb{C}, ad-bc = 1 \tag{2.20}$$

thus L_{-1} generates translations, L_0 generates scalar transformation, and L_{+1} generates special conformal transformations, and in particular $|0\rangle$ is SL_2 invariant. The OPE of $T(z)$ with itself is more complicated. The algebra generated by the vector fields (2.2):

$$[l_n, l_m] = (m-n)l_{n+m} \tag{2.21}$$

is not realised faithfully but projectively. This should be expected because in quantum mechanics the physical states are described by rays on a Hilbert space.

The central extension of (2.21) in the projective representation generated by the energy-momentum tensor is determined by the two point function of $T(z)$. By dimensional analysis and SL_2 invariance of $|0>$ together with the requirement that the energy-momentum tensor should vanish as $z \to \infty$ (no sources at ∞) we obtain:

$$\langle 0|T(z)T(w)|0\rangle = \frac{c/2}{(z-w)^4} \tag{2.22}$$

the constant c depends on the CFT. Hence the OPE between two T's is given by:

$$T(z)T(w) = \frac{c/2}{(z-w)^4} + \frac{2}{(z-w)^2}T(w) + \frac{1}{z-w}\partial_w T(w) + \text{regular} \quad (2.23)$$

implying that under infinitesimal transformation:

$$\begin{aligned} \delta_\varepsilon T(z) &= [T(\varepsilon), T(z)] = \oint_z \varepsilon(w)T(w)T(z) \\ &= \left(\varepsilon(z)\frac{d}{dz} + 2\varepsilon'(z)\right)T(z) + \frac{c}{12}\varepsilon'''(z) \end{aligned} \quad (2.24)$$

or in terms of L_n commutators.

$$[L_n, L_m] = (n-m)L_{n+M} + \frac{c}{12}(n^3 - n)\delta_{n+m,o} \quad (2.25)$$

and we obtain the Virasoro algebra. It is important to notice that the central terms in (2.25) vanishes for $n = 0, \pm 1$. This is reflected in the SL_2 invariance of the ground state. Equivalently, the inhomogeneous terms in (2.24) vanishes for $\varepsilon = 1, z, z^2$. This is useful in determining the finite transformation rule of $T(z)$ under conformal maps. Using the properties of the fractional linear transformations, one can show that under a finite transfromation $z = f(w)$:

$$\begin{aligned} T'(w) &= (\frac{dz}{dw})^2 T(z) + \frac{c}{12}S(z,w) \\ S(z,w) &= \frac{(z'z''' - \frac{3}{2}(z'')^2)}{z'^2} \end{aligned} \quad (2.26)$$

S is the schwarzian derivative and it vanishes when z is a Möbius transformation of w. As an example, we can compute the relation between T in the cylinder $T_{\text{cyl}}(w)$ and $T(z)$. Substituting (2.1) in (2.26) we obtain:

$$\begin{aligned} T_{\text{cyl}}(w) &= z^2 T(z) - \frac{c}{24} \\ T_{\text{cyl}}(w) &= \sum_{n\in Z} L_n z^{-n} - \frac{c}{24} = \sum_{n\in Z}\left(L_n - \frac{c}{24}\delta_{n,0}\right)e^{-nw} \\ (L_0)_{\text{cyl}} &= L_o - \frac{c}{24} \end{aligned} \quad (2.27)$$

and c can be interpreted as a Casimir energy. Hence if we want to compute the physical partition function, we learn from (2.27) that:

$$z = \mathrm{tr} \quad e^{-\beta(L_0+\bar{L}_0)_{\mathrm{cyl}}} = \mathrm{tr} e^{-\beta(L_0+\bar{L}_0-c/12)} , \tag{2.28}$$

The simplest examples of conformal theories are free scalars and free Weyl-Majorana fermions. The energy-momentum tensor of a free scalar is:

$$T(z) = -\frac{1}{2} \; : \partial\phi(z)\partial\phi(z) : , \partial\phi(z) = \sum_{n\in\mathbb{Z}} \alpha_n z^{-n-1} \tag{2.29}$$

with the basic two-point function:

$$\partial\phi(z)\partial\phi(w) = \frac{-1}{(z-w)^2} + \; : \partial\phi(z)\partial\phi(w) : \tag{2.30}$$

and one easily obtains $c = 1$. For a Weyl-Majorana fermion there are two possibilities. We can have periodic or antiperiodic boundary conditions on the cylinder. Since in this case $h = 1/2$, we learn from (2.7) that a periodic (antiperiodic) fermion on the cylinder becomes an antiperiodic (a periodic) fermion on the plane. These boundary conditions are known respectively as Ramond (R) and Neveu-Schwarz (NS) boundary conditions:

$$\begin{aligned} \psi(e^{2\pi i}z) &= \psi(z) \qquad (NS) \\ \psi(e^{2\pi i}z) &= -\psi(z) \qquad (R) \end{aligned} \tag{2.31}$$

with associated mode expansions:

$$\begin{aligned} \psi(z) &= \sum_{n\in\mathbb{Z}+1/2} \psi_n z^{-n-1/2} \qquad (NS) \\ \psi(z) &= \sum_{n\in\mathbb{Z}} \psi_n z^{-n-1/1} \qquad (R) \end{aligned} \tag{2.32}$$

and we see that in the Ramond sector $\psi(z)$ has a square root cut from 0 to ∞. The canonical anticommulatation relations are:

$$\{\psi_n, \psi_m\} = \delta_{n+m} \; . \tag{2.33}$$

and there is a sero mode in the Ramond sector. The singularly on the OPE is

$$\psi(z)\psi(w) = \frac{1}{z-w} + \text{regular}$$

in both cases,

$$T(z) = \frac{1}{2} : \psi(z)\partial_z \psi(z) : \tag{2.34}$$

and $c = 1/2$.

The symmetry of a CFT under the Virasoro algebra $V \otimes \overline{V}$ means that the operators (or states) of the theory fall into different conformal families, each generated by a primary field. The states generated by the action of $T(z)$ on the primary field (2.11), known as descendent fields; are determined by those of the primary fields through the use of the conformal Ward identity (2.14). As an example we compute:

$$\begin{aligned} &\langle (L_{-n}\phi)(z)\phi_1(z_1)\dots\phi_N(z_N)\rangle \\ &= \langle \oint_z (w-z)^{-n+1} T(w)\phi(z)\phi_1(z_1)\dots\phi_N(z_N)\rangle \end{aligned} \tag{2.35}$$

deforming the contour so that it circles $z_1 \dots z_N$ and using the OPE (2.12) (assuming all $\phi_i(z_1)$ are primaries) one obtains:

$$(-1)^{n-1}\sum_{i=1}^{N}\Big(\frac{(1-n)h_i}{(z-z_i)^n} + \frac{1}{(z-z_i)^{n-1}}\frac{\partial}{\partial z_i}\Big)\langle\phi(z)\phi_1\dots\phi_N\rangle \,. \tag{2.36}$$

Notice also that

$$(L_{-1}\phi)(z) = \frac{\partial\phi}{\partial z} \,. \tag{2.37}$$

Hence to proceed further we need to find the general properties of correlation functions of primary fields. In the next subsection we will use the associativity of the OPE to derive the duality properties of 4-point functions. We conclude this subsection by analyzing the general SL_2 invariance properties of the correlation functions[6]. Since $L_{\pm 1}|0\rangle = L_0|0\rangle = 0$ we immediately obtain

$$\begin{aligned} \sum_{i=1}^{N} \frac{\partial}{\partial z_i}\langle\phi_1(z_1)\dots\phi_N(z_N)\rangle &= 0 \\ \sum_{i=1}^{N} \Big(z_i\frac{\partial}{\partial z_i} + h_i\Big)\langle\phi_1(z_1)\dots\phi_N(z_N)\rangle &= 0 \\ \sum_{i=1}^{N} \Big(z_i^2\frac{\partial}{\partial z_i} + 2h_i z_i\Big)\langle\phi_1(z_1)\dots\phi_N(z_N)\rangle &= 0 \end{aligned} \tag{2.38}$$

and similarly for $L_{\pm 1}, L_0$ (for simplicity of notation we present most of the formulae only for the explicit dependence on $z_1 \dots z_N$). The behavior of the correlation functions under finite SL_2 transformations follows from (2.7) and (2.38):

$$g(z) = \frac{az+b}{cz+d} \qquad \frac{dg}{dz} = \frac{2}{(cz+d)^2}$$
$$\langle \phi_1(g(z_1)) \dots \phi_N(g(z_N)) \rangle = \Big(\prod_{i=1}^{N} (cz_i + d)^{2h_i} \Big) \langle \phi_1(z_1) \dots \phi_N(z_N) \rangle \tag{2.39}$$

given any four points w_1, w_2, w_3, w_4 we can construct an SL_2 invariant in terms of the harmonic ratio:

$$\eta = \frac{w_{12}w_{34}}{w_{14}w_{32}}, w_{ij} \equiv w_i - w_j \tag{2.40}$$

and we can write $\langle \phi_1 \dots \phi_N \rangle$ in terms of a function of N-3 harmonic ratios times a prefactor which takes care of the transformation property (2.39). Using

$$z_i - z_j \to g(z_i) - g(z_j) = \frac{z_{ij}}{(cz_i + d)(cz_j + d)}$$

we write:

$$\langle \phi_1(z_1) \dots \phi_N(z_N) \rangle = \Big(\prod_{i<j} z_{ij}^{-\gamma_{ij}} \Big) f(\eta_a) \quad a = 1, \dots, N-3 \tag{2.41}$$

to determine the γ_{ij} we impose (2.39) on the prefactor. Then:

$$\begin{aligned} \gamma_{ij} &= \gamma_{ji} \\ \sum_{j \neq i} \gamma_{ij} &= 2h_i \, . \end{aligned} \tag{2.42}$$

In particular, the two- and three-point functions are completely determined:

$$\begin{aligned} &\langle \phi_{h,\bar{h}}(z,\bar{z}) \phi_{h' h^{-1}}(zw, \bar{w}) \rangle = a \delta_{h,h'} \delta_{\bar{h}\bar{h}'} (z-w)^{-2h} (\bar{z} - \bar{w})^{-2\bar{h}} \\ &\langle \phi_1(z_1) \phi_2(z_2) \phi_3(z_3) \rangle = C_{123} z_{12}^{-\gamma_{12}} z_{13}^{-\gamma_{13}} z_{23}^{-\gamma_{23}} \cdot (\text{antihol. part}) \\ &\gamma_{ij} = h_i + h_j - h_k \end{aligned} \tag{2.43}$$

The only nontrivial dynamical information is the coefficient C_{123} which determines the operator product expansion coefficients for the CFT. Knowledge of these coefficients is equivalent to knowing the CFT. The equations satisfied by C_{123} are presented next.

2.2 *Conformal blocks and duality*

To find the crossing or duality transformation satisfied by the constants C_{ijk} we study the four-point function. We define the out state created by a primary field as:

$$\langle h, \bar{h}| = \lim_{z,\bar{z}\to\infty} \langle 0|\phi_{h,\bar{h}}(z,\bar{z}) z^{2L_0} \bar{z}^{2\bar{L}_0} \tag{2.44}$$

This is the definition induced by the scalar product defined by the two point function (2.43). With this definition we can simplify (2.43):

$$\langle n|\phi_m(z,\bar{z})|l\rangle = C^l_{nm} z^{h_n-h_m-h_l} \bar{z}^{\bar{h}_n-\bar{h}_m-\bar{h}_l} \tag{2.45}$$

and in the four-point function we can choose three points to be 0,1,∞. Hence we are interested in:

$$\langle |l| \; \phi_l(1,1)\phi_n(x,\bar{x})|m\rangle \; . \tag{2.46}$$

If we consider the OPE of two primary fields:

$$\phi_n(z,\bar{z})\phi_m(0,0) = \sum_p \sum_{\{k,\bar{k}\}} C^{p\{k,\bar{k}\}}_{nm} \cdot \\ z^{h_p-h_n-h_m+\Sigma k} \bar{z}^{\bar{h}_p-\bar{h}_n-\bar{h}_m+\Sigma\bar{k}} \phi_p^{\{k,\bar{k}\}}(0,0) \tag{2.47}$$

The index p runs over the primary fields, and $k, \bar{k}$ label descendants, k is a collection of integers $k_1, \ldots k_N \geq 0$ and

$$\phi_p^{\{k,\bar{k}\}}(0) = L_{-k_1}\ldots L_{-k_N}\ldots L_{-\bar{k}_M}\phi_p(0,0) \tag{2.48}$$

from conformal invariance,

$$C^{p\{k,\bar{k}\}}_{n,m} = C^p_{nm}\beta^{p\{k\}}_{nm}\beta^{-\bar{p}\{\bar{k}\}}_{nm} \tag{2.49}$$

and the β-coefficients can be calculated using the conformal Ward identities in terms of the conformal dimensions and the central extension c of the Virasoro algebra (see [BPZ] for details). Substituting (2.47) in (2.46):

$$\begin{aligned}
&\langle k|\phi_l(1,1)\phi_n(x,\bar{x})|m\rangle \\
&= \langle k|\phi_l(1,1)\sum_p C^p_{nm}\, x^{h_p-h_n-h_m}\bar{x}^{\bar{h}_p-\bar{h}_n-\bar{h}_m} \sum_{\{k,\bar{k}\}} \beta^{p\{k\}}_{nm}\,\beta^{p\{\bar{k}\}}_{nm}\,. \\
&\times x^{\Sigma k}\bar{x}^{\Sigma \bar{k}}\phi_p^{\{k,\bar{k}\}}(0,0)|0\rangle
\end{aligned} \tag{2.50}$$

defining,

$$\mathcal{F}^{lk}_{nm}(p|x) = x^{h_p-h_n-h_m}\sum_{\{k\}}\beta^{p\{k\}}_{nm}\frac{\langle k|\phi_l(1,1)L_{-k_1}\ldots L_{-k_N}|p\rangle}{\langle k|\phi_l(1,1)|p\rangle}\cdot x^{\Sigma k} \tag{2.51}$$

The Green's function becomes:

$$G^{lk}_{nm}(x,\bar{x}) = \sum_p C^p_{nm}C_{klp}\,\mathcal{F}^{lk}_{nm}(p|x)\overline{\mathcal{F}^{lk}_{nm}(p|x)}\,. \tag{2.52}$$

Graphically each function $\mathcal{F}$ is represented as a skeleton diagram

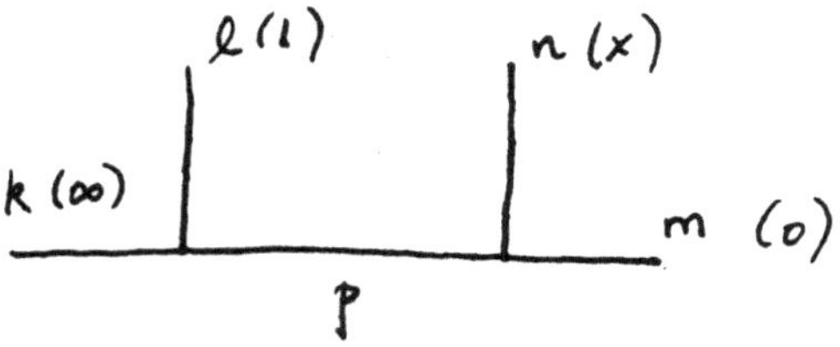

(2.53)

The generally multivalued functions $\mathcal{F}(p|x)$ are known as the conformal blocks. The blocks are normalized so that when $x \to 0$:

$$\mathcal{F}^{lk}_{nm}(p|x) \simeq x^{h_p-h_n-h_m}(1+\ldots) \tag{2.54}$$

This is the convention chosen in Ref. 6. Requiring associativity of the OPE gives us the desired equations for C: the bootstrap, duality or crossing relations. The skeleton in (2.53) indicates the order in which we perform the operator product: we first do the OPE for n with m and then do the expansion with l. Associativity implies that the physical amplitude should not depend on whether we do first the OPE between n, m or l, n. Use first the SL_2 transformation $x \to 1 - x$ to obtain $\langle k|\phi_m(1,1)\phi_n(1-x, 1-\bar{x})|l\rangle$ hence $G^{lk}_{nm}(x,\bar{x}) = G^{mk}_{nl}(1-x, 1-\bar{x})$ similarly with $x \to 1/x$. We thus obtain:

$$G^{lk}_{nm}(x,x) = G^{mk}_{nl}(1-x, 1-\bar{x}) = x^{-2h_n}\bar{x}^{-2\bar{h}_n} G^{lm}_{nk}\left(\frac{1}{x}, \frac{1}{\bar{x}}\right) . \qquad (2.55)$$

now do the OPE between l, n first. Following steps similar to those leading to (2.52) we arrive at:

$$\sum_p C^p_{nm} C_{klp} \mathcal{F}^{lk}_{nm}(p|x)\overline{\mathcal{F}^{lk}_{nm}(p|x)}$$
$$= \sum_q C^q_{nl} C_{kmq} \mathcal{F}^{mk}_{nl}(q|1-x)\overline{\mathcal{F}^{mk}_{nl}(q|1-x)} \qquad (2.56)$$

since $\mathcal{F}, \overline{\mathcal{F}}$ are determined by the conformal dimensions h and the conformal Ward identities; equation (2.56) gives a rather nontrivial set of conditions characterizing the possible operator product coefficients which may appear in CFT. The problem of classifying all CFTs in two dimension is equivalent to solving (2.56) for all possible c (with some extra constraints coming from modular invariance). We can represent (2.56) graphically as:

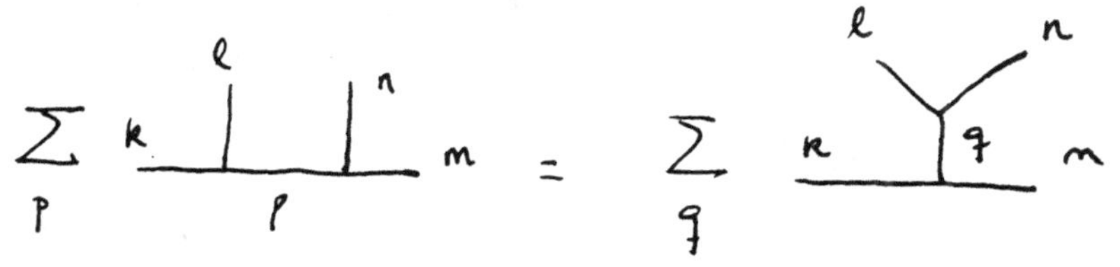

(2.56′)

exhibiting the crossing symmetry implied by (2.56).

It is worthwhile stressing that the two basic properties we have used so far of CFT are: 1) Locality, 2) Associativity of the OPE. Locality implies that although the blocks are in general multivalued, the physical correlation functions have trivial monodromy. Thinking of the blocks as section of a vector bundle, the monodromy invariant metric is constructed in terms of the structure constants of the operator algebra. The general physical correlation function $G_{i_1 \dots i_N}(z_1 \dots z_N)$ can also be expressed in terms of

more general blocks,

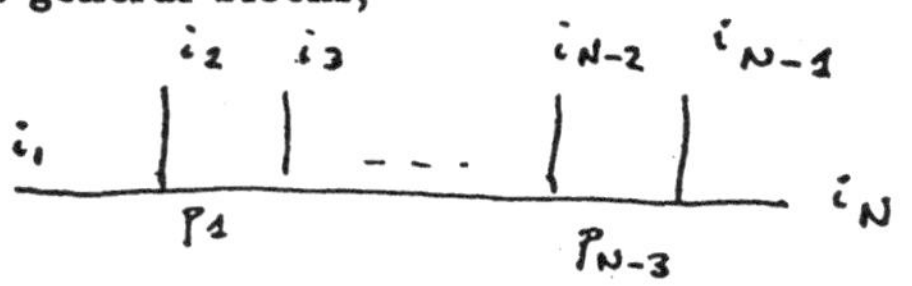

$$= \mathcal{F}^{i_1 \ldots i_N} P_1 \ldots P_{N-3}(z) \tag{2.57}$$

the physical correlator is the monodromy invariant combination of $\mathcal{F}_{\mathrm{p}}$ and $\overline{\mathcal{F}}_{\mathrm{p}'}$:

$$\begin{aligned} &G^{i_1 \ldots i_N}(z_1, \bar{z}_1, \ldots, z_N, \bar{z}_N) \\ &= \sum_{\mathrm{p},\mathrm{p}'} h_{\mathrm{p},\mathrm{p}'} \mathcal{F}_{\mathrm{p}}^{i_1 \ldots i_N}(z_1, \ldots) \overline{\mathcal{F}_{\mathrm{p}}^{\prime i_1 \ldots i_N}(z_1, \ldots)} \end{aligned} \tag{2.58}$$

and the invariant metric $h_{\mathrm{p},\mathrm{p}'}$ is built out of the structure constants C_{ij}^k. Locality implies that G is independent of the order of its arguments. They can be exchanged without changing G. For blocks, the same certainly does not apply. Since $\mathcal{F}$ is in general multivalved we can define the $\mathcal{F}$'s for some ordering of its arguments (for example $|z_1| > |z_2| > \ldots$) and then analytic continuation relates blocks in different domains. This is the question of monodromy and half-monodromy (braids) transformations.

Since analytic continuation depends on the path chosen to perform it, we find that it is the braid group rather than the permutation group which acts on the blocks. The braid group on n strands B_n is defined by $n-1$ generators $\sigma_i, i = 1, \ldots, n-1$ shown in Fig. 1, and two relations[7]:

$$\begin{aligned} \sigma_i \sigma_j &= \sigma_j \sigma_i \quad |i-j| \geq 2 \\ \sigma_i \sigma_{i+1} \sigma_i &= \sigma_{i+1} \sigma_i \sigma_{i+1} \end{aligned} \tag{2.59}$$

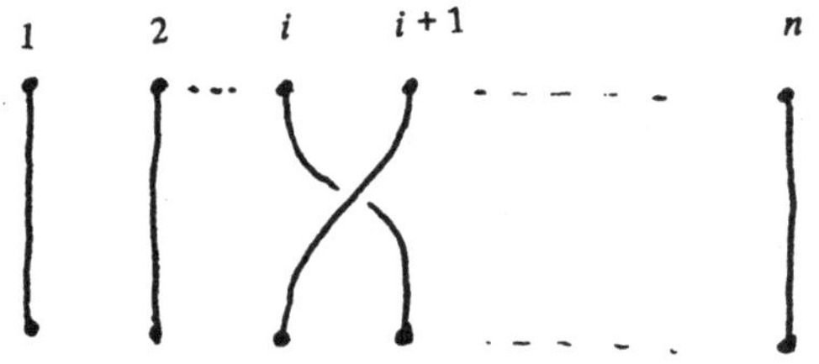

Fig. 1. Generator σ_i of the braid group on n strands B_n.

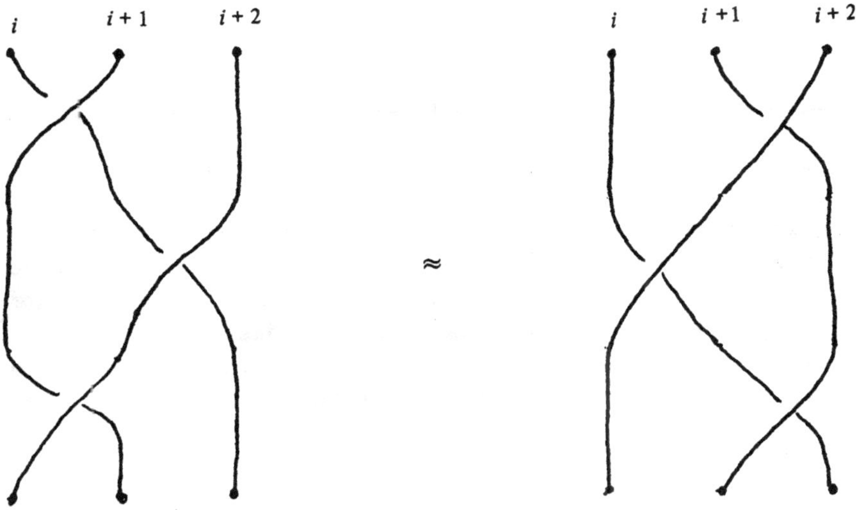

Fig. 2. Graphical representation of the relation $\sigma_i\sigma_{i+1}\sigma_i = \sigma_{i+1}\sigma_i\sigma_{i+1}$.

By looking at Fig. 1 it is fairly obvious that the first relation in (2.5a) is satisfied. The second is shown diagramatically in Fig. 2.

A basis in the space of blocks is equivalent to giving a skeleton representation of the blocks. In Fig. 3 one can see different bases for the 5-point function. The dimension of the space of blocks with fixed states on the external legs can be computed in terms of the fusion rules. The simplest information contained in the structure constants C_{ij}^k is the number of possible couplings between the representations $i, j, k : N_{ij}^k$. The integral matrices $(N_i)_j^k = N_{ij}^k$ define the fusion algebra and later we will review some of its very striking properties[15]. Using N_{ij}^k we can easily count the number of blocks. Take for example the 4-point block in (2.53), then the number of blocks with k, l, n, m fixed is

$$\sum_p N_{lk}^p N_{np}^m = (N_l N_n)_k^m$$
$$(N_i)_j^k \equiv N_{ij}^k \tag{2.59}$$

it is clear that $N_{ij}^k = N_{ji}^k$ and furthermore associativity of the OPE implies (see Sec. 5) $[N_i, N_j] = 0$; and the dimension of the space of blocks is the same regardless of the basis chosen (a consistency condition). We can write

Fig. 3. Different skeletous represent different bases for the space of conformal blocks.

the associativity of the OPE and the locality of the correlation functions in terms of conformal blocks. There are two basic moves that one can make on the space of blocks. One correponds to the action of braid generator on the blocks and the other is a change of basis going from the s-channel to the t-channel. Following Moore and Seiberg[16] we can call these moves respectively B-moves and F-moves (for braiding and fusing), and they are shown in Fig. 4.

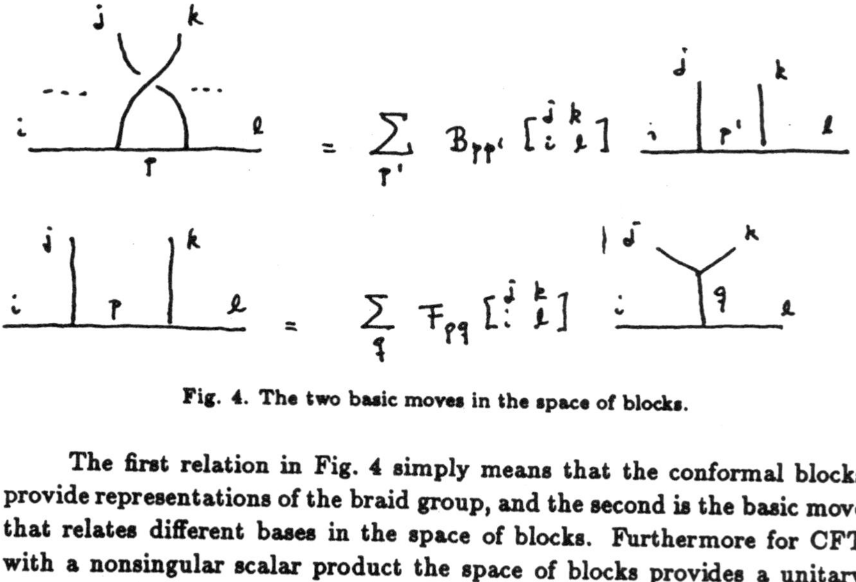

Fig. 4. The two basic moves in the space of blocks.

The first relation in Fig. 4 simply means that the conformal blocks provide representations of the braid group, and the second is the basic move that relates different bases in the space of blocks. Furthermore for CFT with a nonsingular scalar product the space of blocks provides a unitary representation of the braid group. We can now write a set of equations for F and B following from associativity of the OPE and the defining relations of the braid group. These two equations are the hexagon and pentagon identities of Moore and Seibeg[16,17,18]. They are explained graphically, in Figs. 5 and 6. The two equations are not independent, but before we explain this it is convenient at this moment to study in more detail, the braid group

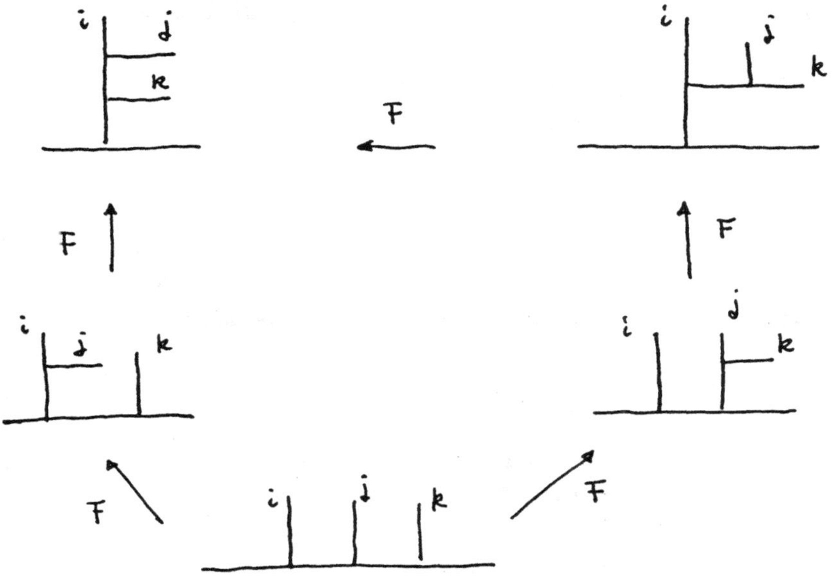

Fig. 5. Pentagon identity for F.

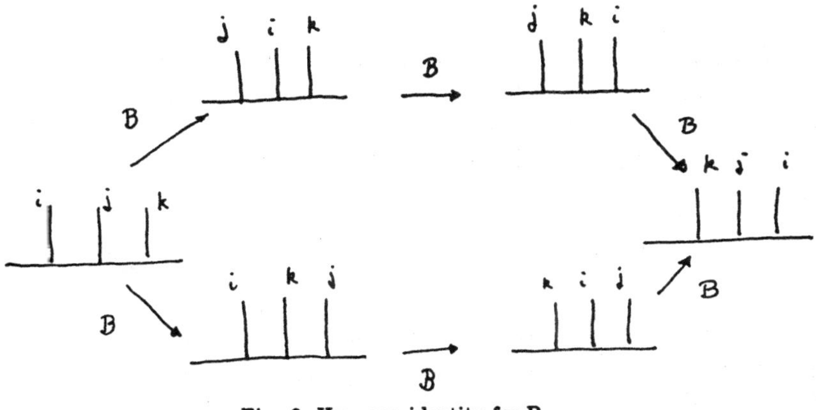

Fig. 6. Hexagon identity for B.

on the sphere and its relation to the mapping class group of the sphere with n distinguished points and local parameters about them.

On the Euclidean plane E^2, the braid group $B_n(E^2)$ is *completely* defined by the relations (2.59). In fact we can construct all the generators in terms of two elements only: $\sigma_1, a = \sigma_1 \ldots \sigma_{n-1}$. this is because

$$\begin{aligned} a\sigma_1 &= \sigma_1\sigma_2\sigma_3 \ldots \sigma_{n-1}\sigma_1 = \sigma_1\sigma_2\sigma_1\sigma_3 \ldots \sigma_{n-1} \\ &= \sigma_2\sigma_1\sigma_2 \ldots \sigma_{n-1} = \sigma_2 a \\ a\sigma_2 &= \sigma_3 a \\ &\ldots\ldots\ldots\ldots\ldots \\ a\sigma_{n-2} &= \sigma_{n-1} a \end{aligned} \tag{2.60}$$

the center of $B_n(E^2)$ is generated by a^n and it is an infinite cyclic group:

$$a^n\sigma_i = \sigma_i a^n \qquad i = 1, \ldots, n-1 \tag{2.61}$$

The braid group on the sphere $B_n(S^2)$ is generated again by $\sigma_1, \ldots, \sigma_{n-1}$ satisfying (2.59) together with the relations:

$$\begin{aligned} &\sigma_1\sigma_2 \ldots \sigma_{n-2}\sigma_{n-1}^2\sigma_{n-2} \ldots \sigma_2\sigma_1 = 1 \\ &(\sigma_1\sigma_2 \ldots \sigma_{n-1})^n = 1 \,. \end{aligned} \tag{2.62}$$

These relations hold as long as the distinguished points do not come with local parameters. In CFT however we always have a local parameter about each point and we have to consider Dehn twist around the distinguished points. These twists are illustrated in Fig. 7 and they are represented in the CFT by the operator

$$e^{2\pi i(L_0 - \bar{L}_0)} \tag{2.63}$$

$e^{2\pi i L_0}(e^{-2\pi \bar{L}_0})$ acting on the holomorphic (antiholomorphic) part of the theory. Hence the effect of the twist on a purely left-(right-) moving state is:

$$\begin{aligned} |h\rangle &\to e^{2\pi i h}|h\rangle \\ |\bar{h}\rangle &\to e^{-2\pi i \bar{h}}|\bar{h}\rangle \end{aligned} \tag{2.64}$$

and this implies that in CFT on the sphere (or any other Riemann surface) with marked points and local parameters we do not obtain representations

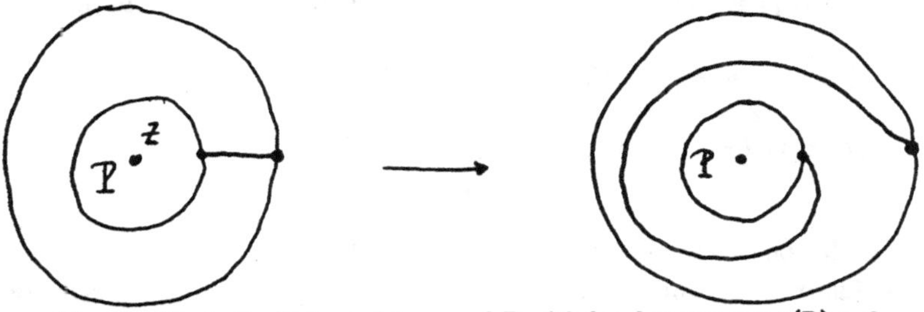

Fig. 7. Effect of a Dehn twist around P with local parameter $z(P) = 0$.

of $B_n(S^2)$ but rather of its extension by the Dehn twists around each puncture.

There are several ways of finding the modification of the relations (2.62) due to Dehn twists. The easiest one is to consider the explicit form of the n-point functions (2.41) due to SL_2 invariance[22]. When the local parameter is included one can think of braiding "ribbons" rather than braids, thus an operation which ends in the identity for braids may end up with twisted ribbons. Braiding the points i and j is expressed analytically as $z_{ij} \to e^{i\pi} z_{ji}$. Hence:

$$z_{ij}^{-\gamma_{ij}} \to e^{-i\pi\gamma_{i,j}} z_{ji}^{-\gamma_{ij}} \,. \tag{2.65}$$

Hence the braiding of (2.41) will have two contributions: one coming from SL_2 invariant quantities and the other from the prefactor. The first contribution gives a faithful representation of the braid group. If we perform the braiding operation $\sigma_1 \ldots \sigma_{n-2}\sigma_{n-1}^2\sigma_{n-2} \ldots \sigma_1$ and consider the phase generated by the prefactor we obtain:

$$\begin{aligned} &e^{-i\pi\gamma_{12}} e^{-i\pi\gamma_{13}} \ldots e^{-i\pi\gamma_{1n}} e^{-i\pi\gamma_{n1}} e^{-i\pi\gamma_{n-1}} \ldots e^{-i\gamma_{21}} \\ &= \exp -2\pi i \sum_{j\neq 1} \gamma_{1j} = e^{-4\pi i h_1} \,. \end{aligned} \tag{2.66}$$

Next, braiding according to $(\sigma_1 \ldots \sigma_{n-1})^n$ we obtain a phase:

$$\begin{aligned} &e^{-i\pi\gamma_{12}} \ldots e^{-i\pi\gamma_{1n}} \\ &e^{-i\pi\gamma_{23}} e^{-i\pi\gamma_{24}} \ldots e^{-i\gamma_{2n}} e^{-\pi\gamma_{21}} \\ &e^{-i\pi\gamma_{n1}} e^{-i\pi\gamma_{n2}} \ldots e^{-i\pi\gamma_{nn-1}} = \prod_i e^{-2\pi i h_i} \,. \end{aligned} \tag{2.67}$$

Therefore if $R(\sigma_i)$ is the representation of σ_i on blocks we conclude:

$$R(\sigma_1| \ldots R(\sigma_{n-2}) R(\sigma_{n-1}^2) R(\sigma_{n-2}) \ldots R(\sigma_1) = e^{-4\pi i h_1}$$
$$(R(\sigma_1) R(\sigma_2) \ldots R(\sigma_{n-1})^n = \prod_{i=1}^{n} e^{-2\pi i h_i} \tag{2.68}$$

where h_i is the L_0 eigenvalue of the field in the ith strand. Since the dimensions of fields in the same conformal family differ by integers, the relations (2.68) hold regardless of which representatives (primary or descendants) we choose in each conformal family. This is a general consequence of the fact that the braiding properties depend on the conformal families and not on their representatives (exercise: prove it).

The mapping class group of the sphere with n distinguished points is closely related to the braid group[7] (2.59) and (2.62). It is generated by Dehn twists around pairs of points and single points. Given two points P_i, P_{j+1} the Dehn twist is constructed by considering a simple closed curve C enclosing P_i, P_j and no other distinguished points. We cut the surface along C, twist one of the lips by 2π and glue back again. Its action on blocks can be constructed in terms of $R(\sigma_i)$. If h_i, h_{i+1} and the dimensions of the fields at the points $i, i+1$ it follows from the definition that the Dehn twist $\tau_{i,i+1}$ about P_i, P_{i+1} is represented by:

$$\tau_{i,i+1} = e^{2\pi i(h_i + h_{i+1})} R(\sigma_i^2) \tag{2.69}$$

and any other twist can be obtained from these ones and twists about single points. If we consider the particular case of the sphere with 4 points, the generators of the mapping class group are $\tau_{12}, \tau_{13}, \tau_{23}, \tau_1, \tau_2, \tau_3, \tau_4$ and using (2.68) and (2.69) one can show:

$$\tau_{12}\tau_{13}\tau_{23} = \tau_1\tau_2\tau_3\tau_4 \ . \tag{2.70}$$

This relation has been used by Vafa[19] to show that for theories with a finite number of primary fields (rational CFT), the dimensions h_i are all rational numbers.

To exhibit the connection between the matrices B and F, we notice that if we braid the i, j legs in Fig. 8, we obtain:

$$R(\sigma_{ij}) W_k^{ij} = e^{i\pi(h_k - h_i - h_j)} W_k^{ji} \tag{2.71}$$

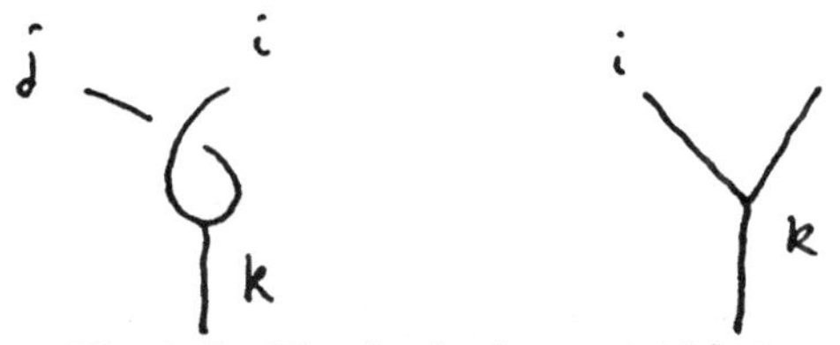

Fig. 8. Braiding in the three point block.

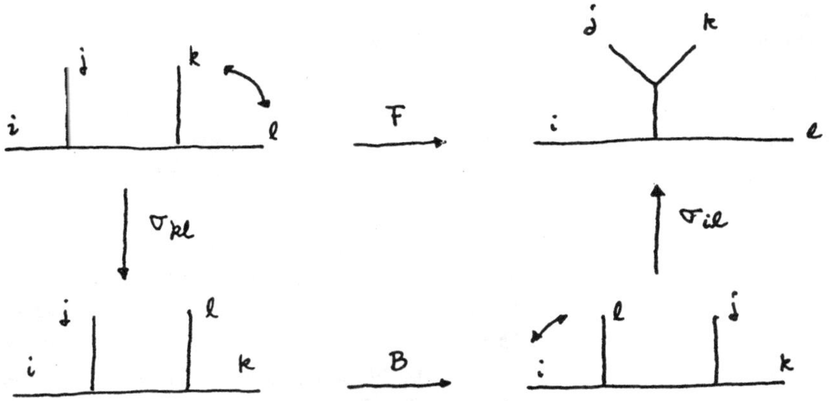

Fig. 9. Relation between F and B matrices.

only a phase appears. If we now consider the set of moves in Fig. 9, it is easy to see from (2.71) that B and F are related by phases.

To conclude, the duality (crossing) properties of a CFT can be rewritten in terms of blocks using the B and F matrices. The content of the duality conditions reduces to the pentagon or hexagon identities plus the σ-move (2.71). These equations are the tree level polynomial equations.[16,18] If the CFT has only a finite number of primary fields one says that the theory is a Rational Conformal Field Theory (RCFT), and the hexagon and pentagon identities becomes a finite set of conditions. When one includes the constraints arising from genus one modular invariance one obtains the complete set of equations characterizing the duality properties of a RCFT[16,18]. The classification of RCFT can be phased in part as the solution of these polynomial equations. A difficult (and interesting) open problem is to what extent a RCFT can be reconstructed from a solution to these equations. After building more machinery we will analyze this question in more detail. We have reviewed quickly the general constraints of duality at the level of conformal blocks to set the framework for futher discussion. Later, in

Sec. 5 we will study in more details, the fusion algebra and the rationality constraints imposed by (2.68) on the set of conformal dimensions.

3. Degenerate Conformal Families. Coulomb Gas Repretation

3.1 *Generalities*

In this subsection we present general properties of conformal theories with null vectors. In the next subsection we will study an explicit formulation of these theories in terms of a coulomb gas representation of the conformal blocks. This will allow us to prove (or at least make plausible) some of the general results. In particular we will derive the representation of conformal blocks in terms of contour integrals, the fusion rules for the minimal conformal theories[6,8], the Rocha-Caridi[10] formulae for the Virasoro characters and a few sample calculations of monodromy invariants and structure constants of the operator algebra.

To understand the meaning of null vectors, we go back to ordinary group theory. When we study the finite dimensional representations of simple or semisimple groups we are accustomed to a general property of the representation theory of this group which implies that any reducible representation can always be brought to block diagonal form i.e. reducible representations are fully reducible. This however is not true in general. To assess whether this is the case we need to use the Brauer-Weyl theory which follows from Schur's lemma. Given an irreducible representation V of some algebra A (a Lie algebra for example) we can take arbitrary tensor products $V \otimes V \otimes \ldots \otimes V = V^{\otimes n}$. If A is the Lie algebra of SU(N) and V is the fundamental representation, we can decompare $V^{\otimes n}$ into irreducible representations labelled by Young tableaux defining different symmetries for a tensor of rank n. The steps in these decompositions are as follows: We first construct the centraliser algebra B of the action of A on $V^{\otimes n}$. In other words, we look for all operators on $V^{\otimes n}$ which commute with the representation of A in $V^{\otimes n}$. For SU(N) the centraliser in $V^{\otimes n}$ is the permutation group S_n exchanging the different copies of V in $V^{\otimes n}$ if $n \leq N$. When $n > N$ we have S_n with the extra constraint that the $N+1$ row antisymmetriser vanishes; there are no Young tableaux with more than N rows. The question of complete reducibility of $V^{\otimes n}$ reduces then to the question of whether the centralizer algebra is semi-simple. For the case of SU(N) this is the case as well as for other semi-simple groups. When this is not the case one finds representation in $V^{\otimes n}$ which generically contains null vectors (i.e. more than one highest weight state) which are not fully reducible. We will find some finite dimensional examples when we come to study the representation theory of quantum groups.

What is the analogous situation for the Virasoro algebra? We have seen in the previous section that the braiding properties of the conformal blocks are independent of the representative fields chosen for each conformal family on the block. Since changing representatives in each family is equivalent to acting with the Virasoro algebra on the primary fields, we find that the centralizer of the Virasoro action on a collection of conformal blocks contains at least the braid group (for RCFT this should be the whole thing). If we find null vectors in the Verma modules associated with our CFT this spells trouble in the representations of the duality transformations acting on the conformal blocks. Thus, to obtain a sensible theory (with a nondegenerate scalar product which guarantees that the space of blocks provides a fully reducible representation of the duality and braid transformations) we have to eliminate the null vectors and consider the reduced theory.[a] This leads to a collection of nontrivial differential equations for the blocks which in some cases completely determine the correlation functions of the theory. To be more precise, let $V(c,h)$ be a Verma module of the Virasoro algebra Vir, with highest weight vector $|h\rangle, L_0|h\rangle = h|h\rangle$, then:

$$L_0\big(L_{-n_1}\dots L_{-n_N}|h\rangle\big) = (h+\sum_i n_i)(L_{-n_1}\dots L_{-n_N}|h\rangle) \tag{3.1}$$

the integer $\sum n_i$ is the level of the state and $|h\rangle$ has zero level. If the representation is unitary, $h>0$,

$$\|L_{-1}|h\rangle\|^2 = \langle h|L_{+1}L_{-1}|h\rangle = 2h > 0\ . \tag{3.2}$$

Furthermore by taking the 2×2 matrix of scalar products of $L_{-N}|h\rangle$ and $L_{-1}^N|h\rangle$ and requiring it to be positive we can obtain that $c>0$. If we can find a state $|\chi\rangle \in V(c,h)$ such that:

$$\begin{aligned} L_0|\chi &= (h+N)|\chi\rangle \\ L_n|\chi\rangle &= 0 \qquad n>0 \end{aligned} \tag{3.3}$$

---

[a] When we study the fusion algebra in more detail (Sec. 5) we will see that for sensible physical theories the space of blocks should provide a fully reducible representation of the braid group.

we say that $|\chi\rangle$ is a null vector and the representation is degenerate. If we have a scalar product such that $L_n^+ = L_{-n}$, then one shows that $\langle\psi|\chi\rangle = 0$ for any state of the theory. Hence we can set $|\chi\rangle = 0$ which is equivalent to taking the quotient by the null vectors and the submodules they generate. If $\phi_h(z)$ is the primary field generating the family we say that the family is degenerate and that it has a null vector at level N. As an example we consider null vectors a level two:

$$(aL_{-2} + bL_1^2)|h\rangle \tag{3.4}$$

since all the $L_n n > 0$ are generated by L_1 and L_2 we only need to impose the annihilation of (3.4) by L_1 and L_2 to obtain the conditions for a null vector. This gives a constraint for given c of what modules can have a level two null state. We obtain:

$$\begin{aligned} |\chi\rangle &= \left(L_{-2} - \frac{3}{2(2h+1)} L_{-1}^2\right)|h\rangle \\ h &= \frac{1}{16}(5 - c \pm \sqrt{(c-1)(c-25)}) \ . \end{aligned} \tag{3.5}$$

Requiring the decoupling of $|\chi\rangle$ from every correlation function leads to a second order differential equation for correlation functions containing $\phi_h(z)$. If $\chi_h(z)$ is the descendant field creating (3.5), we have:

$$\begin{aligned} &\langle \chi_h(z)\phi_1(z_1)\dots\phi_N(z_N)\rangle \\ &= \left(\frac{3}{2(2h+1)}\frac{\partial^2}{\partial z^2} - \sum_{i=1}^N \frac{h_i}{(z-z_i)^2} - \sum_{i=1}^N \frac{1}{z-z_i}\frac{\partial}{\partial z_i}\right)\langle\phi_h(z)\phi_1\dots\phi_N\rangle = 0 \end{aligned} \tag{3.6}$$

where (2.36) and (2.37) have been used. For the 4-point function, using SL_2 invariance this leads to a hypergeometric differential equation whose solutions will be given later in terms of contour integrals. V.Kac has given the set of representations of Vir with null vectors. To obtain this set one has to consider for a given $V(c,h)$ and any level N the determinant of the scalar products between all the states at level N. The seroes of this Kac determinant give the values of (c,h) of representations with a level N null vector. To write the result in a simple form, we introduce two numbers $\alpha_+\alpha_-$ defined by:

$$\begin{aligned} c &= 1 - 24\alpha_0^2 \\ \alpha_0 &= \alpha_+ + \alpha_- \ , \alpha_+\alpha_- = -1 \ . \end{aligned} \tag{3.7}$$

Then the representations with null vectors depend on two integers $n, m \geq 1$ and have dimensions:

$$h(n,m) = -\frac{1}{4}\alpha_0^2 + \frac{1}{4}(n\alpha_+ + m\alpha_-)^2 \tag{3.8}$$

and $V_{n,m} = V(c, h_{n,m})$ has a null vector at level nm. This fact has a simple derivation in the Coulomb gas representation.

The null vectors also give a lot of information about the fusion rules of the theory[6]. To exhibit how this is obtained, we consider $h = h(1,2)$ or $h(2,1)$. In this case the null vector appears at level 2, and the differential equation obtained by decoupling them is (3.6). If $\phi_1(z_1)$ has dimension h_1 and we take $z \sim z_1$, we have the OPE:

$$\phi_h(z)\phi_{h_1}(z_1) = \text{coust.}(z - z_1)^k[\phi_{h1}(z_1) + b(z - z_1)\phi_{h1}^{(-1)}(z_1) + \ldots] \tag{3.9}$$

the expansion (3.9) solves (3.6) if

$$\frac{3}{2(2h+1)}k(k-1) - h_1 + k = 0 \tag{3.10a}$$

and scale invariance implies

$$k = h' - h_1 - h\ . \tag{3.10b}$$

Parametrizing dimensions as:

$$h_1 = -\frac{1}{4}\alpha_0^2 + \frac{1}{4}\alpha^2\ , h' = -\frac{1}{4}\alpha_0^2 + \frac{1}{4}\alpha'^2 \tag{3.11}$$

we find

$$\begin{aligned} h &= h(1,2) \qquad \alpha' = \alpha \pm \alpha_- \\ h &= h(2,1) \qquad \alpha' = \alpha \pm \alpha_+ \end{aligned} \tag{3.12}$$

giving the fusion rules

$$\begin{aligned} \phi_{(1,2)}\phi_{(\alpha)} &= [\phi_{(\alpha-\alpha_-)}] + [\phi_{(\alpha+\alpha_-)}] \\ \phi_{(2,1)}\phi_{(\alpha)} &= [\phi_{(\alpha-\alpha_+)}] + [\phi_{(\alpha+\alpha_+)}]\ . \end{aligned} \tag{3.13}$$

The normalization however remains to be determined. Depending on the theory some of the families on the right-hand side may not appear.

Similar arguments can be used to derive the general fusion rules once the null vectors are known explicitly. With the information we have, we can explore a little the operator algebra of degenerate theories. The simplest fusion rules we can write down are:

$$\begin{aligned}\phi_{(1,2)}\phi_{(1,2)} &= [\phi_{(1,1)}] + [\phi_{(1,3)}] \\ \phi_{(2,1)}\phi_{(2,1)} &= [\phi_{(1,1)}] + [\phi_{(3,1)}]\end{aligned} \tag{3.14}$$

the field $\phi_{(1,1)}$ has zero dimension and it is identified as the identity operator. The normalization requires the associativity of the OPE and some properties of the theory under consideration. Applying (3.13) once more we find:

$$\begin{aligned}\phi_{(1,2)}\phi_{(1,m)} &= [\phi_{(1,m-1)}] + [\phi_{(1,m+1)}] \\ \phi_{(2,1)}\phi_{(m,1)} &= [\phi_{(m-1,1)}] + [\phi_{(m+1,1)}] \,.\end{aligned} \tag{3.15}$$

Thus if $\phi_{(1,2)}$ is on the operator algebra so is $\phi_{(1,m)}$ generically and the same for $\phi_{(2,1)}$ and $\phi_{(m,1)}$. Naive application of these rules would however generate dimension $h(n,m)$ with (n,m) zero or negative. In fact these fields drop from the operator algebra, and only $\phi_{(n,m)} n, m > 0$ appear. This truncation phenomenon can be understood as follows:

$$\phi_{(1,2)}\phi_{(2,1)} = c_1[\phi_{(2,0)}] + c_2[\phi_{(2,2)}] \,. \tag{3.16}$$

Using the second relation in (3.13) we have:

$$\phi_{(1,2)}\phi_{(2,1)} = c_1'[\phi_{(0,2)}] + c_2'[\phi_{(2,2)}] \tag{3.17}$$

and the c and c' are the OPE coefficients. Compatibility between (3.16) and (3.17) requires $c_1 = c_1' = 0, c_2 = c_2'$, and therefore:

$$\phi_{(1,2)}\phi_{(2,1)} = [\phi_{(2,2)}] \,. \tag{3.18}$$

It is clear that if $\phi_{(1,2)}$ and $\phi_{(2,1)}$ are in the operator algebra we can generate all the filds $\phi_{(n,m)}$ in the theory via OPE. Moreover, using the truncation phenomenon and the associativity of the OPE we can derive the fusion rules for a theory with degenerate fields $\phi_{(n,m)}$. The result is[6]

$$\phi_{(n_1,m_1)}\phi_{(n_2,m_2)} = \sum_{k=(n_1-n_2)+1}^{n_1+n_2-1} \sum_{l=(m_1-m_2)+1}^{m_1+m_2-1} [\phi_{(k,l)}] \,. \tag{3.19}$$

If $|n_1 - n_2|$ is odd (even) then the sum over k runs only over even (odd) values, and similarly for $|m_1 - m_2|$ and l.

Next one can try to find theories where all the field are degenerate and with a finite number of fields (RCFT). From (3.7) and (3.8) we see that when $c > 25 \alpha_+, \alpha_-$ are purely imaginary, and as n, m grow larger some $h(n,m)$ will become negative. For $25 > c > 1$, the dimensions are generally complex. Only for $c \leq 1$ the dimensions are real in general. For example when $c = 1, \alpha_0 = 0, \alpha_\pm = \pm 1$, and the representations with null vectors are those such that $h = N^2/4$, N an integer. When α_-/α_+ is a rational number it was shown in Ref. 6 that the operator algebra truncates to a finite set. This is because each module now contains an infinite number of null fields (we will show this later in Sec. 3.2. If p, q are relatively prime positive integers, then the minimal model of Belavin Polyakov and Zamolodchikov satisfy:

$$\begin{aligned} \alpha_-/\alpha_+ &= -p/q \qquad c = 1 - \frac{6(p-q)^2}{pq} \\ h(n,m) &= \frac{1}{4pq}[(nq - mp)^2 - (p-q)]^2 \end{aligned} \tag{3.20}$$

The unitary subseries[8] is obtained for $q = p + 1$ and $p \geq 2$

$$c = 1 - \frac{6}{p(p+1)} \, , L_{n,m} = \frac{1}{4p(p+1)}[(n(p+1) - mp)^2 - 1] \tag{3.21}$$

Notice furthermore the reflection property:

$$h(n,m) = h(p-n, q-m) \tag{3.22}$$

and we can restrict the n, m values to the rectangle:

$$0 < n < p \qquad 0 < m < q \qquad , p < q \tag{3.23}$$

The simplest nontrivial example is the Ising model with $p = 3, q = 4$. Now $c = 1/2$ and we have a grid of conformal dimensions shown in Fig. 10, given by:

$$\begin{aligned} h(1,1) &= h(2,3) = 0 \\ h(1,2) &= h(2,2) = \frac{1}{16} \\ h(2,1) &= h(1,3) = \frac{1}{2} \end{aligned} \tag{3.23}$$

$\phi_{(1,1)}$ (or $\phi_{(2,3)}$) is identified with the identity operator, $\phi_{(1,2)}$ (or $\phi_{(2,2)}$) is the spin field σ and $\phi_{(2,1)}(\phi_{(1,3)})$ is identified with the energy density ε. The fusion rules are easy to derive:

$$\begin{aligned}\varepsilon\cdot\varepsilon &= \phi_{(2,1)}\times\phi_{(2,1)} = c_1[\phi_{(1,1)}]+c_2[\phi_{(3,1)}] \\ &= \phi_{(1,3)}\times\phi_{(1,3)} = c_1'[\phi_{(1,1)}]+c_2'[\phi_{(1,3)}]+c_3'[\phi_{(1,5)}]\,. \end{aligned} \tag{3.24}$$

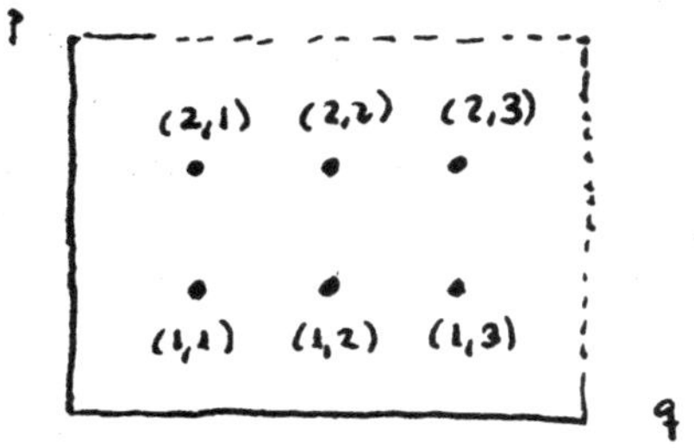

Fig. 10. Conformal weight for p = 3, q = 4.

Hence $c_2 = c_2' = c_3' = 0$, and $\varepsilon\cdot\varepsilon = [1]$. Similarly for the other operator products. We obtain:

$$\begin{gathered}\varepsilon\times\varepsilon = [1] \\ \varepsilon\times\sigma = [\sigma] \qquad \sigma\times\sigma = [1]+[\varepsilon]\end{gathered} \tag{3.25}$$

and similar arguments apply to other cases. This concludes the generalities of degenerate theories. Next we find explicit representations for their correlation functions and characters.

3.2 *Coulomb gas representation. Fusion rules*

The properties of all the minimal models can be obtained in terms of single scalar field l. We follow in part the presentation in Ref. 9. Consider the following energy-momentum tensor:

$$\begin{aligned}T(z) &= -\frac{1}{4}\partial\phi(z)\partial\phi(z)+i\alpha_0\partial^2\phi(z) \\ \langle\phi(z)\phi(w)\rangle &= -2ln(z-w) \\ c &= 1-24\alpha_0^2\,. \end{aligned} \tag{3.26}$$

$T(z)$ follows from the variation of the action of a free scalar field with some background charge: a term propartional to $R\phi$ in the Lagrangian, where R is the two-dimensional scalar curvature. The background charge is $e = -2\alpha_0$. Hence the U(1) current $\partial\phi(z)$ is anomalous:

$$T(z)\partial\phi(w) = \frac{1}{(z-w)^2}\partial\phi(w) + \frac{1}{z-w}\partial^2\phi(w) + \frac{4i\alpha_0}{(z-w)^3} \ . \qquad (3.27)$$

The last term represents the anomaly. The primary fields are exponentials of ϕ, and the background charge changes the conformal dimension with respect to the $\alpha_0 = 0$ case:

$$T(z) : e^{i\alpha\phi(w)} := \frac{\alpha(\alpha - 2\alpha_0)}{(z-w)^2} : e^{i\alpha\phi(w)} + \frac{1}{z-w}\partial_w e^{i\alpha\phi(w)} + \text{regular} \quad (3.28)$$

hence

$$h(: e^{i\alpha\phi} :) = \alpha(\alpha - 2\alpha_0) \qquad (3.29)$$

and the correlation functions of these vertex operators will vanish unless the background charge is screened:

$$\begin{aligned} \langle V_{\alpha_1}(z_1)\dots V_{\alpha_n}(z_n)\rangle &= \prod_{i<j}(z_i - z_j)^{+2\alpha_i\alpha_j} \qquad \text{if} \sum \alpha_i = 2\alpha_0 \\ &= 0 \qquad \text{otherwise} \\ V_\alpha(z) &=: e^{i\alpha\phi(z)} : \ . \end{aligned} \qquad (3.30)$$

From (3.29) we learn that there are always two fields of dimension 1:

$$\begin{aligned} J_\pm \quad &=: e^{i\alpha_\pm\phi} : \\ \alpha_+ + \alpha_- \ &= 2\alpha_0 \ , \quad \alpha_+\alpha_- = -1 \ . \end{aligned} \qquad (3.31)$$

This justifies *a posteriori* the definitions in (3.7). We want to use this construction to represent the primary fields and correlation functions of the minimal models. It is important to notice that both $V_\alpha(z)$ and $V_{2\alpha_0-\alpha}(z)$ have the same dimension. In particular we can write the identity in two ways, 1 and :exp $2i\alpha_0\phi$:. The existence of currents of dimension one allows us to introduce screening operators $Q_\pm$:

$$Q_\pm = \oint_c J_\pm(z)dz \ . \qquad (3.32)$$

Inserting $Q_\pm$ in correlation functions does not change the conformal properties. If we write down the 4-point function of an operator $V_\alpha(z)$.

$$\langle V_\alpha V_\alpha V_\alpha V_{\bar\alpha}\rangle \qquad (3.33)$$

the total charge is $3\alpha + 2\alpha_0 - \alpha = 2\alpha + 2\alpha_0$; hence will vanish generically unless:

$$2\alpha_0 = -n\alpha_+ - m\alpha_- \tag{3.34}$$

in which case we can introduce nQ_+ and mQ_- screening operators to obtain a nonvanishing amplitude. Therefore the spectrum of vertex operators with nonvanishing 4-point functions (3.33) is given by

$$\alpha_{n,m} = \frac{1-n}{2}\alpha_+ + \frac{1-m}{2}\alpha_- \quad , n, m \geq 1 \tag{3.35}$$

and using (3.24) their dimensions are:

$$h(n,m) = -\frac{1}{4}\alpha_0^2 + \frac{1}{4}(n\alpha_+ + m\alpha_-)^2 \ . \tag{3.36}$$

Equations (3.35) and (3.36) reproduce the Kac spectrum of degenerate conformal families. Hence a 4-point block will take the form:

$$\oint_{C_1} dt_1 \ldots \oint_{C_N} dt_N \oint_{C_1'} dt_1' \ldots \oint_{C_m'} dt_m' \langle V_{\alpha'}(z_1) V_{\alpha_2}(z_2) V_{\alpha_3}(z_3) V_{\alpha_4}(z_4)$$
$$J_+(t_1) \ldots J_+(t_N) J_-(t_1') \ldots J_-(t_M')\rangle_{-2\alpha_0} \tag{3.37}$$

where the correlation function is evaluated in terms of a free scalar field with background charge $-2\alpha_0$ (from now on we will remove the subindex $-2\alpha_0$ from $< \ldots >$).

As a first application of this construction we derive the fusion rules (3.19). We want to find the fields $\phi_{(k,l)}$ appearing in the OPE of $\phi_{(m,n)}$ and $\phi_{(r,s)}$. To do this we only need to concentrate on the three-point function. In the Coulomb gas representation, and using Sl_2 invariance, there are three equivalent ways of representing the three point function:

$$\langle V_{\overline{(k,l)}}(\infty) V_{(m,n)}(1) V_{(r,s)}(0) Q_+^{\cdot\cdot} Q_-^{\cdot\cdot}\rangle \tag{3.38a}$$
$$\langle V_{(k,l)}(\infty) V_{\overline{(m,n)}}(1) V_{(r,s)}(0) Q_+^{\cdot\cdot} Q_-^{\cdot\cdot}\rangle \tag{3.38b}$$
$$\langle V_{(k,l)}(\infty) V_{(m,n)}(1) V_{\overline{r,s}}(0) Q_+^{\cdot\cdot} Q_+^{\cdot\cdot}\rangle \tag{3.38c}$$

(the exponents in $Q_\pm$ are not indicated because we will not need them. It suffices to know that one can always screen the charge of the three vertex operators). From (3.35) and (3.38a), the correlation function will not vanish if:

$$\begin{aligned} &k \leq m + r - 1 \quad \text{with} \quad k - m - r - 1 \quad \text{even} \\ &l \leq n + s - 1 \quad \text{with} \quad l - n - s - 1 \quad \text{even} \end{aligned} \tag{3.40}$$

and similarly with (3.38b) and (3.38c), we thus obtain:

$$\left.\begin{array}{l} k\le m+r-1 \\ m\le r+k-1 \\ r\le k+m-1 \end{array}\right\} \quad k{+}m{+}r \quad \text{odd} \quad \left.\begin{array}{l} l\le n+s-1 \\ n\le s+l-1 \\ s\le l+n-1 \end{array}\right\} \quad l{+}n{+}s \quad \text{odd} \tag{3.41}$$

which implies:

$$\phi_{(m,n)}\times\phi_{(r,s)} = \sum_{\substack{k=|m-r|+1 \\ k+m+n=\text{odd}}}^{m+r-1} \quad \sum_{\substack{l=|n-s|+1 \\ l+n+s=\text{odd}}}^{n+s-1} [\phi_{(k,l)}] \,. \tag{3.41}$$

We now consider the minimal series (3.20). We know that $h(n,m) = h(p-n, q-m)$, then:

$$\phi_{(n,m)}\times\phi_{(r,s)} = \sum{}'^{\,n+r-1}_{a=|n-r|+1} \quad \sum{}'^{\,m+s-1}_{b=|m-s|+1} [\phi_{(a,b)}] \tag{3.42a}$$

$$\phi_{(p-n,q-m)}\times\phi_{(p-r,q-s)} = \sum{}'^{\,2p-n-r-1}_{|n-r|+1} \quad \sum{}'^{\,2q-m-s-1}_{|m-s|+1} [\phi_{(a,b)}] \tag{3.42b}$$

where $\sum'$ means that we sum with constraint $k+m+n=0$ etc. (as in (3.41)). Equations (3.42a) and (3.42b) are compatible as long as:

$$\phi_{(n,m)}\times\phi_{(r,s)} = \sum_{|n-r|+1}^{\min(n+r-1,2p-1-n-r)} \quad \sum_{|m-s|+1}^{\min(m+s-1,2q-1-m-s)} [\phi_{(a,b)}]_{\substack{0<n<p \\ 0<m<q}} \tag{3.43}$$

giving the fusion rules for the minimal models. Setting $q = p+1$ we obtain the fusion rules for the unitary series[8]. As we will see in the next section these fusion rules are analogous to the fusion rules for $SU(2)_{p-2}\times SU(2)_{q-2}$ Wess-Zumino-Witten models[11,12,13].

3.3 *Virasoro characters and null vectors*

Before turning into the computation of correlation functions, we can write down the Virasoro characters of the minimal models[10] as a simple application of the Kac formula. This will give us a chance to exhibit the tower of null states in each Verma module appearing in the minimal series. Given a representation of the Virasoro algebra V with highest weight vector h and central extension c, the character of this representation is defined by:

$$\begin{aligned}\chi_V(q) &= q^{h-c/24}\sum_{N=0}^{\infty} a(N)q^N \\ &= \mathrm{tr}_V\, q^{L_0-c/24}\end{aligned} \tag{3.44}$$

where N is the level, and $a(N)$ is the number of states at level N. If there are no null vectors, a basis of V is given by

$$V = \{L_{-n_1}\dots L_{-n_N}|h\rangle n_1 > 0\} \tag{3.45}$$

obtained by applying any polynomial in the L_{-n} $n > 0$ to the highest weight state. Since $[L_0, L_{-n}] = nL_{-n}$ the character of V is easily seen to be:

$$\chi(q) = \frac{q^{h-c/24}}{\prod_{n=1}^{\infty}(1-q^n)} \tag{3.46}$$

when there are null vectors we have to subtract the submodules generated by the null vectors in order to obtain the character of the irreducible representation associated to a given primary field in the conformal theory.

Take p, q to be relatively prime integers. The Verma module $V_{(n,m)}$ obtained as in (3.45) will have plenty of null vectors for $h = h_{(n,m)}$ (3.20). In fact according to Kac formula there is a null vector at level nm with L_0 eigenvalve $h(n,m) + nm = h(n,-m)$. Some formulae useful for the next arguments are:

$$\begin{aligned}h(n,m) &= \frac{1}{4pq}[(nq-mp)^2-(p-q)^2] \\ h(n,m) &= h(p-n,q-m) = h(-n,-m) = h(p+n,q+n) \\ a(k) &\equiv h(n+2pk,m) \quad = h(n,m-2pk) \\ b(k) &\equiv h(n+2pk,-m) = h(n,-m-2qk)\ .\end{aligned} \tag{3.47}$$

If the highest weight vector is $h(n,m) = a(0)$, there are two null vectors

1) $h(n,m) + nm = h(n,-m) = b(0)$
2) $h(p-n,q-m)+(p-n)(q-m) = h(p-n,m-q) = h(n,2q-m) = b(-1)$

the two null vectors have dimensions $b(0), b(-1)$. If we compute the naive character (3.46) for $V_{(n,m)}$, then we have to subtract the contribution of the two null submodules:

$$\frac{q^{-c/24}}{\prod_{n=1}^{\infty}(1-q^n)}\left(q^{a(0)} - q^{b(0)} - q^{b(-1)}\right) . \tag{3.48}$$

This is not the whole story however. The modules with highest weights $b(0), b(-1)$ overlap and have themselves null vectors, so we are not doing the correct subtraction in (3.48). To get a complete picture of the null vector pattern we consider the general case. There are four possibilities:

1. A module with highest weight vector $a(k) = h(n+2pk, m) = h(n, m-2qk) = h(q-n, q+2qk-m), k > 0$. When $k > 0; q-n > 0, q+2qk-m > 0$, then $h(n+2pk, m)$ has a null vector of dimension $h(n+2pk, -m) = b(k)$. There is a second null vector which is obtained by writing $a(k)$ as $h(p-n, q+2kq-m)$, and it has dimension $h(p-n, m-q-2qk) = h(p-n, q+m-2q(k+1)) = h(n, -m+2q(k+1)) = b(-k-1)$. Hence $a(k), k > 0$ has two null submodules with highest weight vectors of dimensions $b(k)$ and $b(-k-1)$.

2. Next take $a(-k), k > 0$. Then:

$$a(-k) = h(n-2pk, m) = h(n, m+2qk) = h(p+2pk-n, q-m) .$$

Since $p+2pk-n > 0, q-m > 0$, we are back into Kac formula, and there is a null vector of dimension

$$\begin{aligned} h(p+2pk-n, m-q) &= h(2p(k+1)-n, m) = h(n-2p(k+1), -m) \\ &= b(-k-1) . \end{aligned}$$

Writing now $a(-k) = h(n, m+2qk)$ we learn again from Kac formula that there is a second null vector with dimension $h(n, -m-2qk) = b(k)$. Hence $a(-k), k > 0$ has two null submodules with highest weight dimensions $b(k), b(-k-1)$.

3. A module with highest weight $h = b(k)$ $k > 0$. Then

$$b(k) = h(n+2pk, -m) = h(n+2pk+p, q-m) .$$

Both entries are positive integers, and by Kac formula there is a null vector of dimension

$$h(n+2pk+p, m-q) = h(n+2p(k+1), m) = a(k+1) .$$

Furthermore:

$$b(k) = h(n + 2pk, -m) = h(n, -m - 2qk) = h(p - n, 2qk + q + m) \ .$$

In the both term both entries are positive, and there is a second null vector with dimension:

$$h(p - n, -2 - qk - q - m) = h(n, 2q(k + 1) + m) = a(-k - 1) \ .$$

Thus $V_{b(k)}$ has two null submodules with dimensions $a(k + 1), a(-k - 1)$.

4. Finally we have $b(-k), k > 0$. Then

$$b(-k) = h(n - 2pk, -m) = h(2pk + p - n, q + m) = h(2pk - n, m)$$

with a null vector $h(n - 2pk, m) = a(-k)$. Similarly: $b(-k) = h(n - 2pk, -m) = h(n, -m + 2qk)$, and there is a second null vector at $h(n, m - 2qk) = a(k)$

We represent all the cases 1–4 in terms of all embedding diagram of modules in Fig. 11. The quantities $a(k), b(k)$ label the Verma modules $V_{a(k)}, V_{b(k)}$ by the L_0 eigenvales of their highest weight vectors.

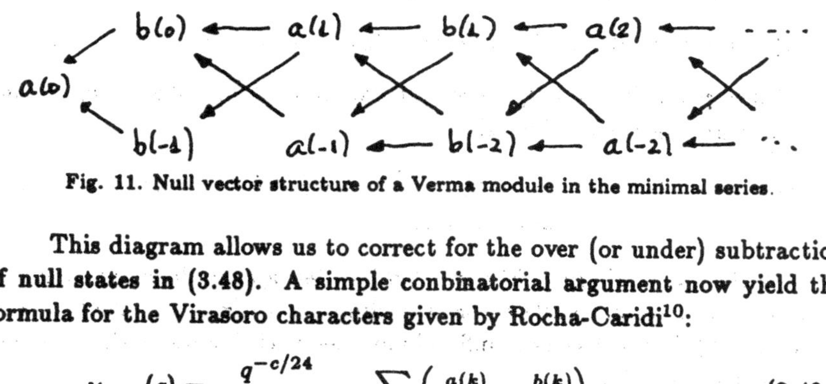

Fig. 11. Null vector structure of a Verma module in the minimal series.

This diagram allows us to correct for the over (or under) subtraction of null states in (3.48). A simple conbinatorial argument now yield the formula for the Virasoro characters given by Rocha-Caridi[10]:

$$\chi_{n,m}(q) = \frac{q^{-c/24}}{\prod_{l=1}^{\infty}(1 - q^l)} \sum_{k \in \mathbb{Z}} \left(q^{a(k)} - q^{b(k)}\right) \tag{3.49a}$$

$$a(k) \equiv h(n + 2pk, m) = \frac{(2pqk + qn - mp)^2 - (p - q)^2}{4pq} \tag{3.49b}$$

$$b(k) \equiv h(n + 2pk, -m) = \frac{(2pqk + qn + mp)^2 - (p - q)^2}{4pq} \ . \tag{3.49c}$$

These characters generate a lot of interesting identities. For instance Euler's pentagonal number theorem is obtained if we take $p = 2, q = 3$. Then we have $c = 0$, and only the identity operator, hence $\chi_{(1,1)} = 1$:

$$\chi_{(1,1)} = 1 = \frac{1}{\prod_{l=1}^{\infty}(1-q^l)} \sum_{k\in\mathbb{Z}} \left(q^{6k^2+k} - q^{6k^2+5k+1}\right)$$

$$\prod_{l=1}^{\infty}(l - q^l) = \sum_{k\in\mathbb{Z}}(-1)^k q^{(3k^2+k)/2} . \tag{3.50}$$

Another example is the character for the spin representation in the Ising model: $p = 3, q = 4, c = 1/2, h_{1,1} = 0, h_{(1,2)} = h_{(2,2)} = 1/16$. Then:

$$\chi_{1/16}(q) = \frac{q^{1/16}}{\prod_{n=1}^{\infty}(1-q^n)} \sum_{k} \left(q^{12k^2+2k} - q^{12k^2+10k+2}\right) = q^{1/16} \prod_{n=1}^{\infty}(1+q^n) \tag{3.51}$$

where the last equality is obtained from the known equivalence between the critical Ising model and a free Weyl-Majorana fermion.

In the character computation just completed we only needed to konw the conformal dimension of the highest weight states in the null submodules. If one wants to explicitly verify that null vectors decouple from correlation functions, we can represent the null vectors explicitly in the Coulomb gas formulation. The naive Verma module (3.45) in this formulation is obtained by first expanding the field ϕ in oscillators. If $V_h(z)$ is the vertex operator creating the highest weight state of dimension h; $|h\rangle, V_h$ is generated by acting on $|h\rangle$ with arbitrary polynomials in the creation operators contained in ϕ. The oscillator representation of the null vectors was given by Fateev and Zamolodchikov[43]. We will need the relations:

$$\begin{aligned}
\alpha_{n,m} &= \frac{1-n}{2}\alpha_+ + \frac{1-m}{2}\alpha_- \\
2\alpha_0 - \alpha_{n,m} &= \frac{1+n}{2}\alpha_+ + \frac{1+m}{2}\alpha_- = \alpha_{-n,-m} \\
2\alpha_0 - \alpha_{n,m} - n\alpha_0 &= \alpha_{n,-m} \qquad \alpha_{n,m} + n\alpha_+ = \beta_{-n,m} \\
2\alpha_0 - \alpha_{n,m} - m\alpha_- &= \alpha_{-n,m} \qquad \alpha_{n,m} + m\alpha_- = \beta_{n,-m} ,
\end{aligned} \tag{3.52}$$

then it is not difficult to show that

$$\chi^{+}_{nm} = \oint_{c_1} \cdots \oint_{c_n} J_+(z_1)\ldots J_+(z_n)V_{\alpha(n,-m)}(z) \tag{3.53a}$$
$$\alpha(+n,-m)$$

$$\chi^{-}_{n,m} = \oint_{c_1} \cdots \oint_{c_m} J_-(z_1)\ldots J_-(z_m)V_{\alpha(-n,m)}(z) \tag{3.53b}$$
$$\alpha(-n,+m)$$

are respectively null vectors of $V_{2\alpha_0} - \alpha_{n,m}$. The proof is simple. From standard vertex operator manipulations:

$$J_+(z_1)\ldots J_+(z_n)V_{\alpha(n,-m)}(0) = \prod_{j<j}(z_i - z_j)^{2\alpha_+^2}\prod_i z_i^{2\alpha+\alpha(n,-m)} \,. \\ : J_+(z_1)\ldots J_+(z_n)V_{\alpha(n,-m)}(0) : \tag{3.54}$$

$$[L_n, J_\pm(z)] = \partial_z\left(z^{n+1}J_\pm(z)\right) \,.$$

If we choose the contours $C_1, C_2, \ldots$ so that there is no cut, we obtain:

$$L_k\chi_{n,m} = Q_+^n L_k V_{\alpha(n,-m)}(0) + \oint dz's\, \partial_{z_1}(z_1^{k+1}\ldots) + \ldots$$

and the contour integrals on the right-hand side vanish. Hence

$$L_k\chi_{n,m} = Q_+^n L_k V_{\alpha(n,m)}(0) \tag{3.55}$$

and:

$$\begin{aligned} L_k\chi_{n,m} &= 0 \quad k>0 \\ L_0\chi_{n,m} &= (h_{n,m}+nm)\chi_{n,m} \,. \end{aligned} \tag{3.56}$$

Let us consisder an example with $n = 2$:

$$\oint_{c_1}\oint_{c_2} dz_1dz_2(z_1 - z_2)^{2\alpha_+^2}(z_1z_2)^{2\alpha+\alpha(2,-m)} : J_+(z_1)J_+(z_2)V_{\alpha(n,-m)}(0) : \tag{3.57}$$

From (3.52) and changing the variables to $z_1 = z, z_2 = z\xi$ (3.57) becomes:

$$\oint_{c_z} dz \oint_{c_\xi} d\xi z^{-1-2m}\xi^{-\alpha_+^2-(1+m)}(1-\xi)^{2\alpha_+^2} \,.: J_+(z)J_+(\xi z)V_{n,-m}(0) : \,. \tag{3.58}$$

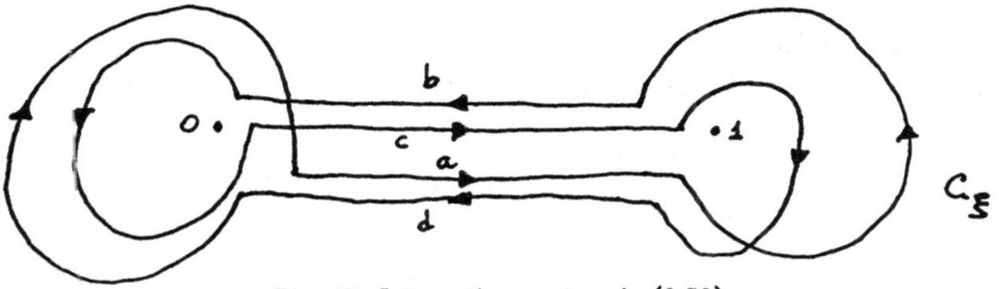

Fig. 12. Integration contour in (3.59).

The z integral will bring down some powers of ξ and some oscillator operators and the coefficients are given by integrals of the form:

$$\oint_{c_\xi} d\xi \; \xi^{-\alpha_+^2-(1+m)+N}(1-\xi)^{2\alpha_+^2} \;, 0 \le N \le nm \; . \tag{3.59}$$

Equation (3.59) is evalvated with the contour in Fig. 12.

When we go around this contour, the integrand comes back to the same value. Keeping track of the phase differences between a, b, c, d we obtain:

$$\begin{aligned} I(\alpha,\beta) &= \oint_{c_\xi} \xi^{\alpha-1}(1-\xi)^{\beta-1} \\ &= \left(1-e^{2\pi i\alpha}\right)\left(1-e^{2\pi i\beta}\right)\oint_0^1 d\xi \; \xi^{\alpha-1}(1-\xi)^{\beta-1} \\ &= -4e^{i\pi(\alpha+\beta)} \sin\pi\alpha \sin\pi\beta \frac{\Gamma(\alpha)\Gamma(\beta)}{\Gamma(\alpha+\beta)} \\ \alpha &= -\alpha_+^2 - m + N \,, \beta = 2\alpha_+^2 + 1 \,. \end{aligned} \tag{3.60}$$

The integrand has four sheets and two branch points, and the genus of the surface is zero by the Riemann-Hurwitz theorem. For more complicated examples the genus of the Riemann surface defined by the integrand will generically be higher than zero.

Using the representation (3.53) for null vectors it is also possible to derive the Kac spectrum of dimensions. Finally a simple consequence of these contour manipulation is that

$$\begin{aligned} Q_+^n V_{\alpha(n,m)}(0) &= 0 \\ Q_-^m V_{\alpha(n,m)}(0) &= 0 \,. \end{aligned} \tag{3.61}$$

Before we considered the module $V_{(2\alpha_0-\alpha_{n,m})} = V_{\alpha_{-n,-m}}$, and found a null vector with the vertex operator $V_{\alpha_{(n,-m)}}$. Now consider the module $V_{(\alpha_{(-n,m)})}$. In the multiple contour integral

$$\oint_{C_1} \cdots \oint_{C_n} J_+(z_1)\dots J_+(z_n) V_{\alpha_{n,m}}(0) \ .$$

we use (3.54) and make the change of variables $z_i = z_1 \xi_i, i = 2, \dots, n; \xi_1 = 1$. Using (3.52) we see that the z_1 integral is well-defined. The power of z_1 in the integrand is z_1^{-1+nm}, hence $Q_+^n V_{\alpha_{n,m}}(0)$ is a null vector in the module $V_{(\alpha_{-n,m})}$. Since the highest weight vector of this module has $h = h_{n,m} + nm$ and $Q_+^n V_{\alpha_{n,m}}(0)$ has $h = h_{n,m}, Q_+^n V_{\alpha_{n,m}}(0)$ must necessarily vanish. A similar argument works for $Q_-^m V_{\alpha_{n,m}}(0)$. The general procedure described up till now is quite useful in the construction of the Kac formula, characters and null vectors for algebras other than Virasoro admitting a Coulomb gas representation.

3.4 *Sample computations: conformal blocks and braiding matrices*

We now come to the explicit computation of correlation functions and fusion and braiding matrices for the minimal theories in this representation[9]. For simplicity we consider correlation functions containing a single screening operator. For instance:

$$\langle \phi_{\overline{(n,m)}} \phi_{(1,2)} \phi_{(1,2)} \phi_{(n,m)} \rangle \tag{3.62a}$$

has charge $2\alpha_0 - \alpha_-$, and therefore it requires one Q_-, and

$$\langle \phi_{\overline{(n,m)}} \phi_{(2,1)} \phi_{(2,1)} \phi_{(n,m)} \rangle \tag{3.62b}$$

has charge $2\alpha_0 - \alpha_+$ and it requires one power of Q_+. From the free field two-point function (3.26) we obtain:

$$\begin{aligned} &\oint dt \langle V_{\alpha_1}(z_1) V_{\alpha_2}(z_2) V_{\alpha_3}(z_3) V_{\alpha_4}(z_4) J_\pm(t) \rangle \\ &= \left(\prod_{i<j} z_{ij}^{2\alpha_i\alpha_j} \right) I(z_i) \\ &I(z_i) = \oint dt \prod_i (z_i - t)^{2\alpha_i \alpha_\pm} \ . \end{aligned} \tag{3.63}$$

We can write this block in a more connonical form using SL_2 invariance

$$\langle\prod_i V_i(z_i)J_\pm(t)\rangle = \prod_{i=1}^{4}(cz_i+d)^{-2h_i}(ct+d)^{-2}$$
$$\langle\prod V_i\left(\frac{az_i+b}{cz_i+d}\right)J_\pm\left(\frac{at+b}{ct+d}\right)\rangle\,.$$

Since t is integrated and $J_\pm$ is a 1-form we can forget about the t-transformation. We now choose a,b,c,d so that $z_1\to\infty, z_2\to 1, z_3\to\eta, z_4\to 0$ where

$$\eta = \frac{z_{12}z_{34}}{z_{13}z_{24}}\,. \tag{3.64}$$

This is achieved by:

$$w = \frac{(z-z_4)(z_1-z_2)}{(z_1-z)(z_2-z_4)}$$

leading to:

$$\oint dt\langle\prod_i V_i(z_i)J_\pm(t)\rangle = \left(\frac{z_{12}z_{14}}{z_{24}}\right)^{h_2+h_3+h_4-h_1}$$
$$\frac{(1-\eta)^{2\alpha_2\alpha_3}\eta^{2\alpha_3\alpha_4}}{z_{12}^{2h_2}z_{13}^{2h_3}z_{14}^{2h_4}}\oint_c dt(1-t)^{2\alpha_2\alpha_\pm}(\eta-t)^{2\alpha_3\alpha_\pm}t^{2\alpha_4\alpha_\pm}\,. \tag{3.65}$$

The integral can be evalvated in terms of hypergeometric functions:

$$F(a,b,c;z) = \frac{\Gamma(c)}{\Gamma(b)\Gamma(c-b)}\oint_0^1 t^{b-1}(1-t)^{c-b-1}(1-tz)^{-a}dt \tag{3.66}$$

and we will need the properties:

$$F(a,b,c;z) = (1-z)^{c-a-b}F(c-a,c-b,c;z) \tag{3.67a}$$

$$F(-n,n,c;z) = \frac{1}{(c)_n}z^{1-c}(1-z)^{n-b+c}\frac{d^n}{dz^n}[z^{n+c-1}(1-z)^{b-c}]$$
$$(c)_n = c(c+1)\dots(c+n-1) \tag{3.67b}$$

$$\cos az = \cos z\, F(\frac{1}{2}+\frac{a}{2},\frac{1}{2}-\frac{a}{2},\frac{1}{2};\sin^2 z) \tag{3.67c}$$

$$\sin az = a\cos z\sin z\, F(1+\frac{a}{2},1-\frac{a}{2},\frac{3}{2};\sin^2 z)\,. \tag{3.67d}$$

In (3.65) there are two possible contours of integration shown in *Fig. 13*. Notice that we are drawing c_1, c_2 as open contours. This requires some explanation. Generically $0, \eta, 1, \infty$ will be branch points of the integrand in (3.65), then the open contour integral and the closed contour integral enclosing a cut differ only by a normalization factor. Since the normalization of the blocks is fixed by the convertion (2.54), we need not worry at this moment about the overall normalization of the integral in (3.65). The difference is quite important however in correlation function. This is very useful, because it gives a different procedure to compute the fusion rules. If the exponenets in the integrand (3.65) conspire in such a way that one of the 2 points in Fig. 13 is not a branch point, then one of the closed contour integrals will vanish. The possible intermediate states in the block (3.62) are determined by the integration contours, hence when one contour gives a vanishing contribution we obtain information about which operators appear in the OPE of the operators in the external legs. Summarizing, even though the contours in Fig. 13 are drawn as open contours, for purpose of determining the fusion rules they should be understood as closed contours on the Riemann surface defined by the integrand.

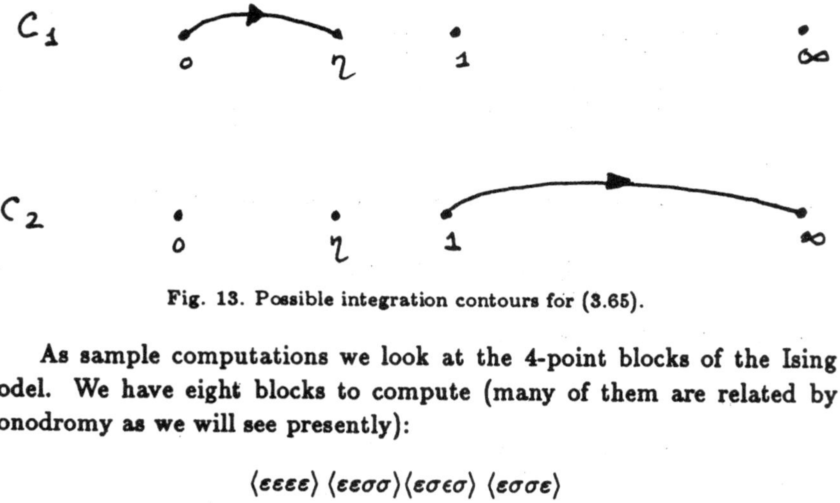

Fig. 13. Possible integration contours for (3.65).

As sample computations we look at the 4-point blocks of the Ising model. We have eight blocks to compute (many of them are related by monodromy as we will see presently):

$$\langle\varepsilon\varepsilon\varepsilon\varepsilon\rangle \ \langle\varepsilon\varepsilon\sigma\sigma\rangle\langle\varepsilon\sigma\varepsilon\sigma\rangle \ \langle\varepsilon\sigma\sigma\varepsilon\rangle$$
$$\langle\sigma\varepsilon\varepsilon\sigma\rangle\langle\sigma\sigma\varepsilon\varepsilon\rangle\langle\sigma\varepsilon\sigma\varepsilon\rangle\langle\sigma\sigma\sigma\sigma\rangle \tag{3.68}$$

for the Ising model:

$$\varepsilon = e^{i\alpha_{2,1}\phi} \qquad c = \frac{1}{2}, p = 3, q = 4$$
$$\sigma = e^{i\alpha_{1,2}\phi} \qquad h_{2,1} = \frac{1}{2}, \alpha_{2,1} = -\alpha_+/2$$

$$h_{1,2} = \frac{1}{16}\alpha_{1,2} = -\alpha_-/2 \tag{3.69}$$

Using (3.65) and (3.66), the integrals along the c_1, c_2 contours needed are:

$$\begin{aligned}\int_0^2 dt_{(c_1)}(1-t)^\alpha(\eta-t)^\beta t^\gamma =& \frac{\Gamma(1+\gamma)\Gamma(1+\beta)}{\Gamma(2+\gamma+\beta)}\eta^{1+\beta+\gamma} \\ & F(-\alpha, 1+\gamma, 2+\gamma+\beta;\eta)\end{aligned} \tag{3.70a}$$

$$\begin{aligned}\int_1^\infty dt(1-t)^\alpha(\eta-t)^\beta t^\gamma =& \frac{\Gamma(1+\alpha)\Gamma(-\alpha-\beta-\gamma-1)}{\Gamma(-\beta-\gamma)} \\ & F(-\beta, -\alpha-\beta-\gamma-1, -\beta-\gamma;\eta)\,.\end{aligned} \tag{3.70b}$$

Start with $\langle\varepsilon\varepsilon\varepsilon\varepsilon\rangle$. The two contours give contributions:

$$\langle\varepsilon\varepsilon\varepsilon\varepsilon\rangle = \frac{1}{z_{13}z_{24}}[\eta(1-\eta)]^{2/3}\begin{cases} [\eta(1-\eta)]^{-5/3}F(-2,-1/3,-2/3;\eta) & 1) \\ F(4/3,3,8/3;\eta) & 2)\end{cases} \tag{3.71}$$

In the second contribution $t=0$ is not a branch point and therefore the second contour does not contribute. Using (3.67b) we obtain:

$$\langle\varepsilon\varepsilon\varepsilon\varepsilon\rangle = \frac{1}{z_{13}z_{24}}\frac{1-\eta+\eta^2}{\eta(1-\eta)} \tag{3.72}$$

satisfying the normalization condition (2.54) and implying that the intermediate state is the identity. Hence the fusion rule is $\varepsilon\times\varepsilon = 1$. In terms of blocks we can write (3.72) as:

$\langle\varepsilon\varepsilon\varepsilon\varepsilon\rangle =$ 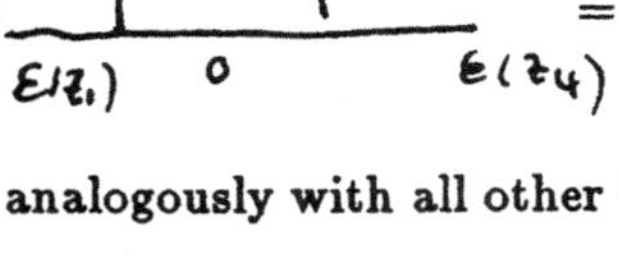

$$= \mathcal{F}_0^{\varepsilon\varepsilon\varepsilon\varepsilon}(z_i) = \frac{1-\eta+\eta^2}{z_{13}z_{24}\eta(1-\eta)}\,. \tag{3.73}$$

Proceding analogously with all other blocks in (3.68) we obtain after some algebra:

$$\mathcal{F}_0^{\varepsilon\varepsilon\varepsilon\varepsilon}(z_i) = \frac{1-\eta+\eta^2}{z_{13}z_{24}\eta(1-\eta)} =$$ 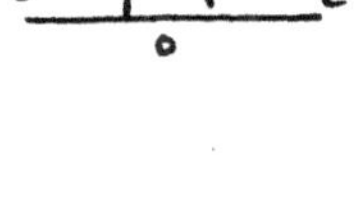

$$\mathcal{F}_0^{\sigma\sigma\sigma\sigma}(z_i) = (z_{13}z_{24})^{-1/8}\frac{(1-\eta)^{3/8}}{\eta^{1/8}\sqrt{2}}\left[\frac{1+\sqrt{1-\eta}}{1-\eta}\right]^{1/2} =$$

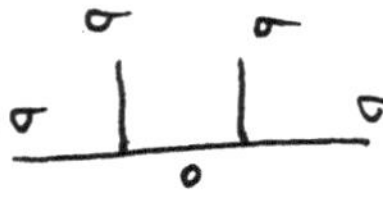

$$\mathcal{F}_\epsilon^{\sigma\sigma\sigma\sigma} = (z_{13}z_{24})^{-1/8}\sqrt{2}\frac{(1-\eta)^{3/8}}{\eta^{1/8}}\left[\frac{1-\sqrt{1-\eta}}{1-\eta}\right]^{1/2} =$$

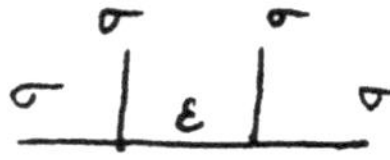

$$\mathcal{F}_0^{\epsilon\epsilon\sigma\sigma} = \frac{(1-\eta)^{-1/2}}{z_{12}z_{34}^{1/8}}(1-\frac{1}{2}\eta)$$

$$=$$

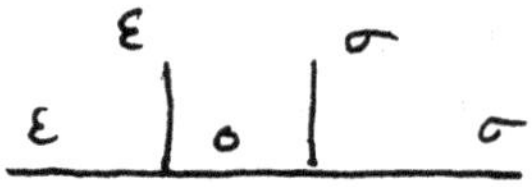

$$\mathcal{F}_\sigma^{\epsilon\sigma\epsilon\sigma} = \frac{1-2\eta}{z_{13}z_{24}{}^{1/8}[\eta(1-\eta)]^{1/2}}$$

$$=$$

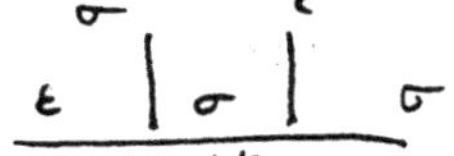

$$\mathcal{F}_\sigma^{\epsilon\sigma\sigma\epsilon} = \left(\frac{z_{14}}{z_{13}z_{24}}\right)^{1/8}\frac{1}{z_{14}}\frac{1+\eta}{\eta^{1/2}(1-\eta)^{1/8}}$$

$$=$$

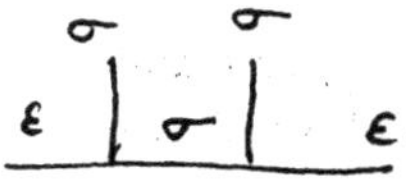

$$\mathcal{F}_0^{\sigma\sigma\epsilon\epsilon} = \frac{1}{z_{12}^{1/8}z_{34}}\frac{1-\eta/2}{(1-\eta)^{1/2}}$$

$$=$$

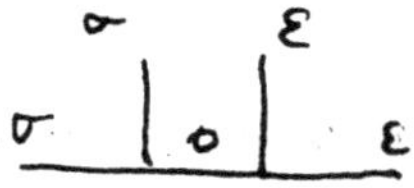

$$\mathcal{F}_\sigma^{\sigma\epsilon\sigma\epsilon} = \frac{[\eta(1-\eta)]^{1/2}}{z_{13}^{1/8}z_{24}}(1-2\eta)$$

$$=$$

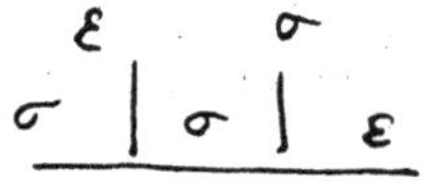

$$\mathcal{F}_\sigma^{\sigma\epsilon\epsilon\sigma} = \frac{(1-\eta)^{-1}\eta^{-1/2}}{z_{14}^{1/8}z_{13}z_{24}}z_{14}(1+\eta)$$

$$= \qquad (3.74)$$

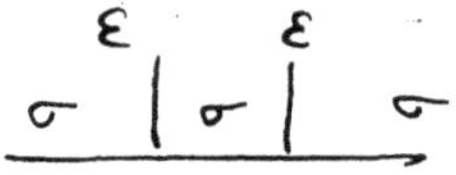

Taking the limit $z_1 = \infty, z_4 = 0$, (3.74) simplifies somewhat:

$$
\begin{aligned}
\eta &= z_3/z_2 \qquad 1-\eta = z_{23}/z_2 \\
\mathcal{F}_0^{\epsilon\epsilon\epsilon\epsilon} &= \frac{z_2^2 - z_2 z_3 + z_3^2}{z_2 z_3 z_{23}} \\
\mathcal{F}_0^{\sigma\sigma\sigma\sigma} &= \frac{1}{\sqrt{2}}(z_2 z - 3 z_{23})^{-1/8}\left(z_2^{1/2} + z_{23}^{1/2}\right)^{1/2} \\
\mathcal{F}_\epsilon^{\sigma\sigma\sigma\sigma} &= \sqrt{2}(z_2 z_3 z_{23})^{-1/8}(z_2^{1/2} - z_{23}^{1/2})^{1/2} \\
\mathcal{F}_0^{\epsilon\epsilon\sigma\sigma} &= \frac{1}{z_2^{1/2} z_{23}^{1/2} z_3^{1/8}}(z_2 - \frac{1}{2} z_3) \\
\mathcal{F}_\sigma^{\epsilon\sigma\epsilon\sigma} &= \frac{z_2 - 2z_3}{z_3^{1/2} z_{23}^{1/2} z_2^{1/8}} \\
\mathcal{F}_\sigma^{\epsilon\sigma\sigma\epsilon} &= \frac{z_2 + z_3}{z_2^{1/2} z_3^{1/2} z_{23}^{1/8}} \\
\mathcal{F}_0^{\sigma\sigma\epsilon\epsilon} &= \frac{z_2 - \frac{1}{2} z_3}{z_2^{1/2} z_3 z_{23}^{1/2}} \\
\mathcal{F}_\sigma^{\sigma\epsilon\sigma\epsilon} &= \frac{z_- 2z_3}{z_3^{1/2} z_2 z_{23}^{1/2}} \\
\mathcal{F}_\sigma^{\sigma\epsilon\epsilon\sigma} &= \frac{z_2 + z_3}{z_2^{1/2} z_3^{1/2} z_{23}}
\end{aligned}
\tag{3.75}
$$

and we can compute the braiding matrices B in Fig. 4. The braiding 2 and 3 is equivalent to the analytic continuation $z_{23} \to e^{i\pi} z_{32}$. For example

$$
\begin{aligned}
\sigma_2 \mathcal{F}_0^{\epsilon\epsilon\epsilon\epsilon}(2,3) &= e^{-i\pi} \mathcal{F}_0^{\epsilon\epsilon\epsilon\epsilon}(3,2) = B_{00}\begin{bmatrix} \epsilon & \epsilon \\ \epsilon & \epsilon \end{bmatrix} \mathcal{F}_0^{\epsilon\epsilon\epsilon\epsilon}(3,2) \\
\sigma_2 \mathcal{F}_0^{\epsilon\epsilon\sigma\sigma}(2,3) &= -\frac{1}{2} e^{-i\pi} \mathcal{F}_\sigma^{\epsilon\sigma\epsilon\sigma}(3,2) \quad B_{0\sigma}\begin{bmatrix} \epsilon & \sigma \\ \epsilon & \sigma \end{bmatrix} = -\frac{1}{2} e^{-i\pi}
\end{aligned}
\tag{3.76}
$$

with only one possible internal state in the block, the computation is rather simple, and we obtain apart from (3.76):

$$
\begin{aligned}
B_{0\sigma}\begin{bmatrix} \epsilon & \sigma \\ \epsilon & \sigma \end{bmatrix} &= -\frac{1}{2} e^{-i\pi/2} \qquad & B_{\sigma\sigma}\begin{bmatrix} \sigma & \sigma \\ \epsilon & \epsilon \end{bmatrix} &= e^{-i\pi/8} \\
B_{\sigma 0}\begin{bmatrix} \sigma & \epsilon \\ \epsilon & \sigma \end{bmatrix} &= -2 e^{-i\pi/2} \qquad & B_{0\sigma}\begin{bmatrix} \sigma & \epsilon \\ \sigma & \epsilon \end{bmatrix} &= -\frac{1}{2} e^{-i\pi/2} \\
B_{\sigma 0}\begin{bmatrix} \epsilon & \sigma \\ \sigma & \epsilon \end{bmatrix} &= -2 e^{-i\pi/2} \qquad & B_{\sigma\sigma}\begin{bmatrix} \epsilon & \epsilon \\ \sigma & \sigma \end{bmatrix} &= e^{-i\pi}
\end{aligned}
\tag{3.77}
$$

the only complicated case comes with braiding in the $\langle\sigma\sigma\sigma\sigma\rangle$ blocks. Using the identities:

$$\begin{aligned}\left[y^{1/2}+i(x-y)^{1/2}\right]^{1/2} &= \frac{e^{i\pi/4}}{\sqrt{2}}\left[x^{1/2}+(x-y)^{1/2}\right]^{1/2}+ \\ &\quad \frac{e^{-i\pi/4}}{\sqrt{2}}\left[x^{1/2}-(x-y)^{1/2}\right]^{1/2} \\ \left[y^{1/2}-i(x-y)^{1/2}\right]^{1/2} &= \frac{e^{-i\pi/4}}{\sqrt{2}}\left[x^{1/2}-(x-y)^{1/2}\right]^{1/2}+ \\ &\quad \frac{e^{-i\pi/4}}{\sqrt{2}}\left[x^{1/2}+(x-y)^{1/2}\right]^{1/2}\end{aligned} \tag{3.78}$$

we easily obtain:

$$\begin{pmatrix} B_{00}\begin{bmatrix}\sigma & \sigma\\ \sigma & \sigma\end{bmatrix} & B_{0\epsilon}\begin{bmatrix}\sigma & \sigma\\ \sigma & \sigma\end{bmatrix} \\ B_{\epsilon 0}\begin{bmatrix}\sigma & \sigma\\ \sigma & \sigma\end{bmatrix} & B_{\epsilon\epsilon}\begin{bmatrix}\sigma & \sigma\\ \sigma & \sigma\end{bmatrix}\end{pmatrix} = \begin{pmatrix} \frac{e^{i\pi/8}}{\sqrt{2}} & \frac{1}{2\sqrt{2}}e^{-3\pi i/8} \\ \sqrt{2}e^{-3\pi i/8} & e^{i\pi/8}/\sqrt{2}\end{pmatrix}. \tag{3.79}$$

The eigenvalves of this matrix are easy to compute if we realize that the F move in Fig. 4 diagonalizes B. Hence the eigenvalves are $e^{i2\pi i h_\sigma}e^{i\pi h_0} = e^{-2\pi i h_\sigma}$ and $e^{-2\pi i h_\sigma}e^{i\pi h_\epsilon}$ or $e^{-i\pi/8}$ and $e^{i\pi 3/8}$.

Finally to compute the physical correlation functions and the coeficients of the OPE we need to combine the left and right blocks onto monodromy invariant combinations. For these blocks where there is only one internal state this is quite straightforward. Choosing standard conventions for the two-point functions:

$$\begin{aligned}\langle\epsilon(z,\bar z)\epsilon(w,\bar w)\rangle &= \frac{1}{|z-w|^2} \\ \langle\sigma(z,\bar z)\sigma(w,\bar w)\rangle &= \frac{1}{|z-w|^{1/4}}\end{aligned} \tag{3.80}$$

we learn that $C_{\epsilon\epsilon} = 1$ and $C^0_{\sigma\sigma} = 1$, to determine the only nontrivial OPE coefficient $C^\sigma_{\epsilon\epsilon}$, we write:

$$\langle\sigma\sigma\sigma\sigma\rangle = A_0|\mathcal{F}_0^{\sigma\sigma\sigma\sigma}|^2 + A_\epsilon|\mathcal{F}_\epsilon^{\sigma\sigma\sigma\sigma}|^2 \tag{3.81}$$

from (2.52) and (2.56), the explicit form of the braiding matrix (3.79) and the normalization (3.80) we easily find

$$C^{\epsilon}_{\sigma\sigma} = 1/2 \ . \tag{3.82}$$

From the general relation between F and B matrices (Fig. 9) it is easy to obtain the F matrix in terms of the results (3.76) and (3.79), and it is left as an exercise.

A useful exercise in this formalism which extends to other algebras is to determine the fusion rules (3.41) using only the properties of the contour integrals defining the conformal blocks. For generic 4-point functions we will have N contours of J_+ and M contours of J_-. Determining when some of these contour integrals contribute is equivalent to determining what internal states can appear in a block and this computes the fusion rules.

Although we have only presented the Coulomb gas representation for the Virasoro algebra, there are similar constructions for more complicated algebras. For instance, this can be done for $N = 1$[44] and $N = 2$[45] superconformal algebras, W-algebras[46], SU(2) Wess-Zumino-Witten theories[47] etc.

To summarize, in this section we have presented some of the general properties of degenerate conformal families, we compute the Virasoro characters and introduced a powerful computational technique based on a Coulomb gas representation with background charge related to the central term c of the Virasoro algebra.

4. Current Algebras and Wess-Zumino-Witten Models

4.1 *Generalities*

In previous sections we have considered conformal theories whose symmetry algebra was Vir$\otimes\overline{Vir}$. In general, for any CFT we will have a symmetry algebra $\mathfrak{A}$, the chiral algebra $\mathfrak{A} = \mathfrak{A}_L \times \mathfrak{A}_R$ with left- and right-handed componenets. In two dimensions we can label tensors as $(p, \bar{p})$ differentials meaning that the quantity $t(z, \bar{z})dz^p d\bar{z}^{\bar{p}}$ is invariant under holomorphic coordinate transformations. The chiral algebras are generated by tensors of the form $(p, 0)$ and $(0, \bar{p})$, i.e. they are either holomorphic or antiholomorphic. For a $(p, 0)$ tensor or p-differential the conservation law can be written as:

$$\bar{\partial} S(z) = 0 \ . \tag{4.1}$$

This immediately implies an infinite number of conserved quantities,

$$\bar{\partial}(z^n S(z)) = 0 \tag{4.2}$$

expanding $S(z)$ in a Laurent series, the conseved quantities are the coefficients of this expansion:

$$S(z) = \sum_{n\in \mathbb{Z}} S_n z^{-n-p} \,. \tag{4.3}$$

In the case of Virasoro $p = 2$, and $S(z) = T(z)$. The OPE of $S(z)$ with itself defines the commutation relations of the chiral algebra. Other examples are provided by the $N = 1$ super-Virasoro algebra where together with $T(z)$ there is a (3/2,0) field and $(S(z), T(z))$ are the components of a two-dimensional chiral superfield. One of the characteristic features of chiral algebras is that $\mathfrak{A}_L(\mathfrak{A}_R)$ always contains the identity operator and the Virasoro $\overline{(Vir)}$ algebra. The purpose of this section is to study in some detail the case when the chiral algebra is generated by (1,0) fields. The simplest example is given by a free scalar field. The U(1) current is $\partial\phi$, and the OPE:

$$\partial\phi(z)\partial\phi(w) = -\frac{1}{(z-w)^2} + \text{regular} \tag{4.4a}$$

$$\partial\phi(z) = \sum_{n\in \mathbb{Z}} \alpha_n z^{-n-1} \tag{4.4b}$$

and the commutation relations are the canonical commutation relations for bosonic oscillators:

$$[\alpha_n, \alpha_m] = n\delta_{n+m,0} \,. \tag{4.5}$$

Another example is provided by a set of Weyl-Majorana fermions $\psi_i(z)$. Given a group G and a real representation $(T^a)ij$, $i, j = 1, 2, \ldots \dim T$ we can construct currents:

$$j^a(z) = \frac{1}{2} : \psi T^a \psi := \sum_{n\in \mathbb{Z}} j_n^a z^{-n-1} \,. \tag{4.6}$$

Since the only nontrivial OPE is:

$$\psi_i(z)\psi_j(w) = \frac{\delta_{ij}}{z-w} + \text{regular} \tag{4.7}$$

the commutation relations of the J_n^a's follow from the OPE:

$$j^a(z)j^b(w) = \frac{\mathrm{tr}T^aT^b/2}{(z-w)^2} + if^{abc}\frac{j^c(w)}{z-w} + \text{reg.} \tag{4.8}$$

where f^{abc} are the structure constants of the group G:

$$[T^a, T^b] = if^{abc}T^c \ . \tag{4.9}$$

In terms of commutators (4.8) becomes:

$$[j_n^a, j_m^b] = if^{abc}j^c n + m + \frac{1}{2}\mathrm{tr}T^aT^b n\delta_{n+m,0} \tag{4.10}$$

It is also useful to compute the OPE between the current and $\psi_i(z)$:

$$j^a(z)\phi_k(w) = -(T^a)_{kj}\frac{\psi_j(w)}{z-w} \ . \tag{4.10}$$

From (4.10) we see that the zero modes j_0^a satisfy the commutation relation of the Lie algebra of G. If we consider complex fermions, the action for the left movers is:

$$S = \frac{1}{\pi}\int b^i\bar{\partial}c_i \qquad i = 1, \dots, N \tag{4.11}$$

with two-point functions:

$$\begin{aligned} b^i(z)c_j(w) &= \frac{\delta_j^i}{z-w} + \text{regular} \\ c_i(z)b^j(w) &= \frac{\delta_i^j}{z-w} + \text{regular} \ . \end{aligned} \tag{4.12}$$

Given representation T^a of G of dimension N we can write

$$j^a(z) = b^i(T^a)_i{}^j c_j = bT^a c \tag{4.13}$$

and

$$j^a(z)j^b(w) = \frac{\mathrm{tr}T^aT^b}{(z-w)^2} + \frac{1}{z-w}if^{abc}J^c(w) + \text{regular} \ . \tag{4.14}$$

Normalizing the roots of G so that the highest root θ is such that $\theta^2 = 2$, and choosing an appropriate basis for the Lie algebra, tr $T^aT^b = k\delta^{ab}$.

Choosing this basis we can always write down the commutation relations of the current algebra as:

$$[j_n^a, j_m^b] = i f^{abc} j_{n+m}^c + n\delta^{ab} k \delta_{n+m,0} \tag{4.15}$$

In the same way that the Virasoro generators provide a projective representation of the algebra of vector fields l_n, the j_n^a are associated to the loop algebra of G(for more details and references on the theory of Kac-Moody algebras and their applications, see Ref. 48):

$$\varepsilon(z) = \sum_{a,n} \varepsilon_n^a T^a z^n \tag{4.16}$$

ε_n^a are the generators of the loop algebra. The algebra (4.16) is a central extension of the loop algebra of G. This is the case when the conserved currents generating the chiral algebra have dimensions (1,0) or (0,1). Their commutation relations always generate a Kac-Moody algebra like (4.16). If the ground state is invariant under the Kac-Moody algebra:

$$j_n^a|0\rangle = 0 \quad n \geq 0 \tag{4.17}$$

the fields of the theory will organize into families each providing an irreducible representation of the Kac-Moody algebra. As with Virasoro we will have primary and descendant fields. A primary field has the simplest transformation properties:

$$j^a(z)\phi_j(w) = -(T^a)_{jk} \frac{\phi_k(w)}{z - w} \tag{4.18a}$$

or:

$$[j_n^a, \phi_j(w)] = -(T^a)_{jk} \phi_k(w) w^n \tag{4.18b}$$

The module generated by $\phi_j(w)$ has a highest weight subspace

$$|j\rangle = \phi_j(0)|0\rangle \tag{4.19}$$

forming a representation of the group G, and the descendants are obtained acting on $|j\rangle$ with j_{-n}^a $n > 0$.

$$V_{(\phi_j)} = \{j_{-n_1}^{a_1} \cdots j_{-n_N}^{a_n} |j\rangle n_1, \ldots, n_N > 0, N \geq 0\} \tag{4.20}$$

The primary fields also have the OPE:

$$T(z)\phi_i(w) = \frac{h}{(z-w)^2}\phi_i(w) + \frac{1}{z-w}\partial_w \phi_i(z) + \ldots \tag{4.21}$$

$$T(z)j^a(w) = \frac{1}{(z-w)^2}j^a(w) + \frac{1}{(z-w)}\partial j^a(w) + \text{regular}\ . \tag{4.22}$$

From the basic OPE between currents:

$$j^a(z)j^b(w) = \frac{k\delta^{ab}}{(z-w)^2} + i f^{abc}\frac{j^c(w)}{z-w} + \text{reg.}\ . \tag{4.23}$$

we can use SL_2 invariance and Bose symmetry to compute the 2,3 and 4-point functions involving only currents:

$$\langle J^a(z)j^b(w)\rangle = \frac{k\delta^{ab}}{(z-w)^2} \tag{4.24a}$$

$$\langle j^a(z)j^b(w)j^c(y)\rangle = \frac{ikf^{abc}}{(z-w)(z-y)(w-y)} \tag{4.24b}$$

$$\begin{aligned}
&\langle j^a(z)j^b(w)J^c(x)j^d(y)\rangle \\
&= k^2\left(\frac{\delta^{ab}\delta^{cd}}{(z-w)^2(x-y)^2} + \frac{\delta^{ac}\delta^{bd}}{(z-x)^2(w-y)^2} + \frac{\delta^{ad}\delta^{bc}}{(z-y)^2(w-x)^2}\right) \\
&- \frac{k}{3}\frac{f^{abe}f^{cde} + f^{ace}f^{bde}}{(z-w)(z-x)(w-y)(x-y)} - \frac{k}{3}\frac{f^{abe}f^{cde} + f^{ade}f^{cbe}}{(z-w)(z-y)(w-x)(x-y)} \\
&- \frac{k}{3}\frac{f^{ace}f^{bde} + f^{ade}f^{bce}}{(z-x)(z-y)(x-w)(w-y)}\ .
\end{aligned} \tag{4.24c}$$

The structure of (4.24c) is fixed by the pole structure forced by the OPE and requiring symmetry under arbitrary exchanges of (a,z), (b,w), (c,x), (d,y).

Given a Kac-Moody algebra one can always obtain a Virasoro algebra. The energy-momentum tensor is constructed using Sugawara's prescription. From the U(1) case $j(z) = \partial\phi(z)$, (see Ref. 48 for more details), we know that $T(z) = -\frac{1}{2} : j(z)j(z) :$. It is reasonable to try and generalize this prescription in the nonabelian case. Define the normal ordering between two currents by:

$$: j^2(z) := \lim_{z\to w}\left(\sum_a j^a(z)j^a(w) - \frac{k\dim G}{(z-w)^2}\right)\ . \tag{4.25}$$

Using (4.25) we learn immediately that for:

$$T(z)j(w) = \frac{j(w)}{(z-w)^2} + \frac{1}{z-w}\partial j(w) + \dots$$

to hold, we need to normalize (4.26) as:

$$T(z) = \frac{1}{2k + c_G} : j^2(z) : \qquad (4.26)$$

where c_G is the quadratic casimir in the adjoint representation. From (4.24c) we can compute the central extension of the Virasoro algebra generated by $T(z)$. If we define the level as $x = 2k/\theta^2$, then

$$\begin{aligned} c &= \frac{x \dim G}{x + \tilde{h}(G)} \\ \tilde{h}(G) &= c_G/\theta^2 \end{aligned} \qquad (4.27)$$

$\tilde{h}$ is the dual Coxetor number of G and it is independent of the way we normalize the roots. Hence in any theory with Kac-Moody symmetry we can have three types of null vectors: a) Purely Kac-Moody. b) Purely Virasoro, c) mixed. The constraints imposed by null-vectors of type a) are derived from the Ward identity[12]

$$\langle j^a(z)\phi_{i_1}(z_1)\dots\phi_{i_N}(z_N)\rangle = -\sum_{j=1}^{N} \frac{(T^a)}{z - z_j}{}_{i_j k_j}\langle\phi_{i_1}\dots\phi_{k_j}\dots\phi_{i_N}\rangle \qquad (4.28)$$

The mixed ones are very important. Their decoupling leads to the Knizhnik-Zamalodchikov equation for the conformal blocks of the theory. Expanding (4.26) in powers of z, L_{-1} is given by:

$$L_{-1} = \frac{1}{2k + c_G} \sum_a \sum_n : j^a_{n-1} j^a_{-n} : \quad . \qquad (4.29)$$

For any primary state then L_{-1} minus the right-hand side of (4.30) always gives a mixed Virasoro Kac-Moody null vector. As any null vector, it should be annihilated by L_1 and j^a_1:

$$\begin{aligned} &\left[L_1, \left(L_{-1} - \frac{1}{2k + c_G} j^b_{-1} j^b_0\right)\right] |i\rangle = 0 \\ &\left[j^a_1, \left(L_{-1} - \frac{1}{2k + c_G} j^b_{-1} j^b_0\right)\right] |i\rangle = 0 \end{aligned} \qquad (4.30)$$

and we have used the highest weight state condition for $|i\rangle$. The second Casimir in a given representation T^a is defined by:

$$\sum_a (T^a T^a)_{ij} = c_2(T)\delta_{ij} \,. \tag{4.31}$$

Then from:

$$[L_n, j^a_m] = -m j^a_{n+m}$$

(a consequence of the OPE $T(z)j^a(w)$) we easily obtain from (4.30) the dimension of the primary state $|i\rangle$[12]:

$$h = \frac{c_2(T)}{2k + c_2(G)} \,. \tag{4.32}$$

In particular, for SU(2) $\theta = \sqrt{2}$, and the highest weights are of the form $j\theta, j = 0, 1/2, 1, \ldots$; then $c_2(j\theta) = 2j(j+1)$ and:

$$h(j) = \frac{j(j+1)}{k+2} \,. \tag{4.33}$$

The general formula for a simple group G is easy to derive. Normalizing the long roots to have length 2, the quadratic Casimir as an operator is:

$$Q = \sum_i H_i^2 + \sum_\alpha \frac{\alpha^2}{2} E_\alpha E_{-\alpha} \tag{4.34}$$

where H_i are the generators of the Cartan subalgebra and α runs over the roots of the algebra. For simplizity take G to be simple laced. Then for a representation with highest weight $|\lambda\rangle$ we have:

$$\begin{aligned} Q|\lambda\rangle &= (\lambda, \lambda + 2\rho)|\lambda\rangle \\ \rho &= \frac{1}{2}\sum_{\alpha>0} \alpha \,. \end{aligned} \tag{4.35}$$

Hence in general if $\phi_\lambda(z)$ is a primary field labelled by a highest weight representation λ its conformal dimension is:

$$h = \frac{(\lambda, \lambda + 2\rho)}{k + \tilde{h}(g)} \,. \tag{4.36}$$

Next we use (4.30) to derive a differential equation satisfied by the correlation functions of primary fields. Let $\phi_1(z_1)\ldots\phi_n(z_n)$ be primary fields,

$$\begin{aligned} |i\rangle &= \phi_i(0)|0\rangle \qquad L_0|i\rangle = \frac{c(i)}{k+\tilde{h}}|i\rangle \\ j_0^a|i\rangle &= -(T^a)_{ij}|j\rangle \qquad L_n|i\rangle = j_n^a|i\rangle = 0 \quad n>0\,. \end{aligned} \tag{4.37}$$

Since

$$\left(L_{-1} - \frac{1}{2k+c_G}J_{-1}^a J_0^a\right)\phi(z) = 0$$

and:

$$(J_{-1}^a\phi)(z) = \oint_z dw \frac{1}{w-z} J^a(w)\phi(z)$$

we obtain:

$$\begin{aligned} \langle\phi_1(z_1)\ldots\left(L_{-1} - \frac{1}{2k+c_G}J_1^a J_0^a\right)\phi_i(z_i)\ldots\phi_N(z_N)\rangle &= 0 \\ \left((2k+c_G)\frac{\partial}{\partial z_i} + \sum_{j\neq i}\frac{T_i^a T_j^a}{z_i - z_j}\right)\langle\phi_i(z_1)\ldots\phi_N(z_N)\rangle &= 0 \end{aligned} \tag{4.38}$$

This is the Knizhnik-Zamalodchikov[12] equation for the conformal blocks.

A field theory which contains all the features we have explained so far, is the Wess-Zumino-Witten (WZW) σ-model[11]. Its action is

$$S = \frac{k}{16\pi}\int_{S^2} \mathrm{tr}\partial_\mu g^{-1}\partial^\mu g d^2\sigma + \frac{k}{24\phi}\int_{D,\partial D=S^2} \mathrm{tr}(g^{-1}dg)^3\,. \tag{4.39}$$

g is a field valued in the group G, D is a three-dimensional disc whose boundary is the two dimension space-time and k is an integer. The action is invariant under the infinite dimension Kac-Moody group $\hat{G}_L \times \hat{G}_k$:

$$g(x^+,x^-) \to \Omega(x^+)g(x^+,x^-)\bar{\Omega}^{-1}(x^-) \tag{4.40}$$

with infinitesimal generators given by

$$\begin{aligned} j^a(z)T^a &= -\frac{ik}{2}\partial_z g\, g^{-1} \\ j^a(\bar{z})T^a &= -\frac{ik}{2}g^{-1}\partial_{\bar{z}} g \end{aligned} \tag{4.41}$$

the central extension is k, and the energy-momentum tensor is a current bilinear given by the Sugawara construction (4.26).

4.2 *Representations and fusion rules*

We can rewrite the defining relations (4.15) in terms of a Cartan-Weyl basis. If G is a simple algebra, H^i is a basis for the Cartan subalgebra $i = 1, \ldots, r = \text{rank } G$, and α labels the roots of G, we can write (for many more details on Kac-Moody algebras see Ref. 49):

$$\begin{aligned}
&[H_n^i, H_m^j] = k\delta^{ij} n\delta_{n+m,0} \\
&[H_n^i, E_m^\alpha] = \alpha^i E_{n+m}^\alpha \\
&[E_m^\alpha, E_n^\beta] = N(\alpha,\beta) E_{m+n}^{\alpha+\beta} \qquad \alpha+\beta \text{ a root} \\
&\qquad\qquad = 0 \qquad \alpha+\beta \text{ not a root} \\
&\qquad\qquad = 2\frac{\alpha^i H_{n+m}^i}{\alpha^2} + \frac{2km}{\alpha^2}\delta_{n+m,0} \qquad \beta = -\alpha \\
&[k, j_n^a] \quad = 0, \quad d = -L_0, [d, j_n^a] = n j_n^a, [d,k] = 0 \,.
\end{aligned} \tag{4.42}$$

We have included the "level" operator d to avoid having an infinite degeneracy for the roots. The maximal set of commutating operators in (4.42) are (H_0^i, k, d), and the roots of the Kac-Moody algebra $\hat{G}$ can be read off from (4.42) to be $(\alpha, 0, n) n \in \mathbb{Z}$, α a root of G. There are $r+1$ simple roots. A root $(\alpha, 0, n)$ is positive if $n > 0$ or if $\alpha > 0$ for $n = 0$. If $\alpha_i i = 1, \ldots, r$ are the simple roots of G, the simple roots of $\hat{G}$ are $(\alpha_i, 0, 0) \equiv a_i$ and $(-\theta, 0, 1) = a_0$, (θ is as before the highest root). The scalar product on the space of weights is:

$$(\boldsymbol{\lambda}_1, a_1, b_1) \cdot (\boldsymbol{\lambda}_2, a_2, b_2) = \lambda_1 \cdot \lambda_2 + a_1 b_2 + a_2 b_1$$

and the Dynkin diagram follows from

$$K_{ij} = 2a_i \cdot a_j / a_j^2 \qquad i, j = 0, 1, \ldots, r \,. \tag{4.43}$$

as for ordinary Lie algebras, we draw a dot on the diagram for each simple root and if $K_{ij} \neq 0$ we draw $K_{ij} K_{ji}$ lives joining dots i and j. One needs the relation

$$\theta/\theta^2 = \sum_{i=1}^{r} m_i \alpha_i / \alpha_i^2 \tag{4.44}$$

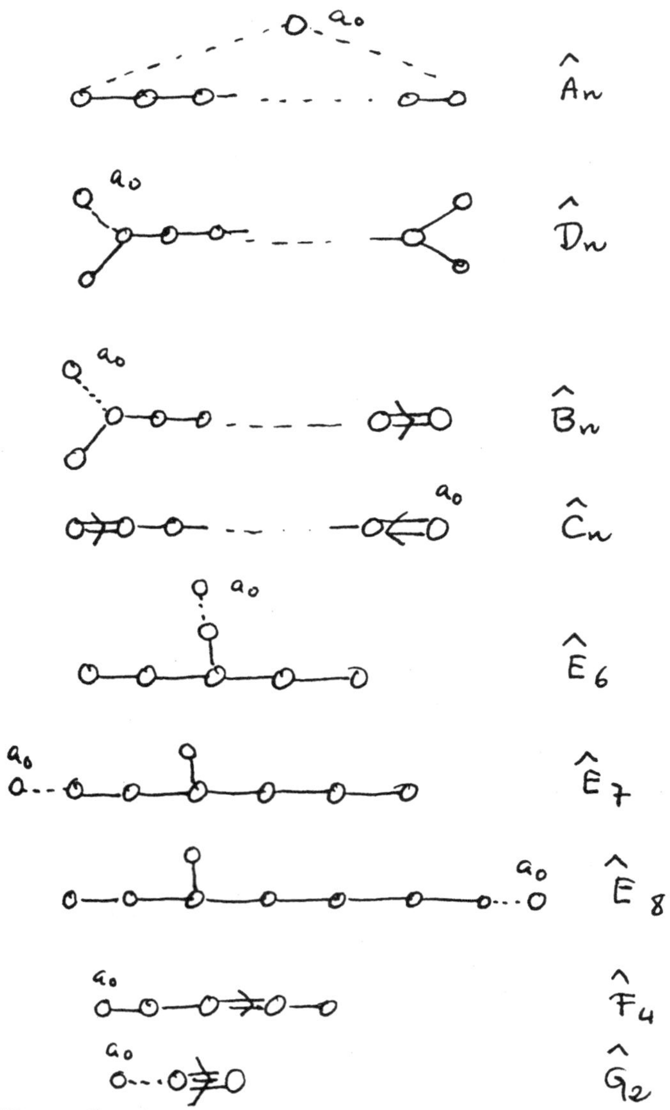

Fig. 14. Dynkin diagrams for untwisted affine Kac-Moody algebras.

to obtain the Dynkin diagrams in Fig. 14.

The highest weight representations are constructed as tensor product of $r + 1$- fundamental representations. The weights are defined as in the classical case:

$$2\Lambda_i \cdot a_j / a_j^2 = \delta_{ij} \quad i,j = 0,1,\ldots,r \; . \tag{4.45}$$

It is not difficult to show that

$$\begin{aligned} \Lambda_i &= \left(\boldsymbol{\lambda}_i, m_i \frac{\theta^2}{2}, 0\right) \qquad i = 1,\ldots,r \\ \Lambda_0 &= \left(0, \frac{1}{2}\theta^2, 0\right) \end{aligned} \tag{4.46}$$

(m_i is defined in (4.44). Any highest weight vector representation of $\hat{G}$ is a positive integer combination of $\Lambda_0, \ldots, \Lambda_r$:

$$\mu = n_0 \Lambda_0 + \sum_{i=1}^{r} n_i \Lambda_i \quad n_0, n_i \geq 0 \tag{4.47}$$

for unitary representations, $k \geq 0$. This is proved by computing:

$$\| E^{\alpha}_{-1} |\mu > \|^2 = \langle \mu | E_1^{-\alpha} E^{\alpha}_{-1} |\mu\rangle = \frac{2}{\alpha^2}(k - \alpha \cdot \mu) \geq 0 \; . \tag{4.48}$$

Since this holds for any root α we conclude that $k \geq 0$. The quantity $2k/\theta^2 \equiv x$ is the level of the Kac-Moody algebra. Important representations of G are those with level one. Normalizing $\theta^2 = 2$, for convenience, we see from (4.46) that Λ_0 (the basic representation) is always of level one. The level one representations depend on the algebra. For example, in $\hat{A}_n$ all the basic representations are level one. In $\hat{D}_n$ we have the basic as well as the two spinors and the fundamental representation all with level 1. For E_8 only the basic representation has level one. This is of great importance in the construction of the heterotic string[50]. To summarize, every highest weight representation of $\hat{G}$ is labelled by a highest weight of the classical algebra and an integer (the level). $\Lambda = (\boldsymbol{\lambda}, k, 0)$.

We come next to determing which representations will appear in the WZW models (or in any unitary CFT whose chiral algebra is $\hat{G}_L \times \hat{G}_k$) and the fusion rules of the operator algebra. We follow closely the presentation in Ref. 13.

To the simple roots of $\hat{G}$ we can associate a Chevalley basis for $\hat{G}$. This simply means that we can construct some SU(2) subalgebras of $\hat{G}$. If $r = \operatorname{rank} G$, r of these subalgebras are ordinary SU(2) subalgebras of G. The more interesting SU(2) is the one associated to a_0; $E_1^{-\theta}$. Then:

$$[E^{\theta}_{-1}, E_1^{-\theta}] = \theta \cdot H_0 - k \; . \tag{4.49}$$

Defining:

$$x^+ \equiv E_1^{-\theta} \quad x^- = E_{-1}^{\theta} \quad x^3 = k - \theta \cdot H_0 \tag{4.50}$$

we obtain:

$$\begin{aligned}[x^3, x^\pm] &= \pm 2x^\pm \\ [x^+ . x^-] &= x^3 \ .\end{aligned} \tag{4.51}$$

If ϕ_λ is the highest weight component of a primary field, then $x^+\phi_\lambda = 0$ and we can construct the SU(2) representation:

$$\begin{aligned}&\phi_\lambda, x^-\phi_\lambda, \ldots \\ &x^3\phi_\lambda = (k - \langle\lambda,\theta\rangle)\phi_\lambda \, ; \langle\lambda,\theta\rangle \equiv 2\lambda\cdot\theta/\theta^2 \ .\end{aligned} \tag{4.52}$$

The representation is finite and integrable if x^3 has integer eigenvalues larger or equal to zero. Hence:

$$k \geq \langle\lambda,\theta\rangle$$

and the dimension of this SU(2) representation is $k - \langle\lambda,\theta\rangle + 1$. Considering instead the operators $E_0^\theta, E_0^{-\theta} H_0^\theta$, they also generate an SU(2) subalgebra, then $(J_0^{-\theta})^{\langle\lambda,\theta\rangle+1}\phi_\lambda = 0$. Since $k \geq \langle\lambda,\theta\rangle$ and $\langle\lambda,\theta\rangle \geq \langle\lambda,\alpha\rangle, \alpha > 0$, we obtain $(J_0^{-\alpha})^{\langle\lambda,\alpha\rangle+1}\phi_\lambda = 0$; and the irreducible representation is integrable iff:

$$(J_0^{-\theta})^{k+1}\phi_\lambda = 0 \ . \tag{4.53}$$

Furthermore, we have a null vector in this integrable representation:

$$(X^-)^{k-\langle\lambda,\theta\rangle+1}\phi_\lambda = 0 \ . \tag{4.54}$$

This null vector allows us to derive the fusion rules of the model. Since

$$(J_{+n}^a\phi)(z) = \oint_z dw(w-z)^n j^a(w)\phi(z)$$

we can write

$$\begin{aligned}0 =& \langle (x^-)^{k-\langle\lambda,\theta\rangle+1}\phi_\lambda(z)\phi_1(z_1)\ldots\phi_n(z_n)\rangle \\ =& \sum_{\substack{l_1,\ldots,l_n \\ l_1+\ldots+l_n=k-\langle\lambda,\theta\rangle+1}} \frac{(k-\langle\lambda,\theta\rangle+1)!}{l_n!l_2!\ldots\ldots l_n!} \\ & \frac{(T_1^\theta)^{l_1}\ldots(T_n^\theta)^{l_n}}{(z-z_1)^{l_1}(z-z_2)^{l_2}\ldots(z-z_n)^{l_n}}\langle\phi_\lambda(z)\phi_1(z_1)\ldots\phi_n(z_n)\rangle = 0 \ .\end{aligned} \tag{4.55}$$

A consequence of this Ward identity is the decoupling of nonintegrable representations from the correlating functions. If $\lambda = id$, the correlation function in (4.55) is independent of z. If $\phi_r(z_r)$ is a nonintegrable primary field, in (4.55) with $\lambda = id$ multiplied by $(z - z_r)^k$ and integrable along a small contour containing z_r and no other z_i's. Then:

$$(T_r^\theta)^{k+1}\langle\phi_\lambda(z)\dots\phi_r(z_r)\dots\rangle = 0 \,. \tag{4.56}$$

If ϕ_λ is integrable, $(J_0^{-\theta})^{k+1}\phi_\lambda = 0$. For nonintegrable ones, if $\phi^{\lambda_r}(z_r)$ is the highest weight state, then there is some state $\tilde{\phi}_r(z_r) = (J_0^{-\theta})^{k+1}\phi^{\lambda_r}(z_r) \neq 0$, taking $\phi_r(z_r) = \tilde{\phi}_r(z_r)$ in (4.56) we find that the correlation functions involving $\phi_r(z_r)$ vanish for ϕ_r a nonintegrable representation. We next use (4.55) for the three-point function:

$$\sum_{\substack{l_1,l_2 \\ l_1+l_2=k-\langle\lambda,\theta\rangle+1}} \frac{(M+1)!}{l_1!l_2!}\,\frac{(T_1^\theta)^{l_1}(T_2^\theta)^{l_2}}{(z-z_1)^{l_1}(z-z_2)^{l_2}}\langle\phi_\lambda(z)\phi_1(z_1)\phi_2(z_2)\rangle = 0 \tag{4.57}$$

Since SL_2 invariance determines the z-dependence of the three-point function we obtain:

$$(T_1^\theta)^{l_1}(T_2^\theta)^{l_2}\langle\phi_\lambda(z)\phi_1(z_1)\phi_2(z_2)\rangle = 0, l_1 + l_2 \geq k - \langle\lambda,\theta\rangle + 1 \,. \tag{4.58}$$

In a highest weight representation $V(\Lambda)$ of G with highest weight $\Lambda, \lambda \in V(\Lambda)$ the depth of λ is defined as the largest integer j such that $\lambda - j^\theta \in V(\Lambda)$. Then the three-point function $\langle\phi_\Lambda(z)\phi_1^i(z_1)\phi_2^j(z_2)\rangle$ vanishes identically unless either $f_{\Lambda_{ij}} = 0$ (f is the Clebsch-Gordan coefficient) or

$$\text{depth}\,(i) + \text{depth}\,(j) \leq k - \langle\lambda,\theta\rangle \tag{4.59}$$

Let $f_{\lambda_{ij}} \neq 0$. If $n = \text{depth}\,(i), m = \text{depth}\,(j)$ and $\langle\phi_\Lambda\phi_i\phi_j\rangle \neq 0$ for $n+m \geq k - \langle\lambda,\theta\rangle + 1$ then we obtain the contradiction:

$$\begin{aligned} 0 &\neq \langle\phi_\Lambda(z)\phi_1^i(z_1)\phi_2^i(z_2)\rangle \\ &= c(T_1^\theta)^n(T_2^\theta)^m\langle\phi_\Lambda\phi_1^{(i-n\theta)}\phi_2^{(j-m\theta)}\rangle = 0 \end{aligned}$$

from (4.58). Hence (4.59) provides the fusion rules for arbitrary WZW models. As an example we consider first $SU(2)_k$. Then $\theta = \sqrt{2}, \lambda = j^\theta, j \in \mathbf{Z}/2$ and the integrable representations satisfy $k \geq \langle\lambda,\theta\rangle = z_j$ i.e. $j =$

$0, 1/2, 1, \ldots, k/2$ and these are $k+1$ irreducible integrable representation. If m is the third component of spin in the spin j representation, then depth $(m) = m + j$. This and (4.59) imply:

$$\left.\begin{array}{l}\mu_1 = m_1\theta \\ \mu_2 = m_2\theta\end{array}\right\} \quad \text{depth}\,(\mu_1) + \text{depth}\,(\mu_2) = m_1 + m_2 + j_1 + j_2 \leq k - 2j_3 \ .$$

Since $m_1 + m_2 + m_3 = 0$ (we are taking Λ in (4.58) to be $j_3\theta$), $(j - 3 - m_3) + (j_1 + j_2 + j_3) \leq k$ leading to

$$j_1 + j_2 + j_3 \leq k \ . \tag{4.60}$$

Hence

$$[\phi_{j_1}] \times [\phi_{j_2}] = \sum_{j_3=|j_1-j_2|}^{\min(j_1+j_2,k-j_1-j_2)} [\phi_j] \tag{4.61}$$

are the fusion rules for SU(2) level k WZW models. With (4.11) it is easy to check that the fusion rules for the minimal models (3.43) are a pair of (4.61) rules. If in the $\phi_{(n,m)}$ field we identify $n = 2j+1, m = 2j'+1$ and $p = k+2, q = k'+2$ the SU(2)$k\times$SU(2)k' fusion rules (4.61) generate (3.43).

For other groups, say SU(N) the computation of the depth of a given weight is a complicated algebraic problem in general. The admissible SU(N) representations are characterized by those Young tableaux $[\lambda_1, \lambda_2, \ldots, \lambda_N], \lambda_1 \geq \lambda_2 \geq \ldots \geq \lambda_N$ such that $\lambda_1 - \lambda_N \leq k$. Examples are given in Fig. 15. The fusion rules are easy to work in some cases. For instance, if one of the representations is the fundamental representation. The weights in the N-dimensional representation of SU(N) are $e_1, \ldots, e_N$ satisfying

$$\begin{aligned} &e_1 + \ldots + e_N = 0 \\ &e_i \cdot e_j = -\frac{1}{N} \quad i \neq j \quad e_i^2 = \frac{N-1}{N} \ . \end{aligned} \tag{4.62}$$

The nonzero roots are:

$$\pm(e_i - e_j) \quad i < j \tag{4.63}$$

and the basic weights are:

$$\Lambda_1 = e_1 \ \ \Lambda_2 = e_1 + e_2, \ldots, \Lambda_k = e_1 + \ldots + e_k \quad k \leq N \tag{4.64}$$

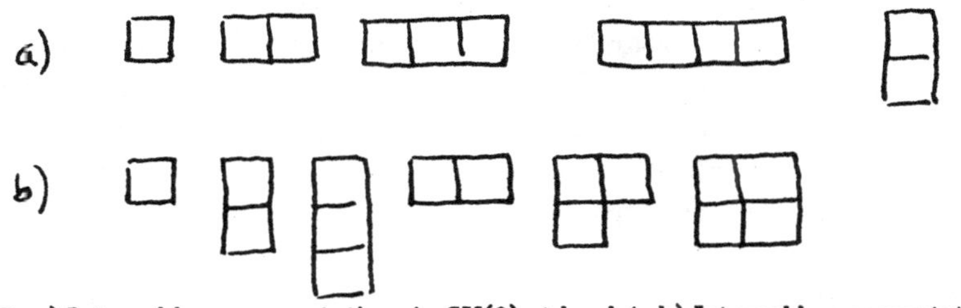

Fig. 15. a) Integrable represenatations in SU(2) at level 4. b) Integrable representations of SU(3) level 2.

Choosing the simple roots as:

$$\alpha_i = e_i - e_{i+1} \quad 1 \le i \le N-1$$

we find that

$$\Lambda_i \cdot \alpha_j = \delta_{ij} \ . \tag{4.65}$$

The weight Λk in (4.64) is represented by a Young tableaux with a single column and k boxes. A general highest weight takes the forms:

$$\Lambda = n_1\Lambda_1 + \ldots + n_{N-1}\Lambda_{N-1} = f_1e_1 + f_2e_2 + \ldots + f_{N-1}e_{N-1} \tag{4.66}$$

represented by a tableaux with f_1 boxes in the first row, f_z in the second etc, $f_1 \ge f_2 \ge \ldots \ge f_{N-1}$. The highest root $\theta = e_1 - e_N$, then $\Lambda \cdot \theta = f_1 \le k$ (for level k in SU(N)). A tableaux is regular if $f_N = 0$. If $f_N \ne 0$, then $[f_1, \ldots, f_N] \simeq [f_1 - f_N, f_2 - f_N, \ldots, f_{N-1} - f_{N,0}]$ because the Young diagrams consisting of a coloumn with N boxes is the trivial representation. The weights of the representation Λk are $e_{i_1} + \ldots + e_{i_k} 1 \le i_1 < i_2 < \ldots < i_k \le N$ and

$$\begin{aligned} &\text{depth}(e_1 + e_{i_2} + \ldots + e_{i\pi}) = 1 \\ &= \text{depth}(\mu) = 0 \qquad \text{otherwise} \end{aligned} \tag{4.67}$$

In SU(N) many of its irreducible representations are complex. The tableaux conjugate to $[f_1, \ldots, f_{N-1,0}]$ is $\overline{[f_1, \ldots, f_{N-1,0}]} = [f_1, f_1 - f_{N-1}, f_1 - f_{N-2}, \ldots, f_1 - f_{2,0}]$. Hence to find out if $[f_i]$ is the OPE $[f_i'] \times [1]$ we have to look at the three-point function:

$$\langle \phi_{\overline{[f_1]}} \phi_{[f_i']} \phi_{[1]} \rangle \ .$$

The conclusion is the following. If $f_1' + 1 \le k$ the fusion rules are identical to the decomposition rules of tensor products of tableaux. When $f_1' = k$, then the decomposition will only include tableaux with $f_1' = k$ and apart from that we follow the standard rules. Since for SU(N) every field can be obtained by decomposing the tensor product of a sufficient number of fundamental fields, it is often useful to collect the fusion rules in Bratelli diagrams. They describe the decomposition of products between the fundamental and other representations. In Fig. 16 we present as an example $\mathrm{SU}(2)_2$ and in Fig. 17 $\mathrm{SU}(3)_2$. The first row in both diagrams is the fundamental representation. The second row represents the result of the OPE of the first row with the fundamental representation; the third row is the result of computing the OPE of the second row with the fundamental etc. Since for finite k the number of integrable representations of $\hat{G}_k$ is finite, the Bratelli diagram will repeat after a finite number of steps. For $\mathrm{SU}(2)_k$ these diagrams can be summarized also in terms of Dynkin-like diagrams. The primary fields have $j = 0, 1/2, 1, \ldots, k/2$. Each one can be associated

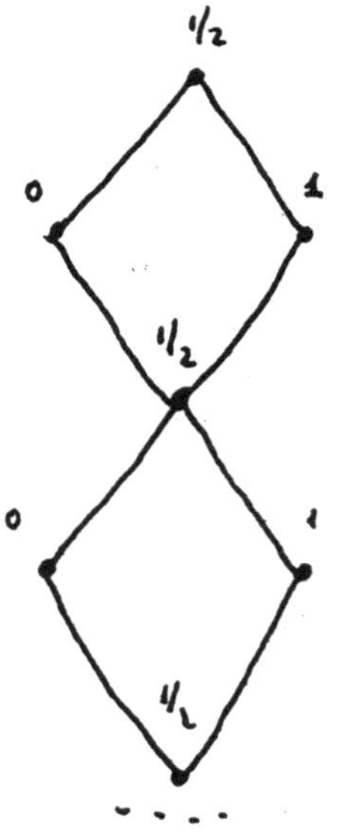

Fig. 16. Bratteli diagram for $\mathrm{SU}(2)_2$.

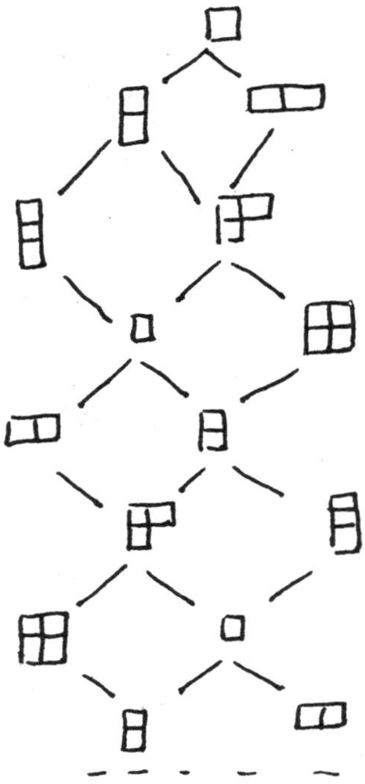

Fig. 17. Bratteli diagram for $\mathrm{SU}(3)_2$.

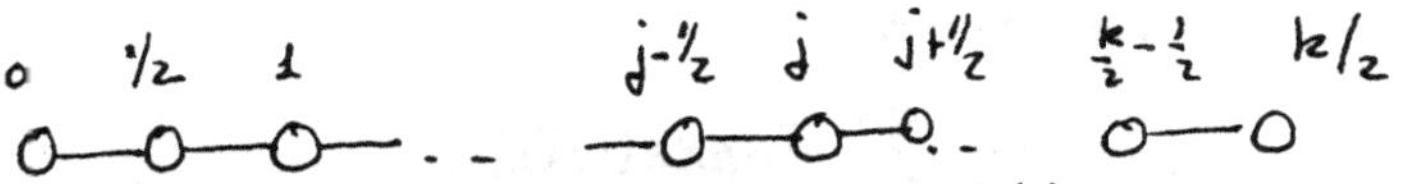

Fig. 18. Dynkin-like representation of the SU(2) fusion rules.

with one of the dots in the Dynkin diagram of A_{k+1} as in Fig. 18. The lines joining different dots represent the result of performing the OPE with the spin 1/2 field. Following the rules of this diagram it is very easy to reconstruct the Bratelli diagram for any level. In cases where the operator algebra is more complicated, we will get multi-dimensional Dynkin-like diagrams. In the minimal models for example we have two generators. $\phi_{(2,1)}, \phi_{(1,2)}$ of the operator algebra leading to two-dimensional graphs and associated Bratelli towers. Other examples of fusion rules will be analyzed in later sections. In our analysis of quantum groups we will show that the fusion rules are a consequence of the properties of tensor products of representations. Before concluding this subsection we would like to add that the braiding and fusing matrices can be obtained in principle by solving the Knizhnik-Zamalodchikov equations (4.28). This is unfortunately a very cumbersome procedure. We will show later that the representation theory of quantum groups provides explicit formulae to compute the duality matrices from simple group theoretic manipulations.

4.3 *Partition functions. modular invariants*

For the minimal theories the partition function is built out from the characters of the Virasoro algebra. For WZW models we can also construct the partition functions in terms of the Kac-Moody characters. So far we have ignored the antiholomorphic dependence of the primary fields. The physical spectrum of the theory however will contain operators with representation $(R, \bar{R}), R(\bar{R})$ with respect the holomorphic (antiholomorphic) Kac-Moody algebra. To find the physical spectrum in a theory where the left and right chiral algebras are isomorphic we have to determine the possible modular (and monodromy) invariant combinations of left and right characters[51]:

$$\begin{aligned} \chi_R(\tau) &= e^{2\pi i\tau\left(\frac{c_R}{2k+c_G} - \frac{k\dim G}{12(2k+c_G)}\right)} \operatorname{tr}_R e^{2\pi i N} \\ &= \operatorname{tr}_R e^{2\pi i(L_0 - c/24)} \end{aligned} \tag{4.68}$$

(and similarly for $\chi_{\bar{R}}(\bar{\tau})$). In (4.68) N is the "number" operator. The partition function takes the form:

$$Z = \sum_{(R,\bar{R})} N_{R\bar{R}} \chi_R(\tau)\overline{\chi_{\bar{R}(\tau)}} \tag{4.69}$$

(we are assuming for simplicity that G is a simple group, the arguments are easily extended to the general case). Since under $\tau \to \tau + 1 \chi_R(\tau)$ changes by a phase, the first requirement on $N_{R\bar{R}}$ is that

$$N_{R\bar{R}} = 0 \quad \text{unless} \quad h_R - h_{\bar{R}} \in \mathbb{Z}\,. \tag{4.70}$$

Invariance under $S : \tau \to -1/\tau$ of Z is more complicated and we need to know the modular properties of the character (or as we will show in the next section the fusion rules for the fields in the theory). The Kac-Moody[49] and minimal model characters[10] can all be expressed in terms of level N ϑ-functions. The simplest ones are:

$$\vartheta_{n,k}(z,\tau;n) = e^{-2\phi_i k_u} \sum_{j\in\mathbb{Z}+n/2k} e^{2\pi i k(j_\tau^2 - jz)} \quad n \in \mathbb{Z}/2k\mathbb{Z}\,. \tag{4.71}$$

Using Poisson resummation one shows that:

$$\vartheta_{n,k}\left(-\frac{z}{\tau}, -\frac{1}{\tau}, u + \frac{z_2}{4\tau}\right) = \left(-\frac{i\tau}{4k}\right)^{1/2} \sum_{e\in\mathbb{Z}/2k\mathbb{Z}} e^{i\pi e n/k} \vartheta_{e,k}(z,\tau;u)\,. \tag{4.72a}$$

Furthermore:

$$\vartheta_{n,k}(z, z+1; u) = e^{i\pi n^2/2k} \vartheta_{n,k}(z,\tau;u)\,. \tag{4.72b}$$

The Rocha-Caridi characters (3.49) with the change of notation:

$$c = 1 - 6\frac{(r-s)^2}{rs}$$

r, s coprime $(r,s) = 1$ becomes:

$$\chi_{p,q}(\tau) = \frac{1}{\eta(\tau)}\left[\vartheta_{n_-,k}(0,\tau;0) - \vartheta_{n_+,k}(0,\tau;0)\right]$$

$$\eta(\tau) = q^{1/24}\prod_{n=1}^{\infty}(1-q^n), q = e^{2\pi i\tau} \tag{4.73a}$$

$$k = rs \quad n_\pm = rp \pm sq\,. \tag{4.74b}$$

Since $(r, s) = 1, 0 < p < r, 0 < q < s$ using (4.72) and some simple arithmetic properties one can show that

$$\chi_{p,q}(\tau + 1) = e^{i\pi n_-^2/2k} \chi_{p,q}(\tau) \tag{4.75a}$$

$$\begin{aligned} \chi_{p,q}(-\frac{1}{\tau}) &= -2\sqrt{\frac{2}{k}} \sum_{p',q'} (-1)^{p'q+pq'} \\ &\quad \sin\left(\frac{ns}{s} pp'\right) \sin\left(\frac{ns}{r} qq'\right) \chi_{p'q'}(\tau) \,. \end{aligned} \tag{4.75b}$$

For Kac-Moody algebras the characters (4.68) are also known[49]. For instance, in the SU(2) algebra at level k the character for the spin j representation is:

$$\chi_j(\tau) = \frac{\vartheta_{-2j-1,k+2} - \vartheta_{2j+1,k+2}}{\vartheta_{-1,2} - \vartheta_{1,2}} \quad (0, \tau; 0) \,. \tag{4.76}$$

this formula is intended as a limit since both numerator and denominator vanish. Using the identity:

$$\eta^3(\tau) = \sum_{m \in \mathbb{Z}} (4m + 1) q^{\frac{1}{8}(4m+1)^2}$$

one shows that

$$\begin{aligned} \chi_j(\tau) &= \eta^{-3}(\tau) \sum_{m \in \mathbb{Z}}^{(2_{j+1} + mn)} q^{(N_m + 2j)^2/2N} \\ N &= 2(k + 2) \,, \end{aligned} \tag{4.77}$$

although in the computation of modular transformations it is more convenient to use the representation (4.76). From (4.72) we obtain:

$$\begin{aligned} \chi_j(-1/\tau) &= \sqrt{\frac{2}{k+2}} \sum_{j'} \sin \frac{\pi(2j+1)(2j'+1)}{k+2} \chi_{j'}(\tau) \\ \chi_j(\tau + 1) &= e^{i\pi\left(\frac{(2j+1)^2}{2(k+2)} - \frac{1}{4}\right)} \chi_j(\tau) \,. \end{aligned} \tag{4.78}$$

For SU(2) and the minimal models Capelli Itzykson and Zuber[52] have given the complete classification of modular invariant combinations of the characters. From the remarks following (4.61) we know that the fusion rules for the minimal models follow from those of SU(2)$p\times$SU(2)p' with $r = p + 2 \quad s = p' + 2$, thus it is reasonable to expect that once the modular invariants for SU(2) are known, we can also construct all the modular invariants for the minimal models. This is indeed the case as shown in Ref. 52.

Using simple algebra it is easy to show that the matrix

$$S_{jj'} = \sqrt{\frac{2}{k+2}} \sin \pi \frac{(2j+1)(2j'+1)}{k+2} \qquad 0 \leq j, j' \leq k/2 \tag{4.79}$$

is orthogonal. Hence S, T (4.78) defines a unitary representation of the modular group $\mathrm{PSL}_2(\mathbb{Z})$ acting on the characters. Hence, for every level k we can always construct the diagonal or A invariant:

$$Z(\tau, \bar{\tau}) = \sum_{j=0}^{k/2} |\chi_j(\tau)|^2 \; . \tag{4.80}$$

The complete classification is intimately connected with the A-D-E classification of simple Lie algebras. To exhibit this relation more clearly we label the characters χ_j as $\chi_{2j+1}(\tau)$. The results of Ref. 52 are:

$$k \geq 1 \quad \sum_{\lambda=1}^{k+1} |\chi_\lambda|^2 \qquad (A_{k+1})$$

$$k = 4\rho, \rho \geq 1, \sum_{\substack{\lambda \mathrm{odd}=1 \\ \lambda \neq 2\rho+1}}^{4\rho+1} |\chi_\lambda|^2 + |\chi_{2\rho}|^2 + \sum_{\lambda \mathrm{odd}=1}^{2\rho-2} (\chi_\lambda \chi^*_{4\rho+2-\lambda} + \mathrm{c.c.}) \qquad (D_{2\rho+2})$$

$$k = 4\rho - 2, p \geq 2 \sum_{\lambda \mathrm{odd}=1}^{4\rho-1} |\chi_\lambda|^2 + |\chi_{2\rho}|^2 + \sum_{\lambda \mathrm{even}=2}^{2\rho-2} (\chi_\lambda \chi^*_{4\rho-\lambda} + \mathrm{c.c.}) \qquad (D_{2\rho+1})$$

$$k + 2 = 12 \quad |\chi_1 + \chi_7|^2 + |\chi_4 + \chi_8|^2 + |\chi_5 + \chi_{11}|^2 \qquad (E_6)$$

$$k + 2 = 18 \quad |\chi_1 + \chi_{17}|^2 + |\chi_5 + \chi_{13}|^2 + |\chi_7 + \chi_{11}|^2 + |\chi_9|^2 \qquad (E_7)$$

$$k + 2 = 30 \quad |\chi_1 + \chi_{11} + \chi_{19} + \chi_{29}|^2 + |\chi_7 + \chi_{13} + \chi_{17} + \chi_{23}|^2 \qquad (E_8)$$

$$\tag{4.81}$$

Note that the values of λ for each invariant coincide with the exponents of the simply-laced algebras on the right-hand side of (4.81). The connection between these invariants, the simply laced algebras, and the crytallographic subgroups of SU(2) is not well-understood. The nondiagonal invariants in (4.81) can be interpreted in a variety of ways. For example the D-invariants $D_{2\rho+2}$ can be thought of a WZW models on SO(3) $=$ SU(2)/$\mathbb{Z}_2$[13]. The possibility of having nondiagonal invariants is also related to the existence of a bigger underlying chiral algebra or that there is a nontrivial automorphism of the fusion rules[53,17]. Consider for example SU$(2)_{4k}$, the primary fields have spin $j = 0, 1/2, 1, \ldots, 2k$ and from the fusion rules (4.61) we know that

$$\begin{aligned} \phi_{2k} \times \phi_{2k} &= 1 \\ \phi_{2k} \times \phi_j &= \phi_{|2k-j|} \;. \end{aligned} \tag{4.82}$$

Furthermore the braiding properties of ϕ_j and $\phi_{|2k-j|}$ are the same as long as j is an integral. The field ϕ_{2k} extends the chiral algebra (the Kac-Moody algebra) with a field of dimension k. If $k = 1$, SU$(2)_1$ algebra[54]. Using the extension of the chiral algebra we see that χ_j and χ_{2k-j} pair up into a single module of the extended algebra for $j \neq k$. Rewriting the $k = 4\rho$ invariant (4.81) as:

$$\sum_{\substack{j\text{integer} \\ j\neq\rho}} |\chi_j + \chi_{2\rho-j}|^2 + 2|\chi_\rho|^2 \,, \tag{4.83}$$

we see that the characters of the extended algebra are of the form $\chi_j + \chi_{2\rho-j}$. (The E_8 invariant can be identified with the G_2 Kac-Moody algebra at level 1[54] and the E_6 invariant with $S_p(4)$ level 1).

Out of the invariants (4.81) we can construct the modular invariant combinations of the minimal models. Write $c = 1 - 6(p-p')^2/pp'$, $(p,p') = 1$ and $k = p-2$, $k' = p'-2$. Since p, p' are coprime, they cannot be both even, hence one of the SU(2) modular invariants must be of A-type. The complete set is[52]:

$$\frac{1}{2} \sum_{r=1}^{p'-1} \sum_{s=1}^{p-1} |\chi_{rs}|^2 \qquad (A_{p'-1}, A_{p-1})$$

$$\begin{matrix} p' = 4\rho + 1 \\ p \geq 1 \end{matrix} \quad \frac{1}{2} \sum_{s=1}^{p-1} \Big\{ \sum_{\substack{r\text{odd} = 1 \\ r \neq 2\rho+1}}^{4\rho+1} |\chi_{rs}|^2 + 2|\chi_{2\rho+1,s}|^2$$

$$+ \sum_{r\,\mathrm{odd}=1}^{2\rho-1} \left(\chi_{rs}\chi^*_{r,p-s} + \text{c.c.}\right) \Big\} \qquad (D_{2\rho+2}, A_{p-1})$$

$$\begin{array}{l} p' = 4\rho \\ p \geq 2 \end{array} \quad \frac{1}{2} \sum_{s=1}^{p-1} \Big\{ \sum_{r\,\mathrm{odd}\,=\,1}^{4\rho-1} |\chi_{rs}|^2 + |\chi_{2\rho,s}|^2$$

$$+ \sum_{r\,\mathrm{even}\,=\,1}^{2\rho-2} \left(\chi_{rs}\chi^*_{p'-r,s} + \text{c.c.}\right) \Big\} \qquad (D_{2\rho+1}, A_{p-1})$$

$$p' = 12 \quad \frac{1}{2} \sum_{s=1}^{p-1} \Big\{ |\chi_{1s} + \chi_{7s}|^2 + |\chi_{4s} + \chi_{8s}|^2 + |\chi_{5s} + chi_{11s}|^2 \Big\}$$

$$(E_6, A_{p-1})$$

$$p' = 18 \quad \frac{1}{2} \sum_{s=1}^{p-1} \Big\{ |\chi_{1s} + \chi_{17s}|^2 + |\chi_{5s} + \chi_{13s}|^2$$

$$+ |\chi_{7s} + \chi_{11s}|^2 + |\chi_{9s}|^2 + [(\chi_{3s} + \chi_{15s})\,\chi_{9s} + \text{c.c.}]$$

$$(E_7, A_{p-1})$$

$$p' = 30 \quad \frac{1}{2} \sum_{s=1}^{p-1} \Big\{ |\chi_{1s} + \chi_{11s} + \chi_{19s} + \chi_{29s}|^2$$

$$+ |\chi_{7s} + \chi_{13s} + \chi_{17s} + \chi_{23s}|^2 \Big\} \qquad (E_8, A_{p-1})$$

$$(4.84)$$

The character formulae for other Kac-Moody algebras is known, and it is easy to show that the diagonal combination of left and right character is always modular invariant. Furthermore, in analogy with SU(2), if $H \subset G$ is a subgroup of the center of $G, H \subset Z(G)$, we can construct the G/H WZW-model using orbifold techniques (as in Ref. 13) to obtain analogues of the D-series. Unfortunately it is not known how to classify all possible modular invariant combinations and whether it has something to do with finite crystallographic subgroups of G. (for some recent progress and references to the literature, see Refs. 55 and 56).

This concludes our quick survey of conformal theories with Kac-Moody symmetry. It is interesting to notice that each module of $\hat{G}_k$ consists of an infinite number of Virasoro modules (G simple). The notion of rationality of a CFT depends on the chiral algebra underlying the theory. From this point of view the classification of rational theories RCFT translates into the classification of chiral algebras and the construction of their associated minimal series.

5. Fusion Algebras. The Verlinde Conjecture

5.1 *The fusion algebra and its properties*

In this section we would like to review in some details the properties of this fusion algebra. The unexpected connection between the fusion rules and the modular transformation properties of the genus 1 characters[15] has generated a lot of work in the classification of rational CFT (RCFT) [16,17,18,22,23,24,25].

Generically, a RCFT is characterized by a chiral algebra $\sigma = \sigma_L \otimes \sigma_R$ and a physical Hilbert space $\mathcal{H}$ which splits into a finite number of irreducible representation (irreps) of $\sigma_L \otimes \sigma_R : \mathcal{H} = \oplus_{(i,\bar{\imath})} \mathcal{H}_i \otimes \mathcal{H}_{\bar{\imath}}$ where $\mathcal{H}_i$ (respectively $\mathcal{H}_{\bar{\imath}}$) is an irrep. of σ_L (respectively σ_R). The primary fields generating $\mathcal{H}_i \otimes \mathcal{H}_{\bar{\imath}}$ are denoted by $\phi_{h,\bar{h}}$. If we consider only the left moves, the OPE yield the fusion rules:

$$[\phi_i] \times [\phi_j] = N_{ij}^k [\phi_k] \ . \tag{5.1}$$

The numbers N_{ij}^k are either zero or positive integers — Defining the matrices $(N_i)_j^k = N_{ij}^k$ the associativity of the OPE implies that:

$$[N_i, N_j] = 0 \ . \tag{5.2}$$

It is obvious from (5.1) that $N_{ij}^k = N_{ji}^k$, hence to every set of fusion rules we can associate a commutative and associative algebra with unit:

$$x_i x_j = N_{ij}^k x_k \qquad i,j,k = 0,1,\ldots,N-1 \ . \tag{5.3}$$

Since the number of primary operators (or conformal families) is finite $\phi_0, \ldots, \phi_{N-1}$, for every ϕ_i we can construct a conjugate field $\phi_{\bar{\imath}}$ such that the OPE $\phi_i \times \phi_{\bar{\imath}}$ contains the identity. The SU(2) examples and minimal models studied in the previous sections are all examples of self-conjugate representations. The "charge conjugation" matrix $C : \phi_i \rightarrow \phi_{\bar{\imath}}$ is given by $N_{ij}^0 = C_{ij}$, and $C^2 = 1$. We can use C_{ij} to raise and lower indices. In particular,

$$N_{ijk} = N_{ij}^l C_{lk} \tag{5.4}$$

is a totally symmetric symbol. In

$$N_{ij}^{l} N_{ml}^{s} = N_{mj}^{l} N_{il}^{s}$$

simply substitute $s = 0$, then:

$$N_{ijm} = N_{mji}$$

which together with $N_{ij}^{k} = N_{ij}^{k}$ implies the desired property. We can analyze the representation theory of the algebra (5.3) using the techniques of finite group theory. First, the existence of a nondegenerate charge conjugation matrix implies that the fusion algebra (5.3) is semi-simple, hence the regular representation $x_i \to N_i$ is both reducible and fully reducible, and there are N one-dimensional inequivalent irreducible representations. Since all the N_i commute, they can be diagonalized simultaneously, and the one-dimensional irreps can be labelled by these eigenvalues:

$$N_i \sim \begin{pmatrix} \lambda_i^{(0)} & & \\ & \ddots & \\ & & \lambda_i^{(N-1)} \end{pmatrix}$$

$$\lambda_i^{(\alpha)} \lambda_j^{(\alpha)} = N_{ij}^{k} \lambda_k^{(\alpha)} \tag{5.5}$$

The matrices N_{ij}^{k} are quite useful in determining the dimension of the conformal blocks of a given theory on an arbitrary Riemann surface. For instance the space of 4-point blocks of the type described in Fig. 19 has dimensions:

$$\sum_p N_{ij}^{p} N_{kp}^{l} = (N_j N_k)_i^{\ l} \ . \tag{5.6}$$

The associativity of (5.3) implies that dimension formulae like the one in (5.6) are independent of the skeleton graph (Fig. 19) we use to compute it. In particular, we can compute the dimension of the space of conformal characters on a Riemann surface of genus g. If $M_{g,o}$ is the moduli space of genus g surface without distinguished points, the characters are sections of a (projectively) flat holomorphic vector bundle over $M_{g,0}$. We can use the skeleton in Fig. 20 to compute the dimension of this vector bundle. This figure looks like two-point function on a genus r surface. The dimension for fixed i is easily computed to be

$$\sum_{\substack{k_1\ldots \\ j_1\ldots \\ l_1\ldots}} N_{ik_1}^{j_1} N_{j_1k_1}^{l_1} N_{l_1k_2}^{j_2} N_{j_2k_2}^{l_2} \cdots N_{l_{r-1}k_r}^{j_r} N_{j_rk_r}^{i} = \sum_{k_1\ldots k_r} (N_{k_1}^2 \cdots N_{k_r}^2)_i^{\ i} \tag{5.7}$$

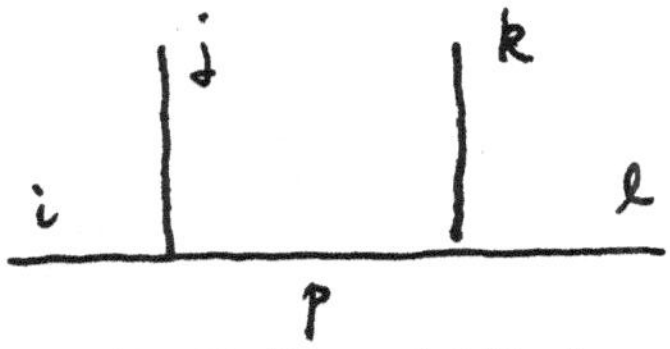

Fig. 19. Four point block.

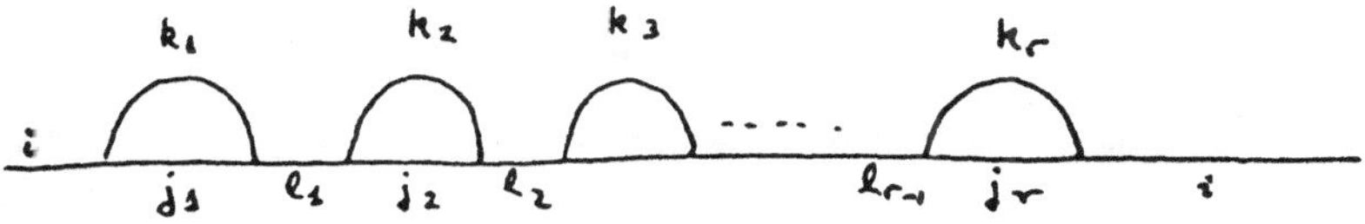

Fig. 20. Skeleton diagram to compute the number of generalized characters at genus g.

we close the loop by summing over i. Hence for a germs g surface, the number of generalized characters is:

$$\dim V_g = \mathrm{Tr}\left(\sum_{k=0}^{n-1} N_k^2\right)^{g-1} . \tag{5.8}$$

At genus 1 we have the standard Virasoro characters:

$$\chi_i(q) = \mathrm{tr}_{[\phi_i]} q^{L_0 - c/24} . \tag{5.9}$$

Under modular transformations:

$$\begin{aligned} T: \quad \chi_i &\to \chi_i\left(qe^{2\pi i}\right) = e^{2\pi i(h_i - c/24)}\chi_i(q) \\ S: \quad \chi_i &\to \chi_i\left(-1/\tau\right) = S_i^j \chi_j(\tau) . \end{aligned} \tag{5.10}$$

T and S generate $\mathrm{PSL}_2(\mathbb{Z})$. If the charge conjugation matrix is not $\mathbb{1}$, then $S^2 = (ST)^3 = C$. In a torus as in Fig. 21 the modular transformation $\tau \to -1/\tau$ acts on the cycles as $(a,b) \to (-b,a)$. Hence $S^2 : (a,b) \to (-a,-b)$ equivalent to charge conjugation. Since $C^2 = 1, S^4 = (ST)^6 = 1$ for the representation of S,T furnished by the CFT. In [15] E. Verlinde introduced a set of operator $\phi_i(a), \phi_i(b)$ defined as follows: $\phi_i(a)$ acts on the characters by first inserting the identity factorized as $\phi_i \times \phi_{\bar{\imath}}$, then we transport the field ϕ_i around the a cycle, and combine it with $\phi_{\bar{\imath}}$ into the identity again. Similarly for $\phi_i(b)$ around the b-cycle. If we think of the a-cycle as space and the b cycle as time, $\phi_i(a)$ acts diagonally on the characters:

$$\phi_i(a)\chi_j = \lambda_i^{(j)}\chi_j \tag{5.11a}$$

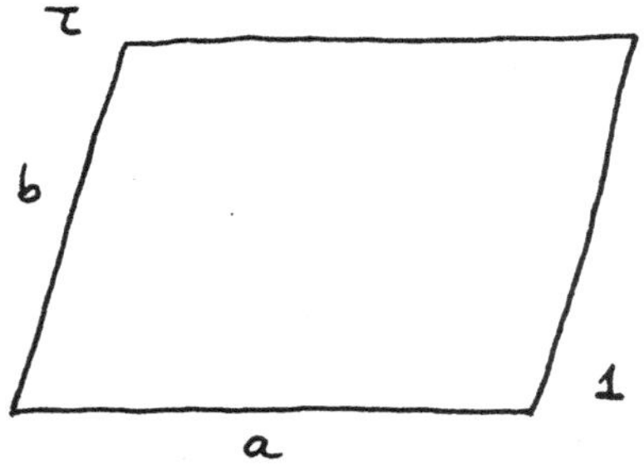

Fig. 21. Fundamental region for a torus $\mathbb{C}/\{\mathbb{Z}+\tau\mathbb{Z}\}$.

(for the time being the $\lambda_i^{(j)}$ are different from the eigenvalues (5.5) although they will be shown to be the same). The operator $\phi_i(b)$ has a less trivial action on the characters. If we start with the character of the identity $\phi_0(\tau)$ it is clear that the action of $\phi_i(b)$ generates $\chi_i(\tau) : \phi_i(b)\chi_0(\tau) = \chi_i(\tau)$ and this also defines the normalization of $\phi_i(b)$. Acting on other characters we will have:

$$\phi_i(b)\chi_j(\tau) = A_{ij}^k \chi_k(\tau) \,. \tag{5.11b}$$

The constants A_{ij}^k satisfy A_{ij}^k, and they also define an associative algebra as a consequence of the associativity of the OPE. By looking at many examples and using some intuitive arguments, E. Verlinde[15] conjectured that $A_{ij}^k = N_{ij}^k$. Later G. Moore and N. Seiberg[16] proved this conjecture for RCFT using the pentagon equation (Fig. 5). Before presenting the proof, we will explore its consequences. First of all, notice that the fact that S acts on the characters implies that there is some notion of completeness of the operator algebra in the CFT. Working at tree level one may satisfy all consistency conditions regarding monodromy invariance etc with a smaller set of fields. For instance in the minimal series we can consider the thermal subalgebra of operators $\{\phi_{(1,m)}\}$. This family closes under operator products and it satisfies all the requirements of locality and duality. One finds that some of the fields are missing when one implements the modular transformation $\tau \to -1/\tau$. It is still an open question whether the requisite on the characters to provide a representation of the modular group can be translated into some set of completeness conditions for the operator algebra at genus zero. However once (5.10) is satisfied and we assume $N_{ij}^k = A_{ij}^k$ we immediately conclude that the matrix S_i^j diagonalizes the fusion matrices N_i. Since the operator S exchanges the a and b cycles we have that:

$$\phi_i(b) = S^{-1}\phi_i(a)S \tag{5.12}$$

as operators on characters. Therefore

$$N_i = S D_i S^{-1} \qquad D_i = \begin{pmatrix} \lambda_i^{(0)} & & \\ & \ddots & \\ & & \lambda_i^{(N-1)} \end{pmatrix} . \tag{5.13}$$

From $N_i S = S D_i$ and $N_{i\cdot}^{\,j} = \delta_i^j$ we obtain:

$$\lambda_i^{(j)} = S_i^j / S_0^j \tag{5.14}$$

and:

$$N_{ij}^k = \sum_m S_j^m \frac{S_i^m}{S_0^m} (S^{-1})_m^k . \tag{5.15}$$

since $S^2 = C, S^{-1} = SC, C^2 = 1$ we obtain:

$$N_{ijk} = \sum_m \frac{S_i^m S_j^m}{S_0^m} S_m^k . \tag{5.16}$$

it is clear that if S is symmetric, N is fully symmetric in (i, j, k). The converse is true. Take $k = 0$ in (5.16)

$$N_{ij0} = C_{ij} = \sum_m S_i^m S_j^m \frac{S_m^0}{S_0^m} = \sum_m S_i^m S_m^j .$$

Hence

$$S_m^j = S_j^m \frac{S_m^0}{S_0^m} .$$

Taking $m = j$ we learn that $S_0^j = S_j^0$ and therefore

$$S^T = S . \tag{5.17}$$

Finally it is also possible to prove the unitarity of S. This is equivalent to showing that $SC = S^*$. The charge conjugation symmetry of the fusion rules $N_{\bar{i}\bar{j}}^{\bar{k}} = N_{ij}^k$ implies that the eigenvalues of $N_{\bar{i}}$ are $\lambda_{\bar{i}}^{(j)} = \lambda_i^{(j)*}$. Then:

$$S_{\bar{i}}^j / S_0^j = S^{*j}{}_i / S_0^{*j} .$$

Since $S_{\bar{i}}^j = (CS)_i^j$, taking $j = 0$ and using $CS = SC$ we obtain $(S_i^0)^* / S_i^0 = (S_0^0)^* / S_0^0$. Hence

$$(CS)_i^j \frac{(S_0^0)^*}{S_0^0} = (S_i^j)^* .$$

Using $S^{*2} = C = S^2$ we obtain $(S_0^0)^*/S_0^0 = \pm 1$. To fix the sign we can use $\tau = i$ as a fixed point of the transformation $S : \tau \to -1/\tau$S. Also at $\tau = i$ the characters are all real $\left(\chi_i(\sqrt{-1}) = v_i\right)$, hence S has an eigenvalue equal to one with a real eigenvector: $S_i^j v_j = v_i$. For $i = 0, S_0^i v_i = v_0$. Since $(S_0^i)^* = \pm S_0^i$, taking complex conjugates in $S_0^i v_i = v_0 \quad v_i^* = v_i$ we avoid a contradiction only if the plus sign is chosen. Thus

$$S_0^i = (S_0^i)^* \quad \text{and} \quad S_i^{j*} = (CS)_i^j \ .$$

we can summarize all the properties of the fusion algebra proved so far:

$$\begin{aligned} S^* &= CS = SC \\ S^T &= S\,, S^2 = C \\ N_{ijk} &= \sum_m \frac{S_i^m S_j^m S_m^k}{S_0^m} \ . \end{aligned} \tag{5.18}$$

These conditions, specially the symmetry of S are surprisingly stringent. It is very easy to construct many examples of associative and commutative algebras (5.3) where structure constants are positive integers, for example the group algebras associated to finite groups. However the symmetry conditions on S rule not almost all examples. If the finite group is Abelian we automatically have a solution as a product $\mathbf{Z}_{N_1} \times \mathbf{Z}_{N_1} \times \ldots \times \mathbf{Z}_{N_m}$ where $\mathbf{Z}_p$ is the group of integers module p. For a single factor $\mathbf{Z}_N$, the rational Gaussian models[5] exhibit a set of fusion rules according to the representation theory of $\mathbf{Z}_N$. The primary fields are labelled by an integer mod N; ϕ_p and the fusion rules are:

$$\phi_p \times \phi_{p'} = \phi_{p+p'} \pmod N \ .$$

For this theory:

$$\begin{aligned} (S\chi)_p &= \frac{1}{\sqrt{N}} \sum_{p'} e^{2\pi i p p'/N} \chi_{p'} \\ S_{pp'} &= \frac{1}{\sqrt{N}} e^{2\pi i p p'/N} \end{aligned}$$

and the modular transformation S is equivalent to finite Fourier transform; then:

$$\lambda_p^{(p')} = e^{2\pi i p p'/N} \qquad C_{pp'} = \delta_{p+p',0} \pmod N \ .$$

As pointed out at the beginning of this section, we can analyze (5.3) using the techniques of the representation theory of finite groups. The regular representation $\phi_i \to N_i$ is fully reducible, and we can derive the orthogonality relations for the characters. The characters of the n one-dimensional representations of (5.3) are:

$$\lambda^{(j)}(x_j) = \lambda_i^{(j)} \equiv \lambda_j^{(i)} \tag{5.19}$$

given the matrix $\Lambda, \Lambda_{ij} = \lambda^{(i)}(x_j)$ let $N, N_{ij} = n_j^{(i)} = n_i(j)$ be its inverse. Then the orthogonality properties of the characters are[59]:

$$\begin{aligned} &\sum_k \lambda_i(k)\overline{\lambda_j(k)} = \frac{1}{\pi_0(i)}\delta_{ij} \\ &\sum_k n_0(k)\lambda_k(i)\overline{\lambda_k(j)} = \delta_{ij} \\ &\overline{\lambda_j(k)} = \lambda_j(\bar{k}) \; . \end{aligned} \tag{5.20}$$

The fusion matrices N_i are diagonalized now by the matrix

$$S_{ij} = \sqrt{n_0(j)}\ \lambda_j(i) \; .$$

If S is symmetric (as required by (5.18)) then from (5.20) one easily derives $S^2 = C$, where C is defined as in CFT.

5.2 *Proof of the Verlinde conjecture*

We now present the proof of $A_{ij}^k = N_{ij}^k$ using the pentagon equation.[16,53] In Sec. 7 we will describe a proof of this conjecture using the properties of quantum groups. The coefficient N_{ij}^k label the number of possible couplings between the conformal families i, j, k. The three-point block can be represented as in Fig. 22 with the extra label counting all possible couplings. In order to avoid making the notation very complicated, we will consider self-conjugate theories and concentrate on one coefficient N_{ij}^k. The arguments are easily generalized to include complex representations. The operator $\phi_i(b)$ is defined as a sequence of manipulations on the chiral blocks of the two-point function on the torus. These operations appear in Fig. 23.

The moves A, B can be written in terms of the F, B moves in Fig. 4.

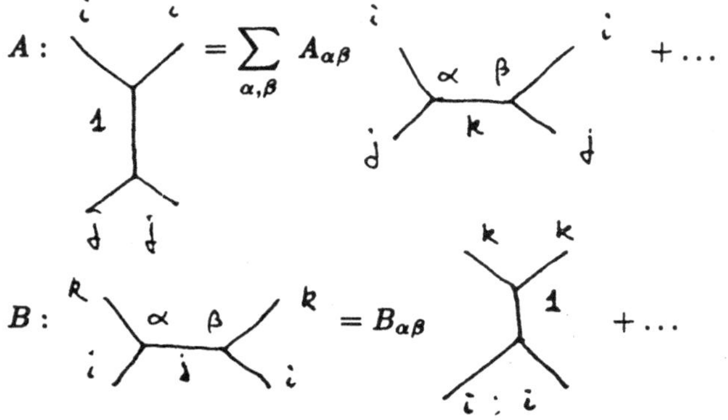

$$\alpha, \beta = 1, \ldots, N_{ij}^{k} \tag{5.21}$$

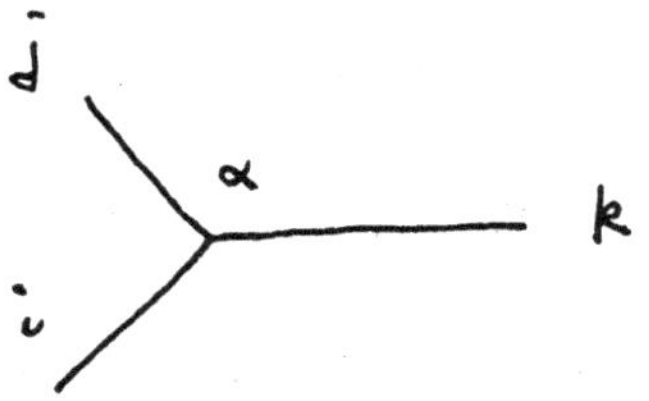

$\alpha = 1, 2, \ldots N_{ij}^{k}$

Fig. 22. An extra label α is unnecessary when $N_{ij}^{k} \neq 1$.

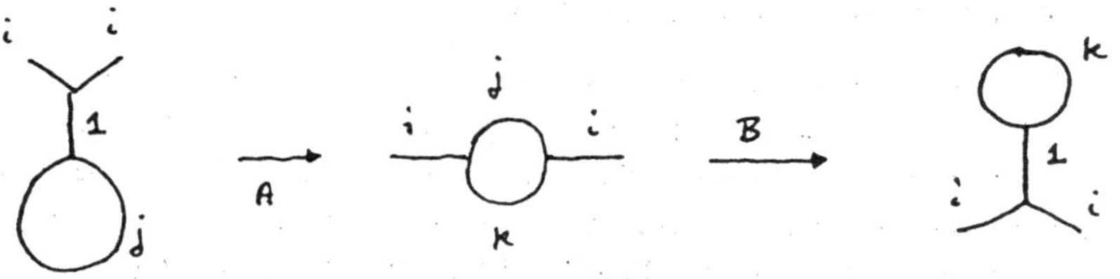

Fig. 23. Operators involved in the definition of the Verlinde operators.

To normalize $\phi_i(b)$ we will also need:

(5.22)

In F, B notation:

$$\begin{aligned} C &= F_{11}\begin{bmatrix} i & i \\ i & i \end{bmatrix} \\ B_{\alpha\beta} &= F_{j1}\begin{bmatrix} k & k \\ i & i \end{bmatrix}_{\alpha\beta} \\ A_{\alpha\beta} &= F_{1k}\begin{bmatrix} j & i \\ j & i \end{bmatrix}^{\alpha\beta} \quad . \end{aligned} \tag{5.23}$$

The dots in (5.21) and (5.22) are the contributions from other representations and they will not be important for the argument. To prove Verlinde's conjecture , we have to show that

$$\sum_{\alpha,\beta=1}^{N_{ij}^{k}} \cdot \frac{A_{\alpha\beta}B_{\alpha\beta}}{c} = N_{ij}^{k} \tag{5.24}$$

according to the moves in Fig. 23. Using the associativity moves which generate the pentagon identity Fig. 5 we obtain the following identity

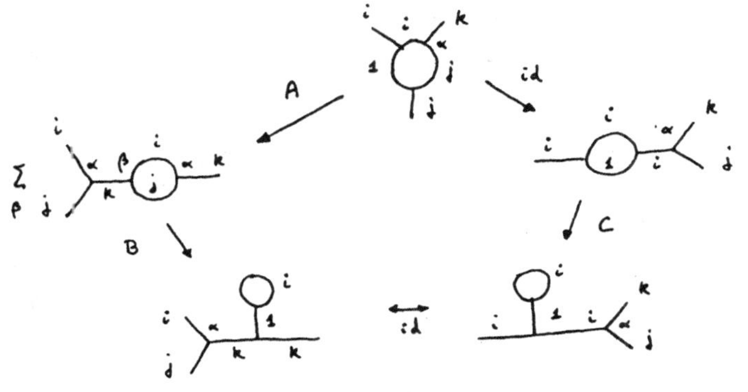

(5.25)

Equations (5.25) become:

$$\sum_{\beta=1}^{N_{ijk}} A_{\alpha\beta}B_{\alpha\beta} = c \; .$$

Since C is independent of α, β we finally obtain (5.24).

As we will see in later sections, the connection between CFT and knot and link invariants follows from the fusion algebra and its properties (5.18). Most of the topological underpinnings of RCFT seem to stem from (5.18).

5.3 *Constraints from rationality*

While we are analyzing general properties of RCFT, it is useful to explore the constraints on the conformal dimension in a RCFT after the fusion rules are known. Using very simple arguments it is possible to show that some fusion rules are not allowed by modular invariance. More important, one can prove the rationality of all the conformal dimensions and the value of c for a RCFT[19,20]. As argued in Sec. 2, the mapping class group of the sphere with n distinguished points can be described in terms of an extension of the braid group B_n by the Dehn twists around the points p_i with local parameters $Z_i, Z_i(P_i) = 0$. Together with the standard defining properties of the braid group we have the relations (2.68). These relations together with (2.69) can be used to derive the relation (2.70). To exhibit the constraints on the possible conformal dimensions, as a consequence of these relations, let us consider an example. We assume that the RCFT has only three primary fields $1, \varepsilon, \sigma$ with fusion rules

$$\varepsilon \cdot \varepsilon = 1 \qquad \sigma \cdot \sigma = 1 + \varepsilon$$
$$\varepsilon \cdot \sigma = \sigma \,. \tag{5.26}$$

We can analyse different conformal blocks depending on the fields ε, σ assigned to the four points p_1, P_2, P_3, P_4 : $\varepsilon\varepsilon\varepsilon\varepsilon, \sigma\sigma\sigma\sigma, \varepsilon\varepsilon\sigma\sigma$. The dimension of the space of blocks $\varepsilon\varepsilon\varepsilon\varepsilon$ is 1, and the eigenvalues of T_{12}, depend only on the dimensions of the fields appearing in the OPE of the fields at p_1 and p_2. The Dehn twist τ_{12} is, in this basis, equivalent to exp $2\pi i L_0$ on the intermediate channels. Since $\varepsilon \cdot \varepsilon = 1$, τ_{ij} are 1×1 matrices with eigenvalues 1. Hence we learn that

$$\alpha_\varepsilon^4 = 1 \qquad \alpha_i = e^{2\pi i h_i} \,. \tag{5.27}$$

With $\sigma\sigma\sigma\sigma$, the block has dimension 2, and the eigenvalues of τ_{ij} are $1, \alpha_\varepsilon$ because $\sigma \cdot \sigma = 1 + \varepsilon$. Taking determinants in

$$\tau_{12}\tau_{13}\tau_{23} = \tau_1\tau_2\tau_3\tau_4$$

leads to:

$$\alpha_\varepsilon^3 = \alpha_\sigma^8 \,. \tag{5.28}$$

$\tau_i = 1,2,3,4$ is proportional to the 2×2 unit matrix. Finally we consider $\epsilon\epsilon\sigma\sigma$. Now τ_{12} has eigenvalue 1, τ_{13} has eigenvalue α_σ and τ_{23} has eigenvalue α_σ and $\tau_1\tau_2\tau_3\tau_4$ yield $\alpha_\epsilon^2\alpha_\sigma^2$. Hence

$$\alpha_\epsilon^2 = 1 \,. \tag{5.29}$$

This implies (5.27) and leads to $\alpha_\epsilon = \alpha_\sigma^8$. Then the possible values of h_ϵ, h_σ are:

$$2h_\epsilon = \text{integer} \qquad 8h_\sigma = h_\epsilon + \text{integer} \,. \tag{5.30}$$

This is obviously satisfied by the Ising model with $h_\epsilon = 1/2, h_\sigma = 1/16$ and for example by the SU(2) WZW theory at level two. The primary fields are 1, $\phi_{1/2}, \phi_1$ with dimensions $h_0 = 0, h_{1/2} = 3/16, h_1 = 1/2$ which clearly satisfies (5.30). In this case the solutions to (5.30) are parameterized by two integers. In general the dimensions $h_i, \ldots, h_n$ of the fields will be determined only mod. 1. It is not known one can obtain extra conditions which put upper bounds on the integers solving equations like (5.27)—(5.29). Using such relations for a general RCFT C. Vafa showed that the dimensions h_i are always rational numbers in a RCFT. Using $(ST)^3 \equiv C$ it is also easy to show that c is a rational number. Taking determinants:

$$\det(ST)^6 = 1 \Rightarrow \det T^6 = \pm 1 \,. \tag{5.31}$$

Since the eigenvalues of T are $\exp 2\pi i(H_i - c/24)$ we obtain:

$$e^{2\pi i c/24} = \pm \prod_{i=1}^{n} \alpha_i^6 \,. \tag{5.32}$$

Hence $c/4 = 6\sum h_i$ mod 1, proving the rationality of c. As a final example of the use of these general algebraic identities we show that a theory with only one field different from the identity: ϕ can only have fusion rules of form $\phi \times \phi = 1$ or $\phi \times \phi = 1 + \phi$. If we try to impose $\phi \times \phi = 1 \times n\phi$ $n \geq 2$, we run into inconsistencies. The argument is very simple. The field ϕ is self-conjugate, $C = 1$ and $(ST)^3 = 1$. The eigenvalues λ of N_ϕ satisfy the quadratic equation

$$\lambda^2 = 1 + n\lambda \tag{5.33}$$

also we can parametrize S and T as

$$S = \begin{pmatrix} \cos\theta & \sin\theta \\ \sin\theta & -\cos\theta \end{pmatrix} \qquad T = e^{-i\pi c/12} \begin{pmatrix} 1 & 0 \\ 0 & \alpha_\phi \end{pmatrix} \alpha_\phi = e^{2\pi i h_\phi} \,. \tag{5.34}$$

The condition that the fusion rules are diagonalised by S implies $\tan\theta = \lambda$; and $(ST)^3 = 1$ leads to

$$\cos 2\pi h_\phi = -\frac{1}{2} n\lambda\,, \qquad 12h - c = 2(\mathrm{mod}\, 8)\,. \tag{5.35}$$

Next consider blocks of the form $\phi\phi\phi\phi$. There are $n^2 + 1$ such blocks. τ_{ij} has n^2 eigenvalues equal to α_ϕ and the remaining eigenvalue τ equal 1. Taking determinants in $\tau_{12}\tau_{12}\tau_{23} = \tau_1\tau_2\tau_3\tau_4$ we arrive at

$$(\alpha_\phi)^{n^2+4} = 1\,. \tag{5.36}$$

The equations:

$$(\alpha_\phi)^{n^2+4} = 1 \qquad \lambda^2 = 1 + n\lambda\,, \quad \alpha_\phi + \alpha_\phi^{-1} = -n\lambda \tag{5.37}$$

are compatible only for $n = 0$ or 1. For $n = 0$ one example is provided the level 1 SU(2) WZW model. With $n = 1$ the examples are the the level 1 G_2 and F_2 WZW models. It is clear from (5.37) and the reality of both h_ϕ, λ that $n \leq 2$. For $n = 0, 1$ we already know of some solutions, so the inconsistency only appears for $n = 2$ because $\lambda = 1 \pm \sqrt{2}$ and $\alpha_\phi^8 = 1$ do not solve the last equation in (5.37).

For a small number of generators it is possible to determine all possible fusion rules. However, once the number of fields different from the identity is larger or equal to 3, the brute force classification seems hopeless and new ideas should come into play if we want to classify the possible fusion rules appearing in RCFT. In the next two sections we will argue that the classification of RCFT is intimately connected to the problem of classified Quasi-Triangular Yang-Baxter algebras (Quantum Groups for short) with some rationality constraint on their representation ring. If this conjecture is true, the fusion rules for RCFT are classified by Clebsch-Gordan decompositions of restricted tensor products of representations of Quantum groups.

We would like to conclude this section with the definition of dimension for a conformal family introduced by Dijkgraaf and Verlinde[53]. Dividing the character $\chi_i(\tau)$ of some family by the character of the identity and taking the $\tau \to 0$ limit gives a regularised version of the dimension of $\mathcal{H}_i$. For finite dimensional spaces, $\chi_i(0) = \dim \mathcal{H}_i$, so this definition appears reasonable:

$$d_i \equiv \lim_{\tau\to 0} \frac{\chi_i(\tau)}{\chi_0(\tau)} \tag{5.38}$$

To compute d_i we use the modular transformation S and the fact that in unitary theories $h_i \geq 0$, and $h_i = 0$ for the identity operator. Then:

$$d_i = \lim_{\tau \to 0} \frac{\sum_j S_i^j \chi_j(-1/\tau)}{\sum_k S_0^k \chi_k(-1/\tau)} = \frac{S_i^0}{S_0^0} \tag{5.39}$$

We will show later how to compute (5.39) for WZW models and how they relate to characters of quantum groups.

6 Duality and Quantum Groups

6.1 *Motivation*

In the first few sections we have reviewed some of the basic properties of CFT. The duality or bootstrap hypothesis provided a set of equations for the OPE coefficients in CFT. Furthermore modular invariance at genus one gives constraints on the possible operators algebra. It is still quite hopeless to try a direct classification of all CFT by solving the bootstrap together with the modular invariance constraints. A resolution of this problem would determine the universality classes of two-dimensional critical phenomena and it would determine completely the classical vacuum manifolds for string and superstring theories. The problem can be drastically simplified if we include a rationality assumption. We want to first classify rational conformal field theories RCFT i.e. those CFT with a finite number of primary fields with respect to some chiral algebra. We have seen that the duality properties of a RCFT are summarized by the hexagon and pentagon identities at genus zero and at genus one we have the relations $S^2 = (ST)^3 = c$. If these conditions are supplemented with the requirement of modular invariance for the one-point functions at genus one, Moore an Seiberg showed that the theory is guaranteed to be modular invariant on a Riemann surface of arbitrary genus. Hence the duality properties are coded in a collection of polynomial equations[16]. Understanding of the space of solutions of the polynomial equations is an important step in the classification of RCFT. In this subsection we analyze the duality properties of a RCFT and argue that their solutions are given by the representation theory of Quantum Groups when the deformation parameters are roots of unity. In Ref. 18 it was shown that the duality properties of these RCFT admitting a classical limit can be described in terms of the representation theory of ordinary groups (finite or continuous). More precisely, the theories with a classical limit fall in sequences of theories with a parameter playing the role of Planck's constant. For instance in WZW theories we have a group G and a level k for the Kac-Moody algebra. The classical limit is obtained as $k \to \infty$. For many of the known sequences the duality properties of their elements can be obtained by quantizing à la Drinf'eld-Jimbo the group appearing in the classical limit. For theories which do not admit a classical limit (if they exist) we may have a situation similar to the classification of exceptional Lie group or sporadic finite simple groups.

Geometrically the conformal blocks can be described as sections of a (projectively) flat holomorphic vector bundle over the moduli space $\mathcal{M}_{g,n}$ of genus g surfaces with n distinguished points (we may also want to include local parameters about the distinguished points to resolve the orbifold singularities in $\mathcal{M}_{g,n}$). At genus zero we have $S^2 - \{P_1, \dots, P_n\}$ and after three points are fixed by $SL_2(C)$ invariance the Teichmüller space can be described in terms of $n-3$ copies of S^2 removing the "diagonal" subsets where two or more points coincide. (The geometrical aspects and their relevance to CFT and string theory has been emphasized by Friedan and Shenker in their attempts to formulate CFT in some Universal Moduli space). The group $\mathrm{Diff}^+(S^+ - \{P_1, \dots P_n\})$ (including maps permuting the points but preserving the orientation) acts on the space of blocks as a consequence of duality. If $\mathrm{Diff}_0^+(S^2 - \{P_1, \dots, P_n\})$ is the subset of diffeomorphisms connected with the identity, the duality properties are related to the representations of the quotient group $\mathrm{Diff}^+/\mathrm{Diff}_0^+$ furnished by the conformal blocks. Since the bundle of conformal blocks is flat, we can use a flat connection (naturally given by the energy-momentum tensor) to identify all the fibers with one of them and study the action of the quotient group on this fiber. Taking into account that the duality properties are the same for primary an secondary fields, the duality properties only involve finite dimensional vector spaces. Formally if $\mathcal{H}$ is the Hilbert space of the theory and σ the chiral algebra we are dealing with $\mathcal{H}/\sigma$. In the decomposition $\mathcal{H} = \oplus_{i,\bar{\imath}} \mathcal{H}_i \times \mathcal{H}_{\bar{\imath}}$ only the highest weight subspaces $V_i \subset \mathcal{H}_i$ survive. The subspace V_i is spanned by all elements in $\mathcal{H}_i$ with the same L_0 eigenvalue as the highest weight vector. If the chiral algebra has a well-defined zero mode subalgebra the spaces V_i are representations of the zero mode algebra.

If $\phi_1, \dots \phi_N$ are the primary fields of the theory, a basis of the space of blocks with the ϕ_i on the external legs is obtained by describing a way of decomposing $V_i \times \dots \times V_N$ according to the fusion rules. This generates a collection of skeleton graphs with labels on the external and internal lines. This decomposition is reminiscent of group theory, albeit not classical group theory. Whatever the algebra structure underlying the decomposition of $V_i \times \ldots \times V_N$ is, it has to contain the duality properties of the CFT if its representation theory is to stand a chance of solving the polynomial equations. To see that this requirement forces us in general to deviate from ordinary group theory, we can think of the three-point block in Fig. 24 as an intertwinner in the algebra A which embodies the extension by the duality operations of the zero mode subalgebra of the chiral algebra. If we exchange (braid) the i, j legs in Fig. 24 in the classical case this will only amount to a sign counting whether the representation k appears symmetrically or antisymmetrically in the tensor product $V_i \otimes V_j$. In CFT this exchange produces this sign together with a phase depending on the conformal weights of the i, j, k fields. If σ_{ij} represents the braiding of i and j (the Ω operation in Refs. 16 and 18) and F_k^{ij} is the block in Fig. 24,

then:

$$\sigma_{ij}(F_k^{ij}) = \varepsilon e^{i\pi(h_k - h_i - h_j)} F_k^{ji} \,. \tag{6.1}$$

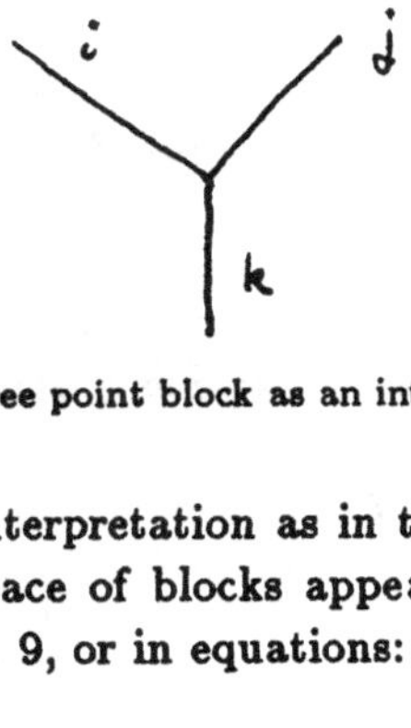

Fig. 24. Three point block as an intertwinner.

ε is a sign with the same interpretation as in the classical case. The two basic moves F, B on the space of blocks appear in Fig. 4 and they are related by the moves in Fig. 9, or in equations:

$$B_{pp'}\begin{bmatrix} j & k \\ i & l \end{bmatrix} = \varepsilon^p_{kl}\varepsilon^i_{p'k} e^{i\pi(h_p + h_{p'} - h_i - h_l)} F_{pp'}\begin{bmatrix} j & l \\ i & k \end{bmatrix} \,. \tag{6.2}$$

We can interpret the B, F operations in (quantum) group theoretic language. Although the precise definitions of quantum groups and quasitriangular algebras will appear in the next subsection, we want to motivate these definitions from the point of view of the duality properties of a RCFT. In ordinary group theory the intertwinners can be used to define the composition of angular momentum. If $K_k^{ij} : V_i \otimes V_j \to V_k$ is an intertwinner and the algebra A acts on V_k via the matrices $\rho^k(a), a \in A$ we can define the action of A in the tensor product $V_i \otimes V_j$ via the diagram:

$$\begin{array}{ccc} V_i \otimes V_j & \xrightarrow{\quad K_k^{ij} \quad} & V_k \\ \rho_i \otimes \rho_j(\Delta(a)) \Big\downarrow & & \Big\downarrow \rho^k(a) \\ V_i \otimes V_j & \xrightarrow{\quad K_k^{ij} \quad} & V_k \end{array} \tag{6.3}$$

Technically this is the same as defining a co-multiplication operations in the algebra $A: \ \Delta : A \rightarrow A \otimes A$, and $\rho^i \otimes \rho^j(\Delta(a)) \equiv \Delta^{ij}(a)$. If A and the tensor product of representations are associative, then the co-multiplication Δ is also associative. As in (6.1) the intertwinner will have non-trivial monodromy, and this is also reflected in the co-multiplication. Once Δ is defined, we can construct the action of A on arbitrary tensor products of representations by iterating the co-multiplication. Exploring further our identification of conformal blocks with the decomposition of the tensor product $V_{j_1} \otimes \ldots \otimes_{i_N}$:

$$V_{i_1} \otimes \ldots \otimes V_{i_N} = \bigoplus_j \mathcal{F}_j^{i_1 \ldots i_N} \otimes V_j \tag{6.4}$$

the space $\mathcal{F}_j^{i_1 \ldots i_N}$ is generated by the blocks $\mathcal{F}_j^{i_1 \ldots i_N}(p_1, \ldots, p_{N-2})$ with one spectator field in the representation j and $p_i, \ldots, p_{N-2}$ are possible intermediate states according to the fusion rules in the basis for the decomposition (6.4) shown in Fig. 25.

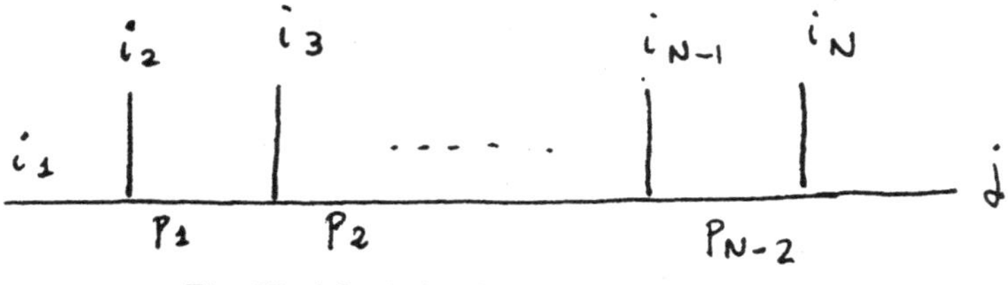

Fig. 25. A basis for the space of blocks in (6.4).

The multiplicity of the representation V_j in (6.4) is given by the dimension of the space $\mathcal{F}_j^{i_1 \ldots i_N}$. There are some properties of (6.4) which follow from the general structure of RCFT. In the decomposition (6.4) we have assumed the representation on the left-hand side to be not only reducible but also fully reducible. This is not the case for general algebras but it holds in

CFT. The space of all conformal blocks can be divided into subsets, each providing an irrep. of the braid group acting on the external legs. From general arguments in representation theory the blocks $\mathcal{F}_j^{i_1 \ldots i_N}$ provide representations of the centralizer of the algebra A acting on the tensor product $V_{i_1} \otimes \ldots \otimes V_{i_N}$. By definition of A, B are two algebras acting on the same space V, B is the centralizer of the action of A on V is the set of operators $b : V \to V$ commuting with the action of A. If A, B are both semisimple and finite dimensional then if B is the centralizer of A in V, A is the centralizer of B. The importance of the centralizer in representation theory originates from Schur's lemma. Consider for example the Lie algebra gl(N), and let V be the defining N-dimensional representation. In the decomposition of $V \otimes \ldots \otimes V$ into irreducible representations we first look for the algebra which centralizes the action of gl(N) on it. If the centralizer algebra B is semisimple, the irreducible representations appearing in $V \otimes \ldots \otimes V$ are labeled by the irreducible representations of B. The centralizer algebra of the action of gl(N) on $V \otimes \ldots \otimes V = V^{\otimes n}$ is generated by the permutations exchanging the factors with the constraint that the $N+1$-row antisymmetrizer vanishes; i.e. this is equivalent to saying that only tableaux with at most N rows appear in the irreducible representations of gl(N).

In CFT we start with (6.4) and for simplicity assume for the time being $V_{i_1} = V_{i_2} = \ldots = V_{i_N} = V_\phi$. The argument extends easily to the general case. These particular blocks are represented in Fig. 26. Since the braiding properties in CFT are independent of descendants, we expecr some representation of the braid group to be the centralizer of the algebra A. If we knew the algebra A explicitly we could in principle find its centralizer B, whose representations would completely characterize the decomposition of the tensor product $V_\phi^{\otimes n}$. The complete reducibility of the braid representations provided by the conformal blocks implies that the centralizer algebra must be semisimple. Later we will see that in the more interesting case (quantum groups whose deformation parameters are roots of unity) some restrictions have to be imposed on their representation theory in order to obtain effectively a semisimple centralizer. (This is equivalent to dividing the centralizer by some Abelian ideals and working with the semisimple quotient). Without going into these complications at the moment let us assume the centralizer to be semisimple. From the duality principle in CFT the braid group B_N acts on the blocks on the right-hand side of (6.4). Therefore we can identify the centralizer of A with some algebra providing a representation of the braid group. Reading (6.4) backwards and using the braid representation on the blocks we can construct a matrix $R^{\phi\phi}$ acting on contiguous spaces in the tensor product. This matrix commutes with the co-multiplication operation i.e. it commutes with $\Delta(a), a \in A$ by the definition of centraliser,

$$R^{\phi\phi}\Delta^{\phi\phi} = \Delta^{\phi\phi}R^{\phi\phi} \,. \tag{6.5}$$

Fig. 26.

If the representations on the external legs are different, the analogue of (6.5) is

$$R^{ij}\Delta^{ij} = \Delta^{ji}R^{ij} \tag{6.6}$$

where:

$$R^{ij} : V^i \otimes V^k \to V^j \otimes V^i \,. \tag{6.7}$$

If $P^{ij} : V_i \otimes V_j \to V_j \otimes V_i$ is the permutation map $P^{ij}(x \otimes y) = y \otimes x$ we can define

$$\begin{aligned} R^{ij} &= P^{ji}R^{ij} : V_i \otimes V_j \to V_j \otimes V_i \\ P^{ij}P^{ji} &= 1 \end{aligned} \tag{6.8}$$

and (6.6) can be interpreted at the level of the algebra A by saying that once A is endowed by a co-multiplication Δ, we can always construct a second one $\Delta' = \sigma \circ \Delta$, where $\sigma = A \otimes A \to A \otimes A$ is the permutation map. Equation (6.6) says that Δ' and Δ are conjugate within $A \otimes A$; the exist a matrix $\mathcal{R} : A \otimes A \to A \otimes A$ so that

$$\Delta'(a) = \sigma \circ \Delta(a) = \mathcal{R}\Delta(a)\mathcal{R}^{-1} \,. \tag{6.9}$$

This can be seen using

$$P^{ij}\Delta'^{ji}P^{ji} = \Delta^{ij} \,. \tag{6.10}$$

Two extra conditions on the co-multiplication are obtained from the fact that in CFT fusing and braiding are compatible. This is illustrated in

Fig. 27. When written in equations the equalities in Fig. 27 imply two of the defining relation for a Quasi-Triangular Yang-Baxter algebra which together with (6.9) provide all but one axiom for this algebraic structure. The missing axiom can be explained in terms of the relation between the braiding and fusing properties of a set of fields and those of their conjugates. A simple consequence of these axioms is the Yang-Baxter equation for the matrix $\mathcal{R}$ and the hexagon and pentagon identities. Furthermore the modular transformation matrix S with all the properties (5.18) can be obtained from the properties of the co-multiplication for these algebras[25].

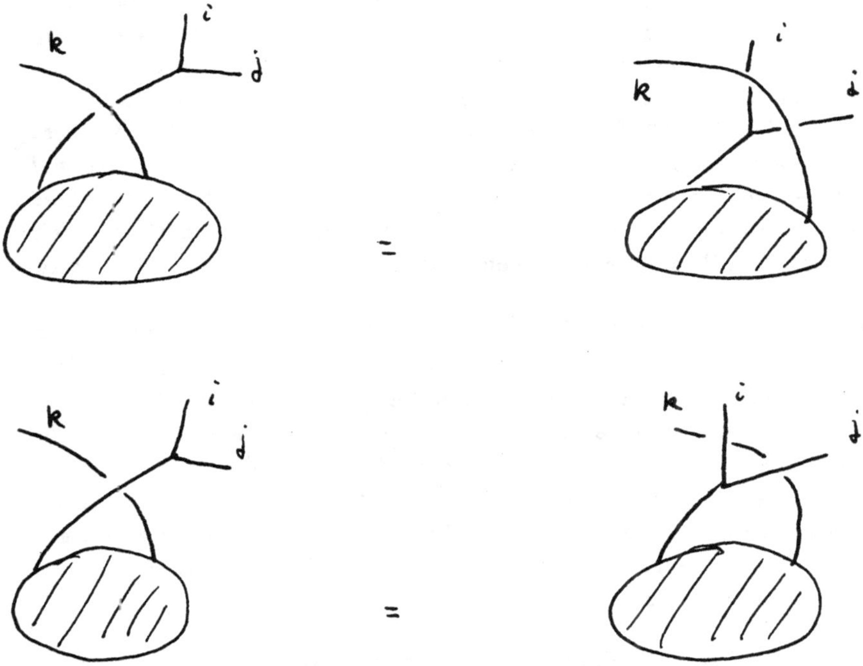

Fig. 27. Compatibility between braiding and fusion.

Although in general we do not yet know whether it is possible to associate a quantum group to every RCFT in some case it is not difficult to derive the quantum groups whose representation theory reproduce their duality properties. Ideally one would like to find an explicit construction of the quantum algebra A in terms of the generators of the chiral algebra σ. There are several ways of constructing quantum group actions on the space of blocks for RCFT[22,24], however it is not known at present how to relate them to σ. The previous arguments are helpful in understanding why such a construction may not be so simple. The duality properties encoded by the quantum group correspond to rather nonlocal operations in terms of the actual blocks (they involve braiding, analytic continuation...). Hence the quantum group generators should be nonlocal functional of the generators of σ. We find it more appealing (and easier) to find the quantum group as the centralizer of the braid group action on the blocks and then use its representation theory to construct explicitly the matrices B, F, S, T satisfyin the polynomial equations. We will work out some examples in details later. This subsection was only intended to motivate the defining relations of quantum groups from the point of view of RCFT. It is also quite clear that the matrices B, F, S, T convey all the topological information contained in a RCFT. This will be exploited and explained in Sec. 9 where we present in some details the connection of CFT with three-dimensional topology and the Chern-Simons approach to both subjects. After these preliminaries we study the general structure of quantum group and their representation theory when the deformation parameters are roots of unity.

6.2 *Quantum groups and their representations for q a root of unity*

Given an associative algebra A with unity, we say that A is a Hopf algebra (general references on Quantum Groups are Refs. 26,27,28,29 and 31) if we can define three operations $\Delta, \gamma, \varepsilon$ on A; $\Delta : A \to A \otimes A$ is the co-multiplication, $\gamma : A \to A$ is the antipodal map and $\varepsilon : A \to \mathrm{C}$ is the counit. C is the field over which A is an algebra. The operations Δ, ε are algebra homomorphisms. $\Delta(ab) = \Delta(a)\Delta(b), \varepsilon(ab) = \varepsilon(a)\varepsilon(b)$ whereas γ is an anti-homomorphism, $\gamma(ab) = \gamma(b)\gamma(a)$. These three operations must satisfy the following axioms:

$$
\begin{aligned}
a, b \in A, \quad (id \otimes \Delta)\Delta(a) &= (\Delta \otimes id)\Delta(a) \\
m(id \otimes \gamma)\Delta(a) &= m(\gamma \otimes id)\Delta(a) = \varepsilon(a)1 \\
(\varepsilon \otimes id)\Delta(a) &= (id \otimes \varepsilon)\Delta(c) = a
\end{aligned}
\tag{6.11}
$$

where m is the multiplication in the algebra; $m : A \otimes A \to A, M(a \otimes b) = a \cdot b$. In words, the first condition is the associativity of the co-multiplication; the second is the definition of the antipode and the third defines the co-unit. If $\sigma = A \otimes A \to A \otimes A$ is the permutation map $\sigma(a \otimes b) = b \otimes a$, it is easy to check that $\Delta' = \sigma \circ \Delta$ is another co-multiplication in A with antipode

$\gamma' = \gamma^{-1}$. A Hopf algebra is a quasi-triangular Yang-Baxter algebra if the co-multiplications Δ, Δ' are related by conjugation:

$$\sigma \cdot \Delta(a) = \mathcal{R}\Delta(a)\mathcal{R}^{-1} \qquad \mathcal{R} \in A \times A \tag{6.12}$$

and the following conditions are satisfied:

$$\begin{aligned} (id \otimes \Delta)(\mathcal{R}) &= \mathcal{R}_{13}\mathcal{R}_{12} \\ (\Delta \otimes id)(\mathcal{R}) &= \mathcal{R}_{13}\mathcal{R}_{23} \\ (\gamma \otimes id)(\mathcal{R}) &= \mathcal{R}^{-1} . \end{aligned} \tag{6.13}$$

Since $(id \otimes \Delta)(\mathcal{R}) \in A \otimes A \otimes A, \mathcal{R}_{13}$ acts as the identity in the second factor and as $\mathcal{R}$ in the first and third, and similarly for $\mathcal{R}_{12}, \mathcal{R}_{23}$. The motivation for these axioms came originally from the theory of integrable models (see Ref. 26 and references therein). In CFT (6.12) comes from 26 the definition of the centralizer and the first two conditions in (6.13) are the precise statement of the compatibility between fusing and braiding represented graphically in Fig. 27. An immediate consequence of these axioms is the Yang-Baxter equation for $\mathcal{R}$. Write

$$\mathcal{R} = \sum_i a_i \otimes b_i .$$

Then

$$\mathcal{R}_{13}\mathcal{R}_{23} = \sum_{i,j} a_i \otimes a_j \otimes b_i b_j$$

and:

$$\begin{aligned} [\sigma \circ \Delta \otimes id]\mathcal{R} &= \sum_i \Delta'(a_j) \otimes b_i = \sum_i \mathcal{R}_{12}\Delta(a_i)\mathcal{R}_{12}^{-1} \otimes b_i \\ &= \mathcal{R}_{12} \sum_i \Delta(a_i) \otimes b_i \mathcal{R}_{12}^{-1} = \mathcal{R}_{12}(\Delta \otimes id(\mathcal{R}))\mathcal{R}_{12}^{-1} \\ &= \mathcal{R}_{12}\mathcal{R}_{13}\mathcal{R}_{23}\mathcal{R}_{12}^{-1} . \end{aligned}$$

However

$$(\sigma \circ \Delta \otimes id)(\mathcal{R}) = \sigma_{12}((\Delta \otimes id)(\mathcal{R})) = \sigma_{12}(\mathcal{R}_{13}\mathcal{R}_{23}) = \mathcal{R}_{23}\mathcal{R}_{13}$$

yielding

$$\mathcal{R}_{12}\mathcal{R}_{13}\mathcal{R}_{23}\mathcal{R}_{12}^{-1} = \mathcal{R}_{23}\mathcal{R}_{13}$$

or:

$$\mathcal{R}_{12}\mathcal{R}_{13}\mathcal{R}_{23} = \mathcal{R}_{23}\mathcal{R}_{13}\mathcal{R}_{12} \ . \tag{6.14}$$

The matrix $\mathcal{R}$ is known as the universal R-matrix for the algebra A.

As an example consider any Cartan matrix (a_{ij}) for an ordinary simple Lie group or untwisted affine Kac-Moody algebra. With it we can associate a deformation of the classical algebra[26]. Working in a Chevalley basis, to every root we associate three generators X_i^+, X_i^-, H_i satisfying the relations:

$$\begin{aligned}
&[H_i, H_j] = 0\\
&[H_j, X_i^\pm] = \pm\langle\alpha_i, \alpha_j\rangle X_i^\pm \quad \langle a, b\rangle \equiv 2a\cdot b/b^2\\
&[X_i^+, X_j^-] = \delta_{ij}\,\frac{q^{H_i/2} - q^{-H_i/2}}{q^{1/2} - q^{-1/2}}\\
&[X_i^\pm, X_j^\pm] = 0 \quad \text{if}\, a_{ij} = 0\\
&\sum_{\nu=0}^{1-a_{ij}} (-1)^\nu \begin{bmatrix} 1-a_{ij} \\ \nu \end{bmatrix}_{q_i} (X_i^\pm)^{1-a_{ij}-\nu} X_j^\pm (X_i^\pm)^\nu q_i^{-\nu(1-a_{ij}-\nu)/2} = 0\\
&i \neq j \qquad q_i = q^{\alpha_i^2/2}
\end{aligned} \tag{6.15}$$

where we use the definitions for q-numbers:

$$\begin{aligned}
[X]_q &= \frac{q^{X/2} - q^{-X/2}}{q^{1/2} - q^{-1/2}}\\
\begin{bmatrix} n \\ m \end{bmatrix} &= \frac{[n]!}{[m]![n-m]!}\,,\, [n]! = [n][n-1]\ldots[2][1]
\end{aligned} \tag{6.16}$$

the operators $\Delta, \gamma, \varepsilon$ are defined by:

$$\begin{aligned}
\Delta(\mathcal{H}_i) &= \mathcal{H}_i \otimes 1 + 1 \otimes \mathcal{H}_i\\
\Delta(X_i^\pm) &= X_i^\pm \otimes q^{\mathcal{H}_i/4} + q^{-\mathcal{H}_i/4} \otimes X_i^\pm\\
\varepsilon(1) &= 1 \qquad \varepsilon(X_i^\pm) = \varepsilon(\mathcal{H}_i) = 0\\
\gamma(X_i^\pm) &= -q^{\rho/2} X_i^\pm q^{-\rho/2}\\
\rho &= \frac{1}{2}\sum_{\alpha>0} \mathcal{H}_\alpha \ .
\end{aligned} \tag{6.17}$$

It is an easy exercise to check that (6.17) satisfies the axioms of a Hopf algebra. Moreover (6.15) also has the structure of a quasi-triangular Yang-Baxter algebra (quantum group for short).[26-29] To keep the discussion as simple as possible consider the SU(2) case. Now we do not have the last relation in (6.15) (the q-analog of the Serre relations) and the defining properties are:

$$\begin{aligned} [X^+, X^-] &= [\mathcal{H}] \\ [\mathcal{H}, X^\pm] &= \pm 2X^\pm \\ \varepsilon(X^\pm) &= \varepsilon(\mathcal{H}) = 0\,, \varepsilon(1) = 1 \\ \gamma(X^\pm) &= -q^{\pm 1/2} X^\pm\,, \gamma(\mathcal{H}) = -H \end{aligned} \tag{6.18}$$

with the universal $\mathcal{R}$ matrix given by[27,29]:

$$\mathcal{R} = q^{\mathcal{H}\otimes\mathcal{H}/4} \sum_{n\geq 0} \frac{(1-q^{-1})^n}{[n]!} q^{-n(n-1)/4} q^{n\mathcal{H}/4} (X^+)^n \otimes q^{-n\mathcal{H}/4} (X^-)^n \tag{6.19}$$

and one can check that it satisfies (6.12) and (6.13).

As long as q is not a root of unity, the representation theory of quantum groups runs parallel with the classical theory. The representations are labelled by the same highest weight vectors and most of the standard expressions valid for classical algebras have q-analogs which often amount to replacing ordinary numbers by q-numbers $[X]$. From the point of view of representation theory the co-multiplication is analogous to the rule for addition of angular momentum. Geometrically for q generic we can understand the representation theory of SU$(2,q)$ as a deformation of the Borel-Weil construction. The fundamental representation of SU(2) has two states: spin up and spin down. We can represent these two states by two variables u, v respectively. If we consider the space of polynomials on these two variable the irreducible representation of SU(2) with spin j is given by the homogeneous polynomials of degree $2j$. The generators of SU(2) become the differential operators:

$$X^+ = u\frac{\partial}{\partial v} \quad X^- = v\frac{\partial}{\partial u} \quad \mathcal{H} = u\frac{\partial}{\partial u} - v\frac{\partial}{\partial v}\,. \tag{6.20}$$

A basis for the spin j representation is given by the monomials:

$$|j, m\rangle = \frac{u^{j+m} v^{j-m}}{\sqrt{(j+m)!(j-m)!}} \qquad m = -j, -j+1, \ldots, j \tag{6.21}$$

orthonormal with respect to the metric:

$$\langle f|g\rangle = f\left(\frac{d}{du}, \frac{d}{dv}\right) g(u,v)|_{u=v=0} \ . \tag{6.22}$$

Geometrically we can describe the sphere S^2 as CP^1 in terms of two complex homogeneous coordinates u, v. In this representation the vectors (6.21) are holomorphic sections of the space of $-j-$ differentials on the sphere (i.e. $f(z,\bar{z})dz^{-j}$ is invariant under holomorphic coordinate changes). According to the Rieman-Roch theorem the space of holomorphic $-j-$ differential $H^{\circ}(S^2, K^{-j})$ has dimension $2j+1$, and in homogeneous coordinate (6.21) provides a basis of $\mathcal{H}^0(S^2, K^{-j})$. The q-deformation of this construction is obtained by using the same polynomial ring in $u, v\mathbb{C}[u,v]$ but now the generators $X^{\pm}$ are represented by q-derivatives. The q-derivative of a function $f(u)$ is defined as:

$$D_u f(u) = \frac{f(q^{1/2}u) - f(q^{-1/2}u)}{(q^{1/2} - q^{-1/2})u} \tag{6.23a}$$

same for D_v:

$$D_v f(v) = \frac{f(q^{1/2}v) - f(q^{-1/2}v)}{(q^{1/2} - q^{-1/2})v} \ . \tag{6.23b}$$

To compute the commutation relations of these q-derivatives we only need to compute them on the basic monomials $u^a v^b$. It is then straightforward to show that the operators:

$$X^+ = uD_v \quad X^- = vD_u \quad \mathcal{H} = u\frac{\partial}{\partial u} - v\frac{\partial}{\partial v} \tag{6.24}$$

satisfy the defining relations of SU(2, q) (6.18). The q-deformed scalar product is:

$$\langle f|g\rangle = f(D_u, D_v)g(u,v)|_{u,v} = 0 \tag{6.25}$$

and with respect to this scalar product an orthonormal basis for the basis j representation is given by:

$$|j,m\rangle = \frac{u^{j+m}v^{j-m}}{\sqrt{[j+m]![j-m]!}} \tag{6.26}$$

and it is easy to derive formulae like:

$$(X^{\pm})^a|j,m\rangle = \sqrt{\frac{[j\mp m]![j\pm m+a]!}{[j\mp m-a]![j\pm m]!}}|j,m\pm a\rangle \ . \tag{6.27}$$

Finally the analogue of the Casimir operator is:

$$C = X^- X^+ + \left[\frac{\mathcal{H}+1}{2}\right]^2 . \tag{6.28}$$

To make contact with the remarks in Sec. 6.1 about the centralizer algebra, we first consider the example SU(N,q). After this example the srategy for an arbitrary group G should be clear. As with SU(2), the fundamental representations of SU(N) and SU(N,q) coincide. Suppose we want to decompose $V_N \otimes \ldots \otimes V_N$ into irreducible components. Defining the action of $\Delta(a)$, $\mathcal{R}$ and R tensor products according to:

$$\Delta^{ij}(a) = \rho^i \otimes \rho^j(\Delta(a)) : V_i \otimes V_j \to V_j \otimes V_i \tag{6.29a}$$

$$\mathcal{R}^{ij} = \rho^i \otimes \rho^j(\mathcal{R}) \tag{6.29b}$$

$$R^{ij} = P^{ij}\mathcal{R}^{ij} : V_i \otimes V_j \to V_j \otimes V_i . \tag{6.29c}$$

Then from (6.12) we obtain

$$R^{ij}\Delta^{ij}(a) = \Delta^{ij}(a)R^{ij} . \tag{6.30}$$

In our example i, j are the fundamental representations and to simplify the notation we will remove in this case, the i, j labels from R^{ij}, Δ^{ij}. Then (6.30) becomes $R\Delta(a) = \Delta(a)R$ and the centralizer of SU(N,q) in $V_N^{\otimes n}$ is generated by:

$$\rho_i = 1 \otimes \ldots \otimes R_{i,i+1} \otimes \ldots \otimes 1 \tag{6.31}$$

where the subindices in R indicate the position of the spaces in the tensor product where ρ_i acts nontrivially. From the definition of comultiplication plus the fact that the fundamental representations of SU(N,q) and SU(N) coincide we can compute R. if e_{ij} is the $N \times N$ matrix unit whose single nonvanishing component is equal to 1 and it is located in position (i,j), the answer is:

$$R = \sum_{i\neq j} e_{ij} \otimes e_{ji} + q^{1/2} \sum_i e_{ii} \otimes e_{ii} + \left(q^{1/2} - q^{-1/2}\right) \sum_{i<j} e_{jj} \otimes e_{ii} \tag{6.32}$$

and a straightfroward computation shows that

$$R^2 = \left(q^{1/2} - q^{-1/2}\right) R + 1 . \tag{6.33}$$

for convenience we define the generators ρ_i by rescaling the R matrix:

$$\rho_i \equiv -q^{1/2} R \ . \tag{6.34}$$

Then ρ_i satisfies the defining relations for the Hecke algebra of type $A_n, \mathcal{H}(q)$:

$$\begin{aligned} \rho_i \rho_{i\pm 1} \rho_i &= \rho_{i\pm 1} \rho_i \rho_{i\pm 1} \\ \rho_i \rho_j &= \rho_j \rho_i \qquad |i-j| \geq 2 \\ \rho_i^2 &= (1-q)\rho_i + q \ . \end{aligned} \tag{6.35}$$

The first equation in (6.35) follows from the Yang-Baxter equation. On $V_i \otimes V_j \otimes V_k$ the Yang-Baxter equation is:

$$\mathcal{R}^{ij} \mathcal{R}^{ik} \mathcal{R}^{jk} = \mathcal{R}^{jk} \mathcal{R}^{ik} \mathcal{R}^{ij} \tag{6.36}$$

and in terms of (6.29c) it becomes:

$$(R^{jk} \otimes 1)(1 \otimes R^{ik})(R^{ij} \otimes 1) = (1 \otimes R^{ij})(R^{ik} \otimes 1)(1 \otimes R^{jk}) \ . \tag{6.37}$$

The centraliser of SU(N,q) in $V_N^{\otimes n}$ satisfies the defining relations of the Hecke algebra $H_n(q)$ (plus one more relation; see below). We can proceed similarly for any representation μ and define a homorphism π^μ from the braid group B_n into the centralizer algebra C_μ^n of SU(N,q) in $V_\mu^{\otimes n}$:

$$\pi^\mu(\sigma_i) = 1 \otimes \ldots \otimes R^{\mu\mu}_{i,i+1} \otimes \ldots \otimes 1 \ . \tag{6.37}$$

The extra relation alluded to above for $C_N^n(V_\mu = V_N)$ is the following: For q generic, the irreps of SU(N,q) are labelled by the same Young tableaux as in the calssical case, the only difference is that the symmetriser associated to a given tableaux is not constructed with the elementary transpositions S_i of the symmetric group S_n but rather with the generator g_i of the Hecke algebra. The extra relations is equivalent to saying that the $(N+1)$—row antisymmetriser vanishes. Only tableaux with at most N rows appear. For example, in the SU$(2,q)$ case the complete set of relations for C_2^n are (6.35) together with:

$$1 - \rho_1 - \rho_2 + \rho_1\rho_2 + \rho_2\rho_1 - \rho_1\rho_2\rho_1 = 0 \ . \tag{6.38}$$

Equations (6.35) and (6.38) ar equivlent to the Temperley-Lieb-Jones algebra defined by $n-1$ projection operators e_i:

$$e_i = \frac{1-\rho_i}{1+q} \tag{6.39}$$

satisfying the relations:

$$\begin{aligned} e_i e_{i\pm 1} e_i &= \beta^{-1} e_i \qquad \beta = 2+q+q^{-1} \\ e_i^2 &= e_i \\ e_i e_j &= e_j e_i \qquad |i-j| \geq 2 \,. \end{aligned} \tag{6.40}$$

This algebra is known as $A_{\beta,n}$ and its properties hae been extensively studied by Jones.[40,41] He has classified the subfactors when q is a root of unity and analyzed the C^* — representations. The centralizers C_N^n of $\mathrm{SU}(N,q)$ in $V_N^{\otimes n}$ define a natural sequence of embeddings $\ldots \subset C_N^n \subset C_N^{n+1} \subset \ldots$ whose properties will be very useful in connecting RCFT and knot invariants (see Sec. 9.2). For q generic C_N^n is semisimple, hence any reducible finite dimensional representation of $\mathrm{SU}(N,q)$ is also fully reducible. From (6.30) we see that the centralizer of C_N^2 defines the co-multiplication operations in $\mathrm{SU}(N,q)$. $\mathrm{SU}(N,q)$ is the centralizer of C_N^n. This reciprocity relation will be useful later when we apply the remarks of Sec. 6.1 to compute the duality properties of some WZW model.

Before dealing with the case when q is a root of unity we present one more property of the general theory[28] that will be very useful in Sec. 9. For any representation μ we have the chain of embeddings $\ldots C_\mu^n \subset C_\mu^{n+1} \subset \ldots$ and the formal direct limit C_μ^∞ describing the braiding operations in $V_\mu^{\otimes n}$ for any n. If $R^{\mu\mu}$ is R^{ij} when $i=j=\mu$, and $\rho = \frac{1}{2}\sum_{\alpha>0} H_\alpha$ (6.30) implies:

$$R^{\mu\mu} q^\rho \otimes \rho^\rho = q^\rho \otimes q^\rho R^{\mu\mu} \tag{6.41}$$

and we can use $(R^{\mu\mu}, q^\rho)$ to define a trace on the algebra C_μ^∞ with the properties:

$$\mathrm{tr}\ (1) = 1 \tag{6.42a}$$

$$\mathrm{tr}\ ab = \mathrm{tr}\ ba \tag{6.42b}$$

$$\mathrm{tr}\ \alpha\rho_n^{\pm 1} = Z_\pm \mathrm{tr}\ \alpha \,. \tag{6.42c}$$

The generators ρ_i are defined in (6.37),

$$Z_{\pm} = \frac{q^{\pm c(\mu)}}{\mathrm{tr}_{\mu} q^{\rho}} \tag{6.43}$$

and $c(\mu)$ is the second Casimir for the classical group in the irrep. Labelled by the highest vector μ. A trace with the property (6.42c) is known as a Markov trace, or a trace with the Markov property. This is crucial in connection with Knot theory. The trace is defined by

$$\mathrm{tr}\, \alpha = \frac{1}{\mathrm{tr}_{\mu} q^{\varrho}} (q^{\varrho} \otimes \ldots q^{\varrho} \alpha) \,. \tag{6.44}$$

(For a proof see Ref. 28). Using the homomorphism (6.37) we get a Markov trace on the braid group.

This trace suggests the definition of a quantum dimension for a representation of a quantum group. If we remove the normalization factor from (6.44) and take the braid element to be the identity we can define the q-dimension of a representation of highest weight μ to be

$$D_q(\mu) = \mathrm{tr} q^{\varrho} \tag{6.45}$$

with tr the standard matrix trace. This definition will be shown to agree with the definition (5.38) and it plays a very important role in the representation theory of quantum groups when q is a root of unity.

We finally come to analyse representations for quantum groups when the deformation parameter is a root of unity. Now the centraliser algebra and the quantum group are not semisimple and we want to find out under what conditions the representation theory can be restricted to a regular set of representations (modular representations of quantum groups appear in Ref. 31. A more physical treatments can be found in Ref. 32. Let us begin with SU(2,q). The basic problem when $q = e^{2\pi i/p}$ is that $(X^{\pm})^p = 0$ and this generates null vectors in some representations. Many of the representations appearing in the decomposition of tensor products of irreducible representations will be reducible but not fully reducible. Instead of using the basis (6.26), we construct a different basis for the representation with highest weight j as

$$\begin{aligned} X^{+}|j\rangle &= 0 \\ |m\rangle &= \frac{(X^{-})^{j-m}}{[j-m]!}|j\rangle \end{aligned} \tag{6.46}$$

then

$$\begin{aligned} X^-|jm\rangle &= [j-m+1]|j,m-1\rangle \\ X^+|jm\rangle &= [j+m=1]|j,m+1\rangle \\ \frac{(X^\pm)^a}{[a]!}|jm\rangle &= \frac{j\pm m+a]!}{[a]![j+m]!}|j,m\pm a\rangle \ . \end{aligned} \tag{6.47}$$

When $q^p = 1, [p] = [2p] = \ldots = [kp] = 0$, however the operators $(X^\pm)^p/[p]!$ are still well-defined as can be seen by taking $a = p$ in (6.47) for generic q and then taking the limit $q \to e^{2\pi i/p}$. It is also clear from the previous equations that $(X^\pm)^p = 0$ and moreover that this property is preserved by the co-multiplication as one should expect:

$$\Delta(X^\pm)^N = \sum_{j=0}^{N} \begin{bmatrix} N \\ j \end{bmatrix} q^{-(N-j)H/4}(X^+)^j \otimes q^{jH/4}(X^+)^{N-j} \ .$$

When $N = p$ the first and the last terms on the right-hand side contain $(X^\pm)^p = 0$ and they vanish, and any of the intermediated terms vanish because they are all proportional to $[p]$. Hence for large enough j the representation will have null vectors. Following Ref. 32 we begin to see the source of the problem if we calculate the eigenvalues of the Casimir operator (6.28) acting on highest weight vectors

$$C|j\rangle = [j+1/2]^2|j\rangle$$

for q generic the eigenvalues of C distinguish between all the different values of j. When $q = e^{2\pi i/p}$, it is easy to see that $j, j+kp, p-1-j+kp$ for any integer k have the same Casimir and if we compute the q-dimensions we find

$$D_q(j) = D_q(j+kp) = -D_q(p-1-j+kp) \ . \tag{6.48}$$

This suggests that if we try to decompose the tensor product $V_{1/2}^{\otimes n}$ for n high enough into irreducible representations, odd things begin to happen. $V_{1/2}^{\otimes n}$ contains in its decomposition reducible but not fully reducible representations which pair up representations that would be distinct irreducible ones for q generic adding up their q-dimensions to zero. For example if $q^3 = 1$ we can try to decompose $V_{1/2} \otimes V_{1/2} \otimes V_{1/2}$. For generic values of q this tensor product decomposes into $V_{3/2} \oplus 2V_{1/2}$. When $q^3 = 1$ the weight states of the spin 3/2 and one of the spin 1/2 representations mix into a reducible but not fully reducible representation due to the presence of null vectors. It is not difficult to check that the state $|\alpha\rangle = X^-|1/2\rangle\otimes|1/2\rangle\otimes|1/2\rangle$ with $m = 1/2$ is annihilated by X^+. Since this state is null, it is orthogonal to itself with respect to the metric (6.25) (notice that this scalar product is the standard one in terms of bras and kets with the prescription that in constructing bras from kets one should not complex conjugate q), and out of the other two states in $V_{1/2}\otimes V_{1/2}\otimes V_{1/2}$ only one of them is orthogonal to $|\alpha\rangle$. The other state $|\beta\rangle$ is not orthogonal to $|\alpha\rangle$ and therefore $X^+|\beta\rangle \propto |1/2\rangle\otimes|1/2\rangle\otimes|1/2\rangle$. The q-dimensions are $D_q(3/2) = [4] = [3-4] = [-1] = -1$ and $D_q(1/2) = [2] = [3-2] = [1] = 1$, where we have used $[p] = 0, [x] = [p-x], [-x] = -[x]$. Thus $D_q(3/2)+D_q(1/2) = 0$. Looking at other tensor product $V_{1/2}^{\otimes n}$ it is possible to prove that this situation is quite general. [32] If we want to have a regular representation theory (so that tensor products of irreducible representations are fully reducible) we have to impose some conditions of the tensor product of representations. To find the integrable or regular representations when $p \geq 0$ we first notice that $V_{1/2}$ has positive q-dimension. Since $[x]$ is a ratio of sine functions, we can solve the equation

$$D_q(j) = [2j+1] = 0 \tag{6.49}$$

to find that the q-dimension vanishes whenever $j = \frac{p-1}{2} + kp$. As we keep on taking tensor products $V_{1/2}\otimes\ldots\otimes V_{1/2}$ we eventually obtain the representation with $j = \frac{p-1}{2}$ with vanishing q-dimension and other with positive q-dimensions. Tensoring another copy of $V_{1/2}$ begins the pairing of representations. We can restrict the representations of SU$(2,q)$ to those with the smallest possible spin and positive q-dimension. This is achieved by requiring that in the tensor product of the fundamental representations we only keep those highest weight vectors annihilated by X^+ and at the same time not in the image of $(X^+)^{p-1}$. This restricts the representations to those with spin smaller or equal to $(p-2)/2$, defining $p = k+2$ we obtain the same restriction as in the SU(2) level k WZW model. In this way we find an alcove in the space of weights where the q-dimension is strictly positive with the lowest possible value of j. Using the scalar product (6.25)

$$\langle jj|\frac{(X^+)^n(X^-)^n}{[n!][n!]}|jj\rangle = \begin{bmatrix} j \\ n \end{bmatrix}$$

it is easy to check the unitarity and positive definiteness of the representation. The representations with $j < (p-1)/2$ are characterised by :1).

They are highest weight representations. 2) The highest weight $|j\rangle$ cannot be written as $(X^+)^{p-1}|anything\rangle$. The highest weights are in the space[32]

$$|j\rangle \in \mathrm{Ker} X^+/\mathrm{Im}(X^+)^{p-1} .$$

Next we can define a restricted tensor product. We have the collection of regular representations $\nu_{\mathrm{reg}} = \{V_j, j \leq (p-2)/2\}$. Given any two regular representations in ν_{reg} we construct the standard tensor product $V_{j_1} \otimes V_{j_2}$. If $j_1 + j_2 \leq (p-2)/2$ the standard tensor product rule for SU(2) holds. As soon as the value of $j_1 + j_2$ is surpassed, the highest weight vector with the largest value of $m_1 + m_2$ will belong to $\mathrm{Im}(X^+)^{p-1}$, if $j = (p-1)/2$ the q-dimension vanishes. For $j > (p-1)/2$ the representation pairs with some other states into a reducible but not fully reducible representations. If we consider the set of highest weight vector $(V_{j_i} \otimes V_{j_2})^+ \in V_{j_i} \otimes V_{j_2}$, we can restrict to those which are not in $\mathrm{Im}(X^+)^{p-1}$ and to the representation they generate. This will be our definition of the restricted tensor product

$$(V_{j_1} \otimes V_{j_2})' \tag{6.50}$$

where the prime stands for the operation of eliminating those representations whose highest weight $\in \mathrm{Im}(X^+)^{p-1}$, in other words eliminating those representations with vanishing q-dimension. Now the scalar product in (6.25) is positive definite and the representation it provides is reducible and fully reducible. Furthermore, the coassociativity of the co-multiplication guarantees this tensor product to be associative

$$((V_{j_1} \otimes V_{j_2})' \otimes V_{j_3})' \cong (V_{j_2} \otimes V_{j_3})')' . \tag{6.51}$$

From the pairing argument described one easily gets the decomposition of (6.50) into irreducible regular representations

$$(V_{j_1} \otimes V_{j_2})' = \bigoplus_{j=|j_1-j_2|}^{\min(j_1+j_2, k-j_1-j_2)} V_j \quad k = p-2 .$$

Since the Bratelli decomposition of an algebra and that of its centralizer are the same, we see that the difinition (6.37) automatically leads to *-representations of the Temperley-Lied-Jones algebra if $j_1 = j_2 = \ldots j_n = 1/2$ or the more complicated centralizer algebras for higher spins in the external legs.

Next we wish to consider a general group G whose quantized commutation relations appear in (6.15). When q is a root of unity we want to find the integrable representations and the conditions on the restricted tensor product. To find the regular alcove in the space of weight we use Weyl's character formula. For a representation with highest weight λ we obtain for the q-dimension the expression

$$D_q(\lambda) = \prod_{\alpha>0} \frac{[\langle \lambda + \varrho, \alpha\rangle]}{[\langle \varrho, \alpha\rangle]} \qquad \varrho = \frac{1}{2}\sum_{\alpha>0} \alpha\,. \tag{6.52}$$

For $q^p = 1$ we first determine when two representations will have equal or opposite q-dimension. The result (6.52) is the same as the standard dimension formula except for the fact that numbers are replaced with q-numbers. For example in SU(N) (6.52) is the same as the rule of hooks to compute dimensions of irreducible representations. For a highest weight Λ (4.66) its representation is given by the tableaux $[f_1, f_2, \cdots, f_{N-1}, 0]$. Using (4.62)—(4.66) we obtain for (6.52):

$$D_q[f_1, \cdots, f_{N-1}, 0] = \prod_{1\le i<j\le N} \frac{[f_i - f_j + j - i]}{[j-i]}$$

and a few examples will convince the reader that this is the q-analogue of the staward formulae:

$$D_q\left(\;\square\;\right) = [N]$$

$$D_q\left(\begin{array}{|c|}\hline \\ \hline \\ \hline\end{array}\right) = \frac{[N][N-1]}{[2]}$$

$$D_q\left(\left.\begin{array}{|c|}\hline \\ \hline \vdots \\ \hline \\ \hline\end{array}\right\} k\right) = \frac{[N][N-1]\ldots[N-k+1]}{[k][k-1]\ldots[2]\cdot 1}$$

$$D_q\left(\begin{array}{|c|c|}\hline & \\ \hline\end{array}\right) = \frac{[N][N+1]}{[2]}$$

$$D_q\left(\overbrace{\begin{array}{|c|c|c|}\hline & \cdots & \\ \hline\end{array}}^{k}\right) = \frac{[N][N+1]\ldots[N+k-1]}{[k][k-1]\ldots[2]\cdot 1}$$

$D_q\Bigg($ 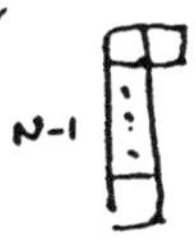 $\Bigg) = \dfrac{[N][N+1][N-1]\ldots[2]}{[N][N-2][N-3]\ldots[1]} = [N+1][N-1]$

etc. We now derive some properties of (6.52). If w_i is the Weyl reflection with respect to the simple root α_i then

$$\prod_{a>0} \frac{[\langle w_i(\lambda) + w_i(\varrho), \alpha\rangle]}{[\langle \varrho, \alpha\rangle]} = -\prod_{a>0} \frac{[\langle \lambda + \varrho, \alpha\rangle]}{[\varrho, \alpha\rangle]}$$

and for any element in the Weyl group:

$$\prod_{a>0} \frac{[\langle w(\lambda + \varrho), \alpha\rangle]}{[\langle \varrho, \alpha\rangle]} = \varepsilon(w) \prod_{a>0} \frac{[\langle \lambda + \varrho, \alpha\rangle]}{\langle \varrho, \alpha\rangle]}$$

where ε is the parity of w. Then we obtain

$$D_q\left(w(\lambda+\varrho) - \varrho + P\sum_{i=1}^{r} n_i\alpha_i\right) = \varepsilon(w) D_q(\lambda), \quad q^P = 1 \tag{6.53}$$

the $\alpha_i, i = 1\ldots, r$ are the simple roots of the classical algebra G. Every positive root can be written as $\alpha = \sum_i n_i\alpha_i, n_i \geq 0$, and $\sum n_i \equiv \text{level}(\alpha)$. For the highest root θ, the level plus one is the dual Coxeter number of the algebra $g = (\theta, \theta + 2\varrho)/\theta^2$. We will normalize the highest root to have length 2. For any weight λ, the largest value of $(\lambda + \varrho, \alpha), a > 0$ is obtained for the highest root θ, $\langle \varrho, \theta\rangle = g - 1$. The denominator of (6.52) can be written as:

$$\prod_{a>0} [\langle \varrho, \alpha\rangle] = \prod_{l(\alpha)=1}^{g-1} [l(\alpha)]^{Nl(\alpha)}$$

$l(\alpha)$ is the level of α and $Nl(\alpha)$ is the number of positive roots with the same level. For $q^p = 1, p > g$ the q-dimension of the generating representations of G are positive and if we consider representations in increasing values of $\langle \lambda, \theta\rangle$ we find that the q-dimension stays positive until $\langle \lambda, \theta\rangle = p - g$, for $\langle \alpha, \theta\rangle = p - g + 1$ the q-dimension vanishes, and beyond this value it can be positive, negative or zero, and we begin to have null vectors, reducible but not fully reducible representations appear, etc. In analogy with SU(2) let $p = k + g$, the integrable (regular) irreducible representations are such that their highest weights are not in

$$\text{Im}(X_\theta^+)^{p-g+1}\ . \tag{6.54}$$

The first one with vanishing q-dimension appears when $\langle\lambda,\theta\rangle = k+1$ and it is such $|\lambda\rangle = (X_\theta^+)^{k+1}|\alpha\rangle$. Hence the condition $|\lambda\rangle \notin (X_\theta^+)^{k+1}$ restricts the representations to those with $\langle\lambda,\theta\rangle \leq k$. Moreover, since $\langle\lambda,\theta\rangle \geq \langle\lambda,\alpha\rangle, \alpha > 0, (X_\alpha^-)^{k+1}|\lambda\rangle = 0, \forall\alpha$. This is the same condition of integrability as for Kac-Moody algebra. Proceeding as with SU(2) one can derive the "depth" rule[13] giving the fusion rules for WZW theories with group G and level k. If V_{reg} is the set of regular representations we can define the restricted tensor product

$$(V_1 \otimes V_2)'$$

as for SU(2) and the co-associativity of the co-multiplication implies the associativity if the restricted tensor product. The representation $(V_1 \otimes V_2)'$ is reducible and fully reducible. If we analyse the decomposition of $(V_1 \otimes V_2 \otimes \ldots V_n)'$ for regular representations, we can also read off the regular representations and the Bratelli diagram of the various centraliser algebras characterizing the braid representations generated by the conformal blocks associated to the decomposition of the tensor product (6.4). In the case of SU(N) and with all but one of the external legs in the fundamental representation we can check that one recovers the results of Wenzl[59]. For other groups or other representations the analysis of Wenzl has not been carried out yet and the restricted tensor product constructed here provides a rather simple way of characterising the regular representations of all the centraliser algebras that appear in the decomposition of representations of quantum groups when q is a root of unity. For SU(N) the representations are labels in terms of Young tableaux $[\lambda_1, \lambda_2, \ldots, \lambda_N], \lambda_1 \geq \lambda_2 \ldots$ where λ_1 is the number of blocks in the first row, λ_2 is the number of boxes in the second row etc. Then $\langle\lambda,\theta\rangle = \lambda_1 - \lambda_N \leq p - g = k$ in Wenzl's notation, only diagrams of type (N,k) appear (at most N rows and the previous condition on the number of columns). The Bratelli diagram of the fundamental representation is easy to work out. Consider a Young tableau Y with $\lambda_1 < k, \lambda_N = 0$. If we take the tensor product with one more copy of the fundamental representation we proceed as with standard SU(N) until $\lambda_1 - \lambda_N = k$. In this case the restricted tensor product eliminates the representation which could be generated by adding one more box to the first row (or any row with the same length as the first one). The arguments is easier to explain with one example. In Fig. 17 we present the Bratelli diagram for the fundamental representation of SU(3,q) with $q^5 = 1$ i.e, level $k = 2$.

To conclude this section we want to show that the quantum dimensions (5.38) for level k WZW models with group G are given by (6.52). Instead of taking the characters $\chi_\lambda(\tau)$ we can consider the characters with arbitrary angles in the Cartan subalgebra

$$\chi_\mu(z,\tau) = \mathrm{tr}_\mu q^{L_0 - c/24} e^{iz\cdot H}$$

z is an r-dimensional vector (r is the rank of G). The modular properties of $\chi_\mu(z,\tau)$ are the same as those of $\chi_\mu(\tau)$.[49] If we take $z = \varrho/(k+g)$ we can use the Weyl-Kac character formula to obtain

$$\frac{\chi_\mu\left(\frac{\varrho}{k+g}|\tau\right)}{\chi_0\left(\frac{\varrho}{k+g}|\tau\right)} = D_q(\mu)\frac{\sum_{\alpha\in M} e^{i\pi\tau((k+g)\alpha+\mu+\varrho)^2/k+g}}{\sum_{\alpha\in M} e^{i\pi\tau((k+g)\alpha+\mu+\varrho)^2/k+g}} \tag{6.55}$$

where M is the lattice generated by the long roots. We can take the $\tau \mapsto 0$ limit by first applying the Poisson resummation formula to the numerator and denominator of the right-hand side in (6.55) to obtain:

$$\lim_{\tau\mapsto 0} \frac{\chi_\mu\left(\frac{\varrho}{k+g}|\tau\right)}{\chi_0\left(\frac{\varrho}{k+g}|\tau\right)} = D_q(\mu) \quad q^{k+g} = 1\,.$$

Hence

$$D_q(\mu) = \frac{S_{\mu 0}}{S_{00}}\,. \tag{6.56}$$

From (5.18) and (5.38) we conclude that the q-dimensions are the eigenvalues of the fusion algebra for WZW models. This agrees also with the derivation of the fusion rules based on the restricted tensor products.

It is intriguing that after the representation theory of the quantum group is restricted as indicated above, the quantum group looks closer to a finite group than to a classical or Kac-Moody group.

Before finishing this subsection we would like to make a few comments in relation with Witten's three-dimensional construction of conformal field theories. The main point in that construction is the identification of the space of blocks of the WZW theory with the physical Hilbert space of the Chern-Simons theory. Considering an S^2 section of the three-dimensional manifold and working in the temporal gauge, the physical Hilbert space for n external charges in representations R_i of the gauge group is given by[33]:

$$\mathcal{H}_{ph}^{R_1R_2\ldots R_n} = \otimes_{i=1}^n \mathcal{H}^{R_i}/(\text{Gauss'law}) = \mathrm{Inv}\left\{\otimes_{i=1}^n \mathcal{H}^{R_i}\right\} \tag{6.57}$$

where the $\mathcal{H}^{R_i}$ are the representation spaces for the irreducible representations R_i and Gauss'law is defined by the constraint:

$$\frac{k}{8\pi}\varepsilon^{ij}F^a_{ij}(x) = \sum_{k=1}^{n} \delta^{(2)}(x - P_i)T^a_{(k)} \ . \tag{6.58}$$

For the fundamental representation, the spaces $\mathcal{H}^{R_1R_2\ldots R_n}_{ph}$ define irreducible representations of the Hecke algebra $H_n(q)$, $q = e^{2\pi i/k+2}$. This follows from the identification of the space of conformal blocks with $\mathcal{H}_{ph}$. We can use the physical Hilbert space defined above to write down a "branching rule" decomposition:

$$\otimes_1^n \mathcal{H}^R = \bigoplus_j \{\mathcal{H}^{R\ldots R,R_j}\} \otimes \mathcal{H}^{R_j}$$

which is equivalent to equation (6.4). From the last equation and the identification of $\mathcal{H}_{ph}$ with the space of blocks we learn that the commutant of the Hecke algebra (i.e. the quantum group) acts on $\otimes^n \mathcal{H}^R$. From the previous arguments we see that this group parametrizes the inequivalent solution to Gauss' law constraint. This connection was already foreseen in Ref. 33.

An alternative way to get the same result is to consider invariants in the quantum group. If we have a tensor product of n representations V^i, the space of invariants is given by

$$\mathrm{Inv}\left(\otimes_{i=1}^n V^i\right) = \left\{v \in \otimes_{i=1}^n V^i ; \varrho_1 \otimes \varrho_2 \ldots \varrho_n(\Delta^{n-1}(a))v = a \ \forall_a \in A\right\} \ .$$

This set is nonempty only when invariant tensors can be constructed out of the representations appearing in the tensor product, and they define interaction vertices according to the fusion rules. For instance, if we work with SU(2, q), $q^3 = 1$ then $\mathrm{Inv}(V^{1/2} \otimes V^{1/2} \otimes V^1)$ is empty. Coming back to (6.57), we know that the space of invariant tensors determined by Gauss'law are the ones allowed by the fusion rules of the SU(2) level k WZW model. In this chapter we have seen that this is precisely what happens for SU(2,q), $q = e^{2\pi i/k+2}$. Thus we can formally identify the generators of (6.58) for n charges in the representations R_i with $\otimes_{i=1}^n \varrho^{R_i}(\Delta^{n-1}(a))$ with a a generator of quantum SU(2).

7. Explicit Examples and Modular Properties

This section will make more concrete the arguments in Sec. 6.1. We will first show how the duality properties of WZW models follow from the representation theory of quantum group[23,25]. Other examples are provided by the rational Gaussian models[25] (Sec. 7.2). The first two subsections will deal mostly with the genus zero properties, B and F matrices, etc. The last subsection deals with one of the more important results obtained in the connection between quantum groups and RCFT, namely that the modular matrix S_{ij} is obtained from the properties of the co-multiplication[25]. This will be explained in detail and we will present a proof of Verlinde's conjecture in the context of quantum groups. The argument will show quite cleary the close relation with knot theory and Witten's use of 3D topological field theories to understand the properties of RCFT.

7.1 *WZW models*

Although in general we do not yet know whether it is possible to associate a quantum group to any RCFT (or to a solution to the polynomial equations), in some cases it is not difficult to derive the quantum group whose representation theory reproduces the duality properties of the RCFT. Ideally one would like to find an explicit construction of the quantum algebra A in terms of the generators of the chiral algebra $\mathcal{A}$ (see Sec. 6.1). There are several ways of constructing quantum group actions on the space of blocks for RCFT[22,24,60], however it is not known at present how to relate them to $\mathcal{A}$. The previous arguments are helpful in understanding why such a construction may not be so simple. The duality properties encoded by the quantum groups correspond to rather nonlocal operations in terms of the actual blocks (as sections of a vector bundle over some moduli space), for example one has to look at analytic continuation from one region to another, etc. Hence the quantum group generators should be nonlocal functionals of the generators of $\mathcal{A}$. We find it more appealing and easier to find the quantum group as a centralizer of the half monodromies, and then use its representation theory to express the B, F, S, T matrices characterizing a solution to the polynomial equations.

As an example consider the $SU(2)_k$ WZW theory, and take $\phi = \phi_{1/2}$ for simplicity. This field generates the operator algebra in the sense that any other primary field $\phi_j, j \leq k/2$ can be obtained in terms of the OPE of a sufficient number of ϕ fields. Consider the blocks in Fig. 26. The dimension of the field ϕ_j is $h_j = j(j+1)/(k+2)$. From the SU(2) fusion rules we know that

$$\phi_{1/2} \times \phi_{1/2} = \phi_0 + \phi_1$$

$$\phi_{1/2} \times \phi_j = \phi_{j-1/2} + \phi_{j+1/2} \qquad \text{if } j \le \frac{k-1}{2}$$
$$\phi_{1/2} \times \phi_{k/2} = \phi_{(k-1)/2} \tag{7.1}$$

Then the Bratelli diagram of Fig. 28 corresponding to ordinary SU(2) will not keep on expanding indefinitely and it will truncate as shown in Fig. 29. We can characterize quite accurately the representations of the braid group furnished by the blocks $F^{(N)}_{\phi,i}$ (as in Fig. 26) from the information we have about the fusion rules (7.1) and the conformal weights h_j of the fields ϕ_j. First in the operation (6.1) we have

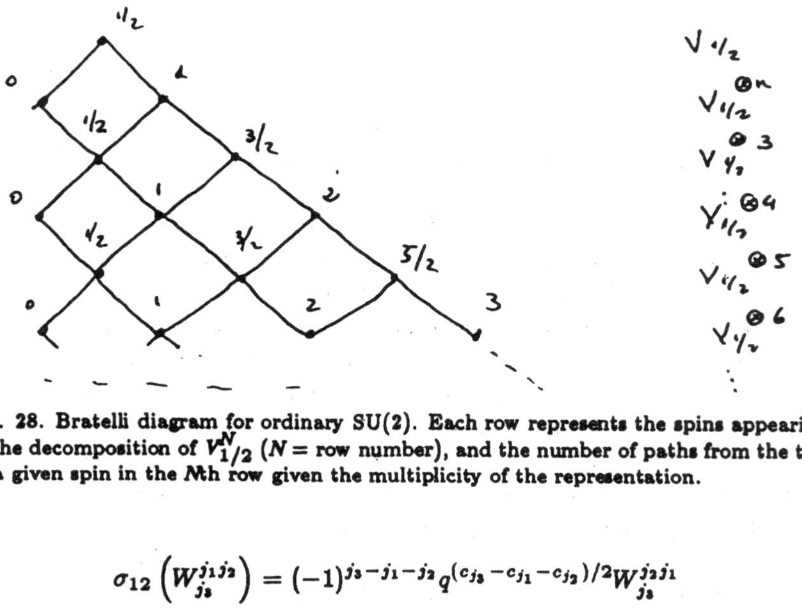

Fig. 28. Bratelli diagram for ordinary SU(2). Each row represents the spins appearing in the decomposition of $V^N_{1/2}$ (N = row number), and the number of paths from the top to a given spin in the *N*th row given the multiplicity of the representation.

$$\sigma_{12}\left(W^{j_1 j_2}_{j_3}\right) = (-1)^{j_3 - j_1 - j_2} q^{(c_{j_3} - c_{j_1} - c_{j_2})/2} W^{j_2 j_1}_{j_3}$$
$$q = e^{\frac{2\pi i}{k+2}} \qquad c_l = l(l+1)\ .$$

Using the F move (Fig. 4) we can diagonalize the matrix ϱ_i braiding the legs $i, i+1$. From (7.1) we learn that all the ϱ_i have the same eigenvalues. By a phase redefinition these eigenvalues can be taken to be $1, q$. Therefore the matrices ϱ_i satisfy a quadratic equation. This together with the braid relations defines the structure of a Hecke algebra of type A_N; $H_N(q)$:

$$\begin{aligned} \varrho_i^2 &= (1-q)\varrho_i + q \\ \varrho_i \varrho_{i+1} \varrho_i &= \varrho_{i+1} \varrho_i \varrho_{i+1} \\ \varrho_i \varrho_j &= \varrho_j \varrho_i \qquad |i-j| \ge 2 \end{aligned} \tag{7.2}$$

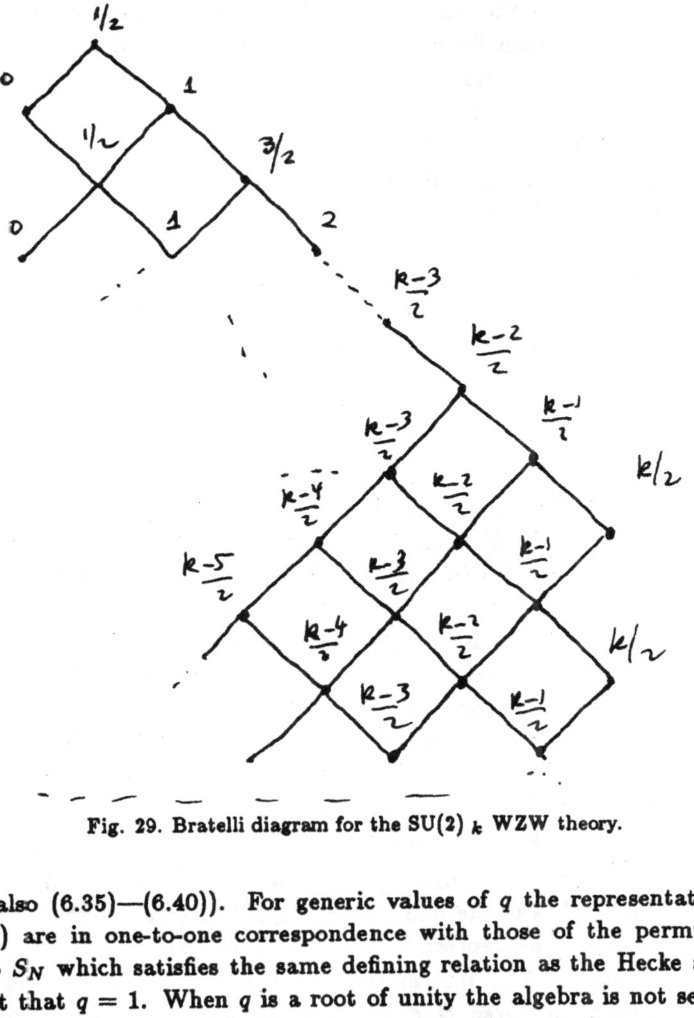

Fig. 29. Bratelli diagram for the SU(2) $_k$ WZW theory.

(See also (6.35)—(6.40)). For generic values of q the representations of $H_N(q)$ are in one-to-one correspondence with those of the permutation group S_N which satisfies the same defining relation as the Hecke algebra except that $q = 1$. When q is a root of unity the algebra is not semisimple and to obtain a semisimple representation theory we have to consider quotients of the algebra by some ideals. Wenzl[59] has studied this question in detail and has classified the necessary restrictions in the representations of $H_N(q)$. From the general properties of conformal theories we know that the algebra representing the action of the braid group must be effectively semisimple because the blocks provide fully reducible representations of B_N. For the blocks considered in this example the semisimple quotient of the Hecke algebra provides the representation of the braid group on the blocks. This can be seen from the fusion rules or the associated Bratelli diagram. The representations of $H_N(q)$ are labelled by the same Young tableau as those of S_N. In the case of SU(2) we can replace the spin labels in Fig. 28 by the Young tableaux. Spin j is represented by a tableau with at most two rows of λ_1, λ_2 boxes each such that $\lambda_1 - \lambda_2 = 2j$. Hence the representation of SU(2) appearing in the WZW model are those satisfying $\lambda_1 - \lambda_2 = 2k$. These are the diagrams of type $(2, k)$ in Wenzl notation and as shown in Ref. 59 they provide the *-representations of $H_N(q)$ when $q = e^{2\pi i/k+2}$ (the *-representations are always semisimple algebras). The absence of tableaux with more than two rows can be written as an extra relation in the Hecke algebra demanding the three row antisymmetrizer to vanish

$$1 - e_1 - e_2 + e_1 e_2 + e_2 e_1 + e_1 e_2 e_1 = 0 \tag{7.3}$$

(since we are working with SU(2) this was expected). Equations (7.2) and (7.3) are equivalent to the Temperley-Lieb-Jones algebra. See (6.35)—(6.40) for more details. These algebras, known as $A_{\beta,N}$ have been extensively studied by Jones[40,41] where he has classified the subfactor when q is a root of unity and analyzed the *-representations. It is remarkable that many of the properties of the semisimple quotients of these algebras can be obtained directly from the theory of the WZW model whose chiral algebra is a Kac-Moody algebra. We have therefore learned that the representations of the braid group provided by the blocks $F^N_{\phi,i}$ in the $\mathrm{SU}(2)_k$ WZW theory are in one-to-one correspondence with the *-representations of the algebra $A_{\beta,n}$. Since we know this algebra, we can ask for its centralizer in order to find the algebra A. As we know from Sec. 6, the centralizer of this algebra is the quantum group SU(2,q), $q = e^{2\pi i/k+2}$ (with some restrictions on its representation theory so that it also becomes a semisimple algebra). This information can now be used to construct explicity the duality matrices B, F in terms of the q-analogues of the $6j$-symbols. The F move (Fig. 4) corresponds to the change of basis implementing the associativity law for the composition of (quantum) angular momentum $(V_1 \otimes V_2) \otimes V_3 \cong V_1 \otimes (V_2 \otimes V_3)$. For generic values of q of the $6j$-symbols were computed and it was shown that they coincide when q is a root of unity with the duality matrices for the SU(2) level k WZW model. The arguments in Ref. 23 were essentially the same as those appearing here. Using Wenzl's analysis of the *-representation of the braid representations of this conformal field theory and then finding the quantum group through the centralizer of the Temperley-Lieb-Jones algebras. The duality matrices were obtained in Ref. 6 by solving the Knizhnik-Zamolodchikov equation[12] corresponding to the decoupling of null vectors of the combined Virasoro and Kac-Moody algebras. The results fo both computations (via the quantum group or via the differential equation) agree up to conjugation by a diagonal matrix. This arbitrariness is used to normalize the blocks according to one's favorite conventions. For instance the prescription in Ref. 6.

Since the previous paragraph may have been a bit too fast, we provide some more details. With the operation (6.1) we have determined that the deformation parameter of SU(2,q) must be a root of unity if we want to make contact with WZW models. From Sec. 6.2 however we know what are the precise conditions for the space of blocks (6.4) to provide fully reducible representations of the braid group. We have to use the modified tensor products (6.50) and (6.51). In general, we know if $q^p = 1$, the set of regular representations ν_{reg}, contains only those representations with highest weights satisfying $\langle\lambda, \theta\rangle \leq p - g \equiv k$. In quantum groups we can define the intertwiner

$$K_k^{ij} : (V^i \otimes V^j)' \mapsto V^k$$

and the universal $\mathcal{R}$ matrix yields the operation (6.1). As a general consequence of the properties of quantum groups[28] we have

$$K_k^{ij} R^{ji} = \epsilon q^{(c_k - c_i - c_j)/2} K_k^{ji} \tag{7.4}$$

where c_i is the quadratic Casimir of the classical algebra for the same representation. Hence, modulo integers we can identify the conformal dimensions of the field generating V^i as

$$h_i = \frac{c_i}{k+g} \,. \tag{7.5}$$

Notice that in (7.4) the intertwiner is defined with respect to the restricted tensor product. In this case we have a positive definite scalar product and we can define the adjoint

$$K_{ij}^k : V^k \mapsto (V^i \otimes V^j)'$$

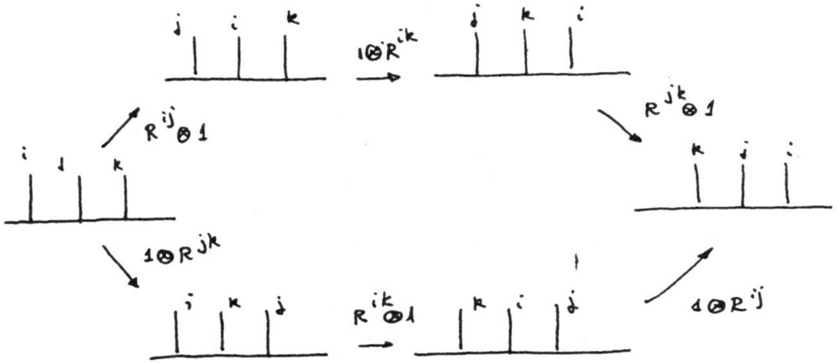

Fig. 30. Yang-Baxter equation in the form (6.37).

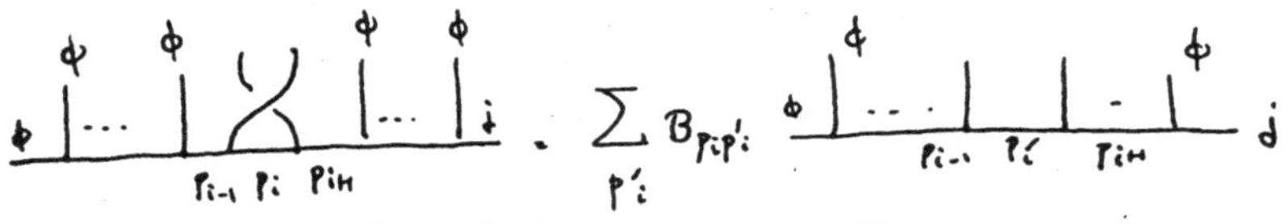

Fig. 31. Braid group action on $F^n_{\phi,j}$.

so that the orthogonality relations now become

$$\sum_k K^k_{ij} K^{lm}_k = \delta^l_i \delta^m_j (l_i \otimes l_j)'$$

with the sum extended only over the representations appearing in the decomposition of the restricted tensor product $(V^i \otimes V^j)'$.

The hexagon identity is a simple consequence of (6.37) which is represented graphically in Fig. 30. The explicit formulae for the braiding and fusing matrices are given in terms of the q-analogues of the $6j$-symbols for SU(2) or their generalisations for other groups. They determine explicity the isomorphism

$$((V_1 \otimes V_2)' \otimes V_3)' \cong V_1 \otimes (V_2 \otimes V_3)')' . \tag{7.6}$$

The pentagon identity is a consequence of the associativity of the restricted tensor product which in turn follows from the co-associativity of the comultiplication.

As argued before, the matrices of this change of basis give the F move in the space of blocks an the B-move is obtained from it using (6.2). For theories where the representation ring is generated by one representation (in Conformal Field Theory this means that there is an operator out of which all other primaries are obtained by decompositing its OPEs) the braiding and fusion matrices of more interest are those of the generating field with arbitrary spectators. Using associativity and the fusion rules one can in principle determine the fusion and braiding matrices of other representations. In the case of SU(N) the F, B matrices are obtained up to conjugation by a diagonal matrix using rather simple arguments and they are shown to be the same as those obtained in WZW theories by solving the Knizhnik-Zamolodchikov equation[12]. For level $k > 1$ there are two irreducible representations appearing in the tensor product of two fundamental representations $V \otimes V, V \otimes V = V_a + V_s$ (the subscripts s, a stand for q-symmetric and q-antisymmetric respectively). Using the F move we diagonalize the braiding matrix of two contiguous representations and find that they have only two distinct eigenvalues. Choosing the common phase of all the braiding matrices conveniently we can make these two eigenvalues equal to $1, q = e^{2\pi i/k+g}$ and therefore the braid generators ϱ_i satisfy a Hecke algebra relation (6.35) together with the condition that the $N+1$ row antisymmetrizer vanishes. Both the quantum group and the WZW theory provide unitary representations of this algebra labelled by the same tableaux. This is a consequence of the relation between the Bratelli diagrams of the quantum group and its centralizer, and in the WZW theory it follows from the fusion rules. Since the basis we have chosen for the blocks have the same quantum numbers as those for the WZW theory, the matrices of these two representations must be the same up to diagonal conjugation[23,62]. In WZW one can determine the explicit form of the representation matrix by solving the Knizhnik-Zamolodchikov equation. In the quantum group we know that the action of the braid group on tensor products is given by

$$\varrho(\sigma_i) = \varrho_i = 1 \otimes \ldots \otimes R_{i,i+1} \otimes \ldots \otimes 1$$

and after the tensor product is decomposed, we have an action on the blocks (see Fig. 31):

$$\varrho_i F_j(\ldots p_i - 1, p_i, p_i + 1 \ldots) = \sum_{p'_i} B_{p_i,p'_i} \begin{bmatrix} \phi & \phi \\ p_{i-1} & p_{i=1} \end{bmatrix} F_j(\ldots p_{i-1}, p'_i, p_{i+1} \ldots)$$

in the SU(2) case $\phi = \phi_{1/2}$. Using (6.2) and the definition of the $6j$-symbols we obtain[23]:

$$F_{jj'}\begin{bmatrix} j_2 & j_3 \\ j_1 & j_4 \end{bmatrix} = \begin{Bmatrix} j_1 & j_2 & j \\ j_3 & j_4 & j' \end{Bmatrix} \tag{7.7}$$

$$B_{jj'}\begin{bmatrix} j_2 & j_3 \\ j_1 & j_4 \end{bmatrix} = (-1)^{j+j'-j_1-j_4} q^{(c_{j_1}+c_{j_4}-c_j-c_{j'})/2} \begin{Bmatrix} j_2 & j_1 & j \\ j_3 & j_4 & j' \end{Bmatrix}$$

where

$$\begin{Bmatrix} j_1 & j_2 & j_{12} \\ j_3 & j & j_{23} \end{Bmatrix} = \Delta(j_1, j_2, j_{12})\Delta(j_3, j, j_{12})\Delta(j_1, j, j_{23})\Delta(j_3, j_2, j_{23})\times$$

$$\sum_{z\geq 0}(-1)^z[z+1]!\frac{1}{[z-j_1-j_2-j_{12}]![z-j_3-j-j_{12}]![z-j_1-j-j_{23}]![z-j_3-j_2-j_{23}]!}$$

$$\frac{1}{[j_1+j_2+j_3+j-z]![j_1+j_3+j_{12}+j_{23}-z]![j_2+j+j_{12}+j_{23}-z]!}$$

$$\Delta(a,b,c) = \sqrt{\frac{[-a+b+c]![a-b+c]![a+b-c]!}{[a+b+c+1]!}} \tag{7.8}$$

implementing the B, F moves for SU(2) WZW theories.

To summarise, the duality properties of a RCFT contain the purely topological information characterizing a CFT. We argued in Sec. 6.1 (and exhibited explicity here) that the braiding and fusing properties of the conformal blocks can be translated into the defining properties of a Quasi-Triangular Yang-Baxter algebra obtained as the centralizer of the action of the braid group on the conformal blocks. With so little information one can go quite far in the explicit computation of the duality matrices in terms of representation theoretic quantities provided by the quantum group. In a more conventional approach this information is obtained only after solving the Knizhnik-Zamalodchikov equations. It is quite remarkable that both computation differ only by conjugation with diagonal matrices (block normalization).

In Sec. 7.3 we will present the computation of the modular transformation matrix S_{ij} from the quantum group point of view.

7.2 *Rational gaussian models*

As a final example we consider the rational theories at $c = 1$: consisting of a single scalar field ϕ taking values on a circle of radius $R = \sqrt{r/s}$

(where s is odd). The chiral algebra is generated by the two fields $\partial\phi(z)$ and $\exp(\pm 2i\sqrt{r/s}\phi(z))$. For details see Ref. 57. The primary fields are vertex operators $\phi_p = e^{ip\phi(z)}$ where the momentum is quantized as

$$p = \frac{k}{2\sqrt{rs}} \quad k \in Z/4rsZ = Z_{4rs} \ .$$

ϕ_p has conformal dimension $h_p = p^2 = k^2/4rs$ and the fusion rules are

$$\phi_k \times \phi_{k'} = \phi_{k+k'} \qquad k, k', k+k' \in Z_{4rs} \ . \tag{7.9}$$

The quantum group description of this RCFT is generated by the operators $p, t_\pm$ in correspondence with $\partial\phi(z)$ and $\exp(\pm 2i\sqrt{rs}\phi(z))$. They satisfy the commutation relations

$$\begin{aligned} [p, t_\pm] &= \pm M t_\pm \\ [t_+, t_-] &= 0 \ . \end{aligned}$$

We want to first show how one endows this algebra with the structure of a quantum group and then show that the duality properties of the rational Gaussian model are recovered from its representation theory. The Hopf algebra axioms are satisfied by the following definitions

$$\begin{aligned} \Delta(p) &= p \otimes 1 + 1 \otimes p & \varepsilon(p) &= 0 & \gamma(p) &= -p \\ \Delta(t_\pm) &= t_\pm \otimes q^p + q^{-p} \otimes t_\pm & \varepsilon(t_\pm) &= 0 & \gamma(t_\pm) &= -q^{\pm M} t_\pm \ . \end{aligned} \tag{7.10}$$

So far the two parameters q, M are independent. The next step is to find under what conditions this Hopf algebra $Q(q, M)$ becomes a Quasi-Triangular Yang-Baxter algebra (6.12,13). Take as the universal $\mathcal{R}$-matrix the ansatz

$$R = q^{ap \otimes p} \tag{7.11}$$

then (6.12) and (6.13) imply the conditions

$$q^{4p} = q^{2Map} = 1 \ . \tag{7.12}$$

It is interesting to see that quasi-triangularity imposes important restrictions on the Hopf algebra Q. Taking q as a phase we obtain the quantization conditions on the spectrum of p. We normalize q^a to $e^{2\pi i}$, which is equivalent to choosing conventions so that $h_p = p^2$. Then (7.12) leads to

$$p = \frac{a}{4}k \quad M = \frac{2\nu_0}{a} \quad k, \nu_0 \in Z \ . \tag{7.13}$$

Since t_+ changes the eigenvalue of p by M units to obtain a representation of the algebra we must have that $p + M$ also takes the form $\frac{a}{4}k'$, hence

$$a = 2\sqrt{\frac{2\nu_0}{N}} \quad N \in Z \ .$$

All this together yields the quantization conditions

$$p = \frac{k}{2\sqrt{rs}} \quad N = 2\nu_0 rs \ . \tag{7.14}$$

To obtain a rational representation theory note first that the irreducible representations of Q are labelled by the eigenstates of p, $|p\rangle$ with $p = (k + mN)/2\sqrt{rs}$, $m \in Z$ obtained from $|k/2\sqrt{rs}\rangle$ by repeated action of $t_\pm$. Since the braiding properties of two elements in the same representation of the chiral algebra are the same we find that for a state in $\{|p_1\rangle\}$ and a state in $\{|p_2\rangle\}$ the braiding is

$$e^{2\pi i p_1 p_2} = e^{2\pi i k_1 k_2/4rs} e^{i\pi(m_1 k_2 + m_2 k_1)\nu_0}$$

implying that to have a consistent representation theory, ν_0 must be an even integer. In fact there is no loss of generality if we choose $\nu_0 = 2$. If $\nu_0 \geq 2$ we can always construct other operators $t'_\pm$ which are "roots" of $t_\pm$ and for them $\nu_0 = 2$. This implies that $k \in Z_{4rs}$. When we enlarge the representations of p as indicated in order to label them only with integers mod. $4rs$ we have to make sure that $t_\pm$ is still compatible with the modding and the similarly for the co-multiplication. The fusion rules are very easy in this case. Assuming the Clebsch-Gordan rule

$$|p_1\rangle|p_2\rangle = |p_1 + p_2\rangle$$

we have

$$t_+|p\rangle = N_p|p + M\rangle \ .$$

From the co-multiplication we can derive the compatibility condition

$$N_{p_1}q^{p_2} + N_{p_2}q^{-p_1} = N_{p_1+p_2}$$

solved by

$$N_p = \sin\frac{k\pi}{2} \quad p = \frac{a}{4}k \quad N_{p+N} = N_p$$

and in the reduced theory $t_\pm$ does not act as the identity

$$t_\pm|k\rangle = \sin\frac{k\pi}{2}|k\rangle \ .$$

t cannot act as the identity because $\Delta(1) = 1 \otimes 1$ and it would not be compatible with the reduction. Therefore the rational Gaussian models can be recovered (or rather their duality properties) by looking at the quantum group Q and imposing a rationality condition on its representation ring (to have only a finite number on representations).

7.3 *Modular properties*

We now construct the matrix S using only quantum group information. Using the relation between quantum groups, RCFT and link invariants detailed in Sec. 9 we will show that our representation of S is as expected from the topological Chern-Simons theory given by the expectation value of two linked Wilson lines on S^3.

From the work of Moore and Seiberg[16,17,18] we know we can represent the Verlinde operators[15] in terms of the duality matrices. In particular, the action of the operators associated to the a-cycle on the characters is diagonal and it is given by the eigenvalues of the fusion matrices $\lambda_k^{(i)}$. The precise relation with the duality matrices is

$$\begin{aligned} \lambda_k^{(j)} &= \frac{1}{F_k}\sum_p B_{op}\begin{bmatrix} k & \bar{j} \\ k & j \end{bmatrix}(+)B_{p0}\begin{bmatrix} \bar{j} & k \\ k & j \end{bmatrix}(+) \\ F_k &= \epsilon^0_{k\bar{k}}F_{00}\begin{bmatrix} k & \bar{k} \\ k & k \end{bmatrix} \end{aligned} \tag{7.15}$$

where the line over an index refers to the conjugate field. (The + argument in the braiding matrices indicates that we are braiding in the positive sense, the $i+1$-st strand goes over the ith strand.) The Verlinde operator acts on characters according to

$$\phi_k(a)\chi_j = \lambda_k^{(j)}\chi_j \tag{7.16}$$

and it can be described as inserting the identity operator factorised as $k, \bar{k}$ on the block in Fig. 20, then we braid the field k around j. These operations result in (7.15). From (6.2) we can write (7.15) as

$$\lambda_k^{(j)} = \frac{1}{F_k} \sum_p \varepsilon^0_{\bar{j}j} \varepsilon^k_{p\bar{j}} \varepsilon^p_{jk} e^{2\pi i(h_p - h_j - h_k)} F_{0p} \begin{bmatrix} k & j \\ k & \bar{j} \end{bmatrix} F_{p0} \begin{bmatrix} \bar{j} & j \\ k & k \end{bmatrix} . \quad (7.17)$$

It is easy to show as a consequence of the pentagon identity that

$$\frac{S_{jk}}{S_{00}} = \sum_p e^{2\pi i(h_p - h_j - h_k)} \frac{S_{0p}}{S_{00}} N^p_{jk} . \quad (7.18)$$

In the case of SU(2) we have

$$S_{jl} = \sqrt{\frac{2}{k+2}} \sin \frac{\pi(2j+1)(2l+1)}{k+2}$$

and (7.18) implies the following identity between q-numbers

$$[(2j+1)(2k+1)] = \sum_p q^{c_p - c_j - c_k} N^p_{jk} [2p+1] \quad c_j = j(j+1) .$$

This relation is certainly different from the addition formula of q-dimensions

$$[(2j+1)][(2k+1)] = \sum_p N^p_{jk} [2p+1] \quad (7.19)$$

which is a particular case of the Verlinde conjecture (see Sec. 5 for details).

In the quantum group there is a useful relation between the trace of the comultiplication $\Delta(a)$ in the tensor product representation $j \otimes k$ and the trace of a in the representations appearing in the decomposition of the tensor product:

$$\mathrm{tr}_{j\otimes k} \Delta(a) = \sum_p N^p_{jk} \mathrm{tr}_p(a) \quad (7.20)$$

where

$$\mathrm{tr}_{j\otimes k} \Delta(a) = \mathrm{tr}\ \varrho^j \otimes \varrho^k(\Delta(a))$$

and ϱ^j is the map from the quantum group A to the representation V^j. Using (6.3) one easily derives (7.20) from the orthogonality properties of the quantum Clebsch-Gordan coefficients. As a trivial example from (7.20) and the comultiplication

$$\Delta(q^{H/2}) = q^{H/2} \otimes q^{H/2}$$

we obtain (7.19). In any Quasi-Triangular Yang-Baxter algebra there is always a distinguished element u[28] defined in terms of the $\mathcal{R}$-matrix and the antipode by

$$\begin{aligned} u &= m(\gamma \circ id)(\sigma(\mathcal{R})) \\ u^{-1} &= m(id \circ \gamma^2)(\sigma(\mathcal{R})) \end{aligned} \tag{7.21}$$

where m is the multiplication in A and we have used the notation in Sec. 6. The element u has the properties

$$uau^{-1} = \gamma^2(a) \qquad \forall_a \in A \tag{7.22}$$

when $A = \mathrm{SU}(2,q)$, u and u^{-1} are given in the representation j by

$$\varrho^j(u) = q^{-c_j} q^{H/2} \qquad \varrho^j(u^{-1}) = q^{c_s} q^{-H/2} \tag{7.23}$$

so that

$$\mathrm{tr}_j u = q^{-c_j}[2j+1] \qquad \mathrm{tr}_j\, u^{-1} = q^{c_j}[2j+1]\,. \tag{7.24}$$

Since the construction of $\lambda_k^{(j)}$ follows from the polynomial equations directly, the relation (7.18) has a very nice interpretation in the quantum group (for the time being we consider SU(2)):

$$\frac{S_{jk}}{S_{00}} = q^{-c_j-c_k}\mathrm{tr}_{j\otimes k}\Delta(u^{-1}) = q^{c_j+c_k}\mathrm{tr}_{j\otimes k}\Delta(u) \tag{7.25}$$

the element u can be identified with the remnant of the Virasoro algebra. The comultiplication rule for the element u is[29]

$$\begin{aligned} \Delta(u) &= \mathcal{R}^{-1}\sigma(\mathcal{R}^{-1})(u \otimes v) \\ \Delta(u^{-1}) &= \sigma(\mathcal{R})\mathcal{R}(u^{-1} \otimes u^{-1})\,. \end{aligned} \tag{7.26}$$

Finally we find the complete form of the matrix S in the quantum group[a]

$$\begin{aligned} \frac{S_{jk}}{S_{00}} &= q^{-c_j-c_k}\mathrm{tr}_{j\otimes k}\sigma(\mathcal{R})\mathcal{R}(u^{-1} \otimes u^{-1}) \\ &= q^{c_j+c_k}\mathrm{tr}_{j\otimes k}\mathcal{R}^{-1}\sigma(\mathcal{R}^{-1})(u \otimes u)\,. \end{aligned} \tag{7.27}$$

[a]In the limit $q \to 1$, $\sigma(\mathcal{R})\mathcal{R} \mapsto 1 \otimes 1$, this is the condition for a quasi-triangular Yang-Baxter algebra to be also triangular [27]. Hence the classical limit of a quantum group is a triangular Yang-Baxter algebra.

We can make contact with the Chern-Simons theory[14]. We know from (6.42) that the quantum group comes endowed with a trace with the Markov property which can be used to define link invariants (see Sec. 9 for more details). As in SU(2), u is propotional in general to q^{ϱ}, hence

$$(u \otimes u)R = R(u \otimes u) \ .$$

To understand what link we relate with S by (7.27), we recall that every link or knot can be obtained by closing a braid (see Sec. 9). Since we have a representation of the braid group whose action is defined in terms of the matrix R, we see that the operation $\sigma(R)R$ used in the definition of S is the braid in Fig. 32a. Taking traces is like closing the braid (identifying the end points of the braid in Fig. 32a to get Fig. 32b) and we prove directly that the modular matrix S is a link invariant associated to Fig. 32. In the Chern-Simons theory this conclusion is obtained by doing surgery on the trivial link (two unknotted unlinked circles on $S^2 \times S^1$) according to the operation S. In other words, in the Chern-Simons theory S_{jk} is the expectation value in S^3 of two linked Wilson lines

$$S_{jk} = Z(S^3, L(R_j, R_k))$$

(see Ref. 33 for more details). In the quantum group this is interpretation is a consequence of the co-multiplication rule for the distinguished element u.

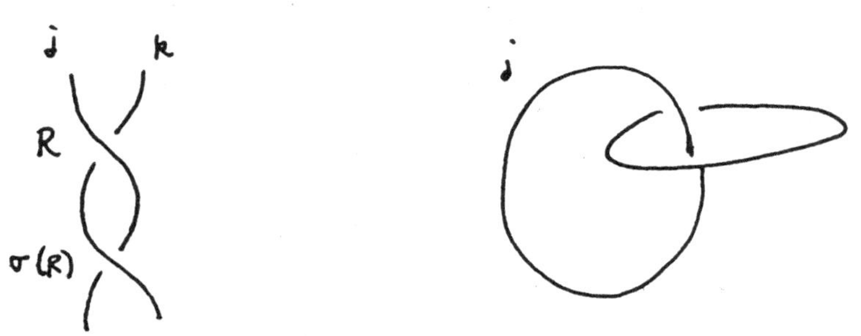

(a) (b)

Fig. 32. Link representation of the matrix S.

We can now proceed along the argument presented in Ref. 33 to prove the Verlinde conjecture. Since here we do not have a functional integral

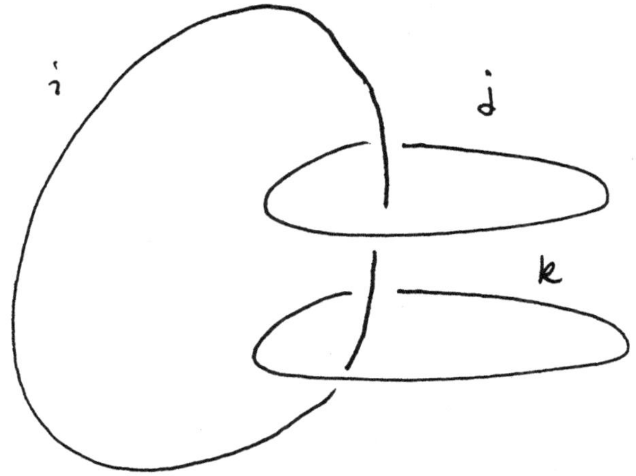

Fig. 33. Link used in the proof of the Verlinde conjecture.

interpretation (only duality properties) the proof will rely on the defining properties of a quantum group. Consider the link in S^3 in Fig. 33. It can be represented by the braid in Fig. 34 which as a representation in terms of traces is

$$\begin{aligned} W_{ijk} &= \mathrm{tr}_{i\otimes j\otimes k}\left(\sigma(R_{12})\sigma(R_{13})R_{13}R_{12}q^{-H/2}\otimes q^{-H/2}\otimes q^{-H/2}\right) \\ &\equiv \mathrm{tr}^{M}_{i\otimes j\otimes k}\left(\sigma(R_{12})\sigma(R_{13})R_{13}R_{12}\right) \end{aligned} \tag{7.28}$$

where in the last step we use the definition of the Markov trace (6.24c). For simplicity we are presenting the argument in detail for SU(2). The generalization to more complicated cases is straightforward. To prove Verlinde's conjecture we compute (7.28) in two ways. First we use the defining properties of a quantum group:

$$\begin{aligned} (1\otimes\Delta)\mathcal{R} &= \mathcal{R}_{13}\mathcal{R}_{12} \\ (1\otimes\Delta)\sigma(\mathcal{R}) &= \sigma(\mathcal{R})_{12}\sigma(\mathcal{R})_{13} \end{aligned}$$

to write

$$W_{ijk} = \mathrm{tr}_{i\otimes j\otimes k}(1\otimes\Delta)\sigma(R)R(q^{-H/2}\otimes q^{-H/2}) \ .$$

Using (7.20) we and (7.27) we obtain

$$W_{ijk} = \sum_m N^m_{jk}\mathrm{tr}_{i\otimes m}\sigma(R)R\left(q^{-H/2}\otimes q^{-H/2}\right) = \sum_m N^m_{jk}\frac{S_{mi}}{S_{00}} \ . \tag{7.29}$$

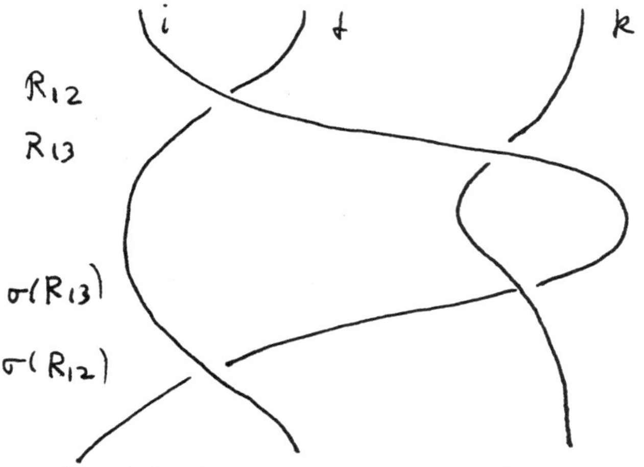

Fig. 34. Braid representation of the link in Fig. 33.

On the other hand the trace (7.28) has the Markov property (6.42c) and next section) and for the braiding we performed in Fig. 34 we have

$$W_{ijk} = \frac{\left(\mathrm{tr}^M_{i\otimes j}\sigma(R)R\right)\left(\mathrm{tr}^M_{i\otimes k}\sigma(R)R\right)}{S_{0i}/S_{00}}$$

$$= \frac{S_{ij}S_{jk}}{S_{00}S_{0i}} \,. \tag{7.30}$$

From (7.29) and (7.30) we obtain the proof of Verlinde's conjecture:

$$\frac{S_{ij}S_{ik}}{S_{0i}} = \sum_m N^m_{jk} S_{mi} \,. \tag{7.31}$$

Since S is obviously symmetric and it diagonalizes the fusion rules, it follows from the general arguments of Sec. 2 that $S^2 = C$ as expected. This now leads to all the formula in (5.18). Although for simplicity we have carried out the argument as though the representations were self-conjugate, the formulae derived hold in the general case. Using (5.18), in particular $S^2 = C, S^* = S^{-1} = SC = CS$ we can rewrite (7.18) as

$$S^j_i = \sum_k N^j_{ki} e^{2\pi i(h_i+h_j-h_k)} S^k_0 \,. \tag{7.32}$$

With respect to the other defining relation of the modular group $(ST)^3 = C$, it can be rewritten as

$$TSTST = S \,. \tag{7.33}$$

Although this equation seems to involve many conditions on the RCFT, it is equivalent to a single equation. Using (5.18), (7.18) and (7.32) one can show that (7.33) is equivalent to

$$e^{2\pi i h_j} S_0^j = e^{i\pi c/4} \sum_k e^{2\pi i h_k} S_0^k S_k^j \,. \tag{7.34}$$

This suggests the definition of a quantum character as

$$\chi_i(q) = e^{2\pi i h_i - i\pi c/8} \frac{S_0^i}{S_0^0}$$

and (7.34) expresses the relation between the character with argument q^{-1} and the characters with argument q. More generally this is a propety of the quantum characters under charge conjugation, and it should be related to the completeness of the set of rational representations of the quantum group. For quantum groups coming from deformations of classical groups (7.34) holds as one can see from explicit computations, it is also quite easy to verify this relation for Gaussian models. In general, however, an argument similar to the one for S in terms of u should ensure this transformation property of quantum characters, although at present this general argument is missing. It seems likely that the Markov property of the trace and its relation to knot and link invariants should provide again the steps to complete the argument in general.

This concludes the proof that quantum groups provide solutions to the polynomial equations. It should be stressed again that the quantum group comes endowed with a Markov trace, and that this is an important property with respect to the relation with knot invariants and also with respect to the partial reconstruction of the space of conformal blocks.

The reconstruction problem in RCFT is to find the extra conditions that determine when a particular representation space of the duality transformations (a solution of the polynomial equations) can be interpreted as the space of blocks of a RCFT. The quantum group approach is a first step in this direction. As we have seen we get the quantum group directly from the duality properties of the conformal theory. What we see at this point is that RCFT are always associated with C^*-irreducible representations of the duality transformations. If these C^*-representations are II_1-factors i.e. they admit a normalized finite trace then we will be able to identify the space of conformal blocks with the Hilbert space obtained by completion of the duality C^*-algebra with respect to the scalar product defined by the trace $(\langle x|y\rangle = \mathrm{tr}(x^* y))$. The extra physical requirement that naturally appears in the quantum group approach is that the conformal theories are in one-to-one correspondence with solutions of the polynomial equations such that:

1) They define irreducible C^*-representations of the duality transformations, and

2) They are equipped with a normalized finite trace satisfying the Markov property (see Sec. 6 and 9). This Markov property which holds under very general circumstances is at the root of the connection between Conformal Field Theory and three-dimensional Topological Field Theories. In general if one begins with a solution to the polynomial equations some of the steps in the previous paragraphs establishing the relation with quantum groups are missing. For instance, in general we will not get the spaces V_i themselves or even their real dimensions, but only their q-dimensions (5.38). In this cases more powerful techniques are required. It is reasonable to expect that an extension of the results of Deligne quoted in Ref. 18 concerning the reconstruction of classical groups may yield ultimately that to any solution of the polynomial equations with some extra condition about the possible q-dimensions can be given in terms of quantum groups. In the next sections we will look in more detail into the structure of quantum groups and show that those groups coming from classical groups do indeed satisfy the polynomial equations and look also into the general situation.

8. Solvable Lattice Models and Conformal Field Theories

8.1 *Vertex models*

Following Baxter[34,36] we will define a vertex model on a generic lattice $\mathcal{L}$. The lattice is defined by the condition that each site must be the endpoint of four edges, where edges are defined as the line segments between sites. These lattices can always be colored in black and white in such a way that two plaqnettes with the same color have no edge in common (Fig. 35). The statistical model on the lattice $\mathcal{L}$ is defined by associating three different interaction coeffcients K_i, K_i', K''_i to each site and spin variables $\sigma_e = \pm 1$ to each face or plaqnette. The Boltzmann weight for a generic site i is (Fig. 36):

$$\exp\left[K_i\sigma_p\sigma_e + K_i'\sigma_g\sigma_m + K''_i\sigma_p\sigma_q\sigma_e\sigma_m\right] . \tag{8.1}$$

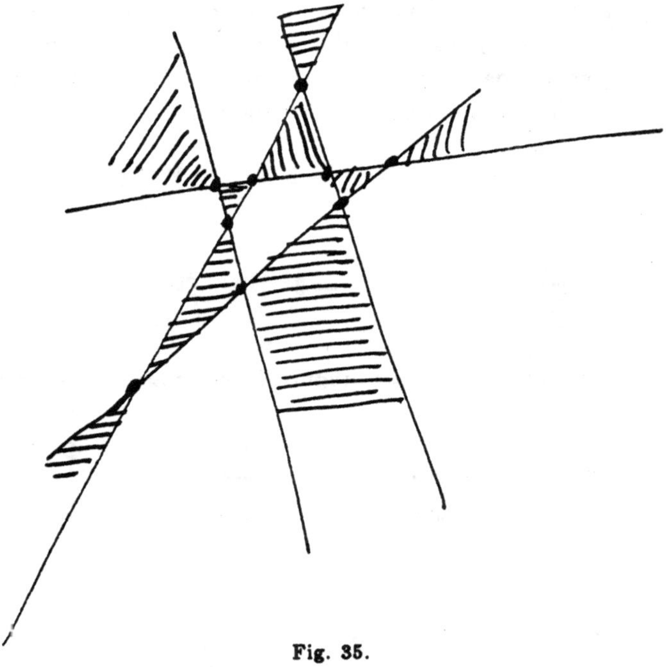

Fig. 35.

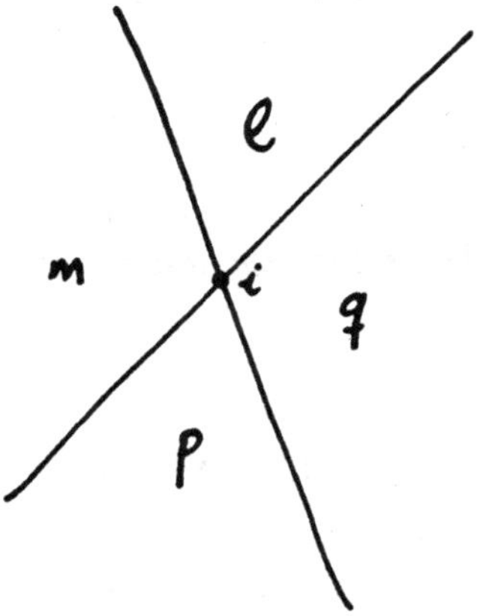

Fig. 36.

Once we have colored our lattice we can describe the different spin configurations *on the plaquettes by drawing arrows on the edges of the lattice.* In fact for each site we have 16 possible choices for the surrounding spins.

These configurations can be divided in two sets of 8, where one can be obtained from the other by reversing all the spins. Notice that reversing all the surrounding spins does not change the value of the Boltzmann weight (8.1). Now we can introduce arrows on the edges of the lattice by the rule:

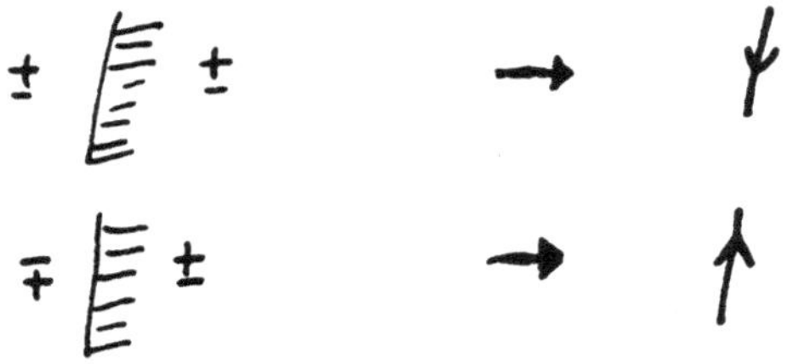

(R)

Using (R) the eight different configurations are associated to the following vertices:

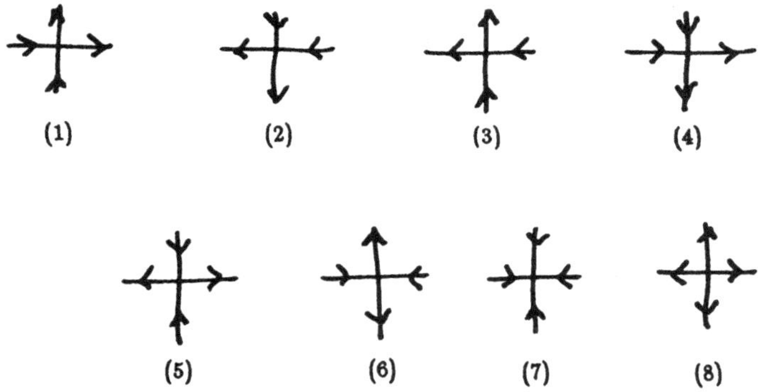

(8.2)

The Boltzmann weights of the eight-vertices can be obtained from (8.1). It is easy to see that they satisfy the following relations:

$$
\begin{aligned}
w_1 &= w_2 = \exp\left(K_i - K_i' - K''_i\right)\\
w_3 &= w_4 = \exp\left(-K_i + K_i' - K''_i\right)\\
w_5 &= w_6 = \exp\left(K_i + K_i' + K''_i\right)\\
w_7 &= w_8 = \exp\left(-K_i - K_i' + K''_i\right)
\end{aligned}
\tag{8.3}
$$

Therefore we have, in principle, four independent variables for each site of the lattice. To solve the model we will consider first the case of a regular lattice and the same interaction coefficients for all sites.

8.1.1 *Regular lattice*

For the Boltzmann weights we will introduce the notation $W(i,j,k,l)$ where i,j,k,l are ± 1 and refer to the directions of the arrows. The partition function of the model can be defined in terms of the transfer matrix:

$$Z = \mathrm{Tr}(T^n) \tag{8.4}$$

where n is the number of rows of the regular lattice. In order to define the transfer matrix consider two consecutive rows and denote by $\alpha = (\alpha_1\alpha_2\dots\alpha_m)$ the configuration of vertical arrows and by $\lambda = (\lambda_1\lambda_2\dots\lambda_m)$ the configuration of horizontal arrows (see Fig. 37). The transfer matrix is then defined by:

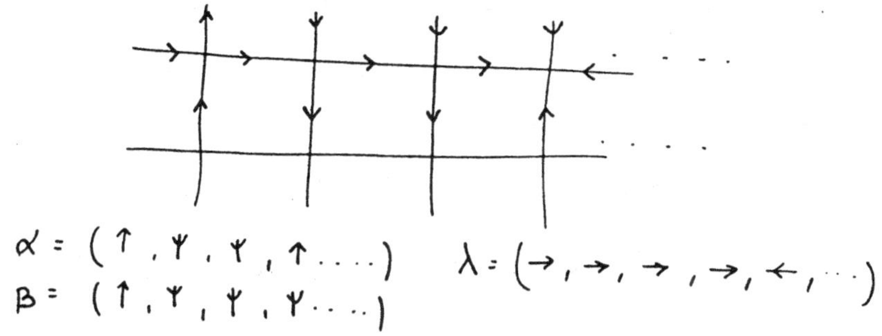

Fig. 37.

$$T_{\alpha|\beta} = \sum_{\lambda's} w(\lambda_1,\alpha_1,\lambda_2,\beta_1)\, w(\lambda_2,\alpha_2,\lambda_3,\beta_2)\dots w(\lambda_m\alpha_m\lambda,\beta_m) \tag{8.5}$$

where the sum is over all possible configurations of horizontal arrows. To solve the model is equivalent to diagonalizing the transfer matrix. Notice that the transfer matrix plays the role of the Hamiltonian and therefore to diagonalize T will be equivalent to finding an infinite set of conserved quantities. To find the conserved quantities we look for a new set of Boltzmann weights. Let us say $w'(i,j,k,l)$ such that the new transfer matrix T' obtained with them commutes with the original one:

$$(TT')_{\alpha\beta} = (T'T)_{\alpha\beta}\ . \tag{8.6}$$

Using (8.5) we can write:

$$(TT')_{\alpha\beta} = \sum_{\mu,nu} \prod_{i} \left(\sum_{\gamma} w(\mu_i \alpha_i \mu_{i+1} \gamma_i) w'(\nu_i \gamma_i \nu_{i+1} \beta_i) \right) . \tag{8.7}$$

Following Ref. 64 we can write (8.7) as:

$$(TT')_{\alpha\beta} = \mathrm{Tr} \left\{ \prod_{i} S(\alpha_i \beta_i) \right\} \tag{8.8}$$

with:

$$[S(\alpha_i \beta_i)]_{\mu\nu/\mu'\nu'} = \sum_{\gamma' s} w(\mu \alpha_i \mu' \gamma) w'(\nu \gamma \nu' \beta_i) . \tag{8.9}$$

To get (8.6) we require the existence of a matrix Q such that:

$$S'(\alpha, \beta) = QS(\alpha, \beta) Q^{-1} . \tag{8.10}$$

Using (8.9) we can write (8.10) as:

$$\begin{aligned} &\sum_{\gamma\mu''\nu''} w(\mu, \alpha, \mu'', \gamma) w'(\nu, \gamma, \nu'', \beta) w''(\nu'', \mu'', \mu', \nu') \\ &= \sum_{\gamma\mu''\nu''} w''(\nu, \mu, \mu'', \nu'') w'(\mu'', \alpha, \mu', \gamma) w(\nu'', \gamma, \nu', \beta) \end{aligned} \tag{8.11}$$

where we denote by w'' the elements of the matrix Q satisfying Eq. (8.10). These elements can be considered as a new set of boltzmann weights. Equation (8.11) is known as the Yang-Baxter equation for vertex models. A solution to Eq. (8.11) can be obtained using the factorized S-matrix models in Ref. 65. Defining the S-matrix by:

$$|\mu; \theta, \alpha; \theta'\rangle = S_{\mu\alpha;\mu''\gamma}(\theta - \theta') \quad |\mu''; \theta, \gamma\theta'\rangle \tag{8.12}$$

where we interpret μ, α, γ as internal quantum numbers and θ as related to the momentum by:

$$P^0 = \mathrm{sh}\ \theta \qquad p = \mathrm{coh}\ \theta . \tag{8.13}$$

The factorisation condition for three particle scattering is[64]:

$$\begin{aligned} &S_{\mu\alpha;\mu''\gamma}(\theta_1 - \theta_2) S_{\mu\gamma;\mu''\beta}(\theta_2 - \theta_3) S_{\nu''\mu'';\mu'\nu'}(\theta_2 - \theta_3) \\ =&S'_{\nu\mu;\mu''\nu''}(\theta_2 - \theta_3) S_{\mu''\alpha;\mu'\gamma}(\theta_1 - \theta_3) S_{\nu''\gamma;\nu'\beta}(\theta_1 - \theta_2) \end{aligned} \tag{8.14}$$

which coincides with Eq. (8.11) if we identify

$$
\begin{aligned}
w(\mu\alpha\mu''\gamma) &= S_{\mu\alpha;\mu''\gamma}(\theta_1-\theta_2) \\
w'(\nu\gamma\nu''\beta) &= S_{\nu\gamma;\mu''\beta}(\theta_1-\theta_3) \\
w''(\nu''\mu''\mu'\nu') &= S_{\nu''\mu'';\mu'\nu'}(\theta_2-\theta_3)\ .
\end{aligned} \tag{8.15}
$$

The graphic representation of Eq. (8.14) is given in Fig. 38. The solution (8.15) to (8.11) gives us a one-parameter family of Boltzmann weights, where the parameter can be identified with the rapidity of the imaginary scattering process (8.12). In general we can write this one-parameter family of Boltzmann weights as $w(i,j,k,l;u)$. This defines a u-dependent transfer matrix $T(u)$ and Eq. (8.6) becomes:

$$[T(u),T(v)]=0\ . \tag{8.16}$$

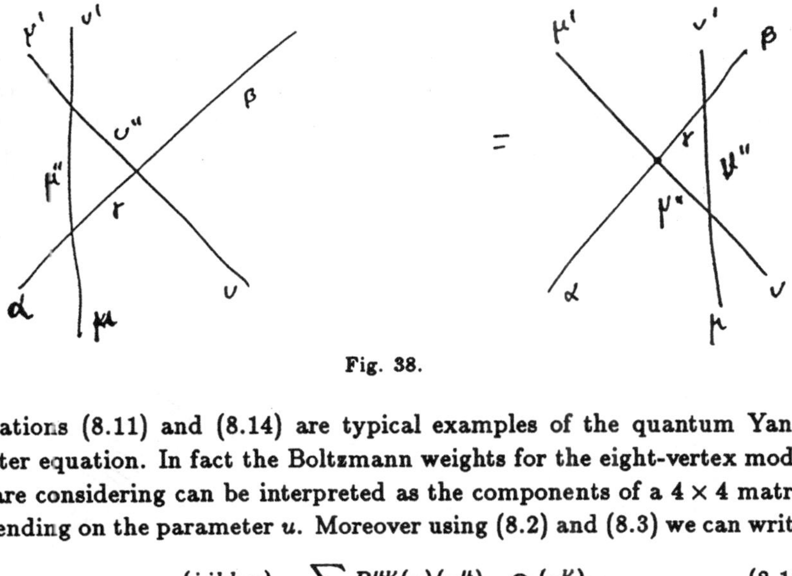

Fig. 38.

Equations (8.11) and (8.14) are typical examples of the quantum Yang-Baxter equation. In fact the Boltzmann weights for the eight-vertex model we are considering can be interpreted as the components of a 4×4 matrix depending on the parameter u. Moreover using (8.2) and (8.3) we can write:

$$w(ijkl;u)=\sum_{\mu,\nu}R^{\mu\nu}(u)(\sigma^{\mu})_{ij}\otimes(\sigma^{\nu})_{kl} \tag{8.17}$$

where the σ's are the Pauli basis on the space of 2×2 matrices. Using the representation (8.17) Eq. (8.11) becomes:

$$R^{(1,2)}(u)R^{(13)}(u+v)R^{(2,3)}(v)=R^{(2,3)}(v)R^{(1,3)}(u+v)R^{(1,2)}(u) \tag{8.18}$$

where

$$R^{(1,3)}(u+v) = \sum_{\mu,\nu} R^{\mu\nu}(u+v)(\sigma^{\mu}) \otimes \mathbb{1} \otimes (\sigma^{\nu}) \qquad (8.19)$$

etc. Notice that (8.18) is simply (8.14) with the upper indices indicating the particles that enter in the scattering process and where we have changed variables: $(\theta_1 - \theta_2) = u$, $(\theta_1 - \theta_3) = u + v$ and $(\theta_2 - \theta_3) = v$.

We now come back to the more general problem of an irregular lattice and different interaction coefficient on the sites of the lattice.

8.1.2 Nonregular lattice. Z-invariance

The question we want to investigate is under what conditions on the interaction coefficients can we solve a vertex model on a generic lattice. The solution to this problem will provide a geometric interpretation of the Yang-Baxter equation. Consider two lattices as in Fig. 39. They differ only by the translation of the line AC. The first thing we will study is when these two lattice are physically equivalent. We assume that the interaction coefficients on both lattices are the same. The Boltzmann weights for the triangles in (Fig. 40) are:

Fig. 39.

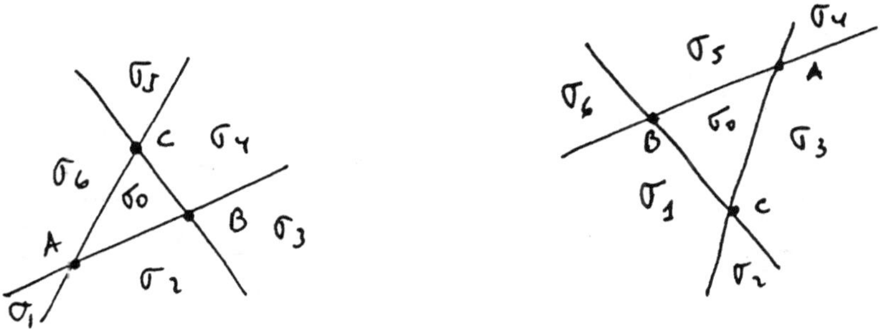

Fig. 40.

$$\begin{aligned} W_1 =&2 \exp(K'_1\sigma_6\sigma_2 + K'_2\sigma_2\sigma_4 + K'_3\sigma_4\sigma_6) \\ &\times \cos h(K'_1\sigma_1 + K_2\sigma_3 + K_3\sigma_5 + K''_1\sigma_1\sigma_6\sigma_2 \\ &+K''_2\sigma_3\sigma_2\sigma_4 + K''_3\sigma_5\sigma_4\sigma_6) \end{aligned} \tag{8.20}$$

$$\begin{aligned} W_2 =&2 \exp(K'_1\sigma_3\sigma_5 + K'_2\sigma_5\sigma_1 + K'_3\sigma_1\sigma_3) \\ &\times \cos h(K_1\sigma_4 + K_2\sigma_6 + K_3\sigma_2 + K''_1\sigma_4\sigma_3\sigma_5 \\ &+ K''_2\sigma_6\sigma_5\sigma_1 + K''_3\sigma_2\sigma_1\sigma_3) \end{aligned} \tag{8.21}$$

where the summation on σ_0 has been performed. The invariance of the partition function under the shift of lines requires that W_1 must equal W_2. This requirement is very close to the Yang-Baxter equation, in fact it is the Yang-Baxter equation when we identify the Boltzmann weights of the vertex model with the components of a factorized S-matrix (see Eq. (8.14) and Fig. 38). The condition:

$$W_1 = W_2 \tag{8.22}$$

becomes to six equations:

$$\exp(2K'_j + 2K'_k) = \frac{\cos h(K_1 + K_2 + K_3 + K''_i - K''_j - K''_k)}{\cos h(-K_i + K_j + K_k - K''_i + K''_j + K''_k)} \tag{8.23}$$

$$\exp(2K'_j - 2K'_k) = \frac{\cos h(K_i - K_j + K_k - K''_i + K''_j + K''_k)}{\cos h(K_i + K_j - K_k - K''_i + K''_j + K''_k)} \tag{8.24}$$

where (i, j, k) is any permutation of $(1,2,3)$. A solution of (8.23) and (8.24) is:

$$K''_1 = K''_2 = K''_3 = K'' \tag{8.25}$$

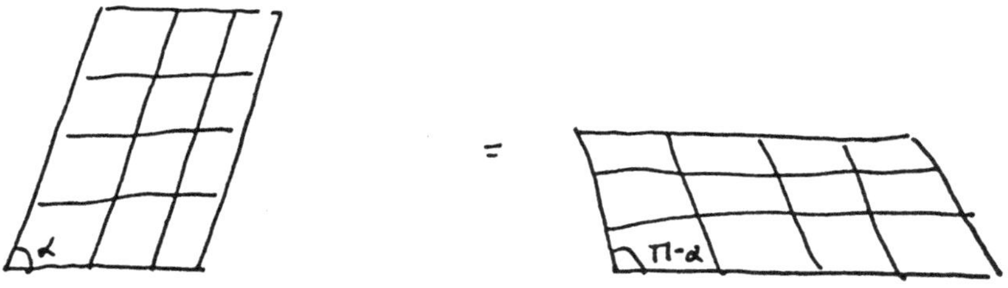

Fig. 41.

and

$$\Delta_1 = \Delta_2 = \Delta_3 = \Delta \tag{8.26}$$

where

$$\Delta = -\sin h2K_i \sin h2K_i' - \tan h2K'' \cos h2K_i \cos h2K_i' \, . \tag{27}$$

The condition of integrability of a model defined on a generic lattice $\mathcal{L}$ is that any triangle of $\mathcal{L}$ must satisfy Eq. (8.22). This condition is known as Z-invariance.. Therefore for an integrable model the interaction coefficient K'' must be the same for any site triangle. Once we fix the values of Δ and K'' the possible values of the interaction coefficients are determined by (8.27). It is possible to prove that the solutions of (8.27) for Δ and K'' fixed can be parametrised in terms of the angles at the sites. In general a nonregular lattice is integrable if the Boltzmann weights depend only on the angles of the lattice and the system is Z-invariant. For the eight-vertex model we have in addition to Z-invariance, invariance under crossing. In fact if we represent the Boltzmann weights in terms of a factorized S-matrix this one will depend, for a nonregular Z-invariant lattice, only on the angles at the sites. Crossing symmetry means that:

$$S_{ij,kl}(\alpha) = S_{ik,jl}(\pi - \alpha) \, . \tag{8.27}$$

This symmetry is true for a generic Z-invariant Baxter model. It is important to point out that crossing symmetry is a consequence of Z-invariance and the special form of the Boltzmann weights (see Eq. (8.1)) we have used. This symmetry is very close to modular invariance in the continuum limit. In particular it means that the partition functions of the two lattices in Fig. 41 must be the same. Moreover if we take the thermodynamic limit, mantaining the ratio between the number of rows and columns constant we will get in the continuum limit, a modular invariant partition function.

8.2 *RSOS-models*[65]

We begin by defining a RSOS model associated to a Coxeter diagram of type A[35]:

(8.28)

The model is parametrized by the number r, the Coxeter number of the diagram (8.28). On a regular lattice the model is defined by associating heights l_i to each site with range:

$$1 \le l_i \le r-1 \tag{8.29}$$

and such that

$$|l_i - l_j| = 1 \tag{8.30}$$

for adjacent sites. Notice that the Coxeter diagram (8.28) contains this informations. For each plaquette we define the Boltzmann weight $W(l_1 l_2 l_3 l_4)$ that must satisfy symmetry under reflections:

$$W(l_1 l_2 l_3 l_4) = W(l_3 l_2 l_1 l_4) = W(l_1 l_4 l_3 l_2) \ . \tag{8.31}$$

The partition function of the model is given by:

$$Z = \sum_{\text{cong}} \prod W(l's) \ . \tag{8.32}$$

To solve the model we proceed in the same way as with the vertex models. First we represent the partition function in terms of the transfer matrix:

$$Z = \mathrm{Tr}(T^N) \tag{8.33}$$

and then look for the conditions under which the transfer matrix can be diagonalized. The transfer matrix is defined by:

$$T_{\bar{l},\bar{l}'} = \prod \{W(l_1 l_2 l_2' l_1') W(l_2 l_3 l_3' l' l_2) \ldots\ldots\} \tag{8.34}$$

where $l = (l_1 \ldots l_m)$ (see Fig. 42). Using different Boltzmann weights, let us say w', we obtain

$$T'_{ll'} = \prod \{w'(l_1 l_2 l_2' l_1') \ldots\} \ . \tag{8.35}$$

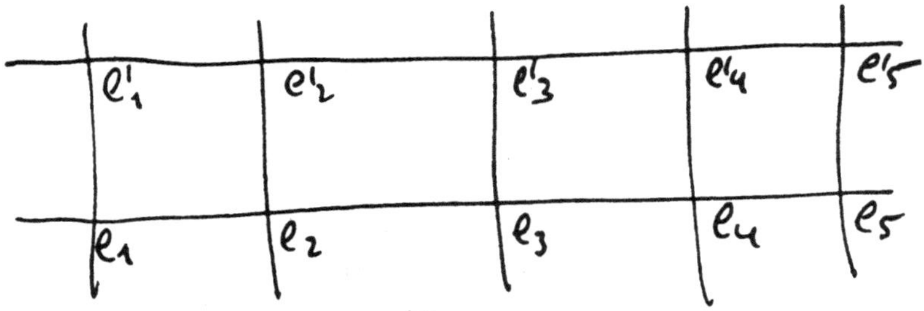

Fig. 42.

The condition of integrability is:

$$(TT')_{TT'} = (TT')_{T'T} \tag{8.36}$$

Using techniques similar to these in the previous case, we can write (8.36) as a triangular equation for three different Boltzmann weights w, w' and w'':

$$\sum_g w(abgf)w'(fgde)w''(gbcd) = \sum_g w''(gcde)w'(abcg)w(fage) \tag{8.37}$$

where $a, b \ldots$ represent the heights l_i. Again we can write (8.37) as an equation for a one-parameter family of Boltzmann weights:

$$\begin{aligned}&\sum_g w(abgf|u)w(fgde|u+v)\,w(gbcd|v)\\ &=\sum_g w(gcde|v)w(abcg|u+v)w(fage|u)\end{aligned} \tag{8.38}$$

The simplest example of SOS-model is the case $r = 4$ which corresponds to the Ising model. In this case the heights are 1,2,3 and any configuration can be separated into two pieces: one even and one odd, each living in a different sublattice (Fig. 43). If we change variables to $\sigma_i = l_i - 2$ and we consider the dual lattice we get the standard Ising model on the white sublattice where we color the lattice as with the Baxter models (Fig. 44). On the white sublattice we have plaqnette variable $\sigma_i = \pm 1$. Now we can use the Boltzmann weights of the RSOS model to define the interaction coefficients. Using the same notation as that for the Baxter models (see Fig. 36 and 46) we arrive at:

$$\exp\left(K_i\sigma_p\sigma_l + K'_i\sigma_q\sigma_m + K''{}_i\sigma_p\sigma_q\sigma_l\sigma_m\right) = W\left(l_pl_ql_ll_m\right) \tag{8.39}$$

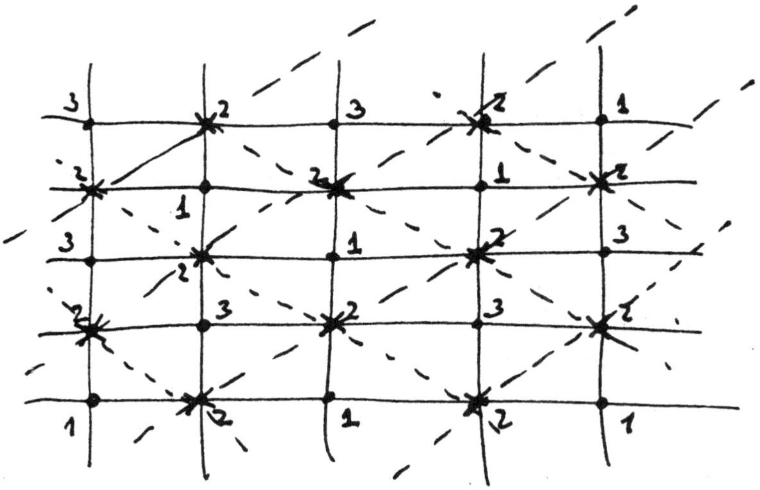
3 2 3 2 1
2 1 2 1 2
3 2 1 2 3
2 3 2 3 2
1 2 1 2 1

Fig. 43.

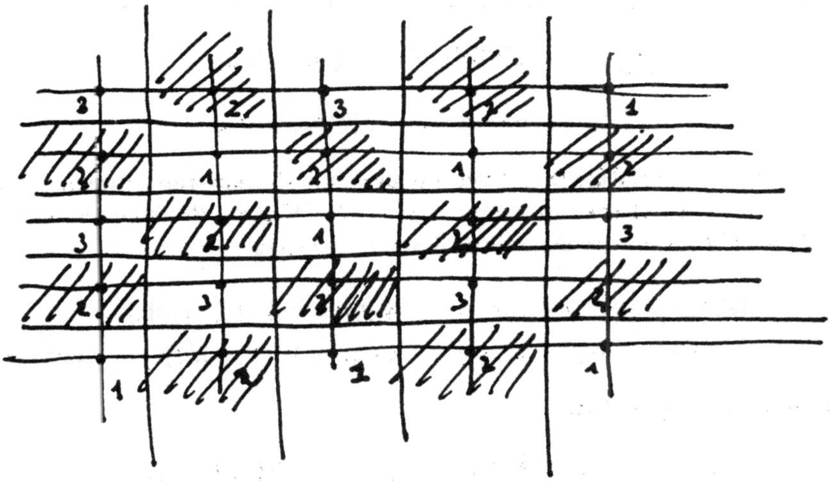
3 2 3 2 1
2 1 2 1 2
3 2 1 2 3
2 3 2 3 2
1 2 1 2 1

Fig. 44.

For the $r=4$ example $K''=0$ and two interaction coefficients, one vertical and another horisontal, are defined by:

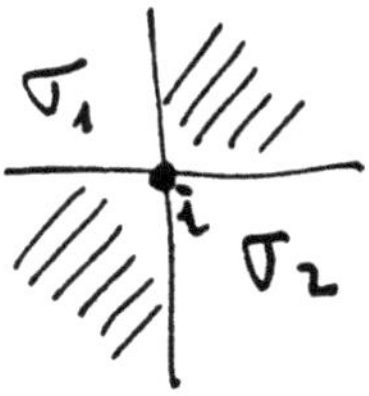

$$= \exp K_i \sigma_1 \sigma_2 = W(2_1\sigma_2 + 2_1 + 2_1\sigma_1 + 2)$$

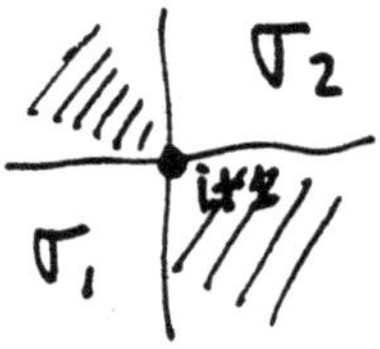

$$= \exp K'_{i+1} \sigma_1 \sigma_2 = W(\sigma_1 + 2_1 + 2_1\sigma_2 + 2_1 2) \tag{8.40}$$

Therefore we see that the RSOS model for $r=4$ is a vertex model with $K''=0$. In general if we include spin variables in the black sublattice and we still impose $K''=0$ we will get two Ising models. Let us now consider the Yang-Baxter equation (8.38) for the special case $r=4$. Now:

$$\begin{aligned}&\sum_g W(a2ag2|u)W(2g2e|u+v)W(g2c2|v)\\&=W(2c2e|v)W(a2c2|u+v)W(2a2e|u)\,.\end{aligned} \tag{8.41}$$

Changing to σ_i variables (8.41) become precisely the equations determining the Z-invariance of the Ising model (Fig. 47). Using reflection symmetry (8.31) we obtain four independent Boltsmann weights:

$$\begin{aligned}W_1(u) &= W(2,1,2,3|u) \qquad & W_3(u) &= W(1,2,3,2|u)\\ W_2(u) &= W(2,1,2,1|u) \qquad & W_3(u) &= W(1,2,1,2|u)\,.\end{aligned} \tag{8.42}$$

The solution to (8.41) is:

$$\begin{aligned}W_1(u) &= \frac{\theta_1(u+\pi/4,p)}{\theta_1(\pi/4,p)} \qquad & W_2(u) &= \frac{\theta_1(\pi/4-u,p)}{\theta_2(\pi/4,p)}\\ W_3(u) &= \frac{\theta_1(u,p)}{\theta_1(\pi/2,p)} \qquad & W_4(u) &= \frac{\theta_1(\frac{\pi}{2}-u,p)}{\theta_1(\frac{\pi}{2},p)}\end{aligned} \tag{8.43}$$

where $\theta_1(u,p)$ is the odd elliptic theta function. Similar arguments apply to other models.

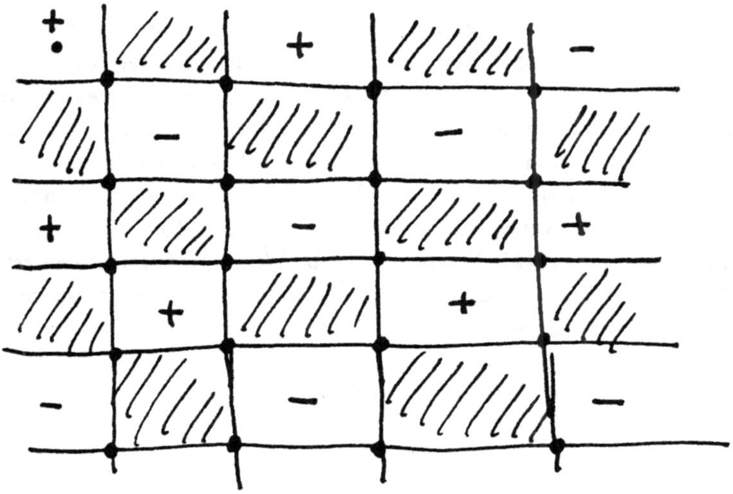

Fig. 45.

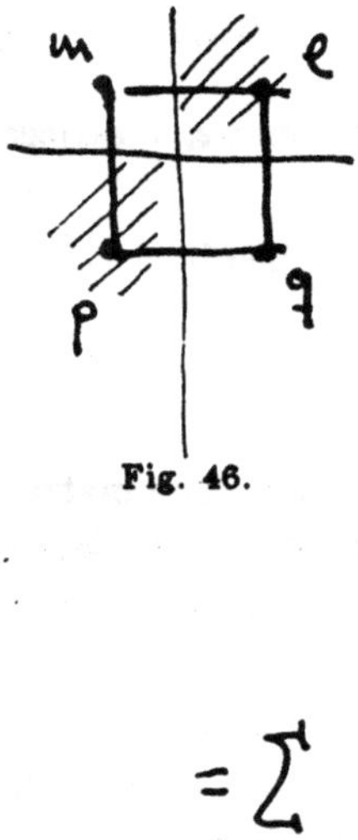

Fig. 46.

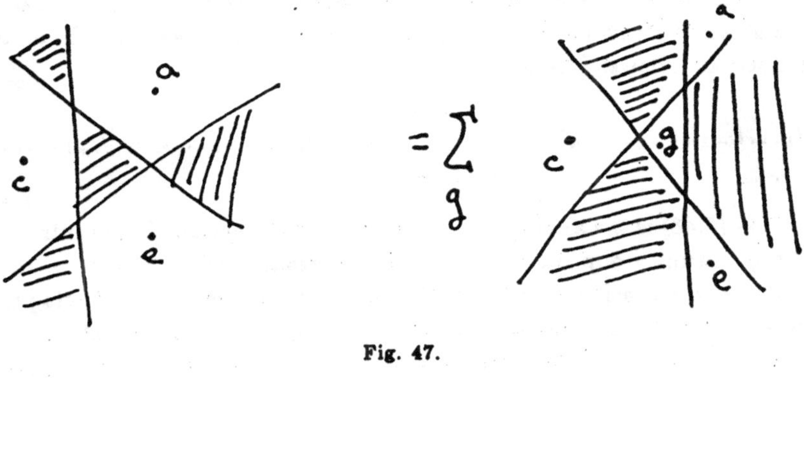

Fig. 47.

8.3 *Conformal field theory*

It is well-known that conformal field theories can be interpreted as a description of the continuum limit of statistical models of the critical point. In particular the minimal models of the $c < 1$ series are in one-to-one correspondence with the critical behavior of restricted SOS models defined on Coxeter diagram of type A_r. A possible way of relating conformal field theory with lattice systems is through quantum groups, namely by identifying the centralizes of the quantum group with some limit, in the spectral parameter, of the Temperley-Lieb algebra of the lattice system.

At the self-dual point (criticality) the partition function of r-SOS models can be written as:

$$Z(u) = \mathrm{Tr}(V(u)W(u))^N \tag{8.44}$$

where

$$V(u) = (1 + f(u)e_1)(1 + f(u)e_3)\dots \tag{8.45}$$

$$W(u) = (1 + f(u)e_2)(1 + f(u)e_4)\dots \tag{8.46}$$

with u, the spectral parameter and e_i matrices satisfying:

$$\begin{aligned} e_i e_j &= e_j e_i \qquad |j - i| \geq 2 \\ e_i e_{i\pm 1} e_i &= \tau e_i \\ e_i^2 &= e_\nu \end{aligned} \tag{8.47}$$

τ depending on the spenfic model. The matrices e_i are defined in terms of Boltzmann weights. More precisely they define a particular representation of the algebra (8.97) on the space of states of the lattice system. The action is defined as follows: (see Fig. 48)

$$e_i|l_1 \dots l_{i-1} l_i l_{i+1} \dots l_m\rangle = \sum_{l'_i} \mathop{\square}\limits^{l_{i-1}\ l'_i}_{l_i\ l_{i+1}} \delta(l_{i-1}, l_{i+1}) \cdot |l_1 \dots l_{i-1} l'_i l_{i+1} \dots l_m\rangle \tag{8.48}$$

Equation (8.48) is the analog of the one defining the action of the centralizer of the quantum group on the space of conformal blocks. There we know that the representation is given by the quantum $6j$-symbols. Following this formal analog it is natural to look for a more precise relation between the Boltzmann weights of the lattice model and $q - 6j$-symbols. This relation

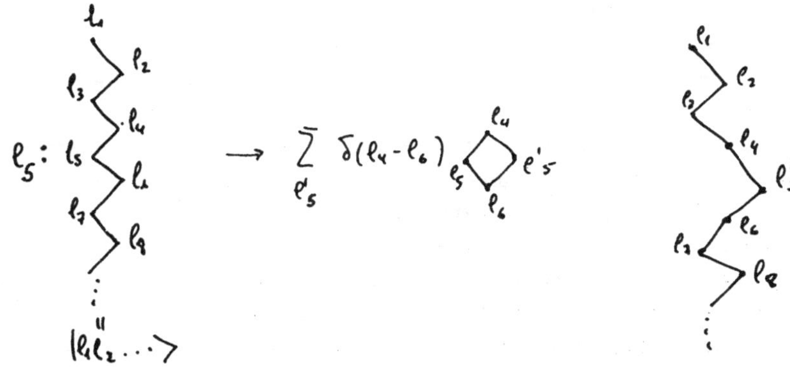

Fig. 48. Graphc representation of the action of an element of the TL algebra.

actually exist (see Ref. 60 for details), for the $u \to \infty$ limit of $\frac{1}{f(u)} W(\ldots|u)$. The reason we need to divide by $f(u)$ is in order to make the representation of the algebra unitary. The same function was used by Zamolodchikov to get from the Boltzmann weights unitary factorized S-matrices. The reason for the $u \to \infty$ limit is less clear from a physical point of view. The other natural limit $u \to 0$ will give us the identity matrices. A qualitative argument for motivating the $u \to \infty$ limit is interpreting this limit as the one corresponding to working with the lattice model on a degenerate torus with modula paramter $\tau = 1$. This is a natural quantum group limit at the level of characters. In fact quantum characters can be thought of as a regularized version of the Kac-Moody characters in the limit $\tau = 1$. These considerations are by no means rigorous but they provide some intuition about the physical content of the $u \to \infty$ limit.

9. Knots Conformal Field Theories and Statistical Mechanics

9.1 *Knot theory*

Knot theory studies the embeddings of S^1 in arbitrary 3-manifolds.[37,38] Until recently the theory was well-establish for embedding in S^3. During the last year important progress has been achieved with the work of Witten to extend the theory to arbitrary three manifolds using Topological gauge theories. In physics knot theory is related to CFT, statistical mechanics and the theory of Integrable Systems.

The simplest way to build a knot is to start with a ϕ^4-diagram (the knot universe) and replace each vertex by a nonplanar crossing (Fig. 49). With the rules in Fig. 49 we can associate to an n-vertex graph 2^n knots. This of course contains an enormous overcounting. A more algebraic way of constructing a knot is by closing braids. Fig. 50. In this way we can write knots in terms of words in the braid group whose classes generate them (the knot in Fig. 50 comes from closing $\sigma_1\sigma_2^{-1}$, and iterating this operating $(\sigma_1\sigma_2^{-1})^n$ produces a girl's braid). This procedure is once again very redundant. From a mathematical point of view we are only interested in topologically equivalent classes of knots. Two knots are equivalent if we can transform one into the other by a continuous deformation. A theorem in knot theory guarantees that two equivalent knots can be transformed into each other via Reidemeister moves. There are three types of moves shown in Fig. 51:

In particular applying a type I move we see that for any element $g \in B_n$ the knots $\hat{g}$ and $\hat{g\sigma}_{n+1}$ (the carat indicates the closure of the braid) are topologically equivalent (Fig. 52).

After the knots have divided into equivalence classes according to the Reidemeister moves we would like to characterise each class by some topological invariants. As with characteristic classes we can define these invariants axiomatically, and these axioms define the link polynomials. We now describe three link polynomials. The simplest one is the Alexander-Conway polynomial given by the axioms:

(A1). To each knot K there is associated a one-variable polynomial $P_k(t)$ with integer coefficients such that if K and K' are equivalent they have the same polynomials.

(A2). The polynomial is normalised by requiring that for the unknot (unknotted circle) $P_{s^1}(t) = 1$.

(A3). Alexander Conway skein rule: Given three knots K^+, K^-, L which only differ in one crossing as:

(9.1)

the corresponding polynomials obey:

$$P_{K^+}(t) - P_{K^-}(t) = tP_L(t) \ . \qquad (9.2)$$

K_+ is an overcrossing. K_- an undercrossing and L is the splicing of the crossing. By applying the skein move (9.1) a sufficient number of times we eventually transform any knot or link into collection of unknotted circles.

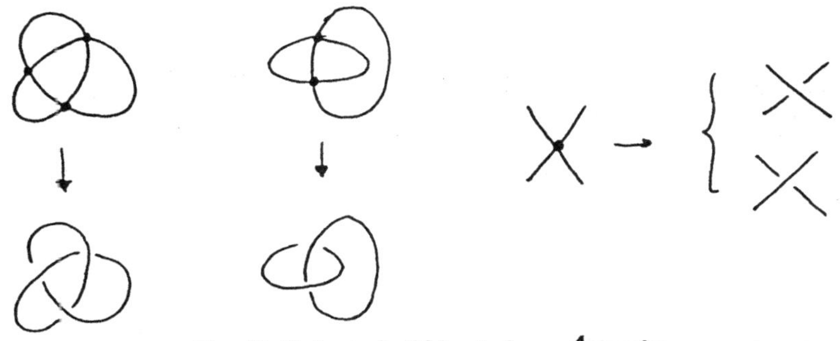

Fig. 49. Rules to build knots from ϕ^4-graphs.

Fig. 50. Knot generated by closing braids.

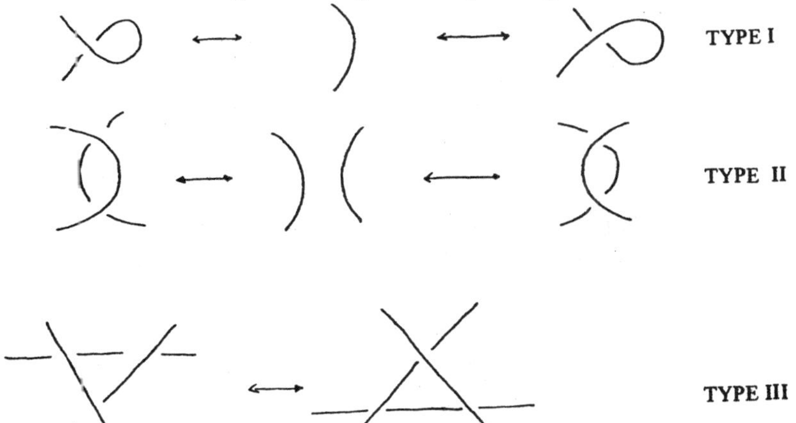

Fig. 51. Reidemeister moves.

Notice however that many inequivalent knots may have the same polynomial. For instance, the Alexander-Conway polynomial is unable to distinguish between a knot and its mirror image even when they are topologically distinct. Progress in the construction of knot invariants can be measured by how precisely the knot is characterized by the invariant. Some simple computations of the Alexander-Conway polynomial $P_K(t)$ following easily

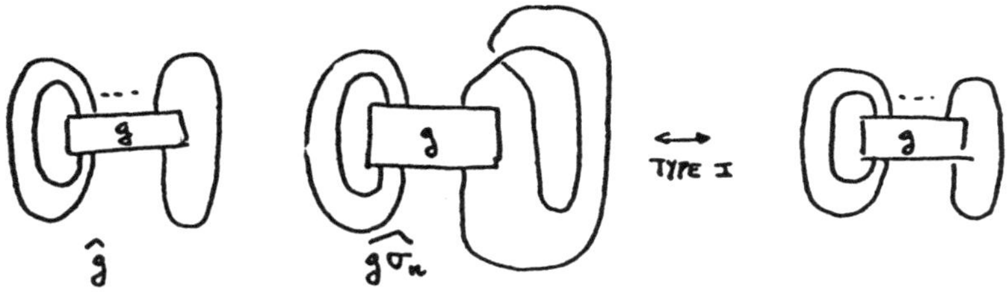

Fig. 52. Equivalence of $\hat{g}$ and $\hat{g\sigma}_n$.

from the skein relation are:

$$P(\bigcirc) = 1 \quad P(\bigcirc \ldots \bigcirc) = 0 \tag{9.3}$$

$P(L) = 0$(if L splits into several pieces)

for a link:

K_+ K_- L_0

Since $P_{K^-} = 0$ (9.3), (9.2) imply $P_{K^+}(t) = t$. Next for the trefoil:

then

$$P_T(t) = 1 + t^2 \tag{9.4}$$

the link with $2n$-crossings

P 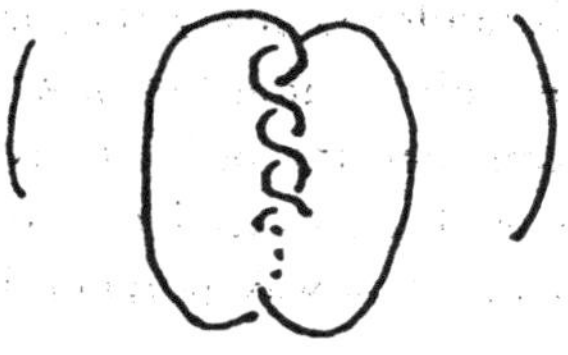 $= nt$ (9.5)

$2n - crossings$

The skein axiom is more difficult to motivate than the other two. It is very specific to knots in S^3. In the skein moves we always consider the crossing as contained in some ball B_3 and this is certaily always the case in S^3. After a finite number of skein moves any knot is transformed into a collection of unknotted unlinked circles. As a final example of how the skein relation works, we show in Fig. 53 how to prove that the split link has a vanishing polynomial.

Fig. 53. The alexander-Conway polynomial vanishes for a split link.

Before studying generalizations of $P_K(t)$, it is instructive to consider the skein relation from the point of view of the braid group characterization of knots. The knots K^+, K^-, L_0 are related by

$$\begin{aligned} K^+ &= \sigma_1^2 K^- \\ L &= \sigma_1 K^- \end{aligned} \tag{9.6}$$

This is illustrated in Fig. 54.

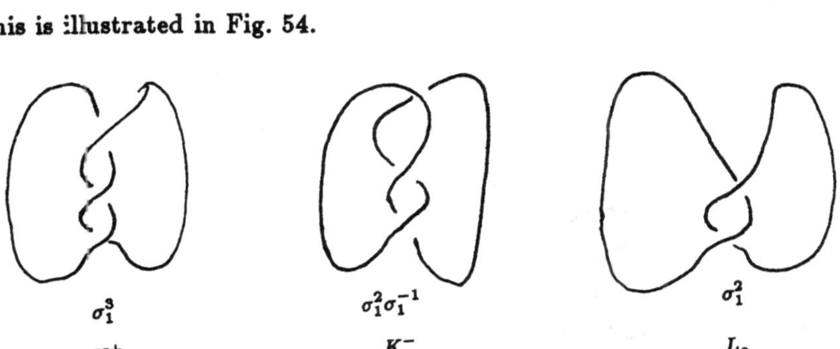

σ_1^3 $\qquad$ $\sigma_1^2\sigma_1^{-1}$ $\qquad$ σ_1^2

K^+ $\qquad$ K^- $\qquad$ L_o

Fig. 54. Example of equation (9.6).

Equation (9.6) is very appealing from the point of view of the previous sections. If we work with Hecke algebras instead of braid groups (9.6) will give us a particular type of skein rule. From the defining relation of a Hecke algebra of type A_n:

$$\sigma_1^2 = (q-1)\sigma_i + q$$

and:

$$K^+ - qK^- = (q-1)L \tag{9.7}$$

very close to the skein rule in (A3). This indicates that the skein rule is related to the Hecke representations of the braid group.

We now consider a two-variable generalization of the previous polynomial.[66] In some convenient normalizations the new polynomial is defined by the skein rule

$$\begin{aligned}(\lambda^{1/2}q^{1/2})^{-1}P_{K^+}(q,\lambda) - (\lambda^{1/2}q^{1/2})P_{K^-}(q,\lambda) \\ = (q^{1/2} - q^{-1/2})P_L(q,\lambda)\ .\end{aligned} \tag{9.8}$$

A particular case of (9.8) is the Jones polynomial[41] defined by

$$V_K(q) = P_K(q,q)\ . \tag{9.9}$$

To motivate the new polynomials it is convenient to have a more algebraic representation of the Reidemeister moves. They correspond to the following two transformations on the braid representation (the second one appears graphically in Fig. 55)

$$\text{MARKOV I} \quad \alpha\cdot\beta \to \beta\cdot\alpha \quad \alpha,\beta\in B_n \tag{9.10a}$$

$$\text{MARKOV II} \quad \alpha\cdot\sigma_n^{\pm1} \to \alpha \quad \alpha\in B_n\ . \tag{9.10b}$$

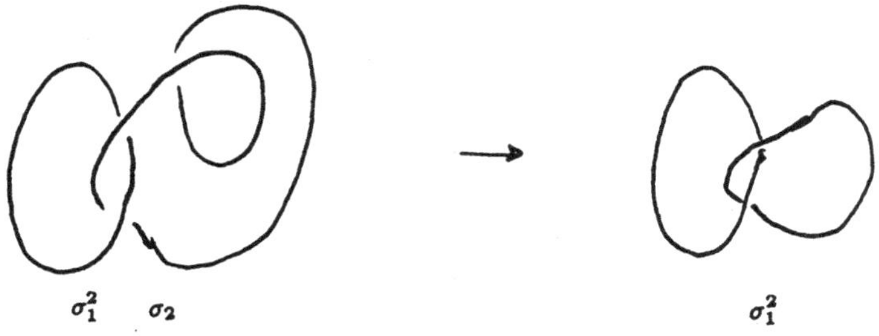

Fig. 55. Markov move of type II.

The transformations (9.10) on braid groups are known as Markov transformations. In general any invariant polynomial must be defined by some functional on $\bigcup_{n=0}^{\infty} B_n$ invariant under the two Markov moves (9.10). Such an invariant will automatically satisfy axiom 1. To have a constructive way of defining such a functional, we need to characterize how this functional changes under the set of transformations which unknot any close braid. For knots inS^3 we can use skein moves in the unknotting procedure. The skein move on the other hand requires a Hecke representation of the braid group, $g_i^2 = (q-1)g_i + q$. Thus to define invariant polynomials of links in S^3 (and with a skein relation as above) we need:

(i). A Hecke representation $H_n(g)$ of B_n.

(ii). A functional defined on $H_\infty = \cup H_n(q)$ invariant under the Markov moves (9.10) for the particular embedding defined in (1).

There is a trace on H_∞ which will allow us to construct a knot invariant. This is the Ocneanu trace defined by,

(1). $\mathrm{tr}(ab) = \mathrm{tr}(ba)$

(2). $\mathrm{tr}\ (1) = 1$

(3). $\mathrm{tr}\ (xg_n) = z\,\mathrm{tr}\, x, x \in H_n, g_n \in H_{n+1}$ the complex parameter z characterises the trace. We now use the Ocneanu trace to derive the polynomials (9.8) and (9.9). First we need the representation $B_n \to H_n(q)$. Following Jones[41] we define:

$$\begin{aligned} \pi_\lambda :& B_n \to H_n(q) \\ & \sigma_i \to \pi_\lambda(\sigma_i) = \sqrt{\lambda} g_i, \lambda = \frac{1-q+z}{qz} \end{aligned} \tag{9.11}$$

(the elements g_i are the $H_n(q)$ generators). It is now easy to verify that:

$$P_{\hat{\alpha}}(q,\lambda) \equiv \left(-\frac{1-\lambda q}{\sqrt{\lambda}(1-q)}\right)^{n-1} (\sqrt{\lambda})^e \,\mathrm{tr}\pi_\lambda(\alpha) \tag{9.12}$$

$\alpha \in B_n$ and $\hat{\alpha}$ is the knot obtained closing α which is invariant under (9.10). The term $(\sqrt{\lambda})^e$ is there to normalise the polynomial so that $P_{\bigcirc}(q,\lambda) = 1$ and the Hecke algebra relation implies a skein rule. The trace in (9.12) satisfies the rules (1)–(3) for the value of z appearing on the representation (9.11). Using the Hecke algebra relation:

$$(\sqrt{\lambda}\sqrt{q})^{-1} P_{K^+}(q,\lambda) - (\sqrt{q}\sqrt{\lambda}) P_{K^-}(q,\lambda) = (q^{1/2} - q^{-1/2}) P_L(q,\lambda)$$

and Jones polynomial is a particular case of this one ($\lambda = q$). In CFT it is the Jones polynomial which appears more naturally and we would like to explain it in more detail. The first significant difference between the Jones polynomial and (9.8) is that one uses the Temperley-Lieb algebra instead of the Hecke algebra. This algebra was discovered in physics in the study of critical phenomena of some solvable models. We know from Sec. 6 that this algebra appears as the centraliser of SU(2,q) for the fundamental representation. Mathematically this algebra is given by a set of projections $e_1, \ldots, e_n$ satisfying:

$$\begin{aligned} e_i e_j &= e_j e_i \qquad |i-j| \geq 2 \\ e_i^2 &= e_i \\ e_i e_{i\pm1} e_i &= \tau e_i \end{aligned} \tag{9.13}$$

for some τ which defines the algebra. Out of (9.13) we can construct a Hecke algebra representation:

$$\begin{aligned} \tau^{-1} &= 2 + q + q^{-1} \\ g_i &= q e_i (1 - e_i) \; . \end{aligned} \tag{9.14}$$

It is straightforward to verify that g_i satisfies that Hecke relations together with:

$$g_i g_{i+1} g_i - g_i g_{i+1} - g_{i+1} g_i + g_i + g_{i+1} - 1 = 0 \tag{9.15}$$

This constraint explains the relation with SU(2,q), it means that the three row antisymmetriser vanishes automatically (see Secs. 6 and 7). The restriction from (9.8) to the Jones polynomial (9.9) corresponds to the representation of B_n into the algebra (9.13).

$$\phi(\sigma_i) = q e_i - (1 - e_i) \; . \tag{9.16}$$

Using the skein rule (9.8) it is easy to compute the polynomial $P(q, \lambda)$ for the trefoil knot:

$$P_{\text{trefoil}}(q, \lambda) = \frac{1}{\lambda} + \frac{1}{\lambda q^2} + \frac{1}{\lambda q} + \frac{1}{\lambda^2 q^2} \tag{9.17}$$

(the mirror image corresponds to changing $\sqrt{\lambda q}$ to $-(\sqrt{\lambda q})^{-1}$ and obviously (9.17) changes. The Jones polynomial, or its two variable extension has succeeded in distinguishing some knots from their mirror images, thus going beyond the Alexander-Conway polynomial). At the end on this section we will give a derivation of these invariants from RCFT using the fusion rules and the duality properties.

9.2 *State models: Turaev and Kauffman invariants*

Given a Yang-Baxter matrix $R : V \otimes V \to V \otimes V$, we can define a representation of the braid group B_n into the space Aut $(V^{\otimes n})$. This was done before in our study of centralizers for quantum groups. Recall that

$$\pi(\sigma_i) = R_i = 1 \otimes \ldots \otimes R_{i,i+1} \otimes \ldots \otimes 1 : V^{\otimes n} \to V^{\otimes n} \tag{9.18}$$

and from the Yang-Baxter equation the R_i provides a representation of B_n. Using this representation we will define link polynomials. We will obtain a function on Aut$(V^{\otimes n})$ invariant under Markov moves. Following Turaev[39] (see also Ref. 67) we define an extended Yang-Baxter system characterized by a Yang-Baxter matrix R and an isomorphism $\mu : V \to V$ transforming a basis $\{\nu_i\}$ into $\{\mu_i \nu_i\}$ satisfying

$$\begin{aligned} (\mu_i \mu_j - \mu_k \mu_l) R_{ij}^{kl} &= 0 \\ \sum_j R_{ij}^{kj} \mu_j &= ab\delta_i^k \\ \sum_j (R^{-1})_{ij}^{kj} \mu_j &= a^{-1} b\delta_i^k \end{aligned} \tag{9.19}$$

for some a, b. From these condition it follows that

$$T(\alpha) = a^{-w(\alpha)} b^{-n} \mathrm{tr}(\pi(\alpha)\mu^{\otimes n}) \tag{9.20}$$

is invariant under Markov moves. Here ε is the wraith number for the word $\alpha \varepsilon B_n$, defined by $w(\sigma_i) = -w(\sigma_i^{-1}) = 1, w(\alpha\beta) = w(\alpha) + w(\beta)$. ($\pi(\alpha)$ is the representation (9.18)). Then:

$$\begin{aligned} T(\alpha\beta) &= T(\beta\alpha) \qquad \alpha\, \beta \in B_n \\ T(\alpha\sigma_n^{\pm 1}) &= T(\alpha) \qquad \alpha \in B_n\ . \end{aligned} \tag{9.21}$$

The interest of this invariant is that it can be automatically defined in the context of quantum groups where the R_i are the generators of the centralizer of the group in $V^{\otimes n}$. As explained in Sec. 7 for quantum groups there is always one element u in the algebra satisfying (9.19) where b is the quantum dimension of the representation V considered. Different representations define different polynomials.

In order to visualise the polynomial defined in (9.20) it is useful to have its skein rule representation. An interesting theorem of Turaev states that for any extended Yang-Baxter system there is always a skein rule of the form:

$$\sum_{i=-p}^{+p} a : t^i T(\alpha_i) = 0 \tag{9.22}$$

where the knots α_i appear in Fig. 56 and the coefficients a_i depend on the extended system. The skein relation (9.22) once again indicates that the Turaev invariants defined so far are naturally related to knots in S^3.

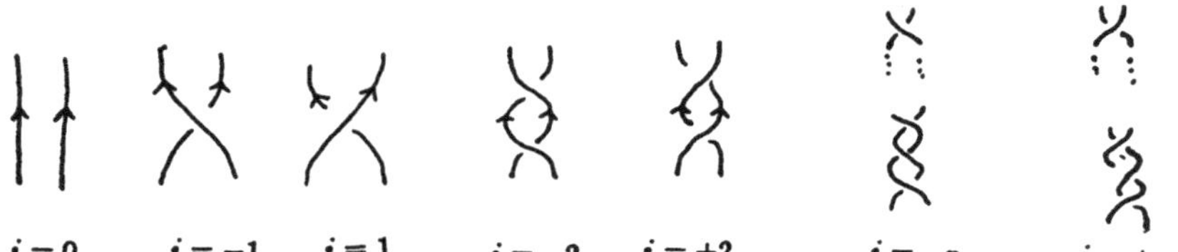

$i=0$ $\quad i=-1$ $\quad i=1$ $\quad i=-2$ $\quad i=+2\ldots$ $\quad i=-p$ $\quad i=+p$

Fig. 56. Crossing entering in the skein rule (9.22).

The invariant (9.20) can be expressed more explicitly in the language of quantum groups. As an example take $G = \mathrm{SU}(2,q) V = V_{1/2}$ the fundamental representation. Then $\pi(\alpha)$ is defined in terms of $R^{1/21/2}$(6.32) and we represent the blocks in (6.4) $\mathcal{F}_j^{1/21/2\ldots1/2} \equiv W_j =$ space of blocks with $N\phi_{1/2}$ fields and one spectator field with spin j, where the representation of $\alpha \in B_n$ in W_j is defined in terms of the $6j$-symbols;:

$\rho^j(\alpha)$ (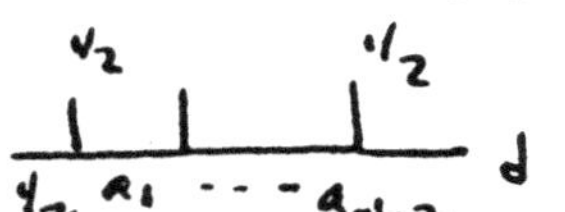) $\otimes e^j_m$

$$= \sum_{m's} \begin{Bmatrix} 1/2 & 1/2 & a_1 \\ m_1 & m_2 & m \end{Bmatrix}_q \cdots \begin{Bmatrix} a_{N-2} & 1/2 & j \\ m_{N-2} & m_N & j \end{Bmatrix}$$
$$\times R(\alpha) \left[e^{1/2}_{m_1} \times \ldots \times e^{1/2}_{m_N}\right] \tag{9.23}$$

where e^j_m is a basis of V^j and $R(\alpha)$ is the representation of the word $\alpha \in B_n$ using $R^{1/21/2}$. The invariant (9.20) now becomes

$$T(\alpha) = \sum_j w_j \mathrm{tr}\, \rho^j(\alpha)\,. \tag{9.24}$$

In the next subsection we will explain detail how to compute the w_j coefficients starting from a more general point of view. This will tie rather nicely different approaches to the computation of knot and link invariants (RCFT, quantum groups, Chern-Simons actions etc.)

9.3 *Rational conformal field theories and knots invariants*

The aim of this section is to show that using the basic imgredients of a RCFT, namely the fusion matrices N_{ij}^{k}, the modular matrix S_{ij} and the braid matrices $B_{pp'}\begin{bmatrix} j & k \\ i & l \end{bmatrix}$, one can construct invariants of links.[25]

Our construction will follow the lines of Jone's original approach.[40]

Let us start with a RCFT by considering the conformal blocks $\mathcal{F}_{\phi,i}^{(n)}$ in Fig. 26 where ϕ is a given primary field which appears n times in the block, and the index i runs over all possible primary fields of the theory.

The number of conformal blocks of the type $\mathcal{F}_{\phi,i}^{(n)}$ is given in terms of the fusion matrix $N_{\phi j}^{k}$ by:

$$d_{\phi,i}^{(n)} = \sum_{l_1,\dots,p_{n-2}} N_{\phi p_1}^{i} N_{\phi p_2}^{p_1} \dots N_{\phi\phi}^{p_{n-2}} = \left(N_{\phi}^{n-1}\right)_{\phi}^{i} \tag{9.25}$$

where

$$(N_\phi)_i^j \equiv N_{\phi i}^{j} \, .$$

From this equation one finds the recursion formula

$$\begin{aligned} d_{\phi,i}^{n+1} &= \sum_j d_{\phi,j}^{n} (N_\phi)_j^i \\ d_{\phi,i}^{2} &= N_{\phi\phi}^{i} \end{aligned} \tag{9.26}$$

where both $N_{\phi i}^{j}$ and $d_{\phi,i}^{n}$ are non-negative integers. It is convenient to organize the integers $d_{\phi,i}^{n}$ into a Bratteli diagram. The nth row of a Bratteli diagram is given by the set of numbers $d_{\phi,i}^{n+1}$ belonging to the $(n+1)^{\text{th}}$ row and the number $d_{\phi,j}^{n}$ of the nth row we draw one arrow whenever $N_{\phi j}^{i} = 1$, two arrows whenever $N_{\phi j}^{i} = 2$, etc. This is represented in Fig. 57.

The arrows in a Bratteli diagram indicates the way the dimensions of the conformal blocks with n-external ϕ fields sum up to reproduce the corresponding conformal blocks with $(n+1)$ extend ϕ fields. Conversely the different paths in the Bratteli tree from the top to a given element j in the nth row gives the different internal states in the block $\mathcal{F}_{\phi,j}^{(n)}$.

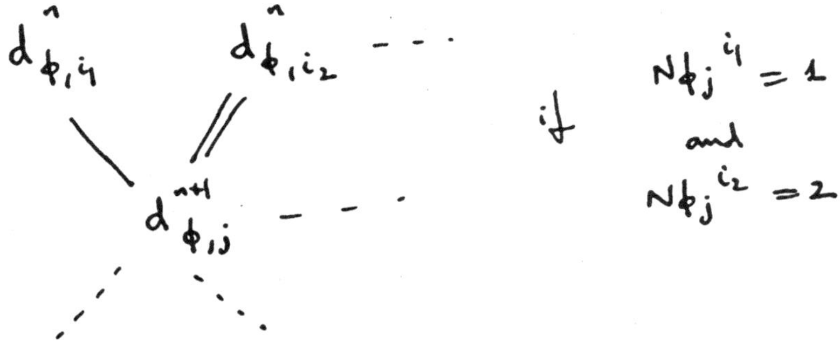

Fig. 57. Pattern for Bratteli diagrams in RCFT's.

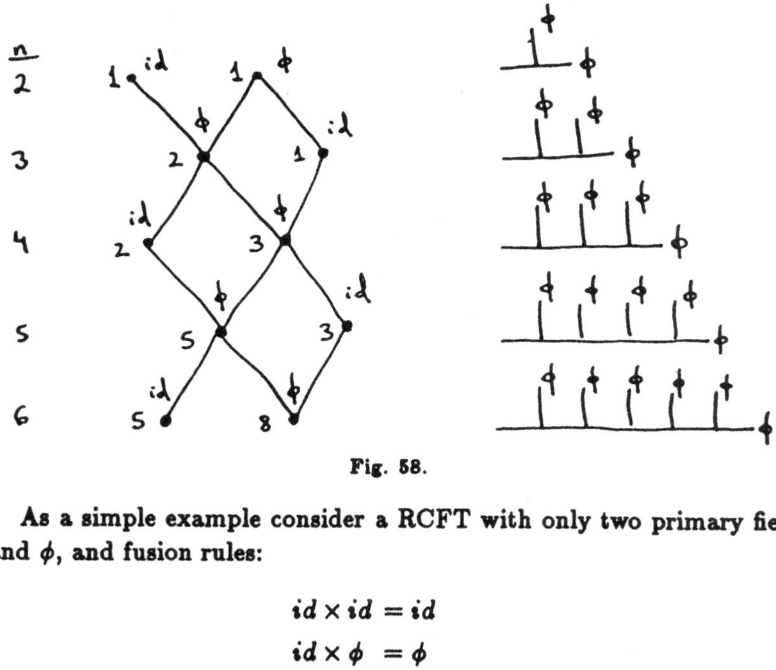

Fig. 58.

As a simple example consider a RCFT with only two primary fields: id and ϕ, and fusion rules:

$$\begin{aligned} id \times id &= id \\ id \times \phi &= \phi \\ \phi \times \phi &= id + \phi \end{aligned} \tag{9.27}$$

which yield the following fusion matrices

$$N_{id} = \begin{pmatrix} 1 & 0 \\ 0 & 1 \end{pmatrix} \qquad N_{\phi} = \begin{pmatrix} 0 & 1 \\ 1 & 1 \end{pmatrix} \tag{9.28}$$

The Bratteli diagram in this case is a Fibonacci tree, and the associated conformed block are as in Fig. 58.

Let us now return to the conformal blocks $\mathcal{F}^{(n)}_{\phi,i}$. The duality properties of RCFT's imply that a braid operation on the ϕ legs leaving the spectator field fixed mixes the internal channels in $\mathcal{F}^{(n)}_{\phi,i}$.

If $g \in B_n$ one obtains in this way a representation of B_n on $\mathcal{F}^{(n)}_{\phi,i}$, of dimension $d^n_{\phi,i}$. Explicitly:

g 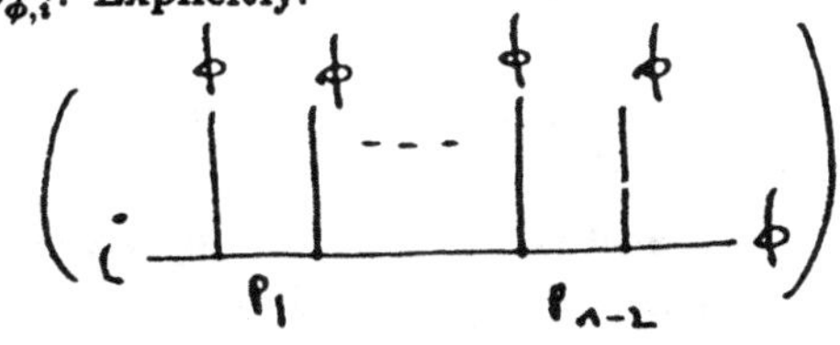 $= \sum_{p'_1,\dots,p'_{n-2}}$

$B^{(i)}_{p_1\dots p_{n-2},p'_1\dots p'_{n-2}}(g)$ 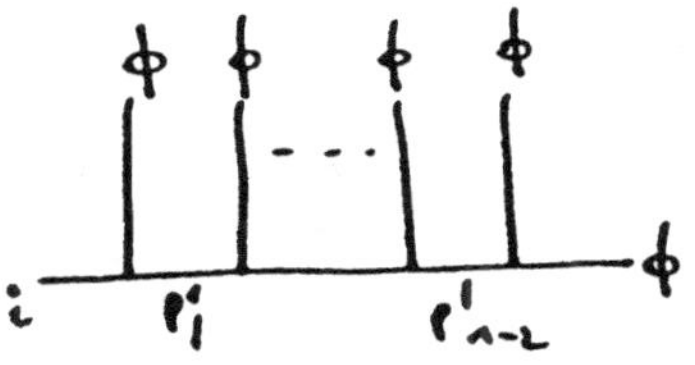 (9.29)

where $B^{(i)}(g)$ is the braid matrix associated to the element $g \in B_n$ and has dimension $d^n_{\phi,i}$.

In fact since B_n is generated by the transpositions $\sigma_1,\dots,\sigma_{n-1}$ it is enough to know the following braid coefficients (Fig. 4 and 8):

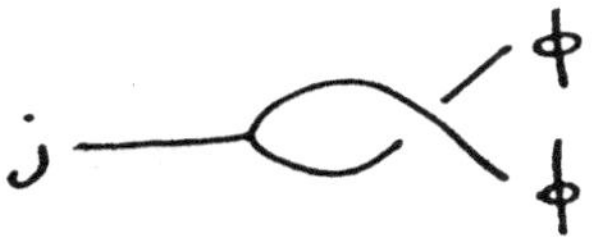 $= \varepsilon^j_{\phi\phi} e^{i\pi(-2h_\phi+h_j)}$  (9.30a)

 $= \sum_{p'} B_{pp'}\begin{bmatrix}\phi & \phi\\ i & j\end{bmatrix}$  (9.30b)

The whole set of braiding matrices $B^{(i)}(g)$ appearing in Eq. (9.29), for a fixed expectator belongs to the space $M_{d^n_{\phi,i}}(\mathbb{C})$ of complex square matrices of dimension $d^n_{\phi,i}$. Taking into account all possible expectators leads us to consider the direct sum:

$$A^n_\phi = \oplus_i M_{d^n_{\phi,i}}(\mathbb{C}) \tag{9.31}$$

which is a finite-dimensional C^*-algebra.

In fact what we obtain is an infinite sequence of C^*-algebras $\{A_\phi^n\}_{n=2,3,\ldots}$ whit an inclusion relation:

$$A_\phi^2 \subset A_\phi^3 \subset \ldots \subset A_\phi^n \subset A_\phi^{n+1} \subset \ldots \tag{9.32}$$

given by (9.26). In other words the inclusion matrix for each step in (9.32) is the fusion matrix N_ϕ of the RCFT.

We have thus arrived at the situation presented by Jones in Ref. 40.

More generally, we have two finite matrix algebra A, B and $B \subset A$. Then

$$\begin{aligned} A &= \oplus_i M_{n_i}(\mathbb{C}) \\ B &= \oplus_\alpha M_{m_\alpha}(\mathbb{C}) \end{aligned} \tag{9.33}$$

and the inclusion is completely specified by the inclusion matrix $\Lambda_{i\alpha}$ whose entries are non-negative integers relating the dimensions of the irreducible representations of A, B

$$n_i = \sum_\alpha \Lambda_{i\alpha} m_\alpha \; . \tag{9.34}$$

The inclusion matrix for the sequence (9.32) of our RCFT is $(N_\phi)_i^j$.

In Jones construction once we have the inclusion $B \subset A$ then one can generate an infinite sequence as in (9.32):

$$B \subset A \subset A \otimes_B A \subset A \otimes_B A \otimes_B A \subset \ldots \tag{9.35}$$

where $A \otimes_B A$ is defined by the condition $xb \otimes y = x \otimes by (b \in B; x, y \in A)$. If Λ is the inclusion matrix for $B \subset A$, then Λ^T is the inclusion matrix for $A \subset A \otimes_B A$, and so on.

The Λ matrix for the Jones algebra $A_{\beta,n}$ described in Sec. 7 with $\beta = 4\cos^2 \pi/m$ can be constructed as the incidence matrix of a partition of the Dinkin diagram of A_{m-1} into two disjoint subsets, such that no pair in the same subset are adjacent in A_{m-1}.

If m is odd then we have the Dinkin diagram

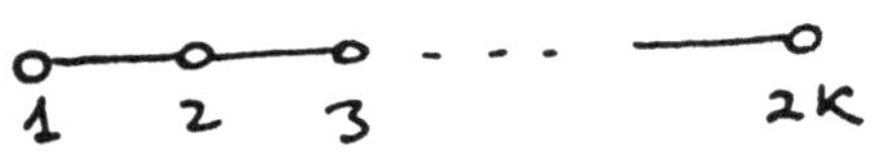

$2k = m - 1$

and the associated incidence matrix

$$\Lambda = \begin{array}{c|cccccc} & 2 & 4 & \ldots\ldots\ldots & & 2k \\ \hline 1 & 1 & 0 & & & \\ 3 & 1 & 1 & & & \\ 5 & & 1 & 1 & & \\ \vdots & & & & \ddots & \\ 2k-1 & & & & & 1 \\ & & & & & 1\,1 \end{array} \tag{9.36}$$

Notice $\Lambda \in M_{\frac{m-1}{2}}(\mathbb{N})$.

If m is even the Dinkin diagram is

$2k = m$

yielding:

$$\Lambda = \begin{array}{c|cccc} & 2 & 4\ldots\ldots & & 2k-2 \\ \hline 1 & 1 & 0 & & \\ 3 & 1 & 1 & & \\ \ldots & & & & \\ & & & 1 & \\ 2k-1 & & & 0 & 1 \end{array} \tag{9.37}$$

$\Lambda \in M_{\frac{m}{2},\frac{m-2}{2}}(\mathbb{N})$.

It can be shown that the largest eigenvalue of the matrix $\wedge^T \wedge$ is $\beta = 4\cos^2 \pi/m$. Notice that m is the Coxeter number of A_{m-1}, and in fact the same result is true for incidence matrices built up from the Dinkin diagrams of $D_m (m \geq 4), E_6, E_7$ and E_8.

Comparing the sequences (9.32) and (9.35) one may wonder about the relations between them. A complete understanding of this point is not yet available, but its elucidation will shed same light in the classification of fusion rules of RCFT. We shall limit ourselves here to exhibit the connection between (9.32) and (9.35) in some particular cases.

In order to interpret the inclusion matrices (9.36) and (9.37) as the fusion matrices of any RCFT we shall permute the Column labels. Let us take for simplicity the case where m is odd so that Λ is a square matrix.

The permuted incidence matrix which we shall call $\wedge_\phi$ is

$$
\wedge_\phi \qquad
\begin{array}{c|ccc}
 & 2k & & 42 \\
\hline
1 & 0 & \cdots\cdots & 1 \\
3 & 0 & & 11 \\
5 & \vdots & \iddots & \vdots \\
\vdots & \vdots & & \vdots \\
\vdots & \vdots & 1 & \vdots \\
2k-1 & 1 & 1\ldots\ldots & 0
\end{array}
\tag{9.38}
$$

Notice that $\wedge_\phi^T = \wedge_\phi$ and that the highest eigenvalue of $(\wedge_\phi)^2$ is given by $\beta = 4\cos^2 \pi/m$. Now we can look for RCFT's with $\wedge_\phi$ as the fusion matrix of same field ϕ. We have already encountered one case given by Eq. (9.27) and (9.28). The matrix N_ϕ of (9.28) is identical to $\wedge_\phi$ for $m = S$. The case $m = 7$ would yield a fusion matrix:

$$
N_\phi \equiv \wedge_\phi = \quad
\begin{array}{c|ccc}
 & id & \psi & \phi \\
\hline
id & 0 & 0 & 1 \\
\psi & 0 & 1 & 1 \\
\phi & 1 & 1 & 0
\end{array}
\tag{9.39}
$$

which requires the existence of 3 primary fields id, ϕ and ψ. An example of (9.39) is provided by the minimal series with $c = -\frac{68}{7}$. Notice that the fusion matrix N_ϕ can be derived from theat of N_ϕ:

$$
\phi \times \phi = id + \psi \mapsto N_\psi = N_\phi^2 - \mathbb{1} = \begin{pmatrix} 0 & 1 & 0 \\ 1 & 1 & 1 \\ 0 & 1 & 1 \end{pmatrix}
\tag{9.40}
$$

Sometimes the fusion matrices of a RCFT are related to incidence matrices in a more complicated way. Take for instance the SU(2)$_k$ WZW model for k odd. The fusion matrix $N_{1/2}$ can be derived from the fusion rules:

$$
\begin{aligned}
\phi_{1/2} \times \phi_0 &= \phi_{1/2} \\
\phi_{1/2} \times \phi_j &= \phi_{j-1/2} + \phi_{j+1/2} \quad j < k/2 \\
\phi_{\frac{1}{2}} \times \phi_{\frac{k}{2}} &= \phi_{\frac{k}{2}-1}
\end{aligned}
\tag{9.41}
$$

then:

$$
N_{1/2} = \begin{array}{c|ccc|ccc}
 & 0 & 1.. & \frac{k-1}{2} & \frac{k}{2} & .. & \frac{1}{2} \\
\hline
0 & & & & & & 1 \\
1 & & & & & 1 & 1 \\
\vdots & & & & & \cdot^{\cdot^{\cdot}} & \\
 & & & & & 1 & \\
\frac{k-1}{2} & & & & 1 & 1 & \\
\hline
\frac{k}{2} & & & 1 & & & \\
\vdots & & 1 & 1 & & & \\
\vdots & & \cdot^{\cdot^{\cdot}} & & & & \\
 & & 1 & & & & \\
\frac{1}{2} & 1 & 1 & & & & \\
\hline
\end{array}
\tag{9.42}
$$

Hence:

$$
N_{1/2} = \begin{pmatrix} 0 & \wedge_{1/2} \\ \wedge_{1/2} & 0 \end{pmatrix} \tag{9.43}
$$

where $\wedge_{1/2}$ is the incidence matrix (9.38) for $m = k + 2$. (The block structure form of $N_{1/2}$ display the graded structure of the representation of $SU(2)_k$).

Notice that in the cases we have considered the largest eigenvalue of $\wedge^T \wedge$ is equal to $(S_{\phi 0}/S_{00})^2$, for example for $SU(2)_k$.

$$
\beta^{1/2} = 2\cos\frac{\phi}{k+2} = \frac{S_{1/20}}{S_{00}} = [2]_q \quad q = e^{2\pi i/k+2}
$$

We see once again the ubiquity of the q-numbers. One should perhaps remark that not all the fusion matrices of a given RCFT have such a nice description as incidence matrices of Dinkin diagram. That seems to be a property of the fusion rules of certain primary fields in the theory which play somehow the rule of the fundamental representation for the corresponding RCFT.

After all the previous examples we want a trace on the sequence of algebras A_ϕ^n with the Markov properties.

Since each factor in (9.33) has the ordinary trace for square matrices we define a "global" trace for A and B as:

$$\begin{aligned}\mathrm{tr}^A(x) &= \sum_i t_i^A \mathrm{tr}(x_i)\\ \mathrm{tr}^B(y) &= \sum_\alpha t_\alpha^B \mathrm{tr}(y_\alpha)\end{aligned} \tag{9.44}$$

where $x \in A, x_i \in M_{n_i}(\mathbf{C})$, similarly for B.

The condition that tr^A descends to tr^B through the inclusion (9.34) implies the constraint

$$\sum_i t_i^A \Lambda_{i\alpha} = t_\alpha^B \tag{9.45}$$

which is dual to (9.34). This time however t^A and t^B are not forced to the integers. If t_i^A and t_α^B are positive numbers we shall say that tr^A and tr^B are positive traces.

In our case we want to construct traces on A_ϕ^n with the Markov properties. Following (9.44) we define

$$\mathrm{tr}^{(n)}(x) = \sum t_{\phi,i}^n \mathrm{tr}(x_i) \tag{9.46}$$

where $x \in A_\phi^n$ and $x_i \in M_{d_{\phi,i}^n}$. Then (9.45) reads

$$\sum_j (N_\phi)_i^j t_{\phi,j}^{n+1} = t_{\phi,i}^n \tag{9.47}$$

Using the Verlinde theorem we can find a solution to (9.47):

$$t_{\phi,j}^n = \frac{S_{j0}}{S_{00}}\left(\frac{S_{00}}{S_{\phi 0}}\right)^n . \tag{9.48}$$

Since $\dfrac{S_{j0}}{S_{00}} \geq 0$, we obtain a positive trace on the whole family $A_\phi = \oplus_n A_\phi^n$.

From the previous discussion we have shown that for each primary field ϕ we can define the maps.

$$\begin{array}{ccccc} B_n & \xrightarrow{\text{RCFT}} & A_\phi^n & \xrightarrow{\text{RCFT}} & \mathbf{C}\\ g & \longrightarrow & \left\{B^{(i)}_{p_1\cdots p_{n-2},p'_1\cdots p'_{n-2}}\right\} & \longrightarrow & \mathrm{tr}^{(n)}_{(g)}\end{array} \tag{9.49}$$

where

$$\mathrm{tr}^{(n)}(g) = \left(\frac{S_{00}}{S_{\phi 0}}\right)^n \sum_j \sum_{p_1 \ldots p_{n-2}} \frac{S_{j0}}{S_{00}} B^{(j)}_{p_1 \ldots p_{n-2}, p_1 \ldots p_{n-2}}(g) \tag{9.50}$$

This trace obviously satisfies the properties $\mathrm{tr}^{(n)}(\alpha\beta) = \mathrm{tr}^{(n)}(\beta\alpha)$ for $\alpha, \beta \in B_n$. As a consequence of Verlinde theorem it also satisfies

$$\mathrm{tr}^{(n)}(1) = 1 \,. \tag{9.51}$$

Hence to assert that $\mathrm{tr}^{(n)}$ is a Markov trace means that it satisfies:

$$\mathrm{tr}^{(n+1)}(\alpha\sigma_n^{\pm 1}) = z_{\pm}\mathrm{tr}^{(n)}(\alpha) \qquad \alpha \in B_n \tag{9.52}$$

for some $z_{\pm} \in \mathbb{C}$.

For RCFT's where duality properties can be represented in terms of quantum groups this equation is indeed satisfied. If one does not want to rely on quantum groups, we need to invoke a result of Jones which guarantees under very general circumstances that for chains of embeddings of algebras $\ldots \subset A_n \subset A_{n+1} \subset \ldots$ The trace defined here has the Markov property (9.52). Independently from these general arguments we can describe the condition imposed by (9.52) on the braiding matrices.

First of all the value of $z_{\pm}$ is found easily from:

$$z_{\pm} = \mathrm{tr}^{(2)}\sigma_1^{\pm 1} = \frac{e^{\mp 2\pi i h_\phi}}{(S_{\phi 0}/S_{00})} \tag{9.53}$$

Then (9.52) is equivalent to:

$$\sum_j S_{j0} B_{pp} \begin{bmatrix} \phi & \phi \\ j & k \end{bmatrix} (\pm) = e^{\pm 2\pi i h_\phi} S_{p0} N^P_{\phi k} \,. \tag{9.54}$$

This equation is the analog of the Turaev's conditions (9.19) and implies in particular (choosing $k = 0$):

$$\sum_j S_{j0} N^k_{\phi\phi} \epsilon^j_{\phi\phi} e^{i\pi h_j} = S_{\phi_0} e^{4\pi i h_\phi} \tag{9.55}$$

It would be interesting to show dierctly that (9.54) follows from the general principle of RCFT. If this is the case then we would have a Markov trace and therfore link and knots invariants/for any RCFT.

Now it is straightforward to construct a polynomial invariant under the Markov moves $g \to \alpha g \alpha^{-1}$ and $\alpha \to \alpha\sigma_n^{\pm 1}$ where $\alpha, g \in B_n, \sigma_n \in B_{n+1}$.

If $g \in B_n$, define

$$T(g) = \left(\frac{S_{\phi 0}}{S_{00}}\right)^n e^{2\pi i h_\phi w(g)} \mathrm{tr}^{(n)} g \qquad (9.56)$$

The coefficient $w(g)$ is the wraith of the braid g (9.20).

It is clear that $T(g)$ satisfies the condition $T(\alpha g \alpha^{-1}) = T(\alpha)$, and T is also invariant under the Markov move of type II:

$$\begin{aligned} T(g\sigma_n) &= \left(\frac{S_{\phi 0}}{S_{00}}\right)^{n+1} e^{2\pi i h_\phi w(g\sigma_n)} \mathrm{tr}^{(n+1)}(g\sigma_n) \\ &= \left(\frac{S_{\phi 0}}{S_{00}}\right)^{n+1} e^{2\pi i h_\phi} e^{-2\pi i h_\phi} \left(\frac{S_{00}}{S_{\phi 0}}\right) \mathrm{tr}^{(n)}(g) = T(g) \ . \end{aligned}$$

Similarly

$$T(g\sigma_n^{-1}) = T(g)$$

As an example, consider n unlinked, unknotted circles. T is the case gives

$$T = \left(\frac{S_{\phi 0}}{S_{00}}\right)^n \ .$$

In RCFT, the dimensions of all primary fields are rational numbers[19,20] and we can always write the dimensions as $h_j = p_j/N$ mod 1. This implies that:

$$\begin{aligned} T(\sigma_1^{2N}) &= e^{2\pi i h_\phi 2N} \sum_j \frac{S_{j0}}{S_{00}} N_{\phi\phi}^j e^{i\pi 2N(2h_\phi - h_j)} \\ &= \sum_j \frac{S_{j0}}{S_{00}} N_{\phi\phi}^j = \left(\frac{S_{\phi 0}}{S_{00}}\right)^2 \end{aligned}$$

by the properties of S. Graphically this is interpreted by saying that the link shown in Fig. 59 is not distinguished in this theory from two unlinked, unknotted circles.

Since all the braid group generators are related to σ_1 by conjugation, this implies that

$$T(\sigma_i^{2N}) = \left(\frac{S_{\phi 0}}{S_{00}}\right)^2$$

σ_1^{2N}

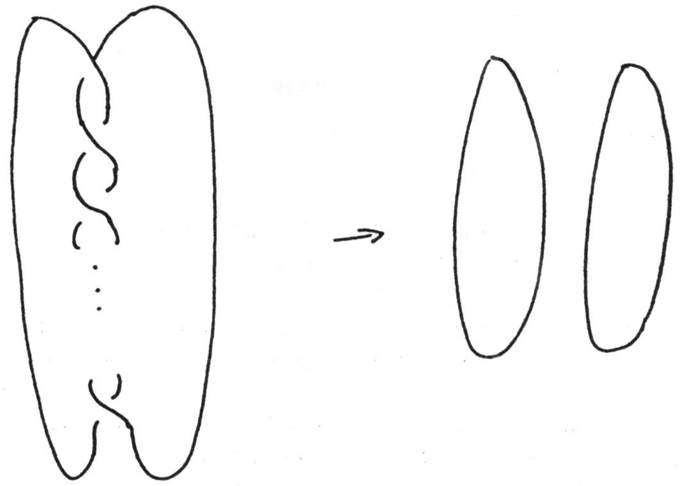

Fig. 59.

gives some constraints on the level of complexity of knots and links that can be distinguished by a given RCFT.

As a final application we consider the SU(2)$_k$ WZW theory. Recalling that $S_{j0}/S_{00} = [2j+1]_q$ with $q = e^{2\pi i/k+2}$ we deduce for $\phi = \phi_j = 1/2$ the polynomial:

$$T(g) = q^{3/4w(g)} \sum_j [2j+1]\ \mathrm{tr}_j(g) \qquad g \in B_n \tag{9.57}$$

where j is constrained to be in the range determined by the fusion rules of SU(2)$_k$ in the decomposition of the tensor product $V_{1/2}^{\otimes n}$. Similar computations can be carried out for other groups and it gives a conformal field theory derivation of the Ocneanu decompositions in terms of Young tableaux of the Jones polynomials.

9.4 *Knots on arbitrary 3– manifolds*

In this final paragraph we describe some aspects of Witten's construction of knot on arbitrary 3-manifolds.[33] This extension is based on a 3–dimensional topological gauge theory whose Lagrangian is the three-dimensional Chern-Simons invariant. The invariant polynomials are represented by expectation values of Wilson lines on the knots or links. This model uses explicitly the topology of the three-manifold and in a single stroke provides an intrinsic geometrical interpretation of knots in arbitrary 3-manifolds.

The metric independent Chern-Simons action is

$$S[A] = \frac{ik}{4\pi} \int_M \mathrm{tr}(AdA + \frac{2}{3}A^3) \tag{9.58}$$

and the first step in Witten's construction is to use (9.58) to construct a modular functor[68] on the moduli space of Riemann surface with distinguished points. One can think about this construction by extending the operator formalism to three dimensions. In two dimensions the operators formalism associate a ray to any Riemann surfact Σ with parametrized boundaries $\partial\Sigma = c_1 \cup c_2 \cup \ldots \cup C_n, |\Sigma_i C_j, \ldots, C_n\rangle$ (Fig. 60) in the tensor product $\otimes^n \mathcal{H}_{c_i}$ of Hilbert spaces associated to each component c_i of $\partial\Sigma$. In three dimension Σ is replaced by a 3-manifold M and $\partial M = c_i \cup \ldots \cup c_n$ is a set of 2D surfaces. Hence to (M, Σ) with $\Sigma = \partial M$ we associate a state (M, Σ) in $\mathcal{H}_\Sigma$ (Fig. 61). We have two maps:

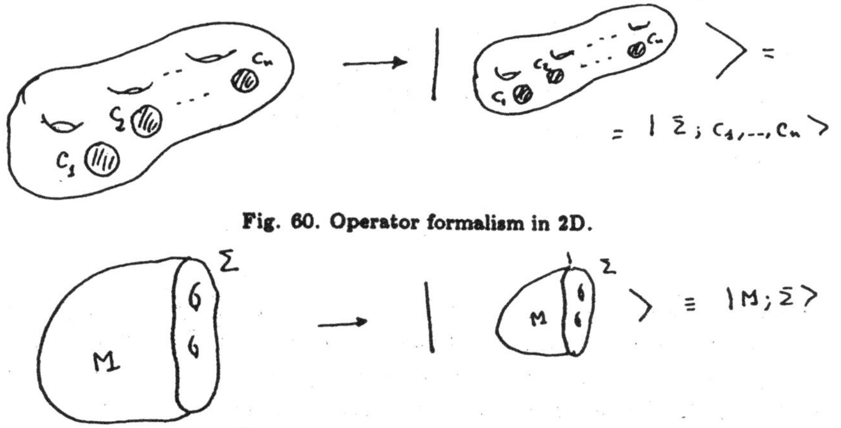

Fig. 60. Operator formalism in 2D.

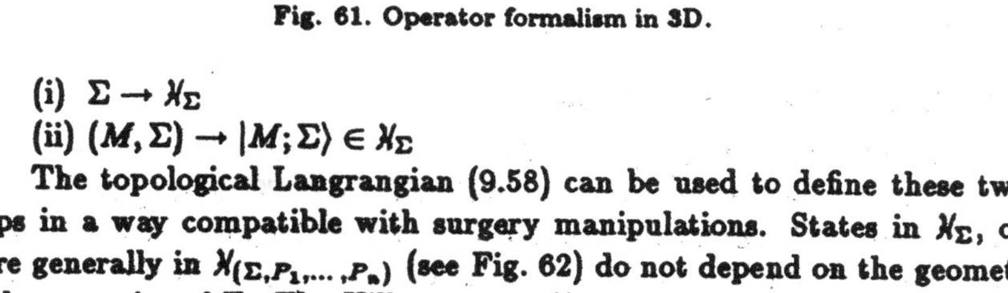

Fig. 61. Operator formalism in 3D.

(i) $\Sigma \rightarrow \mathcal{H}_\Sigma$

(ii) $(M, \Sigma) \rightarrow |M; \Sigma\rangle \in \mathcal{H}_\Sigma$

The topological Langrangian (9.58) can be used to define these two maps in a way compatible with surgery manipulations. States in $\mathcal{H}_\Sigma$, or more generally in $\mathcal{H}_{(\Sigma, P_1, \ldots, P_n)}$ (see Fig. 62) do not depend on the geometrical properties of Σ. The Hilbert space $\mathcal{H}_{(\Sigma, P_1, \ldots, P_n)}$ defines a flat vector bundle on the moduli space of Riemann surfaces with punctures. Using the Schrodinger representation for the states in $\mathcal{H}_{(\Sigma,P_1,\dots,P_n)}$ the states $|M,\Sigma,P_1,\dots,P_n\rangle$ are functionals on the space of gauge equivalent classes of connections satisfying Gauss' law for external charges at the punctures $P_1,\dots,P_n$:

$$\frac{k}{8\pi}\varepsilon^{ij}F^a_{ij}=\sum_i \delta(X-X_i)T^a_i\ . \tag{9.59}$$

Fig. 62. An element of $\mathcal{H}_{(\Sigma;P_1\dots P_4)}$.

The spaces $\mathcal{H}_{(\Sigma,P_1\dots P_n)}$ are representation spaces for the mapping class group of $\Sigma-\{P_1,\dots,P_n\}$. In particular we can braid the points $P_1,\dots,P_n$. This operations lifts to a transformation on the fiber of the Hilbert bundle defined by $\mathcal{H}_{(\Sigma,P_1,\dots,P_n)}$ (Fig. 63).

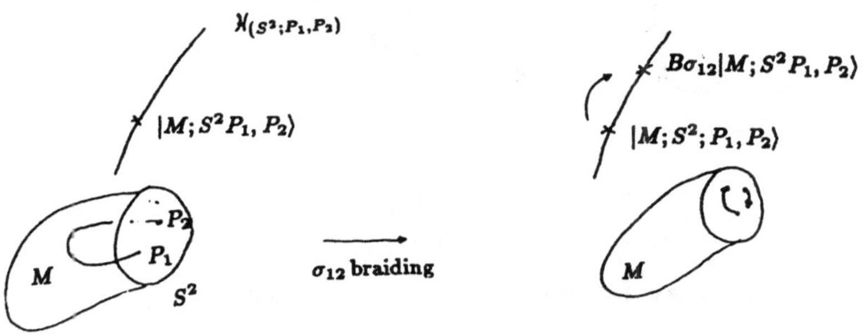

Fig. 63. Lift of the braiding transformation to the Hilbert bundle.

The spaces $\mathcal{H}_{(\Sigma,P_1\dots P_n)}$ when we associate to the punctures representations R_i of the gauge group are finite dimensional and they define irreducible representations of the braiding transformations. Moreover these spaces can be identified with the spaces of conformal blocks for a level k WZW model. Here we present heuristic argument for this identification.

First, the bundle $\mathcal{H}_{(\Sigma;P_1\ldots P_n)}$ is flat over the moduli space of Riemann surfaces with punctures. The same thing happens in conformal field theory where the conformal blocks can be interpreted as a basis of sections of a finite dimensional flat bundle defined on the moduli space. Second let us consider the space $\mathcal{H}_D$ where D is a disc. Using the temporal gauge and Gauss'law $\varepsilon^{ij}F^a_{ij} = 0$ we see that $\mathcal{H}_D$ is the geometric quantization of LG/G, more explicitly the space of sections of a line bundle defined on the base space LG/G. This is the Borel-Weyl approach to representation theory where for each line bundle on a flag manifold we get an irreducible representations. In this case we get a representation of the loop group LG. Therefore if we consider a Riemann surface Σ with n-punctures (Fig. 64) we can associate to each disc around the punctures a particular Hilbert space $\mathcal{H}_{R_i}$ corresponding to the representation R_i of LG. This means that once we include the constraint (9.59) what we are obtaining is the set of invariants states in $\otimes_{i=1}^{n} \mathcal{H}_{R_i}$. For $\Sigma = S^2$ the set of invariants in $\bigotimes_{i=1}^{n} \mathcal{H}^{R_i}$ can be interpreted as the conformal blocks with external legs saturated by primmary fields in the representations R_i:

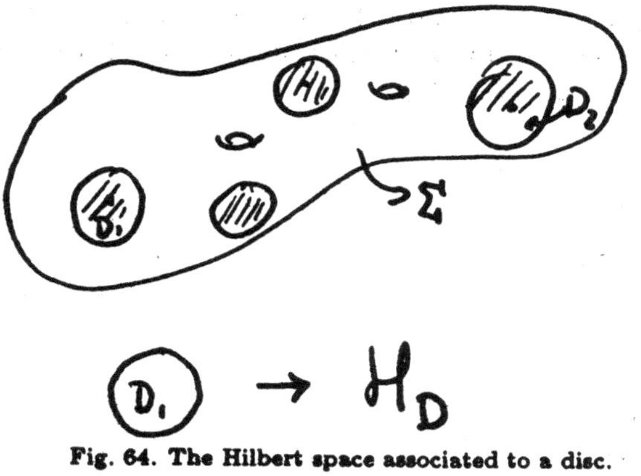

Fig. 64. The Hilbert space associated to a disc.

$$\mathcal{H}_{(\underset{S^2}{\Sigma \,\shortparallel}\,;\; R_1\ldots R_n)} = \text{Imv} \otimes_{n=1}^{n} \mathcal{H}_{R_i} =$$ 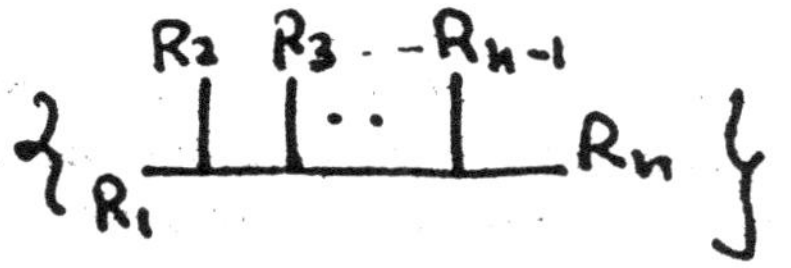

(9.60)

From (9.60) we get[33]

$$\begin{aligned} \dim\ \mathcal{H}(S^2;R) &= 0 \qquad R \neq id\,. \\ \dim\ \mathcal{H}(S^2;R_i,R_j) &= \delta_{i,\bar{j}} \\ \dim\ \mathcal{H}(S^2;R_i,R_j,R_k) &= N_{ij}^k\,. \end{aligned} \tag{9.61}$$

This result, already mentioned in the chapter on quantum groups indicates a deep connection between Gauss's law (9.59) and the fusion algebra of the WZW models.

After these preliminaries, we define Knot invariants. We know that the space of conformal blocks in WZW models are irreducible representations of the braid transformations. From this fact and from (9.60) we can easily get a Skein rule.

Consider for instance the case of Fig. 65 and for simplicity take the group to be SU(2). Choose arbitrary a crossing point and consider it as living inside a 3-ball. The common boundary is S^2. If on the knot we have running a "quark" in the fundamental representation then the Hilbert space at the boundary will be $\mathcal{H}_{(S^2;1/2\ldots 1/2)}$. For $k > 1$ we know that $\mathcal{H}_{(S^2;1/2\ 1/2\ 1/2\ 1/2)}$ is a two dimensional space and therefore a representation space of the Hecke algebra. The expectation value of the Wilson line on k will be:

$$\langle 1/2\ldots 1/2 S^2; M_1 | M_2; S^2; 1/2\ldots 1/2\rangle\,. \tag{9.62}$$

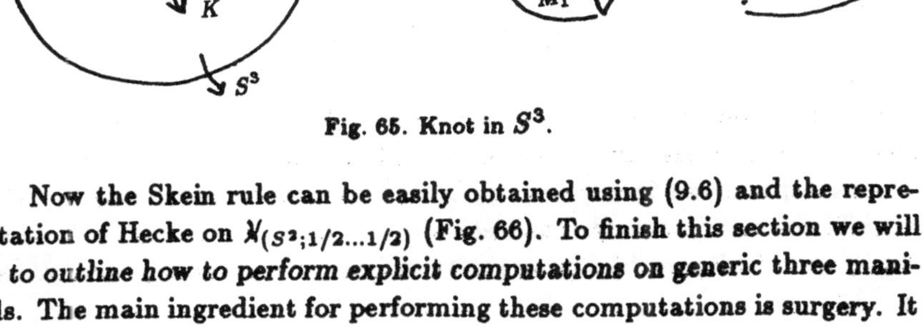

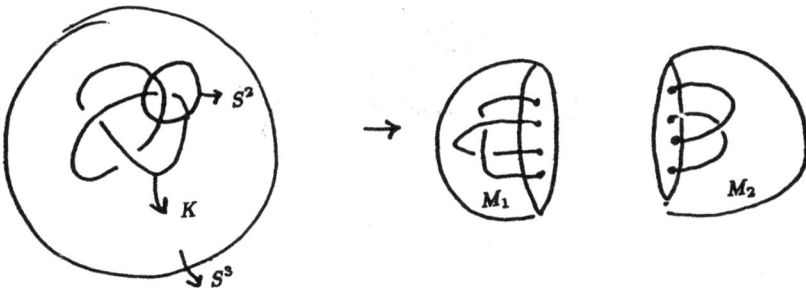

Fig. 65. Knot in S^3.

Now the Skein rule can be easily obtained using (9.6) and the representation of Hecke on $\mathcal{H}_{(S^2;1/2\ldots 1/2)}$ (Fig. 66). To finish this section we will like to outline how to perform explicit computations on generic three manifolds. The main ingredient for performing these computations is surgery. It is well-known that any three manifolds can be obtained by repeated surgery on Knots in S^3. The procedure consists of cutting a solid torus around the Knot and to glue it back after a diffeomorphism on its boundary. The partition function for the new manifold can be obtained as follows. First we divide S^3 in two pieces M_1, M_2, with M_1 the solid torus around the knot Fig. 67. This defines two states $|M_1; \partial M_1\rangle$ $|M_2; \partial M_2\rangle$. Second, we perform a diffeomorphism ρ on the boundary of the solid torus. This diffeomorphism can be represented by an operator acting on $H_{\partial M_1}$, say K_ρ. The new partition function is given by $\langle M_2; \partial M_2 | K_\rho | M_1; \partial M_1\rangle$.

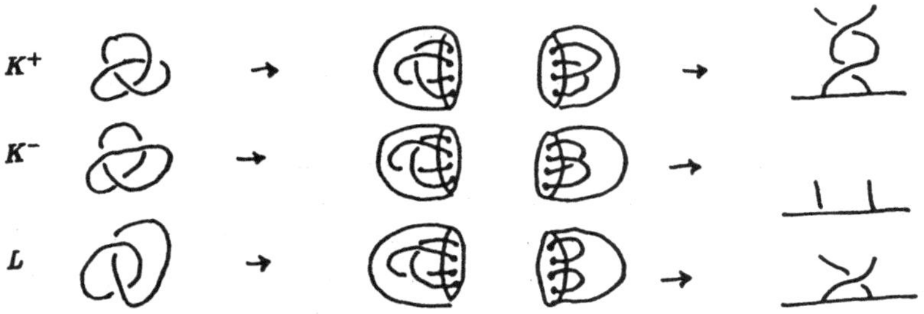

Fig. 66. The Skein rule for the trefoil. The r.h.s are the corresponding states in $\mathcal{H}(S^2 1/2 \ldots 1/2)$

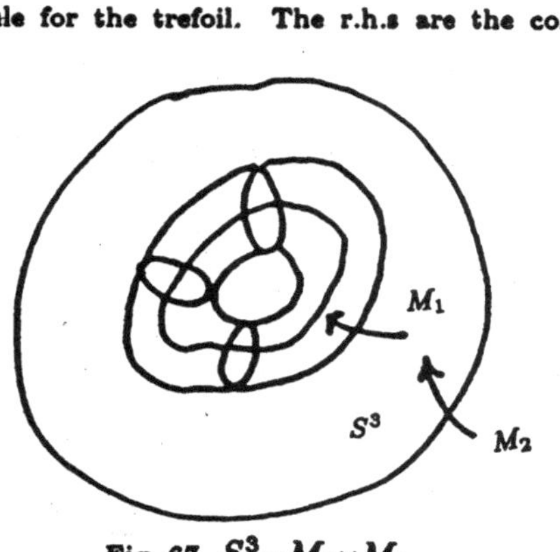

Fig. 67. $S^3 = M_1 \cup M_2$.

Let us consider as an example the two 3-manifolds S^3 and $S^2 \times S^1$. These two manifolds can be obtained by gluing two solid torus. For S^3 this is the case presented in Fig. 67. $S^1 \times S^1$ can also be obtained by gluing two solid torus. If we represent the solid torus by $D \times S^1$ with D a disc, then $S^2 \times S^1$ obtained by gluing the two discs through the boundary. (see Fig. 68).

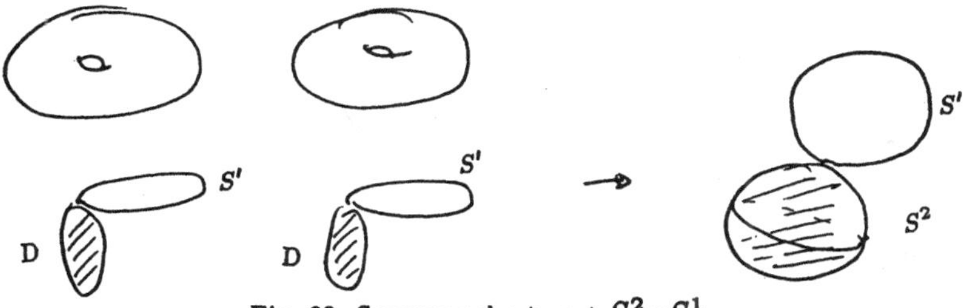

Fig. 68. Surgery rules to get $S^2 \times S^1$.

To get $S^2 \times S^1$ from S^3 we first consider the unknot in S^3, relative to this unknot we divide S^3 in two solid torus and we glue them back again after a modular transformation $\tau \to -1/\tau$ on the boundary of the solid torus that it is surrounding the unknot. To compute the partition function we need to represent modular transformations on $\mathcal{H}_T$, the Hilbert space of a torus. A basis of $\mathcal{H}_T$ is given by the set of character $\mathcal{H}_i$ of the WZW model. The state $|M_1; T\rangle$ will be identified with $\mathcal{H}_0$ the character of the vacuum. If inside M_1 we have a Wilson living in a representation R_i the corresponding state will be $\mathcal{H}_i$. Using the representation of modular transformations on the space of characters we get that the partition function on S^3 is $\Sigma_i S_{oi} Z(S^2 \times S^1; R_i)$ with S the modular transformation matrix, and $z(S^2 \times S^1; R_i)$ is the partition function of $S^2 \times S^1$ with one puncture in S^2 in the representation R_i. (see Fig. 69) From these surgery manipulations we can get an interesting expression for the invariant polynomials in S^3. In fact consider any link in S^3. Divide now S^3 in two solid tori in such a way that the link is contained in one of these Tori. Now cut the second torus away and glue it back again after a modular transformation. What one obtains is $S^2 \times S^1$ with the knot represented by a set of braiding transformations and one spectator Wilson line that is not braided with the original knot. More explicitly one obtains the following representation of the link invariant:

$$\sum_i S_{0i} Z(S^2 \times S^1; \alpha, R_i) \tag{9.63}$$

where α is the braiding transformation defining the original knot. (see Fig. 70).

Equation (9.63) is very interesting. In fact we can compute $Z(S^2 \times S^1; \alpha, R_i)$ by simply computing the trace of the representation of the braiding α on $\mathcal{H}(S^2; R_1 \ldots R_n; R_i)$; the braid group $\alpha \in B_n$. From (9.60) this is equivalent to performing the trace of element α in the space of conformal

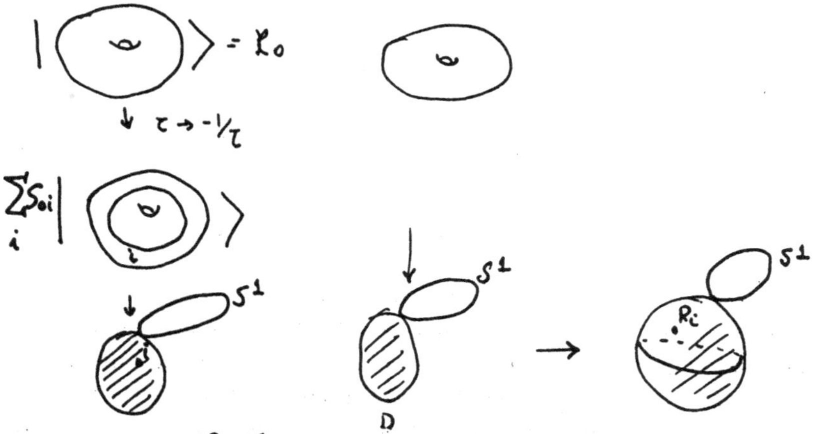

Fig. 69. $S^2 \times S^1$ with an insection in the representation R_i.

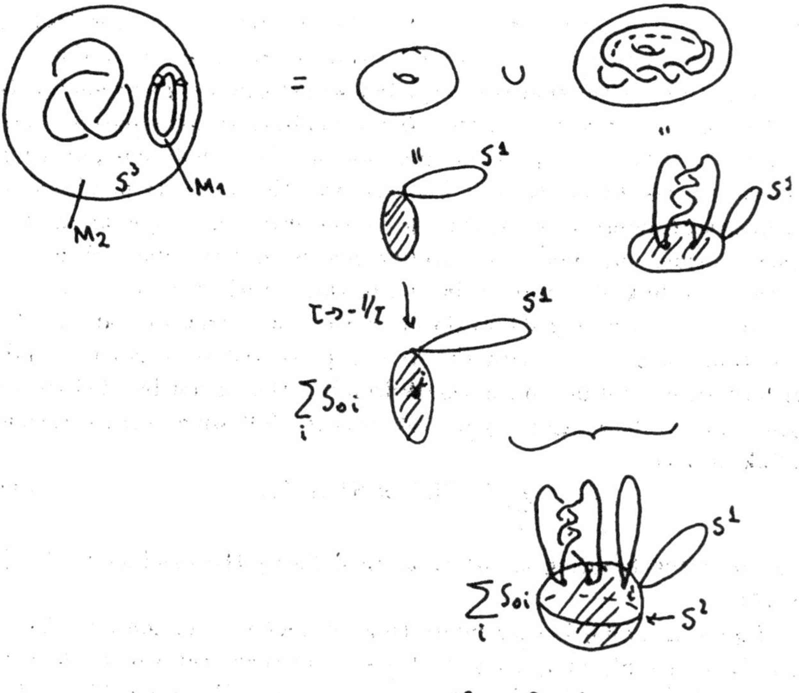

Fig. 70. Knots from S^3 to $S^2 \times S^1$.

blocks with $n+1$ external legs, the last one in the representation R_i. For simplicity let us assume all the representations $R_1 \ldots R_n$ are equal to the fundamental representation. In this case we obtain

$$Z(S^2 \times S^1; \alpha; R_i) = \mathrm{Tr}\,[B_{(\alpha)}]_{\mathcal{H}(S^2;1/2\ldots 1/2;R_i)} \tag{9.64}$$

where $B(\alpha)$ is the representation of $\alpha \in B_n$ in the space of conformal blocks

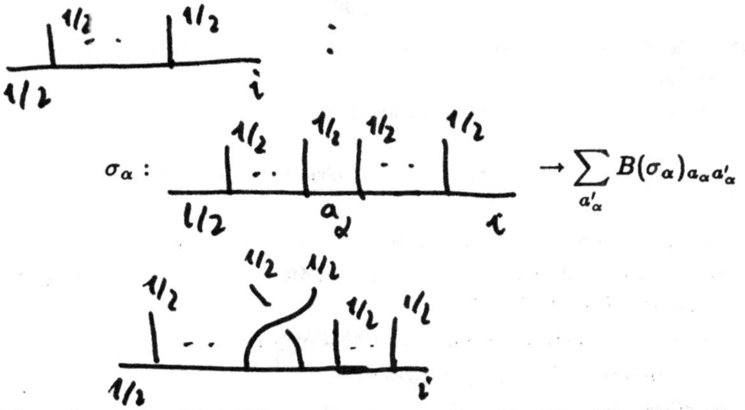

(9.65)

After these considerations we can come back to Eq. (9.24) and compare it with (9.63). We have the natural identification of W_j with S_{0j}. Normalizing (9.63) with respect to the partition function on S^3, we obtain:

$$\frac{1}{S_{00}} \sum_i S_{0i} Z(S^2 \times S^1; \alpha, R_i) \tag{9.66}$$

and therefore we expect:

$$W_j = \frac{S_{0j}}{S_{00}} . \tag{9.67}$$

For SU(2)$_k$ WZW model the value of $\frac{S_{0j}}{S_{00}}$ is $[2j+1]_q$ with $q = e^{2\pi i/k+2}$. This is precisely the value we get for the expansion of the Jones (or Turaev) polynomial in the expansion (9.24). This result is extremely interesting and indicates that the modular properties and the fusion algebra of the conformal field theory are deeply related with the Markov property of the traces used in the definition of invariant polynomials.

This concludes our brief account of Witten's approach to knots on arbitrary 3-manifolds.[33]

References

1. P. Ginsparg, eds. *Les Houches Summer School 1988*, Z. Brezin and J. Zinn-Justin.
2. J. Cardy, eds. *Les Houches Summer School 1988*, E-Brezin and J. Zim-Justin.
3. M. Green, J. Schwarz and E. Witten, *String theory* (Cambridge University Press, 1986), Vols. I and II.
4. M. Kaku, *String theory* (springer Verlag, 1988).
5. D. Lüst on S. Theisen *Lectures and String Theory. Lectures Notes in Physics*. (Springer-Verlag, 1989).
6. A. Belavin, A.M. Polyakov and A. Zamolodchikov, *Nuel Phys.* **B241** (1984) 33.
7. J. Birman, *Links, Braids and Mapping Class Groups* (Princeton University Press, 1974).
8. D. Friedan, Z. Qiu and S. Shenker, *Phys. Rev. Lett.* **51** (1984) 1575.
9. V.S. Dotsenko and V.A. Fateev, *Nucl. Phys.* **B240** (1984) 312; *ibid.* **B251** (1985) 691, *Phys. Lett.* **154B** (1985) 291.
10. A. Rocha-Caridi, in *Vertex Operators in Math Physics*.
11. E. Witten, *Commun. Math. Phys.***92** (1984) 83.
12. V.I. Knizhnik and A-Zamolodchikov, *Nucl. Phys.* **B247** (1984) 83.
13. D. Gepner and E. Witten, *Nucl. Phys.* **278B** (1986) 493.
14. A. Cappelli, C. Itzykson and J.B. Zuber, *Commun. Math. Phys.* **113** (1987) 1.
15. E. Verlinde, *Nucl. Phys.* **B300** (1988) 360.
16. G. Moore and N. Seiberg, *Phys. Lett.* **212B** (1988) 451.
17. G. Moore and N. Seiberg, "Naturalness in conformal field theories", IASSNS-HEP-88/31.
18. G. Moore and N. Seiberg, "Classical and quantum conformal field theories", IASSN-HEP-88/35.
19. C. Vafa, *Phys. Lett.* **B206** (1988) 421.
20. G. Anderson and G. Moore, *Commun. Math. Phys.* **117** (1988) 441.
21. D. Friedan and S. Shenker, *Nucl. Phys.* **B281** (1987) 509.
22. L. Alvarez Gaume, C. Gomez and G. Sierra, *Nucl. Phys. B*, to appear.
23. L. Alvarez Gaume, C. Gomez and G. Sierra, *Phys. Lett.* **220** (1989) 142.
24. G. Moore and N. Yu. Reshetikhin "A comment in quantum groups symmetry in conformal field theory", IASSNS-HEP-89/18.
25. L. Alvarez Gaume, C. Gomez and G. Sierra, "Duality and quantum groups", CERN TH 5369-89.

26. V.G Drinfel'd, *Sov. Math. Doke.* **32** (1985) 254; M. Jimbo, *Lett. Math. Phys.* **10** (1985) 63; **11** (1986) 247.
27. V.G. Drinfel'd, *Quantum Groups*; Berkelg Col 1986.
28. N. Yu. Reshetikhin, "Quantized universal enveloping algebras and invariants of links", I,II LOMI-3-4-87, E-17-87.
29. A.N. Kirillov and N. Yu. Reshetikhin, "Representations of the algebra Uq(SL(2)) q-orthogonal polynomials and invariant of "links", LOMI-E-9-88.
30. E. Witten, "Gauge theories, vertex models and quantum groups", IASSNS-HP-89/32.
31. G. Lusztig, "Modular representations of quantum groups", MIT preprint 1988.
32. V. Pasquier and H. Saleur, "Common structures between finite systems and conformal field theories through quantum groups", Saclay-SPHT-89/031.
33. E. Witten, *Commun. Math. Phys.* **121** (1989) 351.
34. R. Baxter, *Exactly Solved Models in statistical Mechanics* (Academic Press, 1982).
35. V. Pasquier, *Nucl. Phys.* **B 285** (FS19) (1987) 162; *J.Phys.* **A 20** (1987) 5707.
36. R. Baxter, *Proc. Royal Society* **289A** (1978) 1359.
37. J. Rolfsen, *Knots and Links* (Publish or Perish, Boston, 1976).
38. L. Kauffman, *On Knots* (Princeton University Press) 1988.
39. V.G. Tureev, *Invent. Math.* **92** (1988) 527.
40. V. Jones, *Invent. Math.* **72** (1983) 1.
41. V. Jones, *Ann. Math* **126** (9187) 335.
42. S. Ferrara, A. Grillo and R. Gatto, *Nuovo Cimento* **12** (1972) 959.
43. V.A. Fatteev and A. Zamolodchikov, *Nucl. Phys.***B 280** [FS 18] (1987) 644.
44. M. Berhadsky, V. Knizhnik and M. Teitelman, *Phys. Lett.* **B 151** (1985) 31.
45. P.Di Vecchia, V.G. Knizhnik, J.L Petersen and P. Rossi, *Nucl. Phys.* **B 253** (1985) 701.
46. V. A. Fatteev and Lykiarnov.
47. A. Zamolodchikov, Montreal Lectures 1988; A. Gerasimov, A. Marshakov, A. Morozov, M. Olshanetsky and S. Shatashvili, ITEP preprint 1989; J. Distler and Z. Quio, CLNS 89/911; D. Nemeschansky, USC-89/012.
48. P. Goddard and D. Olive, *Int. J. Mod Phys* **A1** (1986) 1.
49. V. Kac, *Infinite Dimensional Algebras* (Cambridge University Press).
50. D. Goss, J. Harvey, E. Martinec and R. Rohm, *Nucl. Phys.* **B 256** (1985) 253 and *Nucl. Phys.* **B 267** (1985) 75.

51. J. Cardy, *Nucl. Phys.* **B270** [FS 16] (1986) 186.

52. A. Capelli, C. Itzykson and J.B. Zuber, *Nucl. Phys.* **B 280** (1987) 445 and *Commun. Math. Phys.* **113** (1987) 1.

53. R. Dijkgraaf and E. Verlinde, "Fusion algebras and conformal field theories", centribution to the Annecy Conference on Conformal Field theory, 1988.

54. P. Bowknegt and W. Nalim, *Phys. Lett.* **184** B (1987) 359.

55. C. Itzykson, Proceedings of Annecy Workshop; *Nucl. Phys.* 5B (1988) 150.

56. B. Schellekens and S. Yankielowicz, CERN TH 5344/89.

57. R. Dijkgraaf, E. Verlinde and H. Verlinde, *Commun. Math. Phys.* **115** (1988) 649.

58. T. Kawai, University of Tokyo Komaba preprint 21340 (88).

59. H. Wenzl, *Invent. Math.* **92** (1988) 349.

60. G. Felder, J. Fröhlich and G. Keller, ETH preprint (February 1989).

61. A. Tsuchiya and Y. Kanie, *Lett. Math. Phys.* **13** (1987) 303.

62. J. Lacki and P. Zaugg, Geneve University preprint 1989.

63. R. Baxter, *Aun. Phys.* **70** (1972) 193.

64. A. Zamolodchikov, *Sov. Phys. Rev.* **2** (1980).

65. G.F. Andrews, R. J. Baxter and P.J. Forrester, *J. Stat Phys.* **35** (1984) 193; E. Date, M. Jimbo, A. Kuniba, T. Miwa and M. Okada, *Nucl. Phys.* **240** [FS 17] 1987 (231), *Adv. Stud. Pure Math.* **16** (1987).

66. P. Freyd, d. Yetter, J. Horte, W. Lickorish, K. Millet and A. Ocneanu, *Bull A M S* **12** (1985) 239.

67. L. Kauffman, *Topology* **26** (1987) 395.

68. G. Segal; Atiyah-Semminar 1989. and paper in preparation.

Representations of $U_q(\mathfrak{gl}(n,\mathbf{C}))$ at $q=0$ and the Robinson-Shensted correspondence

Etsuro Date[1)], Michio Jimbo[2)] and Tetsuji Miwa[3)]

1) Department of Mathematics, College of General Education
Kyoto University, Kyoto 606, Japan
2) Department of Mathematics, Faculty of Science
Kyoto University, Kyoto 606, Japan
3) Research Institute for Mathematical Sciences
Kyoto University, Kyoto 606, Japan

Abstract Let V be the n dimensional irreducible $U_q(\mathfrak{gl}(n,\mathbf{C}))$–module and let v_μ ($\mu = 1,\cdots,n$) be its natural basis. There is an irreducible decomposition $V^{\otimes N} = \oplus_T V(T)$ where T is a standard tableau with N nodes. It is shown that in the $q^{\pm 1} \to 0$ limit $V(T)$ is spanned by decomposable vectors $v_{\mu_1} \otimes \cdots \otimes v_{\mu_N}$, and that the correspondence $T \leftrightarrow v_{\mu_1} \otimes \cdots \otimes v_{\mu_N}$ is given by the Robinson-Shensted correspondence.

1. Introduction

Let $\mathfrak{g}$ be a finite dimensional complex simple Lie algebra. With each $\mathfrak{g}$ one can associate the quantized universal enveloping algebra $U_q(\mathfrak{g})$ — a one–parameter Hopf algebra deformation of the universal enveloping algebra $U(\mathfrak{g})$ [1,2]. Recently there is growing attention to $U_q(\mathfrak{g})$, not only because of its intrinsic mathematical interest but because it seems to underlie the issues centering round solvable lattice models, link invariants and conformal field theory.

Lusztig [3] and Rosso [4] showed that, for generic values of the parameter q, the usual theory of highest weight representations for $U(\mathfrak{g})$ carries over to $U_q(\mathfrak{g})$ without essential change. Of particular interest is the case when q is a root of unity, where one sees subtleties reminiscent of the modular representations [5]. The present paper is concerned with another special value: $q = 0$. As the algebra $U_q(\mathfrak{g})$ loses its meaning for $q = 0$, let us make precise what we mean by this. To be specific we consider here the analog of $U_q(\mathfrak{g})$ for $\mathfrak{g} = \mathfrak{gl}(n, \mathbf{C})$. Let $V = \mathbf{C}^n$ be the n dimensional irreducible representation and v_μ $(\mu = 1, \cdots, n)$ be its natural basis. Since the weights of V are multiplicity free, for any irreducible module V_Y the tensor module $V_Y \otimes V$ decomposes with multiplicity 1. Repeating this decomposition for the tensor power of V one obtains

$$V^{\otimes N} = \oplus_T V(T).$$

The irreducible components $V(T)$ are labeled by standard tableaux T with N nodes and at most n rows — T represents the "growth process of the Young diagram". In this paper we show that in the limit $q \to 0$ $V(T)$ is spanned by decomposable vectors $v_{\mu_1} \otimes v_{\mu_2} \otimes \cdots \otimes v_{\mu_N}$, and that the rule of correspondence

$$T \longleftrightarrow (\mu_1, \ldots, \mu_N)$$

is described by the Robinson-Shensted correspondence [6] (see Theorem 2.7).

Our motivation for this study comes from the computation of one-point functionsin solvable lattice models. Let $R(x, q)$ be the trigonometric solution of the Yang-Baxter equation corresponding to the vector representation of $U_q(\mathfrak{gl}(n, \mathbf{C}))$ [7]. This defines a face model (=an IRF model [8]) on the dual lattice, where the fluctuation variables on the sites take values in the set of level 1 integral weights of the affine Lie algebra $A^{(1)}_{n-1}$ (see e.g. [9]). The model has ground states indexed by fundamental weights $\Lambda_j = \Lambda_{j+n}$. By the one-point function we mean the probability P_a that a particular site assumes a given value a in the ground state, say, Λ_0.

The R–matrix has the spectral decomposition

$$R(x,q) = (x-q^2)P_+ + (1-xq^2)P_-$$

where $P_\pm$ signify the projectors onto the q–analog of the symmetric/antisymmetric parts. At $q=0$ they simplify to

$$\mathrm{Im}\ P_+ = \sum_{\mu\geq\nu} \mathbf{C}v_\mu \otimes v_\nu, \quad \mathrm{Im}\ P_- = \sum_{\mu<\nu} \mathbf{C}v_\mu \otimes v_\nu, \tag{1.1}$$

v_μ being the natural basis of $V = \mathbf{C}^n$. Hence

$$R(x,0)v_\mu \otimes v_\nu = x^{H(\mu,\nu)} v_\mu \otimes v_\nu \tag{1.2}$$

with

$$\begin{aligned} H(\mu,\nu) &= 0 \qquad \text{if } \mu<\nu, \\ &= 1 \qquad \text{if } \mu \geq \nu. \end{aligned} \tag{1.3}$$

The diagonal nature of $R(x,0)$ (1.2) is crucial in applying Baxter's corner transfer matrix method [8]. Consider sequences (called 'paths') of the form $\eta = (\eta_0, \eta_1, \cdots)$, $\eta_j \in \{1,2,\cdots,n\}$. Let $\bar\eta$ be the path such that $\bar\eta_j \equiv j \bmod n$, and set $\mathcal{P}(\Lambda_0) = \{\eta \mid \eta_j = \bar\eta_j \ \text{ for } j \gg 0\}$. Baxter's method entails that, up to a factor, P_a is given by

$$\sum_{k\geq 0} \sharp\mathcal{P}(\Lambda_0)_{a-k\delta} q^{nk}. \tag{1.4}$$

Here δ is the null root, and we set $\mathcal{P}(\Lambda_0)_\mu = \{\eta \in \mathcal{P}(\Lambda_0) \mid \iota_\eta - \omega(\eta)\delta = \mu\}$ with

$$\iota_\eta = \Lambda_j - \sum_{k=0}^{j-1}(\Lambda_{\eta_k} - \Lambda_{\eta_k - 1}) \qquad (j \gg 0),$$

$$\omega(\eta) = \sum_{k=1}^{\infty} k\left(H(\eta_{k-1},\eta_k) - H(\bar\eta_{k-1},\bar\eta_k)\right).$$

As was shown in [9] the number of paths in the set $\mathcal{P}(\Lambda_0)_\mu$ is equal to the multiplicity of μ in the integrable highest weight module $L(\Lambda_0)$ of $A_{n-1}^{(1)}$:

$$\sharp\mathcal{P}(\Lambda_0)_\mu = \dim L(\Lambda_0)_\mu. \tag{1.5}$$

Hence (1.4) is the string function [10] for $L(\Lambda_0)$.

In this paper we calculate the spectral decomposition of the R–matrix associated with the N–th symmetric tensor representation $V_{[N]}$ (eq. (3.6)), and determine its H–function by using the Robinson-Shensted description (Proposition 2.8 and eq. (3.4)). In a previous paper [11] we have proved the corresponding analog of (1.5) for arbitrary level N dominant integral weights. Therefore, along with the result in this paper it is established that the one-point functions for the N–th symmetric tensor model are expressible in terms of the level N string functions.

The text is organized as follows. In section 2, we recall the definition of $U_q(\mathfrak{gl}(n,\mathbf{C}))$, the Gelfand-Tsetlin basis and the q–Wigner coefficients following [12],[13]. We then calculate the limit $q \to 0$ of the latter. This leads to the Robinson-Shensted correspondence mentioned above. In section 3 we describe the fusion procedure of the R–matrix starting from the one for the vector representation. Using the q–Wigner coefficients we derive the spectral decomposition for the R–matrix associated with the pair $(V_Y, V_{[N]})$ where V_Y is an arbitrary irreducible module. Combining the results in sections 2 and 3 we then obtain the H–function for the R–matrix of type $(V_{[N]}, V_{[N]})$.

2. Representation of $U_q(\mathfrak{gl}(n,\mathbf{C}))$ at $q=0$.

2.1. $U_q(\mathfrak{gl}(n,\mathbf{C}))$.

We begin by recalling the definition of $U_q = U_q(\mathfrak{gl}(n,\mathbf{C}))$ [1],[7]. In what follows we fix a positive integer $n \geq 2$. For a complex number $q \neq 0, \pm 1$, U_q is defined to be an associative $\mathbf{C}$–algebra with 1, with generators $\{X_j^{\pm}\}_{1\leq j\leq n-1}, \{q^{\pm\varepsilon_j/2}\}_{1\leq j\leq n}$ and the following relations:

$$
\begin{aligned}
q^{\varepsilon_i/2}q^{-\varepsilon_i/2} &= q^{-\varepsilon_i/2}q^{\varepsilon_i/2} = 1, \\
q^{\varepsilon_i/2}q^{\varepsilon_j/2} &= q^{\varepsilon_j/2}q^{\varepsilon_i/2}, \\
q^{\varepsilon_i/2}X_j^{\pm}q^{-\varepsilon_i/2} &= q^{\pm 1/2}X_j^{\pm} \quad \text{for} \quad i=j, \\
&= q^{\mp 1/2}X_j^{\pm} \quad \text{for} \quad i=j+1, \\
&= X_j^{\pm} \qquad \text{otherwise}, \\
[X_i^{+}, X_j^{-}] &= \delta_{ij}\frac{q^{H_i}-q^{-H_i}}{q-q^{-1}}, \\
(X_i^{\pm})^2 X_j^{\pm} - (q+q^{-1})X_i^{\pm}X_j^{\pm}X_i^{\pm} + X_j^{\pm}(X_i^{\pm})^2 &= 0 \quad \text{for} \quad |i-j|=1, \\
X_i^{\pm}X_j^{\pm} &= X_j^{\pm}X_i^{\pm} \quad \text{for} \quad |i-j|\geq 2.
\end{aligned}
$$

Here $H_i = \varepsilon_i - \varepsilon_{i+1}$.

In fact U_q has a Hopf algebra structure. For our purpose we need the comultiplication $\Delta : U_q \longrightarrow U_q \otimes U_q$ given by

$$
\begin{aligned}
\Delta(q^{\varepsilon_j/2}) &= q^{\varepsilon_j/2} \otimes q^{\varepsilon_j/2}, \\
\Delta(X_j^{\pm}) &= X_j^{\pm} \otimes q^{-H_j/2} + q^{H_j/2} \otimes X_j^{\pm}.
\end{aligned} \tag{2.1}
$$

Throughout this paper we shall assume that $0 < q < 1$. The reality of q is imposed merely for certain sign convention; the essential assumption is that q is not a root of unity. In the rest of this paper we employ the following notation:

$$[\nu]_q = \frac{q^{\nu}-q^{-\nu}}{q-q^{-1}}.$$

2.2. Gelfand-Tsetlin pattern.

A Gelfand-Tsetlin (GT) pattern is an array of integers

$$|m\rangle = (m_{ij})_{1\le i\le j\le n} = \begin{matrix} m_{1n} & m_{2n} & \cdots & m_{nn} \\ m_{1n-1} & \cdots & m_{n-1n-1} & \\ & \cdots & & \\ m_{12} & m_{22} & & \\ m_{11} & & & \end{matrix}$$

such that

$$m_{i\,j+1} \ge m_{ij} \ge m_{i+1\,j+1} \qquad \text{for all } 1 \le i \le j \le n-1. \tag{2.2}$$

Henceforth we shall assume that $m_{nn} \ge 0$. The first row $[m_{1n}, m_{2n}, \cdots, m_{nn}]$ is then a Young diagram, to be called the shape of $|m\rangle$. We denote by $GT(Y)$ the set of GT patterns of shape Y.

Let $Y = [f_1, \cdots, f_n]$ $(f_1 \ge \cdots \ge f_n \ge 0)$ be a Young diagram. A semi-standard tableau of shape Y is an array of integers $S = (s_{ij})_{1\le i\le n, 1\le j\le f_i}$ such that

$$s_{ij} \in \{1, 2, \cdots, n\}, \quad s_{ij} < s_{i+1\,j} \quad \text{and} \quad s_{ij} \le s_{i\,j+1} \quad \text{for all } i, j.$$

We denote by $\mathcal{S}(Y)$ the set of semi-standard tableaux of shape Y.

The GT patterns are in one-to-one correspondence with the semi-standard tableaux. Let $a = (a_0, a_1, \cdots, a_n)$ be a sequence of integers satisfying $0 = a_0 \le a_1 \le \cdots \le a_n$. For each i with $a_{i-1} < a_i$ write the letter i $a_i - a_{i-1}$ times, starting with $i = 1, 2, \cdots$. The resulting sequence $\nu = (\nu_1, \cdots, \nu_f)$, $f = a_n$, is called the rearrangement of a. For example:

$$a = (0\ 2\ 2\ 3\ 5) \quad \leftrightarrow \quad \nu = (1\ 1\ 3\ 4\ 4),$$

$$a = (0\ \overbrace{f\ f\ \cdots\ f}^{n}) \quad \leftrightarrow \quad \nu = (\overbrace{1\ 1\ \cdots\ 1}^{f}).$$

Lemma 2.1. *The following is a bijection:*

$$\tau : GT(Y) \longrightarrow \mathcal{S}(Y), \qquad |m\rangle \mapsto \tau(|m\rangle),$$

where

$$j\text{-th row of } \tau(|m\rangle) = \text{the rearrangement of } (\overbrace{0,\cdots,0}^{j}, m_{jj}, m_{j\,j+1},\cdots,m_{jn}).$$

For example:

$$|m\rangle = \begin{matrix} 3 & & 2 & & 0 \\ & 3 & & 1 & \\ & & 1 & & \end{matrix} \quad \mapsto \quad \tau(|m\rangle) = \begin{matrix} 1 & 2 & 2 \\ 2 & 3 & \end{matrix}$$

2.3. Irreducible representations of $U_q(\mathfrak{gl}(n,\mathbf{C}))$.

Finite dimensional irreducible representations of $U_q = U_q(\mathfrak{gl}(n,\mathbf{C}))$ are parametrized by Young diagrams with at most n rows. Explicitly they are realized as follows [12]. For each Young diagram $Y = [f_1,\cdots,f_n]$ let V_Y be the C–span of GT patterns $|m\rangle \in GT(Y)$. We equip V_Y with a C–bilinear form $(\ ,\ )$ such that $(|m\rangle, |m'\rangle) = \delta_{m\,m'}$. Define the action of the generators of U_q by

$$q^{\epsilon_j/2}|m\rangle = q^{w_j(m)/2}|m\rangle, \qquad w_j(m) = \sum_{i=1}^{j} m_{ij} - \sum_{i=1}^{j-1} m_{i\,j-1}, \tag{2.3a}$$

$$X_j^+|m\rangle = \sum_{m'}{}^{(j)} c_j(m',m)|m'\rangle, \qquad X_j^-|m\rangle = \sum_{m'}{}^{(j)} c_j(m,m')|m'\rangle. \tag{2.3b}$$

Note that $(X_j^+|m'\rangle, |m\rangle) = (|m'\rangle, X_j^-|m\rangle)$. The symbol $\sum^{(j)}$ signifies the sum restricted to such m' that $m'_{ab} = m_{ab}$ if $b \neq j$. The coefficient $c_j(m,m')$ is zero unless for some i $m'_{ij} = m_{ij} - 1$ and $m'_{ab} = m_{ab}$ for any $(a,b) \neq (i,j)$. Non-zero coefficients are given as follows. Set

$$\begin{aligned} l''_k &= m_{k\,j+1} - k & 1 \le k \le j+1, \\ l_k &= m_{kj} - k & 1 \le k \le j, \\ l'_k &= m_{k\,j-1} - k & 1 \le k \le j-1. \end{aligned}$$

Then

$$c_j(m,m') = \left(- \frac{\prod_{k=1}^{j-1} [l'_k - l_i]_q \prod_{k=1}^{j+1} [l''_k - l_i + 1]_q}{\prod_{\substack{k=1 \\ k \neq i}}^{j} [l_k - l_i]_q [l_k - l_i + 1]_q} \right)^{\frac{1}{2}}.$$

If one of $|m\rangle$, $|m'\rangle$ violates the condition (2.2) and the other does not, then $c_j(m,m')=0$.

We note that the 'weight' $w_j(m)$ in (2.3) is the number of the letter j in the semi-standard tableau $\tau(|m\rangle)$. The GT pattern

$$v_Y = \begin{matrix} f_1 & f_2 & \cdots & f_{n-1} & f_n \\ & f_1 & f_2 \cdots & f_{n-1} & \\ & & \cdots & & \\ & & f_1 & & \end{matrix} \in V_Y \tag{2.4}$$

represents the highest weight vector :

$$q^{e_j/2} v_Y = q^{f_j/2} v_Y, \qquad X_j^+ v_Y = 0.$$

The n dimensional representation on $V_{[1]} \cong \mathbf{C}^n$ is referred to as the vector representation. In the sequel we set

$$\begin{aligned} V = V_{[1]} &= \oplus_{j=1}^n \mathbf{C} v_j, \\ v_j &= |m_j\rangle, \\ (m_j)_{ik} &= \delta_{i1} \quad \text{if} \quad j \le k \le n, \\ &= 0 \quad \text{if} \quad 1 \le k \le j-1. \end{aligned}$$

2.4. Wigner coefficients.

Recall that the tensor product of two representations $\pi_i : U_q \longrightarrow \mathrm{End}(V_i)$ is defined via the comultiplication (2.1) : $(\pi_1 \otimes \pi_2)\Delta : U_q \longrightarrow \mathrm{End}(V_1 \otimes V_2)$. These are (in fact all finite dimensional representations are) completely reducible [4]. The branching rule takes the same form as in the Lie algebra case. When one of the representations is the vector representation we have

$$V_Y \otimes V \cong \oplus_W V_W, \tag{2.5}$$

where W runs over Young diagrams obtained by adding one node to Y. Suppose that the node is added at the μ-th row. This is represented by the symbol

$$Y \underset{\mu}{\longrightarrow} W.$$

The embedding $V_W \subset V_Y \otimes V$ is described by Wigner coefficients [13]. Let $|m\rangle \in GT(W)$. For $(i_n, \ldots, i_j)$ such that $i_n = \mu$ and $1 \le i_k \le k$ $(j \le k \le n)$, we define a GT pattern $|m'\rangle = |m; i_n, i_{n-1}, \ldots, i_j\rangle \in GT(Y)$ by

$$\begin{aligned} m'_{ik} &= m_{i_k k} - 1 \quad \text{if } j \le k \le n \text{ and } i = i_k, \\ &= m_{ik} \qquad\quad \text{otherwise.} \end{aligned} \tag{2.6}$$

The formula reads as

$$\begin{aligned} |m\rangle = \sum_{j=1}^{n} \sum_{(i_n=\mu, i_{n-1}, \ldots, i_j)} & w_q(m; i_n, i_{n-1}, \ldots, i_j) \\ & \times |m; i_n, i_{n-1}, \ldots, i_j\rangle \otimes v_j. \end{aligned} \tag{2.7}$$

The Wigner coefficients $w_q(m; i_n, i_{n-1}, \ldots, i_j)$ are given as follows. Suppose that $f'_1 \ge f_1 \ge f'_2 \ge f_2 \ge \cdots \ge f'_{k-1} \ge f_{k-1} \ge f'_k$. We define the reduced Wigner coefficients by

$$\begin{aligned} w_q^{(1)}\left(\begin{matrix} f'_1 & \cdots & f'_k \\ f_1 & \cdots & f_{k-1} \end{matrix} \middle| \, i \right) &= q^{(\sum_k l_k - \sum_{k \ne i} l'_k - k + 1)/2} \\ & \times \left(\frac{\prod_{a \le k-1} [l_a - l'_i]_q}{\prod_{a \le k, a \ne i} [l'_a - l'_i + 1]_q} \right)^{\frac{1}{2}}, \end{aligned} \tag{2.8a}$$

$$\begin{aligned} w_q^{(2)}\left(\begin{matrix} f'_1 & \cdots & f'_k \\ f_1 & \cdots & f_{k-1} \end{matrix} \middle| \begin{matrix} i \\ j \end{matrix} \right) &= S(j-i) q^{(l_j - l'_i)/2} \\ & \times \left(\prod_{a \le k, a \ne i} \frac{[l'_a - l_j + 1]_q}{[l'_a - l'_i + 1]_q} \prod_{a \le k-1, a \ne j} \frac{[l_a - l'_i]_q}{[l_a - l_j]_q} \right)^{\frac{1}{2}}. \end{aligned} \tag{2.8b}$$

Here $l_a = f_a - a$ $(1 \le a \le k-1)$, $l'_a = f'_a - a$ $(1 \le a \le k)$ and

$$\begin{aligned} S(x) &= 1 \qquad \text{if } x \ge 0, \\ &= -1 \quad\ \text{if } x < 0. \end{aligned}$$

Then we have

$$\begin{aligned} & w_q(m; i_n, i_{n-1}, \ldots, i_j) \\ & = w_q^{(1)}\left(\begin{matrix} m_{1j} & \cdots & m_{jj} \\ m_{1\,j-1} & \cdots & m_{j-1\,j-1} \end{matrix} \middle| \, i_j \right) \prod_{k=j+1}^{n} w_q^{(2)}\left(\begin{matrix} m_{1k} & \cdots & m_{kk} \\ m_{1\,k-1} & \cdots & m_{k-1\,k-1} \end{matrix} \middle| \begin{matrix} i_k \\ i_{k-1} \end{matrix} \right). \end{aligned}$$

If we consider $V \otimes V_Y$ instead of $V_Y \otimes V$, the same formula is valid provided we replace $q \to q^{-1}$.

2.5. Paths.

Let us define paths of Young diagrams inductively as follows: By definition any Young diagram is a path. If P is a path and $1 \leq \mu \leq n$, then the ordered pair (P, μ) or (μ, P) is also a path. We denote (P, μ) and (μ, P) by $(P) \underset{\mu}{\otimes} \Box$ and $\Box \underset{\mu}{\otimes} (P)$, respectively. We may abbreviate them to $P \underset{\mu}{\otimes} \Box$ and $\Box \underset{\mu}{\otimes} P$ when there is no notational ambiguities.

For a path P we define a vector space $V(P)$ and its vector $v(P)$ as follows: If $P = Y$ is a Young diagram, then $V(P) = V_Y$ and $v(P) = v_Y$ as given by (2.4). Now suppose that $V(P)(\neq \{0\})$ is an irreducible highest weight U_q–module such that $V(P) \simeq V_Y$, and suppose that $Y \underset{\mu}{\longrightarrow} W$. Then we define $V(P \underset{\mu}{\otimes} \Box)$ (resp. $V(\Box \underset{\mu}{\otimes} P)$) to be the component in $V(P) \otimes V$ (resp. $V \otimes V(P)$) that is isomorphic to V_W, and the highest weight vector $v(P \underset{\mu}{\otimes} \Box)$ (resp. $v(\Box \underset{\mu}{\otimes} P)$) by the formula (2.7) (resp. by (2.7) with q replaced by q^{-1}). If $V(P) = \{0\}$ or if there is no such Young diagram W that $Y \underset{\mu}{\longrightarrow} W$ then we define both $V(P \underset{\mu}{\otimes} \Box)$ and $V(\Box \underset{\mu}{\otimes} P)$ to be the zero space $\{0\}$, and $v(P \underset{\mu}{\otimes} \Box)$ and $v(\Box \underset{\mu}{\otimes} P)$ to be the zero vector. Note that $v((\Box) \underset{1}{\otimes} \Box) = v(\Box \underset{1}{\otimes} (\Box))$ but that $v((\Box) \underset{2}{\otimes} \Box) = -v(\Box \underset{2}{\otimes} (\Box))$.

A path P is called admissible if $V(P) \neq \{0\}$. A sequence of Young diagrams $\{Y_j(P)\}_{M(P) \leq j \leq N(P)}$ is associated with an admissible path P as follows. If $P = Y$, then $M(P) = N(P) = \sharp(Y)$ and $Y_N(P) = Y$ where $N = N(P)$. Suppose that $Q = P \underset{\mu}{\otimes} \Box$ or $\Box \underset{\mu}{\otimes} P$. Then $M(Q) = M(P)$, $N(Q) = N(P) + 1$, $Y_j(Q) = Y_j(P)\big(M(P) \leq j \leq N(P)\big)$ and $Y_{N+1}(Q)$ is such that $Y_N(Q) \underset{\mu}{\longrightarrow} Y_{N+1}(Q)$. We define three sequences of integers $\{\mu_j(P)\}_{M(P)+1 \leq j \leq N(P)}$, $\{\nu_j(P)\}_{M(P)+1 \leq j \leq N(P)}$, and $\{\kappa_j(P)\}_{M(P)+1 \leq j \leq N(P)}$ in such a way that $Y_j(P)$ is obtained from $Y_{j-1}(P)$ by adding a node on the $\mu_j(P)$–th row and the $\nu_j(P)$–th column, and $\kappa_j(P) = \nu_j(P) - \mu_j(P)$.

2.6. Limit of the Wigner coefficients.

Let us study the behavior of the reduced Wigner coefficients (2.8) in the limit $q^{\pm 1} \to 0$.

Lemma 2.2.

(i) As $q \to 0$ the reduced Wigner coefficients vanish except in the following cases.

$$\lim_{q\to 0} w_q^{(1)}\begin{pmatrix} f'_1 & \cdots & f'_k \\ f_1 \cdots f_{k-1} & & \end{pmatrix}\!\!\left|\begin{matrix} i \\ \ \end{matrix}\right) = 1 \quad \text{if} \quad f_a = f'_{a+1} \quad (i \le a < k),$$

$$\lim_{q\to 0} w_q^{(2)}\begin{pmatrix} f'_1 & \cdots & f'_k \\ f_1 \cdots f_{k-1} & & \end{pmatrix}\!\!\left|\begin{matrix} i \\ j \end{matrix}\right) = 1 \quad \text{if} \quad i < j \text{ and } f_a = f'_{a+1} \quad (i \le a < j),$$

$$= 1 \quad \text{if} \quad i = j.$$

(ii) As $q^{-1} \to 0$ the reduced Wigner coefficients vanish except in the following cases.

$$\lim_{q^{-1}\to 0} w_q^{(1)}\begin{pmatrix} f'_1 & \cdots & f'_k \\ f_1 \cdots f_{k-1} & & \end{pmatrix}\!\!\left|\begin{matrix} i \\ \ \end{matrix}\right) = 1 \quad \text{if} \quad i = 1,$$

$$\lim_{q^{-1}\to 0} w_q^{(2)}\begin{pmatrix} f'_1 & \cdots & f'_k \\ f_1 \cdots f_{k-1} & & \end{pmatrix}\!\!\left|\begin{matrix} i \\ j \end{matrix}\right) = 1 \quad \text{if} \quad i = j \text{ and } f_i = f'_i,$$

$$= -1 \quad \text{if} \quad i = j + 1.$$

Let W, Y and $|m\rangle$ be as in 2.4. From Lemma 2.2 one finds that at $q^{\pm 1} = 0$ all terms but one vanish in the right hand side of (2.11). The surviving term $|m; i_n, i_{n-1}, \cdots, i_j\rangle \otimes v_j$ is determined inductively by the following procedure. Set $i_n = \mu$. Suppose that $i_n, \cdots, i_k$ have been fixed.

In the case $q \to 0$, we proceed as follows (Fig.2.1a):

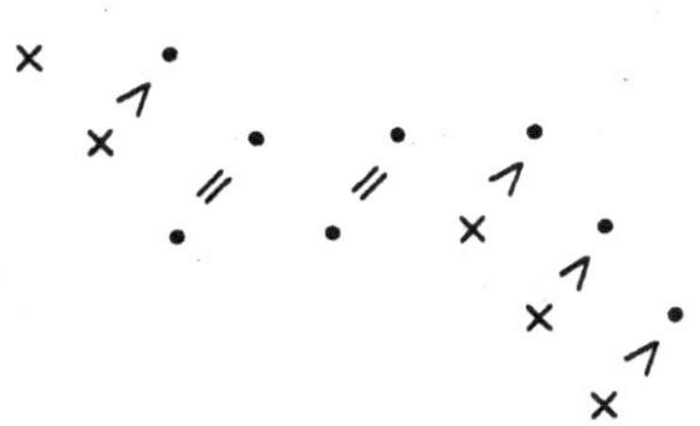

Figure 2.1a.

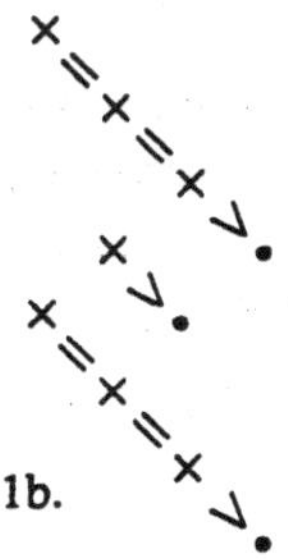

Figure 2.1b.

Fig.2.1. (GT) and $(GT)'$–procedures. Crosses represent the sites on which m_{ij} is decreased.

$(GT1)$ If $i_k = k$, then set $j = k$ and stop.

Suppose that $i_k < k$.

$(GT2)$ If $m_{i_k\, k-1} > m_{i_k+1\, k}$, then set $i_{k-1} = i_k$.

If $m_{i_k\, k-1} = m_{i_k+1\, k}$, then set $l = \max\{a | i_k \leq a \leq k-1, m_{a\, k-1} = m_{a+1\, k}\}$.

$(GT3)$ If $l \neq k-1$, then set $i_{k-1} = l+1$.

$(GT4)$ If $l = k-1$, then set $j = k$ and stop.

Likewise, in the case $q^{-1} \to 0$, we proceed as follows (Fig.2.1b).

$(GT1)'$ If $k = 1$, then set $j = 1$ and stop.

Suppose $k > 1$.

$(GT2)'$ If $m_{i_k k} = m_{i_k k-1}$, then set $i_{k-1} = i_k$.

$(GT3)'$ If $m_{i_k k} > m_{i_k k-1}$ and $i_k > 1$, then set $i_{k-1} = i_k - 1$.

$(GT4)'$ If $m_{i_k k} > m_{i_k k-1}$ and $i_k = 1$, then set $j = k$ and stop.

2.7. Insertion–deletion.

Let Y be a Young diagram with N nodes. For $S \in \mathcal{S}(Y)$ and $x \in \{1, \cdots, n\}$, we define new semi-standard tableaux with $N+1$ nodes $x \to S$ and $S \downarrow x$ [6].

Insertion $x \to S$.

(I1) If $y < x$ for all letters y on the first column, add x to the bottom of the first column.

(I2) Otherwise pick the upmost node that carries a letter $x_2 \geq x$. Replace x_2 on that node by x.

(I3) Do the same for x_2 and the second column, then proceed to the third column, etc..

Insertion $S \downarrow x$.

(I1)′ If $y \leq x$ for all letters y on the first row, add x to the right of the first row.

(I2)′ Otherwise pick the leftmost node that carries a letter $x_2 > x$. Replace x_2 on that node by x.

(I3)′ Do the same for x_2 and the second row, then proceed to the third row, etc..

Example:

$$1 \to \begin{matrix} 1 & 1 & 2 & 3 \\ 2 & 4 & & \end{matrix} = \begin{matrix} 1 & 1 & 1 & 2 & 3 \\ 2 & 4 & & & \end{matrix}.$$

$$\begin{matrix} 1 & 1 & 2 & 3 \\ 2 & 4 & & \end{matrix} \downarrow 2 = \begin{matrix} 1 & 1 & 2 & 2 \\ 2 & 3 & & \\ 4 & & & \end{matrix}.$$

The inverse procedure is described as follows. Suppose that $Y \underset{\mu}{\longrightarrow} W$ and that the node added to W is on the ν–th column. Take $R \in \mathcal{S}(W)$.

Deletion $\leftarrow_\nu R$.

(D1) Remove the letter x_ν on the bottom of the ν–column.

(D2) On the $(\nu-1)$-th column, pick the downmost node which carries a letter $x_{\nu-1} \le x_\nu$. Replace $x_{\nu-1}$ on that node by x_ν.

(D3) Repeat (D2) for the $(\nu-2)$-th column, and so on.

If $R = x \to S$ for some $S \in \mathcal{S}(Y)$ and x, then $S = \leftarrow_\nu R$. Conversely if $S = \leftarrow_\nu R$ and x is the letter thrown away from R, then $R = x \to S$.

Deletion $R \uparrow_\mu$.

(D1)′ Remove the letter x_μ on the rightend of the μ–th row.

(D2)′ On the $(\mu-1)$-th row, pick the rightmost node which carries a letter $x_{\mu-1} < x_\mu$. Replace $x_{\mu-1}$ on that node by x_μ.

(D3)′ Repeat (D2)′ for the $(\mu-2)$-th row, and so on.

If $R = S \downarrow x$ for some $S \in \mathcal{S}(Y)$ and x, then $S = R \uparrow_\mu$. Conversely if x is the letter thrown away from R, then $R = S \downarrow x$.

2.8. Interpretation of the $q^{\pm 1} \to 0$ limit by semi-standard tableaux.

We now translate the procedures in 2.6 in the language of semi-standard tableaux. Let W, Y and $|m\rangle$ be as in 2.4. We set $R = \tau(|m\rangle) = (r_{ij}) \in \mathcal{S}(W)$.

Lemma 2.3. *The procedure (GT1)-(GT4) (resp. (GT1)′-(GT4)′) for $|m\rangle$ is equivalent to the deletion $\leftarrow_\nu R$ (resp. $R \uparrow_\mu$).*

Proof. We prove that in the notation of (GT1)-(GT4) the process of decreasing $m_{i_k k}$ by one $(j \le k \le n, i_n = \mu)$ is equivalent to the deletion process (D1)-(D3).

Divide the interval $\{k \mid j \le k \le n\}$ into $I_{x_1} \cup I_{x_2} \cup \cdots \cup I_{x_q}$ in such a way that $x_1 > x_2 > \cdots > x_q$ and $I_{x_p} = \{k \mid m_{i_k k} = x_p\} \ne \phi$. Set $\lambda_p = i_k$ for $k \in I_{x_p}$. This is independent of $k \in I_{x_p}$. Note that $\lambda_1 \le \lambda_2 \le \cdots \le \lambda_q$. Fix $p (< q)$ and set $a = \min I_{x_{p-1}}, b = \min I_{x_p}$ and $c = \min I_{x_{p+1}}$. (If $p = 1$, then $a = \infty$.) We also set $\mu = \lambda_p$ and $\nu = \lambda_{p+1}$. Accordingly, from the μ-th to the ν-th column, R is as follows: $r_{ik} = b$ if $\mu \le i \le \nu$ and $m_{i\,b-1} + 1 \le k \le m_{ib}$, and $r_{\nu k} = c$ if $m_{\nu c} + 1 \le k \le m_{\nu\, b-1}$. Now it can be seen that decreasing $m_{i_k k}$ by one for I_{x_p} and $I_{x_{p+1}}$ is equivalent to the deletion process for the x-th column of R where $x_{p+1} \le x \le x_p$. In fact, $r_{\mu x_p}$ changes from b to a (if $a = \infty$, this means to remove the (μ, x_p)-th node from R), $r_{\nu x_{p+1}}$ changes from c to b and other $r_{\lambda x}$ for $x_{p+1} \le x \le x_p$ are unchanged. This proves the Lemma. ∎

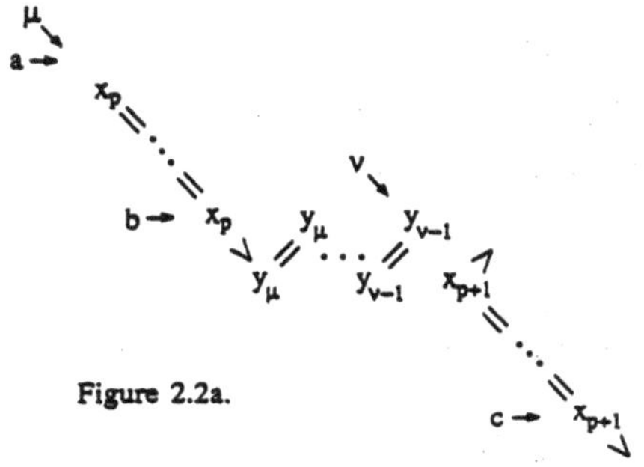

Figure 2.2a.

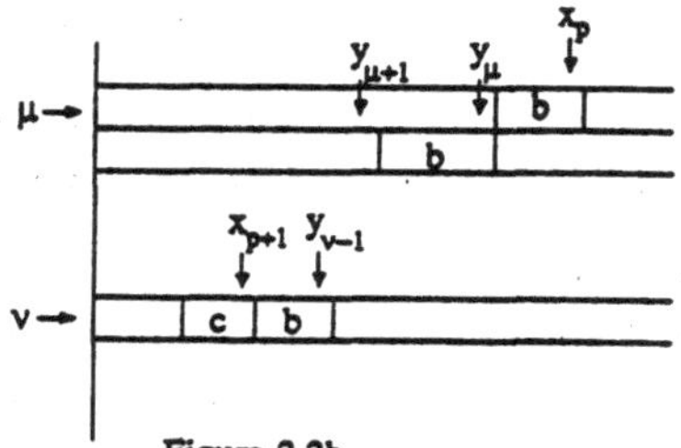

Figure 2.2b.

Fig.2.2. Correspondence between (GT) and (D)–procedures.

We conclude as follows.

Proposition 2.4. *Let $Y \underset{\mu}{\longrightarrow} W$, ν and $R \in \mathcal{S}(W)$ be as in 2.7, and set $|m\rangle = \tau^{-1}(R)$. Consider the limit $q^{\pm 1} \to 0$ in (2.11), and define*

$$|m'\rangle = \tau^{-1}(\leftarrow_\nu R) \quad \text{for} \quad q \to 0,$$
$$= \tau^{-1}(R \uparrow_\mu) \quad \text{for} \quad q^{-1} \to 0.$$

Suppose that the deletion from R throws away the letter j. Then

$$\lim_{q^{\pm 1} \to 0} |m\rangle = \lim_{q^{\pm 1} \to 0} (\pm 1)^{\mu - 1} |m'\rangle \otimes v_j.$$

2.9. Standard tableaux and irreducible decomposition of $V^{\otimes N}$.

Let us fix a positive integer N. Let $Y = [f_1, \cdots, f_n]$ $(f_1 \ge \cdots \ge f_n \ge 0)$ be a Young diagram with N nodes. An array of integers $T = (t_{ij})_{1 \le i \le n, 1 \le j \le f_i}$ is called a standard tableau of shape Y if

$$t_{ij} \in \{1, \cdots, N\}, \quad t_{ij} \ne t_{i'j'} \quad \text{if} \quad (i,j) \ne (i',j'),$$
$$t_{ij} < t_{i+1,j} \quad \text{and} \quad t_{ij} < t_{i,j+1},$$

for all i, j. Determine μ_k and ν_k by $t_{\mu_k \nu_k} = k$, and consider the paths

$$(\cdots((\phi \underset{\mu_1}{\otimes} \square) \underset{\mu_2}{\otimes} \square) \cdots \underset{\mu_N}{\otimes} \square$$

and

$$\square \underset{\mu_N}{\otimes} (\cdots \square \underset{\mu_2}{\otimes} (\square \underset{\mu_1}{\otimes} \phi) \cdots).$$

They are denoted by T_R and T_L, respectively.

Applying the decomposition rule (2.5) successively, we have

Lemma 2.5. *As a U_q-module*

$$V^{\otimes N} = \oplus_T V(T_R), \tag{2.9a}$$

$$= \oplus_T V(T_L), \tag{2.9b}$$

where T runs over the set of standard tableaux with N nodes and at most n rows.

2.10. The Robinson-Shensted correspondence.

Let $\mathcal{W} = \{1, 2, \cdots, n\}^N$. An element of $\mathcal{W}$ is a word $w = \lambda_1 \cdots \lambda_N$ on the letters $1, 2, \cdots, n$. Given a word $w \in \mathcal{W}$ we define a semi-standard tableau $S(w)$ by

$$S(w) = \lambda_N \rightarrow (\lambda_{N-1} \rightarrow \cdots (\lambda_1 \rightarrow \phi) \cdots),$$

ϕ being the empty tableau. Let μ_k be the number of the row where a node is added at the insertion $\lambda_k \rightarrow$. We define $T(w)$ to be the standard tableau obtained by adjoining successively the letter k at the right end of the μ_k-th row for $k = 1, \cdots, N$, starting with ϕ.

Example: If $w = 2\ 1\ 1\ 3\ 2$, then the insertion gives

$$2 \longrightarrow 1\ \ 2 \longrightarrow 1\ \ 1\ \ 2 \longrightarrow \begin{matrix} 1 & 1 & 2 \\ 3 & & \end{matrix} \longrightarrow \begin{matrix} 1 & 1 & 2 \\ 2 & 3 & \end{matrix},$$

hence

$$S(w) = \begin{matrix} 1 & 1 & 2 \\ 2 & 3 & \end{matrix}, \qquad T(w) = \begin{matrix} 1 & 2 & 3 \\ 4 & 5 & \end{matrix}.$$

Likewise we define $S'(w), T'(w)$ by using the insertion $S \downarrow x$ in place of $x \rightarrow S$. Obviously $S(w)$ and $T(w)$ (resp. $S'(w)$ and $T'(w)$) have the same shape.

Proposition 2.6. *([6]) The following maps are bijections.*

$$\begin{aligned} \mathcal{W} &\longrightarrow \amalg_Y \mathcal{S}(Y) \times \mathcal{T}(Y) \\ (\mathrm{RS}):\ w &\mapsto (S(w), T(w)) \\ (\mathrm{RS})':\ w &\mapsto (S'(w), T'(w)). \end{aligned}$$

Here Y runs over Young diagrams with N nodes and at most n rows.

The bijections (RS), (RS)$'$ are known as the Robinson-Shensted correspondence.

2.11. Decomposition of $V^{\otimes N}$ at $q^{\pm 1} = 0$.

Consider now the decomposition (2.9) of the tensor space $V^{\otimes N}$. Repeated application of Proposition 2.4 leads to the following description of the spaces $V(T_R), V(T_L)$ in the limit $q^{\pm 1} \to 0$.

Theorem 2.7. *In the limit $q \to 0$ (resp. $q^{-1} \to 0$) the $V(T_R)$ (resp. $V(T_L)$) are spanned by decomposable vectors*

$$v(w) = v_{\lambda_1} \otimes \cdots \otimes v_{\lambda_N}, \qquad w = \lambda_1 \cdots \lambda_N \in \mathcal{W}.$$

At $q = 0$, $v(w)$ belongs to $V(T_R)$ if and only if $T = T(w)$. At $q^{-1} = 0$, $v(w)$ belongs to $V(T_L)$ if and only if $T = T'(w')$, where $w' = \lambda_N \lambda_{N-1} \cdots \lambda_1$.

Suppose that $v = v_{\mu_1} \otimes \cdots \otimes v_{\mu_N} \otimes v_{\nu_1} \otimes \cdots \otimes v_{\nu_N}$ is contained in the $q \to 0$ limit of $V_{[N]} \otimes V_{[N]} \subset V^{\otimes 2N}$. Then it is immediate that $\mu_1 \geq \cdots \geq \mu_N$ and $\nu_1 \geq \cdots \geq \nu_N$. Since $V_{[N]} \otimes V_{[N]}$ decomposes as

$$V_{[N]} \otimes V_{[N]} = \oplus_{k=0}^{N} V_{[2N-k,k]},$$

the vector v in the limit belongs to $V_{[2N-k,k]}$ for some k. The following proposition determines this k in terms of $\mu_1, \ldots, \mu_N$ and $\nu_1, \ldots, \nu_N$.

Proposition 2.8. *Notations being as above, we have*

$$N - k = H_N(\mu_1, \ldots, \mu_N; \nu_1, \ldots, \nu_N)$$

where

$$H_N(\mu_1, \ldots, \mu_N; \nu_1, \ldots, \nu_N) = \min_{\sigma:\text{permutation}} \sum_{i=1}^{N} H(\mu_{\sigma(i)}, \nu_i). \tag{2.10}$$

Proof. Suppose that σ attains the minimum of $\sum_{i=1}^{N} H(\mu_{\sigma(i)}, \nu_i)$. Then, by changing σ if necessary, we can assume that

i) if $1 \le i \le k$ then $\sigma(i) = \min\{j | j > \sigma(i-1), \mu_j < \nu_i\}$ (Here $\sigma(0) = 0$.),

ii) $\{j | j > \sigma(k), \mu_j < \nu_{k+1}\} = \emptyset$.

To see i) and ii), first note that we can assume $\mu_{\sigma(i)} < \nu_i$ if and only if $1 \le i \le k$, and if $1 \le i < j \le k$ then $\sigma(i) < \sigma(j)$. Next determine $\sigma(i)$ succesively by i). Because the minimum is $N-k$, this process terminates at $i = k$, and we have ii). Now consider the process of insertions $\nu_1, \ldots, \nu_N$ in this order, starting from the tableau of shape $[N]$

$$\mu_N \ \mu_{N-1} \ \cdots \ \mu_1.$$

The conditions i),ii) imply that until the insertion of ν_k the node is added to the second row, while iii) implies that after ν_k the node is added to the first row. ∎

3. *R*–matrix in the scaling region

3.1. *R–matrix for symmetric tensor representations.*

We start from

$$R(u) = \sum_{\mu} [1+u]_q E_{\mu\mu} \otimes E_{\mu\mu} + \sum_{\mu<\nu} q^{u} E_{\mu\mu} \otimes E_{\nu\nu} \qquad (3.1)$$
$$+ \sum_{\mu>\nu} q^{-u} E_{\mu\mu} \otimes E_{\nu\nu} + \sum_{\mu\neq\nu} [u]_q E_{\mu\nu} \otimes E_{\nu\mu}.$$

This is the quantum R–matrix corresponding to the vector representation of $U_q = U_q(\mathfrak{gl}(n, \mathbf{C}))$ [7]. It belongs to $\mathrm{End}_{U_q}(V \otimes V)$, i.e.,

$$\left[R(u), \Delta(X)\right] = 0 \quad \text{for} \quad X \in U_q(\mathfrak{gl}(n, \mathbf{C})), \qquad (3.2)$$

and satisfies the Yang-Baxter equation

$$\left(R(u) \otimes 1\right)\left(1 \otimes R(u+v)\right)\left(R(v) \otimes 1\right)$$
$$= \left(1 \otimes R(v)\right)\left(R(u+v) \otimes 1\right)\left(1 \otimes R(u)\right).$$

From (3.1) we get the following scaling behavior

$$\lim_{\substack{u,q \to 0 \\ w=q^{-2u}: fixed}} \frac{R(u)}{[1-u]_q} = \sum_{\mu,\nu} w^{H(\mu,\nu)} E_{\mu\mu} \otimes E_{\nu\nu}, \qquad (3.3)$$

where $H(\mu,\nu)$ is given by (1.3).

Let us recall the fusion procedure [14],[15]. Consider the following operators in $\mathrm{End}_{U_q}(V^{\otimes N}\otimes V^{\otimes N})$.

$$R^{(i,i+1)}(u)=\underbrace{1\otimes\cdots\otimes 1}_{i-1}\otimes R(u)\otimes\underbrace{1\otimes\cdots\otimes 1}_{2N-i-1},$$

$$R^{(i,i+N)}_{[N],[1]}(u-i+1)\prod_{k=2}^{N}[u-i+k]$$
$$=R^{(i,i+1)}(u-i+1)R^{(i+1,i+2)}(u-i+2)\cdots R^{(N+i-1,N+i)}(u-i+N),$$
$$R^{(1,2N)}_{[N],[N]}(u)=R^{(N,2N)}_{[N],[1]}(u-N+1)\cdots R^{(2,N+2)}_{[N],[1]}(u-1)R^{(1,N+1)}_{[N],[1]}(u).$$

Consider the embedding of the N–th symmetric tensor product, $V_{[N]}\subset V^{\otimes N}$, as U_q–module. In fact, $R^{(1,2N)}_{[N],[N]}(u)$ preserves $V_{[N]}\otimes V_{[N]}$. We denote by $R_{[N],[N]}(u)$ the restriction of $R^{(1,2N)}_{[N],[N]}(u)$ to $V_{[N]}\otimes V_{[N]}$. It satisfies the Yang-Baxter equation. The aim of this section is to show

$$\lim_{\substack{u,q\to 0\\ w=q^{-2u}:fixed}}\frac{R_{[N],[N]}(u)}{\prod_{k=1}^{N}[k-u]_q} \tag{3.4}$$
$$=\sum_{\substack{\mu_1\geq\cdots\geq\mu_N\\ \nu_1\geq\cdots\geq\nu_N}} w^{H_N(\mu_1\cdots\mu_N,\nu\cdots\nu_N)}$$
$$\times\left(E_{\mu_1\mu_1}\otimes\cdots\otimes E_{\mu_N\mu_N}\right)\otimes\left(E_{\nu_1\nu_1}\otimes\cdots\otimes E_{\nu_N\nu_N}\right)$$

where H_N is given by (2.10).

Before discussing the general case let us explain our method in the simplest case (3.1). Let

$$V\otimes V=V_{[2]}\oplus V_{[1,1]}$$

be the irreducible decomposition of $V\otimes V$ with respect to the action of $U_q(\mathfrak{gl}(n,\mathbf{C}))$. Because of (3.2) $R(u)$ is a scalar on each component:

$$R(u)=[1+u]_qP_{[2]}+[1-u]_qP_{[1,1]}. \tag{3.5}$$

Here $P_{[2]}$ and $P_{[1,1]}$ denote the projections to $V_{[2]}$ and $V_{[1,1]}$, respectively. Since

$$\lim_{\substack{u,q\to 0\\ w=q^{-2u}:fixed}}\frac{[k+u]_q}{[k-u]_q}=w\quad(k=1,2,\cdots),$$

the scaling behavior (3.3) follows from (3.5) and (1.1) ($P_{[2]} = P_+, P_{[1,1]} = P_-$).

The generalization of (3.5) reads as (cf. [2])

$$R_{[N],[N]}(u) = \sum_{i=0}^{N} \prod_{j=1}^{N-i} [j+u]_q \prod_{j=N-i+1}^{N} [j-u]_q P_{[2N-i,i]}. \tag{3.6}$$

The formula (3.4) follows immediately from this decomposition and Proposition 2.8 which describes the $q \to 0$ limit of $\operatorname{Im} P_{[2N-i,i]} = V_{[2N-i,i]}$ in $V^{\otimes 2N}$.

The rest of this section is devoted to a proof of (3.6).

3.2. Characterization of the space $V(P)$.

Let $Y = [f_1, f_2, \ldots, f_n] = [l_1+1, l_2+2, \ldots, l_n+n]$ be a Young diagram with N nodes. It is immediate that

$$(1 \otimes R(u))v((Y \underset{\mu}{\otimes} \square) \underset{\mu}{\otimes} \square) = [1+u]_q v((Y \underset{\mu}{\otimes} \square) \underset{\mu}{\otimes} \square). \tag{3.7}$$

Proposition 3.1. *([13], see also [16]) If $\mu < \nu$ we have*

$$\begin{aligned}(1 \otimes R(u))\Big(v((Y \underset{\mu}{\otimes} \square) \underset{\nu}{\otimes} \square), v((Y \underset{\nu}{\otimes} \square) \underset{\mu}{\otimes} \square)\Big) \\ = \Big(v((Y \underset{\mu}{\otimes} \square) \underset{\nu}{\otimes} \square), v((Y \underset{\nu}{\otimes} \square) \underset{\mu}{\otimes} \square)\Big) \begin{pmatrix} \frac{[a-u]_q}{[a]_q} & [u]_q \frac{\sqrt{[a-1]_q[a+1]_q}}{[a]_q} \\ [u]_q \frac{\sqrt{[a-1]_q[a+1]_q}}{[a]_q} & \frac{[a+u]_q}{[a]_q} \end{pmatrix},\end{aligned} \tag{3.8}$$

where $a = l_\mu - l_\nu$.

Proof. We expand $v((Y \underset{\mu}{\otimes} \square) \underset{\nu}{\otimes} \square)$ and $v((Y \underset{\nu}{\otimes} \square) \underset{\mu}{\otimes} \square)$ in $V_Y \otimes V \otimes V$ by using (2.7), and compare the coefficients of $v(Y) \otimes v_\mu \otimes v_\nu$ and $v(Y) \otimes v_\nu \otimes v_\mu$ in both sides. Using that

$$\frac{w_q^{(2)}\left(\begin{matrix} f_1 \cdots f_\nu \\ f_1 \cdots f_{\nu-1} \end{matrix} \middle| \begin{matrix} \nu \\ \mu \end{matrix}\right) w_q^{(1)}\left(\begin{matrix} f_1 & \cdots & f_\nu - 1 \\ f_1 \cdot\cdot f_\mu - 1 \cdot\cdot f_{\nu-1} \end{matrix} \middle| \mu \right)}{w_q^{(1)}\left(\begin{matrix} f_1 \cdots f_\nu \\ f_1 \cdots f_{\nu-1} \end{matrix} \middle| \nu \right)} = -\frac{q^a}{[a]_q}, \tag{3.9}$$

$$\frac{w_q^{(1)}\left(\begin{matrix} f_1 \cdot\cdot f_\mu - 1 \cdot\cdot f_\nu \\ f_1 \cdot\cdot f_\mu - 1 \cdot\cdot f_{\nu-1} \end{matrix} \middle| \nu \right)}{w_q^{(1)}\left(\begin{matrix} f_1 \cdots f_\nu \\ f_1 \cdots f_{\nu-1} \end{matrix} \middle| \nu \right)} = \frac{\sqrt{[a-1]_q[a+1]_q}}{[a]_q},$$

we have (3.8). ∎

Fix $T \in \mathcal{T}(Y)$ and set $P = T_R$. We set $\mu_j = \mu_j(P), \nu_j = \nu_j(P)$ and $\kappa_j = \kappa_j(P) (1 \le j \le N)$. Assume that $P \underset{\mu}{\otimes} \square$ is admissible, and set $\kappa = \kappa_{N+1}(P \underset{\mu}{\otimes} \square) =$

$l_\mu + 1$. Suppose that $\gamma_N \underset{\text{def}}{=} \kappa_N - \kappa, \gamma_{N-1} \underset{\text{def}}{=} \kappa_{N-1} - \kappa, \ldots, \gamma_{s+1} \underset{\text{def}}{=} \kappa_{s+1} - \kappa$ are different from ± 1 and $\gamma_s \underset{\text{def}}{=} \kappa_s - \kappa = \pm 1$. Consider the following operator in $\mathrm{End}_{U_q}(V^{\otimes N} \otimes V)$.

$$G(P,\mu) = R^{(s,s+1)}(\gamma_s) R^{(s+1,s+2)}(\gamma_{s+1}) \cdots R^{(N,N+1)}(\gamma_N).$$

Lemma 3.2. *(cf. [15])*

$$V(P \underset{\mu}{\otimes} \square) = \{ v \in V(P) \otimes V \,|\, G(P,\mu)v = 0 \}.$$

Proof. First we show that

$$G(P,\mu)v(P \underset{\mu}{\otimes} \square) = 0. \tag{3.10}$$

Denote by T' the standard tableau obtained from T by removing the last node, i.e., the one on the μ_N-th row and the ν_N-th column. We write $P' = T'_R, \mu' = \mu_N$. Note that $P = P' \underset{\mu'}{\otimes} \square$. If $\mu = \mu'$, then $\gamma_N = -1$ and from (3.7) we have

$$R^{(N,N+1)}(\gamma_N) v((P' \underset{\mu}{\otimes} \square) \underset{\mu}{\otimes} \square) = 0.$$

If $\mu \neq \mu'$, from (3.8) we have

$$R^{(N,N+1)}(\gamma_N) v((P' \underset{\mu'}{\otimes} \square) \underset{\mu}{\otimes} \square) = \sqrt{[\gamma_N - 1]_q [\gamma_N + 1]_q}\, v((P' \underset{\mu}{\otimes} \square) \underset{\mu'}{\otimes} \square).$$

If $\gamma_N \neq \pm 1$, then $P' \underset{\mu}{\otimes} \square$ is admissible and by the induction hypothesis $G(P', \mu) v(P' \underset{\mu}{\otimes} \square) = 0$. Hence follows (3.10).

Now take $\lambda \neq \mu$ such that $P \underset{\lambda}{\otimes} \square$ is admissible. Set $\bar{\kappa} = \kappa_{N+1}(P \underset{\lambda}{\otimes} \square)$. Denote by $T^{(s)}$ the standard tableau obtained from T by removing the last $N - s$ nodes. Set $P^{(s)} = T^{(s)}_R$. Noting $\gamma_N \neq \kappa_N - \bar{\kappa}$ and $\gamma_j \neq \kappa_j - \kappa_{j+1}$ $(s \leq j \leq N-1)$ and keeping track of the coefficients of $v(((P^{(s)} \underset{\mu_{s+1}}{\otimes} \square) \cdots) \underset{\mu_N}{\otimes} \square) \underset{\lambda}{\otimes} \square)$ we have

$$\begin{aligned} &G(P,\mu) v(P \underset{\lambda}{\otimes} \square) \\ &= \left(\prod_{j=s}^{N-1} \frac{[\kappa_j - \kappa_{j+1} - \gamma_j]_q}{[\kappa_j - \kappa_{j+1}]_q} \right) \frac{[\kappa_N - \bar{\kappa} - \gamma_N]_q}{[\kappa_N - \bar{\kappa}]_q} \\ &\qquad \times v(((P^{(s)} \underset{\mu_{s+1}}{\otimes} \square) \cdots) \underset{\mu_N}{\otimes} \square) \underset{\lambda}{\otimes} \square) + \cdots \\ &\neq 0. \end{aligned}$$

This completes the proof of Lemma 3.2 ∎

3.3. Spectral decomposition: The case of $V_Y \otimes V_{[1]}$.

Using the Yang-Baxter equation and the above lemma one can show that the following operator belongs to $\mathrm{Hom}_{U_q}(V(P) \otimes V, V \otimes V(P))$:

$$R_{P,[1]}(u) = \frac{R^{(1,2)}(u+\kappa_1)R^{(2,3)}(u+\kappa_2)\cdots R^{(N,N+1)}(u+\kappa_N)}{\prod_{k=2}^{N}[u+\kappa_k]_q}.$$

Proposition 3.3. *(cf. [14])*

$$R_{P,[1]}(u)v(P\underset{\mu}{\otimes}\square) = [u+\kappa_{N+1}(P\underset{\mu}{\otimes}\square)]_q v(\square\underset{\mu}{\otimes}P). \tag{3.11}$$

Proof. We use an induction on N. Define κ, P' and μ' as in 3.2. We also set $\kappa' = \kappa_N$.

First assume that $\mu < \mu'$. Applying (2.7) we have

$$\begin{aligned} &v((P'\underset{\mu'}{\otimes}\square)\underset{\mu}{\otimes}\square) \\ &= w_q^{(1)}\begin{pmatrix} f_1 & \cdots & f_\mu+1 \\ & f_1\cdots f_{\mu-1} & \end{pmatrix}\Big|\mu\Big) w_q^{(1)}\begin{pmatrix} f_1 & \cdots & f_{\mu'} \\ f_1\cdots f_{\mu'-1} & & \end{pmatrix}\Big|\mu'\Big) v(P')\otimes v_{\mu'}\otimes v_\mu + \cdots \end{aligned}$$

where the terms $+\cdots$ contain neither elements in $V(P')\otimes v_{\mu'}\otimes v_\mu$ nor in $V(P')\otimes v_\mu \otimes v_{\mu'}$. Therefore we have

$$\begin{aligned} v' &= R^{(N,N+1)}(u+\kappa')v((P'\underset{\mu'}{\otimes}\square)\underset{\mu}{\otimes}\square) \\ &= w_q^{(1)}\begin{pmatrix} f_1 & \cdots & f_\mu+1 \\ & f_1\cdots f_{\mu-1} & \end{pmatrix}\Big|\mu\Big) w_q^{(1)}\begin{pmatrix} f_1 & \cdots & f_{\mu'} \\ f_1\cdots f_{\mu'-1} & & \end{pmatrix}\Big|\mu'\Big) [u+\kappa']_q \\ &\quad \times v(P')\otimes v_\mu \otimes v_{\mu'} + \cdots \end{aligned} \tag{3.12}$$

where the terms $+\cdots$ contain no elements in $V(P')\otimes v_\mu\otimes v_{\mu'}$. Suppose that

$$v' = w \otimes v_{\mu'} + \cdots$$

where the terms $+\cdots$ contain no elements in $V(P'\underset{\mu}{\otimes}\square)\otimes v_{\mu'}$. Since v' is a highest weight vector and contains no elements in $V(P'\underset{\mu}{\otimes}\square)\otimes v_{\mu'+1}$, the vector w is also a highest weight vector, i.e., a constant multiple of $v(P' \underset{\mu}{\otimes} \square)$. The constant is determined from (3.12):

$$v' = w_q^{(1)}\begin{pmatrix} f_1 & \cdots f_{\mu'} \\ f_1 \cdots f_{\mu'-1} & \end{pmatrix}\!\!\left|\mu'\right)[u+\kappa']_q v(P' \underset{\mu}{\otimes} \square) \otimes v_{\mu'} + \cdots .$$

Using the induction hypothesis we obtain

$$\begin{aligned} &R_{P,[1]}(u) v((P' \underset{\mu'}{\otimes} \square) \underset{\mu}{\otimes} \square) \\ &\quad = w_q^{(1)}\left(\begin{matrix} f_1 & \cdots & f_{\mu'} \\ f_1 \cdots f_{\mu'-1} \end{matrix}\middle| \mu'\right) w_{q^{-1}}^{(1)}\left(\begin{matrix} f_1 \cdots f_\mu + 1 \\ f_1 \cdots f_{\mu-1} \end{matrix}\middle| \mu\right)[u + \kappa_N(P' \underset{\mu}{\otimes} \square)]_q \\ &\qquad \times v_\mu \otimes v(P') \otimes v_{\mu'} + \cdots . \end{aligned}$$

Noting that

$$\begin{aligned} v(\square \underset{\mu}{\otimes} (P' \underset{\mu'}{\otimes} \square)) = w_q^{(1)}&\left(\begin{matrix} f_1 & \cdots & f_{\mu'} \\ f_1 \cdots f_{\mu'-1} \end{matrix}\middle| \mu'\right) w_{q^{-1}}^{(1)}\left(\begin{matrix} f_1 \cdots f_\mu + 1 \\ f_1 \cdots f_{\mu-1} \end{matrix}\middle| \mu\right) \\ &\times v_\mu \otimes v(P') \otimes v_{\mu'} + \cdots , \end{aligned}$$

and that $\kappa_N(P' \underset{\mu}{\otimes} \square) = \kappa_{N+1}(P \underset{\mu}{\otimes} \square)$ we have (3.11).

If $\mu = \mu'$ a similar argument shows

$$\begin{aligned} &R^{(N,N+1)}(u+\kappa') v((P' \underset{\mu}{\otimes} \square) \underset{\mu}{\otimes} \square) \\ &\quad = w_q^{(1)}\left(\begin{matrix} f_1 \cdots f_\mu + 1 \\ f_1 \cdots f_{\mu-1} \end{matrix}\middle| \mu\right)[u+\kappa'+1]_q v(P' \underset{\mu}{\otimes} \square) \otimes v_\mu + \cdots , \end{aligned}$$

and

$$\begin{aligned} &R_{P,[1]}(u) v((P' \underset{\mu}{\otimes} \square) \underset{\mu}{\otimes} \square) \\ &\quad = w_q^{(1)}\left(\begin{matrix} f_1 \cdots f_\mu + 1 \\ f_1 \cdots f_{\mu-1} \end{matrix}\middle| \mu\right) w_{q^{-1}}^{(1)}\left(\begin{matrix} f_1 & \cdots f_\mu \\ f_1 \cdots f_{\mu-1} \end{matrix}\middle| \mu\right)[u+\kappa'+1]_q \\ &\qquad \times v_\mu \otimes v(P') \otimes v_\mu + \cdots . \end{aligned}$$

Comparing this with $v(\square \underset{\mu}{\otimes} (P' \underset{\mu}{\otimes} \square))$ we have (3.11).

If $\mu > \mu'$ we have

$$\begin{aligned} R^{(N,N+1)}(u+\kappa') v((P' \underset{\mu'}{\otimes} \square) \underset{\mu}{\otimes} \square) &= \Big\{ w_q^{(1)}\left(\begin{matrix} f_1 \cdots f_\mu + 1 \\ f_1 \cdots f_{\mu-1} \end{matrix}\middle| \mu\right) q^{u+\kappa'} \\ &+ w_q^{(2)}\left(\begin{matrix} f_1 \cdots f_\mu + 1 \\ f_1 \cdots f_{\mu-1} \end{matrix}\middle| \begin{matrix} \mu \\ \mu' \end{matrix}\right) w_q^{(1)}\left(\begin{matrix} f_1 & \cdots & f_\mu \\ f_1 \cdot\cdot f_{\mu'} - 1 \cdot\cdot f_{\mu-1} \end{matrix}\middle| \mu'\right)[u+\kappa']_q \Big\} \\ &\times v(P' \underset{\mu'}{\otimes} \square) \otimes v_\mu + \cdots \end{aligned}$$

Using (3.9) (with the substitutions $\nu \to \mu$, $\mu \to \mu'$ and $f_\nu \to f_\mu + 1$) and

$$q^{u+\kappa'} - \frac{q^{\kappa'-\kappa-1}}{[\kappa'-\kappa-1]_q}[u+\kappa']_q = -\frac{[u+\kappa+1]_q}{[\kappa'-\kappa-1]_q},$$

we have

$$\begin{aligned} &R_{P,[1]}(u)v\big((P' \underset{\mu'}{\otimes} \Box) \underset{\mu}{\otimes} \Box\big) \\ &= -w^{(1)}_{q^{-1}}\left(\begin{matrix} f_1 \cdots f_{\mu'} \\ f_1 \cdots f_{\mu'-1} \end{matrix}\middle| \mu'\right) w^{(1)}_q \left(\begin{matrix} f_1 \cdots f_\mu + 1 \\ f_1 \cdots f_{\mu-1} \end{matrix}\middle| \mu\right) \frac{[u+\kappa]_q}{[\kappa'-\kappa-1]_q} \\ &\quad \times v_{\mu'} \otimes v(P') \otimes v_\mu + \cdots . \end{aligned}$$

Comparing this with $v(\Box \underset{\mu}{\otimes} (P' \underset{\mu'}{\otimes} \Box))$ we have (3.11). ∎

3.4. Spectral decomposition: The case of $V_Y \otimes V_{[k]}$.

Define

$$R_{P,[k]} \in \mathrm{Hom}_{U_q}\big(V(P) \otimes V_{[k]}, V_{[k]} \otimes V(P)\big).$$

inductively by

$$R_{P,[k]} = \big(1 \otimes R_{P,[k-1]}(u-1)\big)\big(R_{P,[1]}(u) \otimes \underbrace{1 \otimes \cdots \otimes 1}_{k-1}\big).$$

Let

$$V(P) \otimes V_{[k]} = \oplus_W V_W$$

be the irreducible decomposition of $V(P) \otimes V_{[k]}$. The following proposition gives a description of V_W in $V(P) \otimes V^{\otimes k}$.

Proposition 3.4. *Fix $s = (s_1, \ldots, s_n)$ such that $s_\mu \geq 0$, $\sum s_\mu = k$. Consider*

$$v_{P,s} = \sum_{(\mu_1,\ldots,\mu_k) \in I(P,s)} c_{\mu_1,\ldots,\mu_k} v\big((\cdots(P \underset{\mu_1}{\otimes} \Box) \underset{\mu_2}{\otimes} \Box \cdots) \underset{\mu_k}{\otimes} \Box\big)$$

where

$$I(P,s) = \Big\{(\mu_1,\ldots,\mu_k) \Big| \ (\cdots(P \underset{\mu_1}{\otimes} \Box) \underset{\mu_2}{\otimes} \Box \cdots) \underset{\mu_k}{\otimes} \Box \text{ is admissible} \\ \text{and } \textstyle\sum_{i=1}^k \delta_{\mu_i \mu} = s_\mu \text{ for } \mu = 1, \ldots, n\Big\}.$$

The highest weight vector $v_{P,s}$ belongs to $V(P)\otimes V_{[k]}$ if and only if the following set of linear equations are satisfied:

$$c_{\mu_1,\ldots,\mu_{j-1},\mu,\nu,\mu_{j+2},\ldots,\mu_k}\sqrt{[a_j+1]_q}$$
$$= c_{\mu_1,\ldots,\mu_{j-1},\nu,\mu,\mu_{j+2},\ldots,\mu_k}\sqrt{[a_j-1]_q}, \tag{3.13}$$

where $a_j = l_\mu - l_\nu + \sum_{i=1}^{j-1}(\delta_{\mu_i\mu} - \delta_{\mu_i\nu})$.

Proof. Recall Lemma 3.2. The k–th symmetric tensor product $V_{[k]}$ in $V^{\otimes k}$ is contained in $V_{[k-1]}\otimes V$ and is characterized by $R^{(k-1,k)}(-1)V_{[k]} = 0$. This is nothing but (3.13) with $j = k-1$. Arguing inductively we get (3.13). ∎

Since the multiplicity of V_W in $V(P)\otimes V_{[k]}$ is 1, these equations uniquely determine $\{c_{\mu_1,\ldots,\mu_k}\}$ up to constant multiple. If we compute

$$R_{P,[2]}(u) = \big(1\otimes R_{P,[1]}(u-1)\big)\big(R_{P,[1]}(u)\otimes 1\big)$$

by using Lemma 3.2, we need to know the matrix of the transformation of two different bases of $V\otimes V(P)\otimes V$:

$$\begin{array}{l}(V(P)\otimes V)\otimes V\\ \qquad\downarrow\ R_{P,[1]}(u)\otimes 1\\ (V\otimes V(P))\otimes V\\ \qquad\downarrow\ id\\ V\otimes(V(P)\otimes V)\\ \qquad\downarrow\ 1\otimes R_{P,[1]}(u-1)\\ V\otimes(V\otimes V(P))\end{array}$$

From (2.8) it is immediate that

$$v((\square\underset{\mu}{\otimes}P)\underset{\mu}{\otimes}\square) = v(\square\underset{\mu}{\otimes}(P\underset{\mu}{\otimes}\square)).$$

Lemma 3.5. *For a given μ,ν such that $\mu\neq\nu$ set $a = l_\mu - l_\nu$. Then we have*

$$\Big(v((\square\underset{\mu}{\otimes}P)\underset{\nu}{\otimes}\square), v((\square\underset{\nu}{\otimes}P)\underset{\mu}{\otimes}\square)\Big)$$
$$= \Big(v(\square\underset{\nu}{\otimes}(P\underset{\mu}{\otimes}\square)), v(\square\underset{\mu}{\otimes}(P\underset{\nu}{\otimes}\square))\Big)\begin{pmatrix}\frac{-1}{[a]_q} & \frac{\sqrt{[a-1]_q[a+1]_q}}{[a]_q}\\ \frac{\sqrt{[a-1]_q[a+1]_q}}{[a]_q} & \frac{1}{[a]_q}\end{pmatrix}.$$

The proof is much the same as Proposition 3.1.

Note that

$$\sqrt{[a-1]_q}\, v\big((P \underset{\mu}{\otimes} \square) \underset{\nu}{\otimes} \square\big) + \sqrt{[a+1]_q}\, v\big((P \underset{\nu}{\otimes} \square) \underset{\mu}{\otimes} \square\big)$$

belongs to $V(P) \otimes V_{[2]}$, and also that

$$(R_{P,[1]}(u) \otimes 1) v\big((P \underset{\mu}{\otimes} \square) \underset{\nu}{\otimes} \square\big) = [u + l_\mu + 1]_q v\big((\square \underset{\mu}{\otimes} P) \underset{\nu}{\otimes} \square\big). \tag{3.14}$$

From Lemma 3.5 we have

Lemma 3.6.

$$\begin{aligned} &[u + l_\mu - 1]_q \sqrt{[a-1]_q}\, v\big((\square \underset{\mu}{\otimes} P) \underset{\nu}{\otimes} \square\big) \\ &\qquad + [u + l_\nu + 1]_q \sqrt{[a+1]_q}\, v\big((\square \underset{\nu}{\otimes} P) \underset{\mu}{\otimes} \square\big) \\ &= [u + l_\nu]_q \sqrt{[a-1]_q}\, v\big(\square \underset{\nu}{\otimes} (P \underset{\mu}{\otimes} \square)\big) + [u + l_\mu]_q \sqrt{[a+1]_q}\, v\big(\square \underset{\mu}{\otimes} (P \underset{\nu}{\otimes} \square)\big). \end{aligned} \tag{3.15}$$

Now we state the main result in this section:

Theorem 3.7. *Suppose that $c_{\mu_1,\ldots,\mu_k}$ satisfies (3.13). Then we have*

$$\begin{aligned} &R_{P,[k]}(u) \sum_{(\mu_1,\ldots,\mu_k)\in I(P,s)} c_{\mu_1,\ldots,\mu_k} v\big((\cdots(P \underset{\mu_1}{\otimes} \square) \underset{\mu_2}{\otimes} \square \cdots) \underset{\mu_k}{\otimes} \square\big) \\ &= \prod_{\mu=1}^{n} [u + l_\mu + 2 - k]_q [u + l_\mu + 3 - k]_q \cdots [u + l_\mu + 1 - k + s_\mu]_q \\ &\quad \times \sum_{(\mu_1,\ldots,\mu_k)\in I(P,s)} c_{\mu_1,\ldots,\mu_k} v\big(\square \underset{\mu_k}{\otimes} (\cdots \square \underset{\mu_2}{\otimes} (\square \underset{\mu_1}{\otimes} P) \cdots)\big). \end{aligned} \tag{3.16}$$

Proof. First note that for any $(\mu_1, \cdots, \mu_k) \in I(P, s)$

$$\begin{aligned} &\prod_{\mu=1}^{n} [u + l_\mu + 2 - k]_q [u + l_\mu + 3 - k]_q \cdots [u + l_\mu + 1 - k + s_\mu]_q \\ &= \prod_{j=1}^{k} [u + \kappa_{N+j}((\cdots(P \underset{\mu_1}{\otimes} \square) \underset{\mu_2}{\otimes} \square \cdots) \underset{\mu_j}{\otimes} \square) - k + 1]_q. \end{aligned} \tag{3.17}$$

Using (3.14) and (3.15) we have

$$(R_{P,[1]}(u)\otimes 1)\sum_{(\mu_1,\ldots,\mu_k)\in I(P,s)} c_{\mu_1,\ldots,\mu_k} v((\cdots(P\underset{\mu_1}{\otimes}\square)\underset{\mu_2}{\otimes}\square\cdots)\underset{\mu_k}{\otimes}\square)$$

$$=\sum_{(\mu_1,\ldots,\mu_k)\in I(P,s)} c_{\mu_1,\ldots,\mu_k}[u+\kappa_{N+1}(P\underset{\mu_1}{\otimes}\square)]_q v((\cdots(\square\underset{\mu_1}{\otimes}P)\underset{\mu_2}{\otimes}\square\cdots)\underset{\mu_k}{\otimes}\square)$$

$$=\sum_{(\mu_1,\ldots,\mu_k)\in I(P,s)} c_{\mu_1,\ldots,\mu_k}[u+\kappa_{N+2}((P\underset{\mu_1}{\otimes}\square)\underset{\mu_2}{\otimes}\square)-1]_q$$

$$\times\, v((\cdots\square\underset{\mu_2}{\otimes}(P\underset{\mu_1}{\otimes}\square)\cdots)\underset{\mu_k}{\otimes}\square)$$

$$\cdots$$

$$=\sum_{(\mu_1,\ldots,\mu_k)\in I(P,s)} c_{\mu_1,\ldots,\mu_k}[u+\kappa_{N+k}((\cdots(P\underset{\mu_1}{\otimes}\square)\underset{\mu_2}{\otimes}\square\cdots)\underset{\mu_k}{\otimes}\square)-k+1]_q$$

$$\times\, v(\square\underset{\mu_k}{\otimes}(\cdots(P\underset{\mu_1}{\otimes}\square)\underset{\mu_2}{\otimes}\square\cdots)).$$

By induction and by using (3.17) we have (3.16). ∎

Since the multiplicity of $V_{[N]}$ in $V^{\otimes N}$ is one, (3.6) follows from (3.16). (The constant multiple is determined by evaluating $R_{[N],[N]}(u)$ at $\underbrace{v_1\otimes\cdots\otimes v_1}_{2N}$.)

References.

[1] Drinfeld, V. G., "Quantum groups", ICM proceedings, Berkeley, 798–820, 1986.

[2] Jimbo, M., "A q–difference analogue of $U(\mathfrak{g})$ and the Yang-Baxter equation", *Lett. Math. Phys.* **10**, 63–69 (1985).

[3] Lusztig, G., "Quantum deformation of certain simple modules over enveloping algebras", *Advances in Math.* **70**, 237–249 (1988).

[4] Rosso, M., "Finite dimensional representations of the quantum analog of the enveloping algebra of a complex simple Lie algebra", *Commun. Math. Phys.* **117**, 581–593 (1988).

[5] Lusztig, G., "Modular representations and quantum groups", preprint 1988.

[6] Shensted, C., "Longest increasing and decreasing subsequences", *Canad. Jour. Math.* **13**, 179–191 (1961).

[7] Jimbo, M., "A q–analogue of $U(\mathfrak{gl}(N+1))$, Hecke algebra and the Yang-Baxter equation", *Lett. Math. Phys.* **11**, 247–252 (1986).

[8] Baxter, R. J., "Exactly solved models in statistical mechanics", Academic, London 1982.

[9] Date, E., Jimbo, M., Kuniba, A., Miwa, T. and Okado, M., "One dimensional configuration sums in vertex models and affine Lie algebra characters", to appear in *Lett. Math. Phys.*.

[10] Kac, V. G. and Peterson, D. H., "Infinite dimensional Lie algebras, theta functions and modular forms", *Advances in Math.* **53**, 125–264 (1984).

[11] Date, E., Jimbo, M., Kuniba, A., Miwa, T. and Okado, M., "Chemins, diagrammes de Maya et représentations de $\widehat{\mathfrak{sl}}(r,\mathbf{C})$", *C. R. Acad. Sci. Paris*, **t.308**, 129–132 (1989); "Paths, Maya diagrams and representations of $\widehat{\mathfrak{sl}}(r,\mathbf{C})$", RIMS preprint **642** (1988), to appear in *Adv. Stud. Pure Math.* **19**.

[12] Jimbo, M., "Quantum R–matrix related to the generalized Toda system: an algebraic approach", Lecture Notes in Phys. **246**, 335–361, Springer 1986.

[13] Pasquier, V., "Etiology of IRF models", *Commun. Math. Phys.* **118**, 335–364 (1988).

[14] Kulish, P. P., Reshetikhin, N. Yu. and Sklyanin, E. K., "Yang-Baxter equation and representation theory. I", *Lett. Math. Phys.* **5**, 393–403 (1981).

[15] Cherednik, I. V., "On special bases of irreducible finite–dimensional representations of the degenerated affine Hecke algebra", *Funct. Anal. and Appl.* **20**, 87–89 (1986).

[16] Jimbo, M., Miwa, T. and Okado, M., "Local state probabilities of solvable lattice models. An $A^{(1)}_{n-1}$ family", *Nucl. Phys.* **B300**[FS22], 74–108 (1988).

THE SEWING TECHNIQUE AND CORRELATION FUNCTIONS ON ARBITRARY RIEMANN SURFACES

P. Di Vecchia

NORDITA, Blegdamsvej 17, DK-2100 Copenhagen Ø, Denmark

Abstract

We describe in the case of free bosonic and fermionic theories the sewing procedure, that is a very convenient way for constructing correlation functions of these theories on an arbitrary Riemann surface from their knowledge on the sphere. The fundamental object that results from this construction is the N-point g-loop vertex. It summarizes the information of all correlation functions of the theory on an arbitrary Riemann surface. We then check explicitly the bosonization rules and derive some useful formulas.

1 Introduction

Several methods have been recently proposed for computing correlation functions of conformal field theories on arbitrary Riemann surfaces[1]. In this paper we discuss in some detail the oldest one that is based on the sewing technique and we apply it to various conformal theories. This formalism has been recently used [3] to compute correlation functions of minimal theories on arbitrary Riemann surfaces. Most of the results listed here are contained in the two papers [1,2] in collaboration with M. Frau, K. Hornfeck, A. Lerda, F. Pezzella and S. Sciuto.

The sewing procedure was introduced in the early days of dual theories as a way for computing higher order corrections to the original Veneziano model in order to implement unitarity in a perturbative way in a model with zero width resonances. The Veneziano model was already regarded in those days as the tree approximation of a yet unknown theory and the higher order corrections were constructed sewing external legs together after the insertion of a propagator as one constructs loop diagrams from trees in a quantum field theory. The main new ingredient with respect to a quantum field theory was the presence of the very important property of duality, that must be preserved in the sewing procedure.

Only later on, it was recognized that a stringlike structure was the one underlying the dual theories and that a complete quantum mechanical treatment implied a sum over all possible topologies. The contribution of the lowest topology (sphere with genus equal to zero) gives rise to the original Veneziano model, while that of higher topologies corresponds to the multiloop diagrams, that are necessary to have a unitary theory.

After taking into account that only physical states propagate in the loops or, using a covariant procedure, that the contribution of the unphysical degrees of freedom cancel exactly against the one of the ghosts it has been by now recognized that multiloop amplitudes coming out from the calculation of the functional integral on an arbitrary

[1]For a list of those methods and their references see [1]

Riemann surface are coincident with the ones obtained by applying the old sewing procedure [4].

Such a technique has the advantage with respect to other approaches to give an explicit expression for the measure of integration on moduli space for multiloop amplitudes in string theories. Only the integration region on moduli space must be fixed for the time being by hand.

It has recently been proposed [5,1,2] to use the sewing procedure also for computing correlation functions of conformal invariant quantum field theories on arbitrary Riemann surfaces.

The starting point of the sewing construction is the N-point vertex on the sphere, that has the important property of summarizing all possible correlation functions of the theory on the sphere. It is a kind of "generating functional" for the correlation functions involving N primary fields of the theory on the sphere. They can in fact be obtained by saturating the N-point vertex $V_{N;0}$ with N primary fields according to the formula:

$$< 0|\prod_{i=1}^{N}\varphi_i(z_i)|0 >_{sphere}= V_{N;0}\prod_{i=1}^{N}|\Phi_i >_i \tag{1.1}$$

$V_{N;0}$ is a bra-like operator acting in the direct product of N different Hilbert spaces corresponding to the N external legs and $|\Phi_i >_i$ is a highest weight state corresponding to the transported primary field $\Phi_i(z_i) \equiv [V_i^{-1\prime}(0)]^{\Delta_i}\varphi[V_i^{-1}(0)](\Delta_i$ is the conformal dimension of φ). For the sake of simplicity we consider only the dependence on the left variable z omitting its complex conjugate $\bar{z}$.

Once $V_{N;0}$ for a certain conformal field theory is known, the sewing procedure amounts to compute in a straightforward way $V_{N;g}$ that is the "generating functional" for all the correlation functions on a genus g Riemann surface.

$V_{N;g}$ is given by

$$V_{N;g} = \prod_{\mu=1}^{g} Tr_{(2\mu-1,2\mu)}\left[V^{\dagger}_{(N+2g);0}\prod_{\mu=1}^{g}P(x_\mu)\right] \tag{1.2}$$

where for notational simplicity we label the first N legs of the vertex in the r.h.s. with an index i, running from 1 to N, and divide the remaining $2g$ legs into "odd" legs, labelled by $2\mu - 1$, and "even" legs labelled by 2μ with $\mu = 1, \ldots, g$. The expression (1.2) means that we sum over all states circulating in the loop by taking the trace in the Hilbert spaces $2\mu - 1$ and 2μ, that are identified after the insertion of a sewing operator $P(x_\mu)$.

A correlation function on a genus g Riemann surface with N primary fields $|\Phi_i>$ is obtained from $V_{N;g}$ exactly as the one on the sphere is obtained from $V_{N;0}$. It is given by:

$$< 0| \prod_{i=1}^{N} \varphi_i(z_i)|0>_g = V_{N;g} \prod_{i=1}^{N} |\Phi_i>_i \tag{1.3}$$

In the next two sections we give the result of the sewing procedure for some particular free bosonic and fermionic conformal theories. Finally in the last section we check the bosonization rules on an arbitrary Riemann surface obtaining some interesting formulas.

2 Free Bosonic Theory

A free bosonic scalar field φ coupled to a vacuum charge Q is described by the following action

$$S[\varphi] = \frac{1}{2\pi} \int_\Sigma d^2z[-\bar{\partial}\varphi\partial\varphi - \frac{1}{4} Q\sqrt{g}R^{(2)}\varphi] \tag{2.1}$$

where g and $R^{(2)}$ are respectively, the determinant of the metric and the scalar curvature of the two dimensional world-sheet Σ, on which the theory is defined.

The starting point of our construction is the N-point vertex on the sphere, that in the case of a free scalar theory is given by :

$$V_{N;0} = \prod_{i=1}^{N} [\sum_{n_i} {}_i\!< n_i, O_a||\exp\left[-\frac{1}{2}\sum_{\substack{i,j=1\\ i\neq j}}^{N}\sum_{n,m=0}^{\infty} a_n^{(i)} D_{nm}(U_i V_j) a_m^{(j)}\right] \delta(\sum_{i=1}^{N} N_i + Q) \tag{2.2}$$

It contains N infinite sets of harmonic oscillators $a_n^{(i)}$, that are related to the oscillators

$\alpha_n^{(i)}$ usually used in the literature by:

$$a_0 = \alpha_0 \equiv -N \qquad \alpha_n = a_n\sqrt{n} \quad n > 0 \tag{2.3}$$

The scalar field φ has the following oscillator expansion:

$$\varphi(z) = x + N\log z + \sum_{n\neq 0}\frac{\alpha_n}{n}z^{-n} \tag{2.4}$$

The state ${}_i\!<n_i, O_a|$ is dual to $|O_a, n_i>_i$, which is annihilated by all the a_n's with $n > 0$ and is an eigenstate of N_i with eigenvalue n_i. For the sake of simplicity we assume here that the sum over n_i runs over integers for odd Q and over integers and half integers for even Q. However other choices are possible as shown in the last section of [1]. V_i and $U_i \equiv \Gamma V_i^{-1}$, with $\Gamma(z) = 1/z$ are transformations related to the choice of local coordinates around the punctures z_i's, and finally, the D_{nm}'s form an infinite dimensional representation of the projective group [2] given by

$$D_{nm}(\gamma) = \frac{1}{m!}\frac{\sqrt{m}}{\sqrt{n}}\partial_z^m[\gamma(z)]^n\Big|_{z=0} \quad , \quad n,m \neq 0 \ ; \ D_{00}(\gamma) = \frac{1}{2}\log\gamma'(0)$$

$$D_{n0}(\gamma) = \frac{1}{\sqrt{n}}[\gamma(0)]^n \quad n \neq 0 \ ; \ D_{0m}(\gamma) = \frac{\sqrt{m}}{2m!}\partial_z^m\log[\gamma'(z)]\Big|_{z=0} \quad , \quad m \neq 0 \tag{2.5}$$

where $\gamma(z)$ is an arbitrary projective transformation of the form

$$\gamma(z) = \frac{Az + B}{Cz + D} \qquad AD - BC = 1 \tag{2.6}$$

As it is explained in detail in [1] it is possible to rewrite $V_{N;0}$ in the following suggestive way

$$V_{N;0} \equiv \hat{V}_{N;0}\,\delta(\sum_{i=1}^N N_i + Q)$$

$$= \prod_{i=1}^N[\sum_{n_i}{}_i < n_i, O_a|]\exp\left\{\frac{1}{2}\sum_{\substack{i,j=1\\ i\neq j}}^N\oint_0 dz\oint_0 dy\,\partial\varphi^{(i)}(z)\log[V_i(z) - V_j(y)]\partial\varphi^{(j)}(y)\right\}$$

$$\exp\left\{-\frac{1}{2}\sum_{i=1}^N\oint_0 dz\,\partial\varphi^{(i)}(z)[\alpha_0^{(i)} - Q]\log[V_i'(z)]\right\}\delta(\sum_{i=1}^N N_i + Q) \tag{2.7}$$

---

[2] This is strictly speaking correct only if $n, m \neq 0$.

The two exponentials in (2.7) have a clear physical meaning. The first one contains only the terms $i \neq j$ of the Fourier components of the Green's function of the Laplace operator on the sphere, given by $\log(z-y)$, but written in a frame in which the local coordinates around the punctures z_i's are given by the functions $V_i^{-1}(z)$ with the property that $V_i^{-1}(z_i) = 0$. The second one reproduces what is left from the conformal invariant regularization of the diagonal terms of the Green's function, as explained in [6]. Finally the δ function expresses the conservation of the U(1) charge with the inclusion of the background charge Q and corresponds to the constant zero mode of the Laplace operator.

By means of the sewing procedure, whose main steps have been sketched in the introduction, one can then construct the N-point g-loop vertex that is given by:

$$
\begin{aligned}
V_{N;g} = & \prod_\alpha{}' \prod_{n=1}^{\infty} \left(\frac{1}{1-k_\alpha^n} \right) \prod_{i=1}^{N} [\textstyle\sum_{n_i} < n_i, O_a|] \; \delta(\sum_{i=1}^{N} N_i - (g-1)Q) \\
& \exp\left\{ -\frac{1}{2} \sum_i^N \oint_0 dz \partial\varphi^{(i)}(z) [\alpha_0^{(i)} - Q] \log[V_i'(z)] \right\} \\
& \exp\left\{ \frac{1}{2} \sum_{i,j=1}^{N} \oint_0 dz \oint_0 dy \partial\varphi^{(i)}(z) \log \frac{E(V_i(z), V_j(y))}{V_i(z) - V_j(y)} \partial\varphi^{(j)}(y) \right\} \\
& \exp\left\{ \frac{1}{2} \sum_{\substack{i,j=1 \\ i \neq j}}^{N} \oint_0 dz \oint_0 dy \partial\varphi^{(i)}(z) \log(V_i(z) - V_j(y)) \partial\varphi^{(j)}(y) \right\} \\
& \left[\Theta \begin{pmatrix} \alpha \\ \beta \end{pmatrix} \left(\left[\frac{1}{2\pi i} \sum_{i=1}^{N} \oint_0 dz \partial\varphi^{(i)}(z) [\int_{z_0}^{V_i(z)} \omega_\mu] - Q \Delta_\mu^{z_0} \right] | \tau \right) \right] \\
& \exp\left\{ Q \sum_{i=1}^{N} \oint_0 dz \partial\varphi^{(i)}(z) \log \sigma [V_i(z)] \right\}
\end{aligned}
\tag{2.8}
$$

We have introduced the standard Riemann Θ-function with characteristics α and β, given by

$$
\Theta \begin{pmatrix} \alpha \\ \beta \end{pmatrix} (z|\tau) = \sum_{\{n_\mu\}} \exp 2\pi i \left\{ \sum_{\mu,\nu=1}^{g} \frac{1}{2} (n_\mu + \alpha_\mu) \tau_{\mu\nu} (n_\nu + \alpha_\nu) + \sum_{\mu=1}^{g} (n_\mu + \alpha_\mu)(z_\mu + \beta_\mu) \right\}
\tag{2.9}
$$

with n_μ integer.

We now give the explicit expressions for the various geometrical objects that enter in the vertex written in terms of the Schottky parametrization of the Riemann surface, that is obtained in a natural way through the sewing procedure.

The period matrix $\tau_{\mu\nu}$ is given by

$$\tau_{\mu\nu} = \frac{1}{2\pi i}\left[\log k_\mu\,\delta_{\mu\nu} + {}^{(\mu)}\sum_\alpha{}'^{(\nu)} \log\frac{[\eta_\mu - T_\alpha(\eta_\nu)][\xi_\mu - T_\alpha(\xi_\nu)]}{[\eta_\mu - T_\alpha(\xi_\nu)][\xi_\mu - T_\alpha(\eta_\nu)]}\right] \tag{2.10}$$

The symbol ${}^{(\mu)}\sum_\alpha{}'^{(\nu)}$ means sum over all elements T_α of the Schottky group which contain neither $S_\mu^{\pm n}$ as leftmost factor nor $S_\nu^{\pm m}$ as rightmost factor for arbitrary $n, m \neq 0$. Moreover the prime means that the identity is excluded when $\mu = \nu$.

The g holomorphic differentials of the first kind are given by

$$\omega_\mu(z) = \sum_\alpha{}^{(\mu)}\left(\frac{1}{z - T_\alpha(\eta_\mu)} - \frac{1}{z - T_\alpha(\xi_\mu)}\right)dz \tag{2.11}$$

where the sum is restricted to all the elements of the Schottky group without $S_\mu^{\pm n}$ as rightmost factor for any $n \neq 0$.

The vector of Riemann constants $\Delta_\mu^{z_0}$ in the Schottky representation is given by

$$\Delta_\mu^{z_0} = \frac{1}{2\pi i}\left\{-\tfrac{1}{2}\log k_\mu - \pi i + \sum_{\nu=1}^{g}{}^{(\nu)}\sum_\alpha{}'^{(\mu)}\log\frac{\xi_\nu - T_\alpha(\eta_\mu)}{\xi_\nu - T_\alpha(\xi_\mu)}\frac{z_o - T_\alpha(\xi_\mu)}{z_o - T_\alpha(\eta_\mu)}\right\} \tag{2.12}$$

Finally the prime form E and the function σ are given respectively by

$$E(z,w) = (z-w)\prod_\alpha{}'\frac{[z - T_\alpha(w)]}{[z - T_\alpha(z)]}\frac{[w - T_\alpha(z)]}{[w - T_\alpha(w)]} \tag{2.13}$$

where $\prod'$ means that T_α and T_α^{-1} are counted only once and the identity is excluded from the product, and by

$$\log\sigma(z) = \frac{1}{2(g-1)}\left[\sum_{\mu,\nu=1}^{g}{}^{(\mu)}\sum_{\alpha\neq I}{}^{(\nu)}\log\frac{[\xi_\mu - T_\alpha(\xi_\nu)][z - T_\alpha(z)]}{[z - T_\alpha(\xi_\nu)][\xi_\mu - T_\alpha(z)]}\right.$$

$$\left. + \sum_{\mu\neq\nu}\log\frac{\xi_\mu - \xi_\nu}{(z-\xi_\nu)(\xi_\mu - z)}\right] \tag{2.14}$$

for $g > 1$; whereas in the case of the torus, i.e. $g = 1$, one simply has

$$\log\sigma(z) = -\frac{1}{2}\log[(z-\xi)(\xi - z)]$$

where ξ is the repulsive fixed point of the single Schottky generator of the torus.

Finally the infinite product over the prime classes of the transformations of the Schottky group appearing in front of the vertex in (2.8) corresponds to the determinant of the Laplace operator on a genus g Riemann surface as described in [1].

3 Free Fermionic Theory

A free fermionic system with background charge Q is described by the following action:

$$S(b,c) = \frac{1}{\pi}\int d^2 z b\bar{\partial}c \tag{3.1}$$

It contains two fields $b(z,\bar{z})$ and $c(z,\bar{z})$ having conformal dimensions equal to λ and $1-\lambda$ respectively. λ is a positive integer or half integer number.

The fermionic fields have the following Fourier expansions:

$$b(z) = \sum_{n\in Z+\lambda} b_n z^{-n-\lambda} \qquad\qquad c(z) = \sum_{n\in Z+\lambda} c_n z^{-n+\lambda-1} \tag{3.2}$$

in terms of the oscillators satisfying the anticommutation relations:

$$\{b_n, c_m\} = \delta_{n+m,0} \qquad \{b_n, b_m\} = \{c_n, c_m\} = 0 \tag{3.3}$$

We assume for simplicity that n,m are integer if λ is integer. If λ is half integer different spin structures appear. In this paper we restrict our analysis to the case that n,m are half integer if λ is half integer.

The N-point vertex for the b,c system can be easily written and is given by:

$$V_{N,0} = \prod_{i=1}^{N} [{}_i< q=-Q|] \exp\left[(-1)^{1-\lambda}\sum_{i\neq j}\sum_{n=\lambda}^{\infty}\sum_{1-\lambda}^{\infty} c_n^{(i)} E_{nm}(U_i V_j) b_m^{(j)}\right]$$
$$\prod_{n=1-\lambda}^{\lambda-1}\left[\sum_{i=1}^{N}\sum_{m=1-\lambda}^{\lambda-1} E_{nm}(V_i) b_m^{(i)}\right] \tag{3.4}$$

where $U_i = \Gamma V_i^{-1}$, $\Gamma(z) = \frac{1}{z}$,

$$E_{nm}(\gamma) = \frac{1}{(m+\lambda-1)!}\partial_z^{m+\lambda-1}\left[[\gamma(z)]^{n+\lambda-1}[\gamma'(z)]^{1-\lambda}\right]|_{z=0} \quad ; \quad \gamma(z) = \frac{Az+B}{Cz+D} \tag{3.5}$$

(with $n, m = 1-\lambda, -\lambda, \ldots, \infty$) is an infinite dimensional representation of the projective group with conformal weight $1-\lambda$ and $Q = 1-2\lambda$.

The previous expression can also be rewritten in the following form:

$$V_{N;0} = {}_i < q = -Q| \prod_{n=1-\lambda}^{\lambda-1} \left[\sum_{i=1}^{N} \oint_{z_i} dv B_i(v) v^{n+\lambda-1} \right]$$

$$: \exp\left\{ -\sum_{i=1}^{N} \oint_{z_i,v} du \oint_{z_i} dv C_i(u) \frac{1}{u-v} \left(\frac{v}{u}\right)^{2\lambda-1} B_i(v) \right\}$$

$$\exp\left\{ -\sum_{\substack{i,j=1 \\ i\neq j}}^{N} \oint_{z_i} du \oint_{z_j} C_i(u) \frac{1}{u-v} \left(\frac{v}{u}\right)^{2\lambda-1} B_j(v) \right\} : \tag{3.6}$$

where

$$C_i(u) \equiv [\partial_u V_i^{-1}(u)]^{1-\lambda} c_i[V_i^{-1}(u)] \qquad B_j(v) \equiv [\partial_v V_j^{-1}(v)]^{\lambda} b_j[V_j^{-1}(v)] \tag{3.7}$$

It is easy to see that the diagonal terms in the first exponential of the Vertex contain only the zero modes of the field $B_j^{zero}(v)$. The subscript (z_i, v) in the first integral of the diagonal terms means that the u integration must be performed along a contour surrounding both z_i and v.

Notice that the presence of the $2\lambda - 1$ δ functions allows one to rewrite the previous expression by using the following modified Green's function:

$$\frac{1}{u-v}\left(\frac{v}{u}\right)^{2\lambda-1} \rightarrow \frac{1}{u-v} \prod_{i=1}^{2\lambda-1} \left(\frac{v-a_i}{u-a_i}\right) \tag{3.8}$$

with a_i being arbitrary constants.

Usually the Green's function for a b, c system on the sphere is given by:

$$< q = -Q| b(u) c(v) | q = 0 >= \frac{1}{u-v} \tag{3.9}$$

where the state on the right is the projective invariant vacuum $|q = 0>$, while the one on the left carries a charge equal to $-Q$ in order to have a nonvanishing matrix element. The Green's function (3.9) corresponds to the case in which the background charge is located at infinity and is therefore equal to the l.h.s. of (3.8) in the limit $a_i \rightarrow \infty$. If we want to have the background charge at arbitrary points a_i then, instead of (3.9), we have to compute the matrix element:

$$\frac{< q=0|\prod_{i=1}^{2\lambda-1} c(a_i)b(u)c(v)|q=0>}{< q=0|\prod_{i=1}^{2\lambda-1} c(a_i)|q=0>} = \frac{1}{u-v}\prod_{i=1}^{2\lambda-1}\left(\frac{v-a_i}{u-a_i}\right) \tag{3.10}$$

As in the bosonic theory also here the various terms of the vertex have a simple meaning. The last exponential in (3.6) with the non diagonal terms contains the Green's function for the bc system on the sphere. The other exponential reproduces what is left from the conformal invariant regularization of the diagonal terms of the Green's function. Finally the $|Q|$ fermionic δ functions in(3.6) correspond to the fact that the equation $\bar{\partial}c[b]=0$ for $c[b]$ admits -Q[0] solutions on the sphere in agreement with the Riemann-Roch formula for $g=0$.

Starting from (3.4) and performing the sewing procedure we can construct $V_{N;g}$ as explained in sect. 1. For the sake of simplicity we discuss in detail only the two lowest cases corresponding to $\lambda=1/2$ and $\lambda=1$. The same calculation has been also performed for an arbitrary λ. It is presented in Ref. [2].

1. $\lambda= 1/2$

In the case of a b,c system with $\lambda=1/2$ we get the following N-point g-loop vertex [5]:

$$V_{N;g} = [\det\bar{\partial}]\prod_{i=1}^{N}{}_i < 0|\exp\left\{-\sum_{i,j=1}^{N}\oint_{z_i} du\oint_{z_j} C_i(u)[G(u,v)-\delta_{ij}\frac{1}{u-v}]B_j(v)\right\} \tag{3.11}$$

where

$$C_i(u)\equiv[\partial_u V_i^{-1}(u)]^{1/2}c_i[V_i^{-1}(u)] \qquad B_j(v)\equiv[\partial_v V_j^{-1}(v)]^{1/2}b_j[V_j^{-1}(v)] \tag{3.12}$$

and the determinant of the operator $\bar{\partial}$ on an arbitrary Riemann surface is given by [5]:

$$\det\bar{\partial} = \prod_{\alpha}{}'\prod_{n=1}^{\infty}[1-e^{2\pi i N_\alpha}k_\alpha^{n-1/2}]^2 \tag{3.13}$$

where the product is over the primary classes of the transformations of the Schottky group as explained in detail in [1] and k_α is the multiplier of the Schottky transformation T_α. Finally

$$N_\alpha = \sum_{\mu=1}^{g} N_\mu^\alpha (\frac{1}{2} - \beta_\mu) \tag{3.14}$$

where N_μ^α is the number of times that the generator S_μ appears in the element T_α of the Schottky group counted with its sign;this means that each S_μ is counted with a contribution +1, while each S_μ^{-1} is counted with a contribution -1. Of course if we restrict only to values of $\beta = 0, \frac{1}{2}$, as we do now, the sign does not make any difference. However later on we will consider arbitrary values of β and in this case the contribution to N_μ^α should be counted with the sign as explained above in order to get the right result.

If $\lambda = 1/2$ the equations $\bar{\partial} b = 0$ or $\bar{\partial} c = 0$ do not admit any solution and therefore we do not have to worry about the contribution of the zero modes to the determinant. For the same reason there is no δ function appearing in the vertex.

$G(u, v)$ is the Green's function of a b, c system with $\lambda = 1/2$ on a genus g Riemann surface. In the literature it is called the Szegö kernel. In the Schottky parametrization of a Riemann surface it is given by:

$$G \begin{pmatrix} 0 \\ \beta \end{pmatrix} (u, v) = \sum_{T_\alpha} e^{2\pi i N_\alpha} \frac{[T'_\alpha(u)]^{1/2}}{T_\alpha(u) - v} \tag{3.15}$$

This form of the Szegö kernel was found by Montonen [8] in 1974 for $\beta = 1/2$ and by Pezzella [5] for $\beta = 0, 1/2$.

It has the important property that under an arbitrary transformation of the Schottky group T_β it transforms as an automorphic function of weight 1/2 in both variables u and v :

$$G(u, v) \to G[T_\beta(u), v] = e^{-2\pi i N_\beta} [T'_\beta(u)]^{-1/2} G(u, v)$$

$$G(u,v) \rightarrow G[u, T_\beta(v)] = e^{2\pi i N_\beta}[T'_\beta(v)]^{-1/2} G(u,v) \tag{3.16}$$

From (3.16) it follows that the quantity $G(u,v)[du]^{1/2}[dv]^{1/2}$ is invariant apart from a phase under any transformation of the Schottky group T_β acting on either u or v.

Since our original vertex (3.6) does not contemplate the possibility of the emission of a spin field, we cannot compute from it loops with the exchange of spin field. Therefore the Green's functions (3.15) correspond only to the 2^g spin structures without spin field exchange in the loops.

It is interesting, however, to give in the case of one loop the Green's functions corresponding to all four spin structures including also those corresponding to spin field exchange in the loops. They are given by[3]:

$$G\begin{pmatrix}\alpha \\ \beta\end{pmatrix}(u,v) = \sum_{T_\alpha} e^{2\pi i N_\alpha} \frac{k^{n/2}}{k^n u - v} \frac{1}{2}\left[\left(\frac{v}{k^n u}\right)^\alpha + \left(\frac{k^n u}{v}\right)^\alpha\right] + \delta_{\beta\frac{1}{2}}\delta_{\alpha\frac{1}{2}} \frac{\log\frac{u}{v}}{\log k}\frac{1}{\sqrt{uv}} \tag{3.17}$$

The extra term appearing for the odd spin structure insures the right automorphic properties around the B cycles. In fact if we perform a transformation of the Schottky group the first term alone in (3.17) has naively the right automorphic properties. However one has to be careful and somewhat regularize the sum over the elements of the Schottky group. Once this is done one discovers that one really needs the additional term in (3.17) in order to have a function with the right automorphic properties. No extra term is instead needed for the three even spin structures.

Up to now we have assumed that β takes only the values $0, 1/2$. In the following we want to give the results coming out from the sewing procedure for an arbitrary value of β between 0 and 1. In this case the sewing procedure is done by inserting in each loop together with the propagator, as explained in eq. (1.2), also the

[3]This has been worked out in collaboration with K.Hornfeck and K. Roland.

factor

$$\exp\left\{2\pi i(\frac{1}{2} - \beta_\mu) \sum_{n=1/2}^{\infty} [c_n^{+(2\mu-1)} b_n^{(2\mu-1)} - b_n^{+(2\mu-1)} c_n^{(2\mu-1)}]\right\} \quad (3.18)$$

containing the charge operator of the b, c system, where now β can assume any value between 0 and 1.

By carrying the sewing procedure with this extra factor we get again the expression (3.11) where now the determinant is given by:

$$\det \bar{\partial} = \prod_{\alpha}{}' \prod_{n=1}^{\infty} [1 - e^{2\pi i N_\alpha} k_\alpha^{n-1/2}][1 - e^{-2\pi i N_\alpha} k_\alpha^{n-1/2}] \quad (3.19)$$

that generalizes (3.13) to an arbitrary value of β.

The Green's function is given again by (3.15) where now β can assume an arbitrary value between 0 and 1. N_α appearing in the phase factor is given by (3.14).

2. λ =1

In the case of $\lambda = 1$ we get the following N-point g-loop vertex:

$$V_{N;g}^{\lambda=1} = \prod_{\alpha}{}' \prod_{n=1}^{\infty} (1 - k_\alpha^n)^2 \prod_{i=1}^{N} [_i < q = |Q||] : \exp\left\{-\sum_{i=1}^{N} C_i(a) b_0^{(i)}\right\}$$

$$\exp\left\{-\sum_{i,j=1}^{N} \oint_{z_i} du C^{(i)}(u) \oint_{z_j} dv \left[\hat{G}[u,v] - \delta_{ij}\frac{1}{u-v}\frac{v-a}{u-a}\right] B^{(j)}(v)\right\} :$$

$$[\sum_{i=1}^{g} b_0^{(i)}] \prod_{\mu=1}^{g} \left[\sum_{i=1}^{N} \oint_0 dz c^{(i)}(z) \omega^\mu [V_i(z)]\right] \quad (3.20)$$

where $\oint_{z_i}$ means that the integral is performed around the point z_i.

We have also defined the transported fields:

$$C(u) \equiv c[V_i^{-1}(u)] \qquad B(v) \equiv [V_i^{-1\prime}(v)] b[V_i^{-1}(v)] \quad (3.21)$$

It is interesting to discuss now in an intuitive way the geometrical and topological meaning of the various terms of the vertex.

The factor in front of the vertex containing the infinite product over the prime classes of the transformations of the Schottky group is equal to the non zero mode contribution to the determinant of the operator $\bar{\partial}$ for a b, c system with $\lambda = 1$ on an arbitrary Riemann surface :

$$\det{}'\bar{\partial} = \prod_{\alpha}{}' \prod_{n=1}^{\infty} (1 - k_{\alpha}^{n})^2 \tag{3.22}$$

The fermionic δ function containing a sum over $b_0^{(i)}$ is a consequence of the fact that for the c field with vanishing conformal dimension there is always a constant solution of the equation $\bar{\partial}c = 0$ on an arbitrary Riemann surface.

On the other hand, because of the Riemann Roch theorem, the equation $\bar{\partial}b = 0$ for the b field has g solutions on a genus g Riemann surface. They give rise to the g fermionic δ functions containing the g abelian differentials given explicitly in (2.11) in the Schottky parametrization.

The second exponential contains the function

$$\hat{G}(u, v) = \sum_{T_\alpha} \frac{T_\alpha'(u)}{T_\alpha(u) - v} \left(\frac{v - a}{T_\alpha(u) - a} \right) \tag{3.23}$$

where the sum is over all elements of the Scottky group T_α.

(3.23) is exactly equal to the function $\hat{G}(u, v)$ given in eq.(16) of ref. [7] for $\lambda = 1$.

The contribution of the identity transformation in the function $\hat{G}$ will give rise to an infinity for the terms $i = j$. If one uses, however, a regularization procedure, that preserves conformal invariance, one gets the expression corresponding to the first exponential in (3.20).

The vertex (3.20) contains an arbitrary point a of the Riemann surface but it does not depend on it. In fact the δ functions allow one to change the value of the constant a without changing the vertex.

It is easy to see that $\hat{G}(u, v)$ is an automorphic function of weight 1 in the variable u:

$$\hat{G}(u, v) \rightarrow \hat{G}[T_\beta(u), v] = [T_\beta(u)]^{-1} \hat{G}(u, v). \tag{3.24}$$

It is, however, not an automorphic function of weight 0 in the variable v. Under a transformation of the Schottky group T_β on the variable v we get:

$$\hat{G}(u,v) \to \hat{G}[u, T_\beta(v)] = \hat{G}(u,v) - \hat{G}(u, T_\beta^{-1}(a)) \tag{3.25}$$

A function with the right automorphic properties with respect to both variables u and v is given by the combination [7]:

$$G(u,v) = \hat{G}(u,v) + \frac{1}{2\pi} \sum_{\mu,\nu=1}^{g} \omega_\mu(u)(Im\tau)^{-1}_{\mu\nu} Re[\Omega_\nu(v) - \Omega_\nu(a)] \tag{3.26}$$

where $\tau_{\mu\nu}$ is the period matrix and Ω is the abelian integral

$$\Omega_\nu(v) - \Omega_\nu(a) = \int_a^v \omega_\nu(w) dw \tag{3.27}$$

$G(u,v)$ has all the correct properties for being the Green's function for the b, c system with $\lambda = 1$ on an arbitrary Riemann surface.

Because of the g fermionic δ functions of the c field appearing in (3.20) one can use $G(u,v)$ instead of $\hat{G}(u,v)$ in the exponential term.

4 Bosonization Rules

Let us now use the N-point g-loop vertex in (2.8) to compute the correlation functions in our theory on arbitrary Riemann surfaces of genus g, i.e.

$$< q = 0| \prod_{i=1}^{N} : e^{q_i \varphi(z_i)} : |q = 0 >_g$$

The procedure is exactly the same as we have described in the introduction (See (1.3)). Indeed, we simply saturate $V_{N;g}$ with the highest weight states and get

$$< q = 0| \prod_{i=1}^{N} : e^{q_i \varphi(z_i)} : |q = 0 >_g \equiv V_{N;g} \prod_{i=1}^{N} |q_i >_i =$$

$$\delta(\sum_{i=1}^{N} q_i - (g-1)Q) \prod_\alpha{}' \prod_{n=1}^{\infty} \left(\frac{1}{1 - k_\alpha^n} \right) \prod_{i<j} [E(z_i, z_j)]^{q_i q_j} \prod_{i=1}^{N} [\sigma(z_i)]^{Q q_i}$$

$$\left[\Theta\begin{pmatrix}\alpha\\ \beta\end{pmatrix}\left(\left[\frac{1}{2\pi i}\sum_{i=1}^{N} q_i \int_{z_0}^{z_i}\omega_\mu - Q\Delta_\mu^{z_0}\right]|\tau\right)\right] \tag{4.1}$$

that completely agrees with the expression originally obtained in Ref. [9] using geometrical methods.

In particular from (4.1) we can compute the following correlation functions

$$<q=0|\prod_{i=1}^{N_1} :e^{-\varphi(z_i)}: \prod_{h=1}^{N_2} :e^{\varphi(y_h)}: |q=0>_g=$$

$$\prod_\alpha{}' \prod_{n=1}^{\infty}\left(\frac{1}{1-k_\alpha^n}\right)\frac{\prod\limits_{\substack{i,j=1\\ i<j}}^{N_1} E(z_i,z_j)\prod\limits_{\substack{h,k=1\\ h<k}}^{N_2} E(y_h,y_k)}{\prod\limits_{i=1}^{N_1}\prod\limits_{h=1}^{N_2} E(z_i,y_h)}\frac{\prod\limits_{h=1}^{N_2}\sigma(y_h)^Q}{\prod\limits_{i=1}^{N_1}\sigma(z_i)^Q}\delta(N_1-N_2+(g-1)Q)$$

$$\left[\Theta\begin{pmatrix}\alpha\\ \beta\end{pmatrix}\left(\left[-\frac{1}{2\pi i}\sum_{i=1}^{N_1}\int_{z_0}^{z_i}\omega_\mu + \frac{1}{2\pi i}\sum_{h=1}^{N_2}\int_{z_0}^{y_h}\omega_\mu - Q\Delta_\mu^{z_0}\right]|\tau\right)\right] \tag{4.2}$$

In the bosonic theory with $Q=0$ we can also compute the partition function that is given by:

$$Z_B^{Q=0} = \prod_\alpha{}' \prod_{n=1}^{\infty}\left(\frac{1}{1-k_\alpha^n}\right)\Theta\begin{pmatrix}\alpha\\ \beta\end{pmatrix}(0|\tau) \tag{4.3}$$

It is well known that free bosonic and fermionic theories are equivalent. This means that the partition functions and the corresponding correlation functions must be equal in the two theories. In the two previous sections we have constructed $V_{N;g}$ for the two theories; we can now use it for checking their equivalence. This is based on the bosonization rules that state that the fermion fields b and c are expressed in the bosonic theory by:

$$c(z) = e^{\varphi(z)} \qquad\qquad b(z) = e^{-\varphi(z)} \tag{4.4}$$

The partition function for free complex fermion theory with $Q=0$ is given by:

$$Z_F^{Q=0} = \prod_\alpha{}' \prod_{n=1}^{\infty} \left[1 - e^{2\pi i N_\alpha} k_\alpha^{n-\frac{1}{2}}\right] \left[1 - e^{-2\pi i N_\alpha} k_\alpha^{n-\frac{1}{2}}\right] \tag{4.5}$$

The identity between (4.3) and (4.5) implies the following suggestive formula for the Θ function for arbitrary genus:

$$\Theta\begin{pmatrix} 0 \\ \beta \end{pmatrix}(0|\tau) = \prod_\alpha{}' \prod_{n=1}^{\infty} \left[1 - e^{2\pi i N_\alpha} k_\alpha^{n-\frac{1}{2}}\right] \left[1 - e^{-2\pi i N_\alpha} k_\alpha^{n-\frac{1}{2}}\right] \left[1 - k_\alpha^n\right] \tag{4.6}$$

(4.6) for $\beta = 0, 1/2$ appears in [5].

Taking into account that now α and β are arbitrary and that $\Theta(z|\tau)$ is an analytic function of z we get the following expression for the Θ function:

$$\Theta\begin{pmatrix} \alpha \\ \beta \end{pmatrix}(z|\tau) = \exp\left[i\pi\alpha \cdot \tau \cdot \alpha\right] \exp\left[2\pi i (z + \beta) \cdot \alpha\right]$$

$$\prod_\alpha{}' \prod_{n=1}^{\infty} \left[1 - \exp\left\{2\pi i \sum_{\mu=1}^{g} (\frac{1}{2} - \beta_\mu - (\alpha \cdot \tau)_\mu - z_\mu) N_\mu^\alpha\right\} k_\alpha^{n-\frac{1}{2}}\right]$$

$$\left[1 - \exp\left\{-2\pi i \sum_{\mu=1}^{g} (\frac{1}{2} - \beta_\mu - (\alpha \cdot \tau)_\mu - z_\mu) N_\mu^\alpha\right\} k_\alpha^{n-\frac{1}{2}}\right] \left[1 - k_\alpha^n\right] \tag{4.7}$$

This expression for the genus g Θ function generalizes the Euler identity valid in the case of genus 1. It has also been derived in a recent preprint by Losev [10] using different arguments.

The previous formula allows one to compute the fermion determinant for an arbitrary spin structure in the Schottky parametrization. Since it has been shown [11,12] in general that

$$\det \bar{\partial}_{\lambda=1/2} = \theta\begin{pmatrix} \alpha \\ \beta \end{pmatrix}(0|\tau) \tag{4.8}$$

by using (4.7) we get the following expression for the fermion determinant in the Schottky parametrization of the Riemann surface:

$$\det \bar{\partial}_{\lambda=1/2} = \exp\left[i\pi\alpha \cdot \tau\alpha\right] \exp\left[2\pi i\beta \cdot \alpha\right]$$

$$\prod_{\alpha}{}' \prod_{n=1}^{\infty} \left[1 - \exp\left\{2\pi i \sum_{\mu=1}^{g} (\frac{1}{2} - \beta_\mu - (\alpha \cdot \tau)_\mu) N_\mu^\alpha\right\} k_\alpha^{n-\frac{1}{2}}\right]$$
$$\left[1 - \exp\left\{-2\pi i \sum_{\mu=1}^{g} (\frac{1}{2} - \beta_\mu - (\alpha \cdot \tau)_\mu) N_\mu^\alpha\right\} k_\alpha^{n-\frac{1}{2}}\right] \tag{4.9}$$

for an arbitrary spin structure.

It would be interesting to compare (4.9) with the fermion contribution given in eq. (7.8) by Petersen in Ref. [13].

In the case of the torus we have that $k = \exp(2\pi i \tau)$ and (4.9) reduces to the well known result:

$$\det \bar{\mathcal{C}}^{g=1}_{\lambda=1/2} = k^{\frac{\alpha^2}{2}} e^{2i\pi\beta\alpha} \prod_{n=1}^{\infty} \left[1 + k^{n-1/2-\alpha} e^{-2\pi i\beta}\right] \left[1 + k^{n-1/2+\alpha} e^{+2\pi i\beta}\right] \tag{4.10}$$

In the fermionic theory we have computed the fermionic Green's function on an arbitrary Riemann surface, that is given by (3.15) for those spin structures that do not correspond to any spin field exchange in the loops. It can also be computed in the bosonic theory from (4.2) by taking $N_1 = N_2 = 1, Q = 0$ and by dividing it with the partition function. The result [5] is the well known expression for the Szegö kernel in terms of the Θ functions and prime form:

$$G\begin{pmatrix}\alpha\\ \beta\end{pmatrix}(u,v) = \frac{\Theta\begin{pmatrix}\alpha\\ \beta\end{pmatrix}\left(\frac{1}{2\pi i}\int_u^v \omega_\mu|\tau\right)}{E(u,v)\Theta\begin{pmatrix}\alpha\\ \beta\end{pmatrix}(0|\tau)} \tag{4.11}$$

valid for any even spin structure.

Since the two expressions (3.15) and (4.11) have the same properties that uniquely identify the Szegö kernel they must be equal at least for $\alpha = 0$. It would be interesting to extend our formalism to include the emission of spin fields and therefore by the sewing procedure get a fermionic Green's function that generalizes to an arbitrary α the expression (3.15) valid for $\alpha = 0$.

We thank K. Hornfeck and K. Roland for a critical reading of the manuscript.

References

[1] P. Di Vecchia, F. Pezzella, M. Frau, K. Hornfeck, A. Lerda and S. Sciuto, Nordita preprint 88-47 P, to be published in Nuclear Physics B.

[2] P. Di Vecchia, F. Pezzella, M. Frau, K. Hornfeck, A. Lerda and S. Sciuto, Nordita Preprint 89/29 P.

[3] M. Frau, A. Lerda, J.G. McCarthy and S. Sciuto, MIT preprint CTP-1753 (1989).

[4] J.L. Petersen, K.O. Roland and J.R. Sidenius, Phys. Lett. **B205** (1988) 262

[5] F. Pezzella, Phys. Lett. **B220** (1989) 544.

[6] P. Di Vecchia, R. Nakayama, J.L. Petersen, S. Sciuto and J.R. Sidenius, Nucl. Phys. **B287** (1987) 621.

[7] E. Martinec, Nucl. Phys. **B281** (1987) 157.

[8] C. Montonen, Nuovo Cimento **19A** (1974) 69.

[9] E. Verlinde and H. Verlinde, Nucl. Phys. **B288** (1987) 357.

[10] A. Losev, ITEP preprint (1989).

[11] V.G. Knizhnik, Phys. Lett. **108B** (1986) 247.

[12] L. Alvarez-Gaumé, G. Moore and C. Vafa, Comm. Math. Phys. **106** (1986) 1.

[13] J.L. Petersen, contribution to this volume.

SUPERGRASSMANNIANS, SUPER τ-FUNCTIONS AND STRINGS

S. N. DOLGIKH, A. S. SCHWARZ

Recently infinite-dimensional grassmannians and their super-generalizations were used to study conformal two-dimensional fields and strings.[1] In particular the super Mumford form (holomorphic square root from the superstring measure on moduli space) was expressed in Ref. 2 through super analog of Sato τ-function. In this paper we present results about supergrassmannians and super τ-functions. Some of these assertions generalize the results of Ref. 2; in these cases we omit the proofs. Let us denote by H_N (or simply by H) the space of square integrable functions $f(z, \theta_1, \ldots, \theta_N)$ where z is a complex number, $|z| = 1$ and $\theta_1, \ldots, \theta_N$ are elements of some Grassmann algebra Λ_N. We introduce in H a bilinear scalar product:

$$\langle f, g\rangle = \oint\limits_{|z|=1} f(z, \theta_1, \ldots, \theta_N) g(z, \theta_1, \ldots, \theta_N) dz d^N\theta \qquad (1)$$

and a hermitian scalar product

$$(f, g) = \oint\limits_{|z|=1} \overline{f(z, \theta_1, \ldots, \theta_N)} g(z, \theta_1, \ldots, \theta_N) z^{-1} dz d^N\theta\,. \qquad (2)$$

Note that these scalar products are even for N even and odd for N odd. The standard basis in H consists of the elements $z^n \theta_{i_1} \ldots \theta_{i_k}$ where n is an integer and $1 \leq i_1 < \ldots < i_k \leq N$. The subspace of H spanned by the elements of standard basis with $n < 0$ (with $n \geq 0$) will be denoted by

H_- (by H_+). (All subspaces are assumed to be closed.) We say that the subspace $W \subset H$ belongs to the super-grassmannian manifold $Gr = Gr(N)$ if the natural projection π_- of W into H_- is a Fredholm map and the natural projection π_+ of W into H_+ is a compact map. Remember that P is a Fredholm operator if $\operatorname{Ker} P$ is finite-dimensional and $\operatorname{Im} P$ can be singled out by finite number of linear equations. In such a way if $W \in \mathrm{Gr}$ the space $\mathcal{A}(W) = W \cap H_+ = \operatorname{Ker} \pi_-^W$ and the space $H_-/\operatorname{Im} \pi_-^W$ are finite-dimensional. (We use the notation π_-^W for the operator π_- considered only on the space W.) One can prove that

$$H_-/\operatorname{Im} \pi_-^W = \Pi^N \mathcal{A}^*(W^\perp) . \tag{3}$$

(Here $W^\perp$ denotes the subspace of H, consisting of vectors orthogonal to the space W in the sense of bilinear scalar product (1) and Π denotes the parity reversion.)

We will define the index of $W \in \mathrm{Gr}$ by the formula

$$\begin{aligned} \operatorname{ind} W &= \dim \mathcal{A}(W) - \dim(H_-/\operatorname{Im} \pi_-^W) \\ &= \dim \mathcal{A}(W) - \dim \Pi^N \mathcal{A}(W^\perp) . \end{aligned} \tag{4}$$

(Remember that the dimension of superspace is a pair of numbers.)

We will say that the basis in H_- is admissible if it differs from the standard one by the matrix of the form $1+T$ where T belongs to the trace class. (The set of matrices belonging to the trace class will be denoted by $\mathcal{T}$ and the set of matrices of the form $1+T$ where $T \in \mathcal{T}$ will be denoted by $\mathcal{S}$. Note that the matrix $P \in \mathcal{S}$ has well-defined determinant admitting the usual property $\det(P_1 \cdot P_2) = \det P_1 \cdot \det P_2$.) The basis in $\operatorname{Im} \pi_-^W$ is called admissible if it can be considered as a part of an admissible basis in H_-. The basis $(w_1, \dots, w_k, \dots)$ in W is called admissible if the vectors $\pi_- w_k$ (or part of these vectors) form an admissible basis in $\operatorname{Im} \pi_-^W$. The measure in the infinite dimensional space is by definition a function of weight 1 on the set of admissible bases (i.e. a function satisfying $\mu(\tilde{e}) = \det P \cdot \mu(e)$ if $\tilde{e}_i = P_i^j e_j, P \in \mathcal{S}$). The set of measures in linear space E will be denoted by $m(E)$; this set can be considered as one dimensional space. (Of course in the infinite dimensional case we assume that the set of admissible bases in E is fixed.) It follows from $\operatorname{Im} \pi_-^W = W/\mathcal{A}(W)$ and (3) that there exist canonical isomorphisms

$$\begin{aligned} m(\operatorname{Im} \pi_-^W) &= m(W)/m(\mathcal{A}(W)) , \\ m(\Pi^N \mathcal{A}^*(W^\perp)) &= m(H_-)/m(\operatorname{Im} \pi_-^W) . \end{aligned} \tag{5}$$

Noting that the standard basis in H_- determines a standard measure in this space, we obtain canonical isomorphism

$$m(W) = m(\Sigma(W)) \tag{6}$$

where

$$\Sigma(W) = \mathcal{A}(W) + \Pi^N \mathcal{A}(W^\perp) . \tag{7}$$

In other words for every basis w in $\Sigma(W)$ we can construct a basis $\hat{w}$ in W determined up to unimodular transformation.

Let us denote by Γ the subset of H consisting of even invertible smooth functions. The set Γ can be considered as a supergroup. It is easy to check that the operator of multiplication $f \to Rf$, $f \in H$, $R \in \Gamma$, transforms the subspace $W \in \mathrm{Gr}$ in the subspace $RW \in \mathrm{Gr}$. In such a way Γ acts in Gr. We introduce also the following notations: the set $\Gamma_+(\Gamma_-)$ consists of the functions from Γ having the form $\exp(\phi(z,\theta))$ where $\phi(z,\theta)$ is holomorphic in the domain $|z| \leq 1$ ($|z| \geq 1$) and $\phi(0,0) = 0$. The set of functions from Γ consisting of functions which are holomorphic for $z \neq 0$ and invertible for $|z| \geq 1$ will be denoted by Γ'. These sets are subgroups of the group Γ. Let us consider the subspace $W \in \mathrm{Gr}, F \in \Gamma$, the bases w in $\Sigma(W)$ and w' in $\Sigma(FW)$. The above considerations permit us to define the bases $\hat{w}$ in W and $\hat{w}'$ in FW. Besides the basis $\hat{w}'$ in FW we can consider the basis $F\hat{w}$ in this space. We will define the τ-function as the determinant of the matrix, connecting these two bases in FW:

$$\tau(w, w', W, F) = \det(F\hat{w}|\hat{w}') . \tag{8}$$

Note that the matrix on the right-hand side of (8) does not in general belong to the class $\mathcal{S}$. Therefore we have to explain the meaning of (8). It is proved in Ref. 2 that in the supercase ($N \geq 1$) the right-hand side of (8) is well-defined by appropriate definition of (super) determinant. In the bosonic case ($N = 0$) the determinant is ill defined in general. However one can give unambiguous definition of the τ-function in bosonic case for $F \in \Gamma_+$. Really in this case the matrix $(F\hat{w}|\hat{w}')$ in (8) can be represented in the form $P + Q$, where P is an upper triangular matrix (i.e. $P_{ij} = \delta_{ij}$ for $i \leq j$), P and P^{-1} are bounded, $Q \in \mathcal{J}$. Therefore we can define det $(F\hat{w}|\hat{w}')$ as det $(1 + P^{-1}Q)$. Note that in the case $F \in \Gamma_-$ we can define $\det(F\hat{w}|\hat{w}')$ by means of similar considerations (however we have to use lower triangular matrices in this case). If $F \notin \Gamma_+$, $F \notin \Gamma_-$, for example if F is a constant ($F \neq 1$), the right-hand side of (8) is ill defined in the bosonic case. Note that in the case $F \in \Gamma_+$ the calculation of (8) is trivial. It is easy to check that

$$\tau(w, w', W, F) = \det(Fw|w') \ . \tag{9}$$

(If $F \in \Gamma_+$ we can say that F and F^{-1} are holomorphic for $|z| \leq 1$. Hence the multiplication by F determines an isomorphism between $\mathcal{A}(W)$ and $\mathcal{A}(FW)$, between $\mathcal{A}(FW)^{\perp})$ and $\mathcal{A}(W^{\perp})$ and therefore between $\Sigma(W)$ and $\Sigma(FW)$. The symbol Fw denotes the image of the basis $w \in \Sigma(W)$ by this isomorphism.) Note that the relation (9) holds for $F \in \Gamma_+$ in super case too.

The bosonic τ-function is also well-defined in the case $F \in \Gamma'$. The proof is similar to that for the group Γ_-.

Let us consider the case when $\Sigma(W) = \Sigma(FW) = 0$, i.e. the projections π_-^W, π_-^{FW} can be considered as isomorphisms between W and H_-, FW and H_-. Then we can define $\tau(W, F)$ by the formula:

$$\tau(W, F) = \det(Fe|e') \tag{10}$$

where e and e' are bases in W and FW connected with the standard one in H_- by means of isomorphisms π_-^W and π_-^{FW}. Of course, the above definition can be considered as a special case of (8). However the calculation of general τ-function can be reduced to the special case (10).

The following obvious equations are useful by the calculation of τ-function

$$\begin{aligned} \tau(w, w'', W, F_2F_1) &= \tau(w, w', W, F_1)\tau(w', w'', F_1, W, F_2) \, , \\ \tau(w, w', W, F) &= \tau^{-1}(w', w, FW, F^{-1}) \ . \end{aligned} \tag{11}$$

The τ-function is defined as an infinite dimensional determinant. However the calculation of this function can be reduced to the calculation of finite dimensional determinants. Let us assume at first that the function F is holomorphic for $|z| \leq 1$. In the bosonic case we have to add the requirement $F \in \Gamma'$ i.e. $F = \prod\limits_i (z - a_i), |a_i| < 1$. Under these conditions the multiplication by F determines a map from $\mathcal{A}(W)$ into $\mathcal{A}(FW)$ and a map from $\mathcal{A}((FW)^{\perp})$ into $\mathcal{A}(W^{\perp})$. Let us consider the exact sequence (acyclic complex) $\mathcal{E}$

$$\mathcal{A}(W) \to \mathcal{A}(FW) \to \mathcal{D} \to \Pi^N \mathcal{A}^*(W^{\perp}) \to \Pi^N \mathcal{A}^*((FW)^{\perp}) \, . \tag{12}$$

Here $\mathcal{D}$ denotes the kernel of the map $\pi_- F$ from H_- to H_-. The map $\mathcal{A}(FW) \to \mathcal{D}$ transforms the function $h \in \mathcal{A}(FW)$ into $\pi_- F^{-1} h$, the map $\mathcal{D} \to \Pi^N \mathcal{A}^*(W^\perp)$ transforms $a \in \mathcal{D}$ into the linear functional $\langle a, Fx\rangle$, $x \in \Pi^N \mathcal{A}(W^\perp)$. The multiplication by F generates the other maps in (12). One can prove[2] that under the above condition the τ-function $\tau(w, w', W, F)$ is equal (up to sign) to the Reidemeister torsion of the complex (12):

$$\tau(w, w', W, F) = \pm T \text{ or } \mathcal{E} \ . \tag{13}$$

This result is useful in concrete calculations as well as in general considerations. For example it can be used to prove the relation

$$\begin{aligned} \ln \tau(w, w', W, F) &= (-1)^{N+1} \ln \tau(\Pi^N w', \Pi^N w, (FW)^\perp, F^{-1}) \\ &= (-1)^N \ln \tau(\Pi^N w, \Pi^N w', W^\perp, F^{-1}) \ . \end{aligned} \tag{14}$$

(It is evident that $\Sigma(W^\perp) = \Pi^N \Sigma(W)$ and therefore $\Pi^N w, \Pi^N w'$ can be considered as bases in $\Sigma(W^\perp), \Sigma((FW)^\perp)$.)

Equation (14) follows from (13) if $N \geq 1$ and F is meromorphic for $|z| \leq 1$. (One can represent F as a quotient of two functions holomorphic for $|z| \leq 1$ and use (13) for these functions.) To derive (14) from (13) in the bosonic case we have to suppose that $F = \prod_i (z - a_i) \cdot \prod_j (z - b_j)^{-1}, |a_i| < 1, |b_j| < 1$.

Let us formulate some generalizations of the above definitions and results. We will introduce the space $H(p, q)$ as the space of functions on the circle $|z| = 1$ taking values in the space $\mathbb{C}^{p,q}$ ((p, q)-dimensional linear complex superspace). The space H can be identified with $H(1, 0)$ in the bosonic case, with $H(1, 1)$ for $N = 1$ and with $H(2^{N-1}, 2^{N-1})$ for $N \geq 1$. If $\mathbb{C}^{p,q}$ is provided by a bilinear scalar product we can introduce a bilinear scalar product in $H(p, q)$. (Of course if the scalar product is odd the scalar product in $H(p, q)$ will be odd too.) If W is a subspace of $H(p, q)$ we define $W^\perp$, $\mathcal{A}(W)$ and $\Sigma(W)$ as earlier (in the definition (7) of $\Sigma(W)$ we take $N = 0$ if the scalar product is even and $N = 1$ if the scalar product is odd).

One can define easily the analogs of Gr and identify $m(W)$ and $m(\Sigma(W))$ for $W \in$ Gr. If K is a matrix group acting in $\mathbb{C}^{p,q}$ we define the group $\Gamma(K)$ as a group of smooth matrix functions $g(z)$ taking values in the group K. The group $\Gamma(K)$ naturally acts in $H(p, q)$ and correspondingly in the grassmannian Gr. Note that the group Γ for $N \geq 1$ can be embedded in the group $\Gamma(K)$, where $K = \mathrm{SL}(2^{N-1}, 2^{N-1})$ denotes the group of volume preserving transformations in $\mathbb{C}^{2^{N-1},2^{N-1}}$. The definition of the groups, $\Gamma_+(K), \Gamma_-(K), \Gamma'(K)$ is similar to that of $\Gamma_+, \Gamma_-, \Gamma'$. If $W \in \mathrm{Gr}, F \in \Gamma(K), w, w'$ are bases in W, FW the τ-function $\tau(w, w', W, F)$ can be defined formally as earlier by means of (10). Note that the embedding of Γ into $\Gamma(K)$ can be used to check that the new definition of τ-function contains the old one. It is easy to verify that the τ-function is well-defined in the case $F \in \Gamma_-(K)$ or $F \in \Gamma'(K)$. If π_-^W is an isomorphism (i.e. $\Sigma(W) = 0$), we can consider the basis $e_{n,\alpha}$ in W which transforms by the map π_-^W in the standard basis $\{z^{-n}e_\alpha\}$ of H_-. Let us define the matrix Baker-Akhieser function $B_W(z), W \in \mathrm{Gr}$ by the formula:

$$e_{1,\alpha} = B_W^{\alpha\beta}(z)e_\beta \,. \tag{15}$$

(Here $\{e_\alpha\}$ denotes a basis in $\mathbb{C}^{p,q}$.) In other words one can define the Baker-Akhieser function as a matrix function having the form

$$B_W^{\alpha\beta}(z) = \delta^{\alpha\beta}z^{-1} + \sum_{k=1}^{\infty} b_k^{\alpha\beta}z^{k-1}$$

and satisfying

$$B_W^{\alpha\beta}(z)e_\beta \in W \,.$$

Note that one can replace the argument z of Baker-Akhieser function by a nondegenerate matrix S and consider a matrix function

$$B_W(S) = S^{-1} + \sum_{k=1}^{\infty} b_k S^{k-1} \,.$$

Let us calculate the τ-function in the case when $F(z) = 1 - Az^{-1} \in \Gamma_-(\mathrm{GL}(p,q))$ and the projections π_-^W and π_-^{FW} are isomorphisms. Using (11) we can represent the τ-function under consideration through τ-functions for $F_1 = z - A, F_2 = z, F_1, F_2 \in \Gamma'(\mathrm{GL}(p,q))$. Using (12) we can obtain

$$\tau(W, F) = \det(B_{W^\perp}(A^*) \cdot A^*) \,. \tag{16}$$

(Here * denotes conjugation with respect to the scalar product in $\mathbb{C}^{p,q}$.)

Really, using (11) we obtain

$$\begin{aligned}\tau(W, 1 - Az^{-1}) &= \tau(0, 0, W, 1 - Az^{-1}) \\ &= \tau(0, \tilde{w}, W, z - A) \cdot \tau^{-1}(0, \tilde{w}, W', z) \,. \end{aligned} \tag{17}$$

Here $\widetilde{W} = (z - A)W, W' = (1 - Az^{-1})W, \tilde{w}$ denotes a basis in $\Sigma(\widetilde{W}) = \mathcal{A}(\widetilde{W})$ and 0 denotes the basis in zero dimensional spaces $\Sigma(W), \Sigma(W')$. To calculate $\tau(0, \tilde{w}, W, z - A)$ we use (12). It is easy to see that in the case under consideration only the second and third members in (12) are nonzero. The basis v in the space $\mathcal{D} = \operatorname{Ker} \pi_-(z - A)$ can be written in the form:

$$v_\alpha = \sum_{k=1}^{\infty} A_{\alpha\beta}^{k-1} z^{-k} e_\beta ,$$

$\{e_\alpha\}$ denotes a basis in the space $\mathbb{C}^{p,q}$. It follows from (12) that

$$\tau(0, \tilde{w}, W, z - A) = \det(v|\pi_-(z - A^{-1})\tilde{w}) . \tag{18}$$

Let us take the basis e in W connected with the standard basis in H_- by the isomorphism π_-^W:

$$e_{n,\alpha} = z^{-n} e_\alpha + \sum_{k=1}^{\infty} U_{nk}^{\alpha\beta} z^{k-1} e_\beta . \tag{19}$$

The basis $\tilde{w}$ in $\mathcal{A}(\widetilde{W})$ can be chosen in the form:

$$\tilde{w}_\alpha = \left(\delta^{\alpha\beta} - \sum_{k=1}^{\infty} \sum_{n=1}^{\infty} (A^k U_{kn})^{\alpha\beta} z^{n-1} \right) e_\beta \tag{20}$$

where U_{kn} are defined by (19). It is easy to check that by this choice of basis $\tilde{w}$ we have $\tau(0, \tilde{w}, W, z - A) = 1$. Using (18) for $A = 0$ we obtain

$$\tau(W, 1 - Az^{-1}) = \tau^{-1}(0, \tilde{w}, W', z) = \det(\pi_- z^{-1} \tilde{w}|v) , \tag{21}$$

$$\tau(W, 1 - Az^{-1}) = \det \left(1 - \sum_{k=1}^{\infty} A^k U_{k1} \right) . \tag{22}$$

It is easy to verify that $z^{-1}\left(1 - \sum_{k=1}^{\infty} U_{k1}^* z^{k-1}\right) \in W^\perp$ and therefore coincides with the Baker-Akhiezer function $B_{W^\perp}(z)$. We obtain Eq. (16). Note that the τ-function $\tau(W, (z - A)(z - B)^{-1})$ may be calculated in a similar way. Using (16) we obtain the equation

$$\begin{aligned} \tau(W, F) &= \tau(W, (z - A)(z - B)^{-1}) \\ &= \frac{\det(B_{W^\perp}(A^*) \cdot A^*)}{\det(B_{F^{-1}W^\perp}(B^*) \cdot B^*)} . \end{aligned} \tag{23}$$

By means of (16) and (14) (which can be proved as earlier) we can obtain the equation which connects τ-function and Baker-Akhieser function for the space $W \in \mathrm{Gr}$:

$$\ln \tau(w, w', W, (1 - Az^{-1})^{-1}) = \pm \ln \det(B_W(A^*) \cdot A^*) . \qquad (24)$$

This equation can be considered as a generalisation of the standard relation between the τ-function and Baker-Akhieser function. In particular it is useful in the case $A \in Q$ (Q denotes the subgroup of GL corresponding to the operators of the multiplication by even invertible element of $\Lambda_N = \mathbb{C}^{2^{N-1},2^{N-1}}$). The consideration above shows that in the supercase ($N \geq 1$) the multiplication by $F \in \Gamma$ generates an isomorphism between $m(W)$ and $m(FW)$ where $W \in \mathrm{Gr}$; this isomorphism will be denoted by α_F. (Really the multiplication by F transforms an admissible basis in W into a basis FW which is connected with the admissible basis by a matrix having well-defined determinant.) Using the isomorphisms $m(W) = m(\Sigma(W))$, $m(FW) = m(\Sigma(FW))$ we obtain isomorphisms between $m(\Sigma(W))$ and $m(\Sigma(FW))$; this isomorphism will be denoted by $\tilde{\alpha}_F$. It is easy to check that $\alpha_{F \circ G} = \alpha_G \circ \alpha_F$, $\tilde{\alpha}_{F \circ G} = \tilde{\alpha}_G \circ \tilde{\alpha}_F$. Let us define a determinant bundle over Gr as a line bundle with fibres $m(W)$, $W \in \mathrm{Gr}$. It follows from the above considerations that in the supercase the group Γ acts in the determinant bundle. (Note that in the bosonic case the group Γ does not act naturally in the determinant bundle. However one can define the action of the central extension of Γ.) If two spaces $W, W' \in \mathrm{Gr}$ belong to the same orbit of group Γ then by means of the element $F \in \Gamma$ satisfying $W' = FW$ we can construct an isomorphism between $m(W)$ and $m(FW)$. However, in general this isomorphism depends on the choice of $F \in \Gamma$ and therefore we cannot say that isomorphism between $m(W)$ and $m(W')$ is canonical. If W, W', W'' belong to the same orbit of the group Γ in Gr, i.e. $W' = F'W, W'' = F''W, F', F'' \in \Gamma$ then the space $\widetilde{W} = F'' \cdot F'W$ do not depend on the choice of the F', F'' as well as the isomorphism between $m(W) \otimes m(\widetilde{W})$ and $m(W') \otimes m(W'')$. (We define this isomorphism as tensor product of isomorphisms $\alpha_{F'} : m(W) \to m(W')$ and $\alpha_{F'}^{-1} : m(\widetilde{W}) \to m(W'')$.)

The canonical isomorphism $m(\Sigma(W)) \otimes m(\Sigma(\widetilde{W})) = m(\Sigma(W')) \otimes m(\Sigma(W''))$ can be used to construct the super Mumford form for superstring. Remember that the natural generalisation of the Krichever construction assigns to every $(1|N)$ dimensional compact complex supermanifold $\mathcal{M}$ an element $W(\mathcal{M}) \in \mathrm{Gr}$ and this element satisfies $W^{\perp} = \phi W$ for some $\phi \in \Gamma$. The subset of Gr(N) consisting of such subspaces is denoted by $\mathrm{UMS_N}$. Using the isomorphisms

$$m(\Sigma(\phi^3 W)) \otimes m(\Sigma(W)) = m(\Sigma(\phi^2 W)) \otimes m(\Sigma(\phi W)) ,$$
$$m(\Sigma(\phi^2 W)) \otimes m(\Sigma(W)) = m(\Sigma(\phi W)) \otimes m(\Sigma(\phi W)) ,$$

following from (18) and

$$m(\Sigma(W^{\perp})) = m^*(\Sigma(W)), N \text{ is odd} ,$$

we obtain for $W \in \mathrm{UMS_1}$ that

$$m(\Sigma(\phi^3 W)) = (m^*(\Sigma(W)))^5 . \tag{25}$$

The isomorphism (25) is called a super Mumford form. One can check that for $W = W(\mathcal{M})$ where $\mathcal{M}$ is a superconformal manifold this isomorphism coincides with conventional super Mumford form.[2] This assertion becomes evident if we note that the isomorphism (25) for $W = W(\mathcal{M})$ coincides with the isomorphism used in Ref. 8 by the construction of super Mumford form.

As we said the isomorphism between $m(\Sigma(W)) = m(W)$ and $m(\Sigma(FW)) = m(W)$ in general depends on the choice $F \in \Gamma$. However one can prove that this isomorphism does not depend on the choice of $F \in \Gamma$ in the important case $W^{\perp} = W$ if the projection π_-^W is one-to-one. (This assertion will be proved in the Appendix.) This result remains correct of course for every space W' belonging to the orbit ΓW (i.e. for every space $W' = \phi W, \phi \in \Gamma$). For odd N one can prove that the isomorphism between $m(W)$ and $m(FW)$ does not depend on the choice of $F \in \Gamma$ for every $W \in$ UMS (see Appendix). The same result remains correct in general if $N \geq 2$ is even and index W is equal to $(0|0)$. Then $W = \phi\widetilde{W}$ where $\phi \in \Gamma, \widetilde{W}^{\perp} = \widetilde{W}$. Noting that in general case the projection $\pi_-^{\widetilde{W}}$ is one-to-one we obtain the above assertion. In particular this assertion can be applied to the case $W = W(\mathcal{M})$ where $\mathcal{M}$ is a $(1|N)$-dimensional compact complex manifold. In this case one can give an independent proof. (An equivalent statement was proved by Deligne.) In conclusion we consider briefly the connection between the theory of formal Baker-Akhiezer functions and τ-functions with the generalizations of KP hierarchy. We define at first the matrix Baker-Akhiezer function $B(W, F, z)$ as a matrix function having the form $B(W, F, z) = F^{-1}(z)(z^{-1} + \sum_{k=1}^{\infty} U_k z^{k-1}) = F^{-1}(z)\widetilde{B}(W, F, z)$

and satisfying $B^{\alpha\beta}(W,F,z)e_\beta \in W$ (here $F(z) \in \Gamma, W \in \mathrm{Gr}, U_k$ are linear operators in $\mathbb{C}^{p,q}$). Note that one can define the function $\widetilde{B}(W,F,S)$ depending on the matrix argument S. In particular case $F = 1$ the function $\widetilde{B}(W,F,S)$ coincides with the function $B_W(S)$ defined by (15). Conversely one can express $\widetilde{B}(W,F,S)$ through $B_W(S)$:

$$\widetilde{B}(W,F,S) = B_{FW}(S) \, .$$

It follows from the above considerations that

$$\oint B(W,F,z)B^+(W^\perp,F',z)dz = 0\,, \tag{26}$$

$$\det(\widetilde{B}(W^\perp,(F^T)^{-1},S^*)\cdot S^*) = \frac{\tau(W,(1-Sz^{-1})\cdot F)}{\tau(W,F)}\,, \tag{27}$$

$$\ln \tau(W^\perp,F) = \pm \ln \tau(W,F^{-1})\,. \tag{28}$$

(Equation (26) can be derived from the orthogonality of $B^{\alpha\beta}(W,F,z)e_\beta \in W$ and $B^{\gamma\delta}(W^\perp,F',z)e_\delta \in W^\perp$. Equations (27) and (28) follow from (14) and (16).)

In the scalar case ($N = 0$) one must assume that in the definition of the Baker-Akhiezer function and τ-function we have $F \in \Gamma_-$. Representing $F \in \Gamma_-$ in the form $F(z) = \exp\left(\sum_{k=1}^\infty x_k z^{-k}\right)$ we can consider these functions as depending on the infinite set of complex variables $x = (x_1, \ldots x_k, \ldots)$. Equations (26), (27), (28) then take the form

$$\oint B(W,x,z)B(W^\perp,x',z)dz = 0\,, \tag{29}$$

$$B(W^\perp,-x',z) = \frac{\hat{X}_+(x',z)\tau(W,x')}{\tau(W,x')}\cdot z^{-1}\,, \tag{30}$$

$$\tau(W^\perp,x) = \tau(W,-x) \tag{31}$$

where $\hat{X}_+(x,z) = \exp\left(\sum\limits_{k=1}^{\infty} x_k z^{-k}\right)\exp\left(-\sum_{k=1}^\infty \frac{z^k}{k}\frac{\partial}{\partial x_k}\right)$. Using (27) and (31) we obtain

$$B(W,x,z) = \frac{\hat{X}_-(x,z)\tau(W,x)}{\tau(W,x)}\cdot z^{-1} \tag{32}$$

where $\hat{X}_-(x,z) = \exp\left(-\sum\limits_{k=1}^{\infty} x_k z^{-k}\right)\exp\left(\sum\limits_{k=1}^{\infty}\frac{z^k}{k}\frac{\partial}{\partial x_k}\right)$. Hence (29), (30) and (31) immediately give the bilinear Hirota equations for τ-function (see for example Ref. 7).

$$\oint dz z^{-2} \hat{X}_+(x,z)\tau(x,z)\hat{X}_-(x',z)\tau(W,x') = 0 .$$

Therefore one can consider (26), (27) and (28) as matrix analog of KP hierarchy. (Related questions are studied in Refs. 4,5 and 6.)

Appendix

Let us suppose that $W \in \text{UMS} \subset \text{Gr}(N), N \geq 1, W^{\perp} = W$ and the projection π_-^W is one-to-one. We will prove under these conditions that the isomorphism between $m(W)$ and $m(FW)$ does not depend on the choice of $F \in \Gamma$. As follows from (11) it is sufficient to check that

$$\tau(W, F_0) = \tau(0,0,W,F_0) = 1 \tag{A1}$$

if $F_0 \in \Gamma$ and $W = F_0 W$.

For the proof we consider at first the matrix of the operator of multiplication by the function $F \in \Gamma$. The standard basis $w = (\pi_-^W)^{-1} w_0$ in W where w_0 is the standard basis in H_- can be written in the form

$$e_{n,\alpha} = z^{-n} e_\alpha + \sum_{k=1}^{\infty} U_{nk}^{\alpha\beta} z^{k-1} e_\beta \tag{A2}$$

where $\{e_\alpha\}$ is an orthonormal basis in the Grassmann algebra Λ_N generated by the elements $\theta_1, \ldots, \theta_N$. (The bilinear scalar product in Λ_N is defined by the formula $(f,g) = \int f(\theta) \cdot g(\theta) d\theta$.) The Grassmann algebra Λ_N can be identified with linear superspace $\mathbb{C}^{2^{N-1}, 2^{N-1}}$. For $F \in \Gamma$ we define half-infinite dimensional matrices $F^{--}, F^{-+}, F^{+-}, F^{++}$ by the formulas:

$$F_{n,k}^{--} = \phi_{-n,-k}\,; \quad F_{n,k}^{-+} = \phi_{-n,k-1}\,; \quad F_{n,k}^{+-} = \phi_{n-1,-k}\,;$$
$$F_{n,k}^{++} = \phi_{n-1,k-1} \,.$$

Here $n, k \in \mathbb{N}$, ϕ denotes the matrix of multiplication operator F in the standard basis w_0 of H_-. Matrices $F^{ij}, i,j = +,-$ are connected by the equations

$$(F^{++})^T = F^{--}, \quad (F^{-+})^T = F^{-+}, (F^{+-})^T = F^{+-} \tag{A3}$$

Let $\{e_{n,\alpha}\}$, $\{\tilde{E}_{m,\beta}\}$ be the standard bases in W, FW correspondingly, $e_{n,\alpha}$ is given by (2), $\tilde{e}_{m,\beta} = z^{-m}e_\beta + \sum_{k=1}^{\infty} U^{\beta\gamma}_{mk} z^{k-1} e_\beta$. Then the matrix of the operator F can be written in the form

$$\begin{aligned} F \cdot e_{n,\alpha} &= (F^{--} + UF^{+-})^{\alpha\beta}_{nk} z^{-k} e_\beta + (F^{-+} + UF^{++})^{\alpha\beta}_{np} z^{p-1} e_\beta \\ &= R^{\alpha\beta}_{nk} \tilde{e}_{k,\beta} \end{aligned}$$

where

$$\begin{aligned} R^{\alpha\beta}_{n,k} &= (F^{--})^{\alpha\beta}_{nk} + U^{\alpha\gamma}_{ns}(F^{+-})^{\gamma\beta}_{sk} \,, \\ R^{\alpha\gamma}_{np} U^{\gamma\beta}_{pq} &= (F^{-+})^{\alpha\beta}_{nq} + U^{\alpha\mu}_{ns}(F^{++})^{\mu\beta}_{sq} \,. \end{aligned} \tag{A4}$$

Note that by the definition $\tau(W,F) = \tau(0,0,W,F) = \det R$. In our case $F_0 W = W$ and we obtain from (A4) the equations

$$\begin{aligned} R_0 &= F_0^{--} + UF^{+-} \,, \\ R_0 U &= F_0^{-+} + UF_0^{++} \,. \end{aligned} \tag{A5}$$

If $\tilde{R}_0 = R_0 - 1, \tilde{F}^{--} = F^{--} - 1$, $\tilde{F}^{++} = F^{++} - 1, \tilde{F}^{+-} = F^{+-}$, $\tilde{F}^{-+} = F^{-+}$ we have from (A5)

$$\begin{aligned} \tilde{R}_0 &= \tilde{F}^{--} + U\tilde{F}^{+-} \,, \\ \tilde{R}_0 u &= \tilde{F}^{-+} + U\tilde{F}^{++} \,. \end{aligned} \tag{A6}$$

It is easy to check that

$$\begin{aligned} \tilde{R}_0^k &= A_k^{--} + UB_k^{+-} \,, \quad k \geq 1 \,, \\ A_k^{--} &= \sum_{i_1,\dots i_{k-1}} \tilde{F}_0^{ii_1} \tilde{F}_0^{i_1 i_2} \dots \tilde{F}_0^{i_{k-1} i} \,, \\ & \qquad i = -, i_1, \dots i_{k-1} = +, - \,, \\ B_k^{+-} &= \sum_{j_1,\dots j_{k-1}} \tilde{F}_0^{ij_1} \tilde{F}_0^{j_1 j_2} \dots \tilde{F}^{j_{k-1} j} \,, \\ & \qquad i = +, j = -, j_1, \dots j_{k-1} = +, - \,. \end{aligned} \tag{A7}$$

The proof is based on the recurrent equations

$$\begin{aligned} A_{k+1}^{--} &= \tilde{F}_0^{--} A_k^{--} + \tilde{F}_0^{-+} B_k^{+-} \,, \\ B_{k+1}^{+-} &= \tilde{F}_0^{+-} A_k^{--} + \tilde{F}_0^{++} B_k^{+-} \,, \end{aligned}$$

following from (A6).

Let us consider the set Q of the operators of multiplication by even element from Λ_N acting in $\Lambda_N = \mathbb{C}^{2^{N-1},2^{N-1}}$ (by definition if $a, b \in Q$ then $ab, a+b \in Q$). Obviously $F^{ij}_{kn} \in Q, i, j = +, -$ and therefore $(A_k^{--})_{mn}$, $(B_k^{+-})_{mn}$, $k \geq 1$. One can easily see that for $a \in Q$ the equation $\mathrm{Sp}\, a = 0$ is valid (remember that Sp denotes the supertrace of the matrix).

The condition $W^{\perp} = W$ means that in the basis (A2) the matrix $U = \{U^{\alpha\beta}_{nk}\}$ defined by (A2) is antisymmetric:

$$U^T = -U\ . \tag{A8}$$

Now we can calculate det R_0. Let us use the decomposition

$$\mathrm{Sp}\ln(1+\varepsilon\widetilde{R}_0) = \sum_{k=1}^{\infty}(-1)^k\frac{\varepsilon^k}{k}\mathrm{Sp}(\widetilde{R}_0^k)\ .$$

In our case ($F_0 \in \Gamma$) one can prove that this decomposition is valid in some neighborhood of the point $\varepsilon = 0, \varepsilon \in \mathbb{C}$. We shall prove that Sp $\widetilde{R}_0^k = 0$. Reality it follows from (A7) that Sp $\widetilde{R}_0^k = \mathrm{Sp}(A_k^{--} + uB_k^{+-})$. Using that $(A_k^{--})_{mn} \in Q$ we obtain $\mathrm{Sp}A_k^{--} = 0$. We have from (A3) and (A7) that B_k^{+-} is symmetric:

$$(B_k^{+-})^T = B_k^{+-}\ , \quad k \geq 1\ . \tag{A9}$$

Hence

$$\mathrm{Sp}\widetilde{R}_0^k = \mathrm{Sp}UB_k^{+-} = \mathrm{Sp}(B_k^{+-})^T U^T = -\mathrm{Sp}UB_k^{+-}$$

and

$$\mathrm{Sp}\widetilde{R}_0^k = 0 \qquad \text{for } k \geq 1\ .$$

We see that $\ln\det(1+\varepsilon\widetilde{R}_0) = 0$ for small ε and therefore by analyticity for $\varepsilon = 1$. Hence

$$\tau(W, F_0) = \det R_0 = 1$$

and relation (A1) is proved.

Note that for the case where N is odd, one can give a simple proof of more general assertion than (A1). Namely, if $W \in$ UMS, $F \in \Gamma, FW = W$ and w is a basis in $\Sigma(W) = \Sigma(FW)$ then

$$\tau(w, w, W, F) = 1\ . \tag{A10}$$

Really if $W^{\perp} = \phi W, \phi \in \Gamma$ it follows from (A11), (A14) that

$$\begin{aligned} \ln \tau(w, w, W, F) &= (-1)^N \ln \tau(\tilde{w}, \tilde{w}, W^{\perp}, F^{-1}) \\ &= (-1)^N \ln \tau(\tilde{w}, \tilde{w}, \phi W, F^{-1}) \\ &= (-1)^N \ln \tau(w, w, W, F^{-1}) \end{aligned}$$

where $\tilde{w} = \Pi^N w$ is a basis in $\Sigma(W^{\perp})$. Using that N is odd we obtain (A10).

References

1. A. S. Schwarz, *JETP Lett.* **46** (1987) 340.
2. A. S. Schwarz, *Nucl. Phys.* (in press).
3. G. Segal and G. Wilson, *Publ. Math. IHES* **61** (1985) 1.
4. Yu. I. Manin and A. O. Radul, *Commun. Math. Phys.* **98** (1985) 65.
5. J. Ueno and H. Yamada, *Lett. Math. Phys.* **13** (1987) 59.
6. H. Yamada, *Hiroshima Math. J.* **17** (1987) 377.
7. J. Verdier, Seminaire Bourlaki 34-e annee 1981/82 N 596.
8. A. Voronov, *Funk. Anal. i Priloshen.* **67** (1988) 22(2).

CONFORMAL FIELD THEORY AND PURELY ELASTIC S-MATRICES

V.A. FATEEV and A.B. ZAMOLODCHIKOV

L.D. Landau Institute for Theoretical Physics
Chernogolovka 1989

Particular perturbations of a 2D Conformal Field Theory leading to Integrable massive Quantum Field Theories are examined. The mass spectra and S-matrices for some models, including the field theory of the Ising Model with magnetic field and "thermal" deformations of the tricritical Ising and 3-state Potts models, are proposed. The hidden Lie-algebraic structures of these spectra and their relation to the Toda systems are discussed.

1. Introduction

There is a large class of 2D Quantum Field Theories associated with a Conformal Field Theory. The Conformal Field Theory corresponds to the fixed point of the Renormalization Group and all the field theories corresponding to the RG trajectories going away from this fixed point can be considered as the Conformal Field Theory perturbed by the appropriate relevant operator. So, the field theory is completely defined if one specifies its "CFT data", i.e. the fixed point (i.e. the Conformal Field Theory) and the RG trajectory (i.e. the relevant operator in this CFT). The CFT data contains explicit information about the ultraviolet asymptotics of the field theory, while its long distance properties are the subject of analysis. One can expect that the RG trajectory either ends in another "critical" fixed point (in which case the infrared limit of the field theory is described by another CFT) or goes to a "noncritical fixed point". The last possibility means that the field theory develops finite correlation length i.e. it contains only massive particles.

As is known, the massive field theory is equivalent to the relativistic scattering theory and so it is completely defined by specifying the S-matrix (which of course has to satisfy the standard requirements of unitarity, crossing symmetry, etc[a]). Contrary to the "CFT data" the "S-matrix data" exhibit some information about infrared properties of the theory in an explicit way, while the ultraviolet asymptotics have to be derived.

A link between these two kinds of data would provide a good viewpoint for understanding the general structure of the 2D QFT. In general this problem does not look tractable. Whereas the CFT data can be specified in a relatively simple way once the CFT is known, the general S-matrix is a very complicated object even in 2D. However, there is an important class of 2D S-matrices known as the Purely Elastic (or Factorized) S-matrices (PESM), which can be described in great detail (see e.g. Ref. 1). So, there is the much more tractable "intermediate" problem of relating the PESM to the CFT data. The PESM is characteristic for an integrable QFT. In particular, any PESM is compatible with an infinite set of additive (local) Integrals of Motion (IM). On the other hand, it is possible to analyse the equations of motion in perturbed CFT and show that there are particular perturbations, which preserve a number of nontrivial local IM[2,3]. The IM have the form

$$P_s = \int T_{s+1} dz + \theta_{s-1} d\bar{z} \,, \tag{1.1}$$

where $(z, \bar{z})$ are standard light cone coordinates, T_{s+1} and θ_{s-1} are local field, satisfying the continuity equation

$$\partial_{\bar{z}} T_{s+1} = \partial_z \theta_{s-1} \,, \tag{1.2}$$

and the suffix s indicates the Lorentz spin of the operator P_s; $s \in \{s\}$, where $\{s\}$ is a certain subset of the positive integers (we explicitly mention only the "left" IM P_s; in fact, there is an analogous set of "right" IM $\overline{P}_s$ with negative spins $-s$, which are obtained from (1.1) by spatial reflection; the $\mathbb{P}$-symmetry is always implied here). The set $\{s\}$ is specific for every model; the structure of this set provides important information about the "bootstrap" properties of the corresponding PESM. In some cases this information, together with some "minimality" assumption, allows one to reconstruct the S-matrix[3,4,5].

In this paper we discuss three models of perturbed CFT

$$\hat{H}_p = H_p + \lambda \int \Phi_p(x) d^2x \,; \quad p = 3, 4, 6 \tag{1.3}$$

[a]Here we always imply the unitary Field Theory. How the concept of an S-matrix generalizes to nonunitary Field Theory is an interesting open problem.

where H_p stands for the action of the "minimal" unitary CFT with central charge $c = 1 - 6/p(p+1)$, and Φ_p is the primary field $\Phi_{1,2}$ of this CFT. (See e.g. Ref. 6 for the definitions.) We prove that each of the models (1.3) possesses nontrivial IM (1.1) and compute the first few representatives of the corresponding sets $\{s\}$:

$$s = 1, 7, 11, 13, 17, 19 \quad \text{for } p = 3\ ; \tag{1.4a}$$

$$s = 1, 5, 7, 9, 11, \quad \text{for } p = 4\ ; \tag{1.4b}$$

$$s = 1, 4, 5, 7, 8, 11, \quad \text{for } p = 6\ . \tag{1.4c}$$

We conjecture that the sets $\{s\}$ for these models consist of the exponents of the Lie algebras E_8, E_7, E_6 (for $p = 3, 4, 6$, resp.) repeated modulo the corresponding Coxetter number hp, where $h_3 = 30, h_4 = 18, h_6 = 12$. Relying on some assumptions about qualitative properties of the models (1.3), which come from "physical" reasoning, we propose the PESM for these models as the "minimal" solutions of the bootstrap equations compatible with the IM (1.1), (1.4). The mass spectra of these QFT are listed below.

For $p = 3$ the theory contains $N_3 = 8$ neutral particles $A_1, A_2, \ldots, A_8$ with the masses

$$\begin{aligned} &M_2 = 2M_1 \cos(\pi/5)\ ; && M_3 = 2M_1 \cos(\pi/30)\ ; \\ &M_4 = 2M_2 \cos(7\pi/30)\ ; && M_5 = 2M_2 \cos(2\pi/15)\ ; \\ &M_6 = 2M_2 \cos(\pi/30)\ ; && M_7 = 4M_2 \cos(\pi/5)\cos(7\pi/30)\ ; \\ &M_8 = 4M_2 \cos(\pi/5)\cos(\pi/30)\ . && \end{aligned} \tag{1.5a}$$

For $p = 4$ there are $N_4 = 7$ neutral particles $A_1, A_2, \ldots A_7$ with the masses

$$\begin{aligned} &M_2 = 2M_1 \cos(5\pi/18)\ ; && M_3 = 2M_1 \cos(\pi/9)\ ; \\ &M_4 = 2M_1 \cos(\pi/18)\ ; && M_5 = 2M_3 \cos(2\pi/9)\ ; \\ &M_6 = 2M_2 \cos(\pi/18)\ ; && M_7 = 2M_3 \cos(\pi/18)\ . \end{aligned} \tag{1.5b}$$

For $p = 6$ there are $N_6 = 6$ particles which split into two particle-antiparticle doublets $A_1, \overline{A}_1$ and $A_3, \overline{A}_3$ and two neutral particles A_2 and A_4. The masses are

$$M_3 = 2M_1 \cos(\pi/12)\ ;$$

$$M_2 = 2M_1 \cos(\pi/4)\ ; \quad M_4 = 2M_3 \cos(\pi/4)\ . \tag{1.5c}$$

There is a hidden relationship of these PESM to the root systems E_8, E_7, E_6 respectively. Note that according to our conjecture the spins

(1.4) are just the corresponding exponents repeated modulo h_p and the numbers of particles N_3, N_4 and N_6 coincide with the ranks of these systems. Moreover, under appropriate normalization, squares of the masses (1.5) coincide with the eigenvalues of the matrix

$$M_{ab} = \sum_{i=0}^{N_p} n_i \alpha_i^a \alpha_i^b \tag{1.6}$$

where α^a; $i = 1, 2, \ldots, N_p$, are components of the corresponding positive simple roots in Euclidean space and the integers n_i are defined by the decomposition

$$\alpha_0 = -\sum_{i=1}^{N_p} n_i \alpha_i \tag{1.7}$$

of the maximal positive root $-\alpha_0$. So, the masses (1.5) coincide with the spectra of the E_8, E_7 and E_6 Toda systems. A possible way of understanding the relationship to the Toda systems is discussed in Sec. 7.

2. Integrals of Motion in the Models $\hat{H}_3, \hat{H}_4, \hat{H}_6$

The space of fields in "minimal" unitary CFT H_p is made up of $p(p-1)/2$ irreducible representations $V_{(n,m)}$; $n = 1, 2, \ldots, p-1$, $m = 1, 2, \ldots, p$ of the $c = 1 - 6/p(p+1)$ Virasoro algebra (in fact there are two Virasoro algebras VIR and $\overline{\text{VIR}}$, the conformal blocks being constructed by tensoring the representations V and $\overline{V}$)[6]. Amongst them the two spaces, $V_{(1,1)}$ and $V_{(1,2)}$, are of special importance in our present study. The former space is obtained by applying the Virasoro generators L_n with $n < -1$ to the identity field I while the latter is similarly obtained from the primary field $\Phi_{(1,2)}$. The following contour integral

$$\overline{D}V_{(1,1)}(z, \bar{z}) = \int \Phi_{(1,2)}(\varsigma, \bar{z}) V_{(1,1)}(z, \bar{z}) d\varsigma \tag{2.1}$$

where the contour surrounds the point z, defines the operator $\overline{D} : V_{(1,1)} \to V_{(1,2)}$. Let $\hat{V}_{(1,1)}$ and $\hat{V}_{(1,2)}$ be the factor spaces $V_{(1,1)}/L_{-1}V_{(1,1)}$ and $V_{(1,2)}/L_{-1}V_{(1,2)}$ where the subspaces $L_{-1}V_{(1,1)}$ and $L_{-1}V_{(1,2)}$ contain the total ∂_1 derivatives. Let us define the operator $\hat{D} : \hat{V}_{(1,1)} \to \hat{V}_{(1,2)}$ as the product $\hat{D} = \pi * \overline{D}$, where π is the projector $\pi : V_{(1,2)} \to \hat{V}_{(1,2)}$; this operator is well-defined on the factor space since $\overline{D}$ commutes with $L_{-1} = \partial_z$. Each of the factor spaces V admits the standard decomposition

$$\hat{V}_{(1,1)} = \oplus_{s=0}^{\infty} \hat{V}_{(1,1)}^s \; ; \; \hat{V}_{(1,2)} = \oplus_{s=0}^{\infty} \hat{V}_{(1,2)}^s \tag{2.2}$$

in terms of the eigenspaces of the spin operator $L_0 - \overline{L}_0$. As is shown in Ref. 5, every element of ker $\hat{D}$ (if the kernel exists) gives rise to local IM in the perturbed theory (1.3). Namely, every field $T_{s+1} \in$ ker $\hat{D}(\hat{V}^{s+1}_{(1,1)})$ satisfies (in the perturbed theory) the continuity equation (1.2) with some local field $\theta_{s-1} \in V_{(1,2)}$ and hence gives rise to IM (1.1) of spin s.

In general, the computation of ker $\hat{D}$ appears to be a complicated problem. However for the first few s, one can reveal a nonvanishing kernel just by dimensional counting. The dimensionalities of the factor spaces $V_{(1,1)}$ and $V_{(1,2)}$ are given by the formulae

$$\sum_{s=0}^{\infty} q^s \dim(\hat{V}^s_{(1,1)}) = (1-q)\chi_{(1,1)}(q) + q \; ; \tag{2.3a}$$

$$\sum_{s=0}^{\infty} q^{s+\Delta} \dim\left(\hat{V}^s_{(1,2)}\right) = (1-q)\chi_{(1,2)}(q) \, , \tag{2.3b}$$

where $\chi_{(1,1)}(q)$ and $\chi_{(1,2)}(q)$ are the characters of the corresponding irreducible Virasoro module[6]. Let us analyze the situation separately for the models $\hat{H}_p$; $p = 3, 4$, and 6.

$p = 3$

In this case the characters in (2,3) are given by the formulae[7]

$$\chi_{(1,1)}(q) = 1/2\left\{ \prod_{n=1}^{\infty}(1+q^{n+1/2}) + \prod_{n=0}^{\infty}(1-q^{n+1/2})\right\} ; \tag{2.4a}$$

$$\chi_{(1,2)}(q) = q^{1/16}\prod_{n=0}^{\infty}(1-q^{2n+1})^{-1} = q^{1/16}\prod_{n=1}^{\infty}(1+q^n) \tag{2.4b}$$

Using (2.3) and (2.4) one can check that for $s = 1, 7, 11, 13, 17$ and 19 $\dim(\hat{V}^{s+1}_{(1,1)})$ exceeds by one $\dim(\hat{V}^s_{(1,2)})$. This proves that this model possesses at least five nontrivial IM (1.1) with the spins $s = 7, 11, 13, 17, 19$ (the IM P_1 always coincides with the left light-cone component of momentum; it is considered as a trivial IM). For other values of s the dimensionalities of $\hat{V}^s_{(1,2)}$ are always greater than or equal to those of $\hat{V}^{s+1}_{(1,1)}$. Of course, this does not mean that there are no more IM, but if other IM exist, they cannot be recovered by this simple method. We conjecture that this model possesses an infinite set of IM (1.1) for all integers s relatively prime to 30.

$p = 4$

For $p = 4$ the characters in (2.3) are given by the following expressions[7]

$$\chi_{(1,1)}(q) = \frac{1}{2}\{ \prod_{n=0}^{\infty} (1 + q^{5n+3/2})(1 + q^{5n+5/2})(1 + q^{5n+7/2}) + \prod_{n=0}^{\infty} (1 - q^{5n+3/2})(1 - q^{5n+5/2})(1 - q^{5n+7/2})\} \Big/ \prod_{n=0}^{\infty} (1 - q^{5n+2})(1 - q^{5n+3}), \tag{2.5a}$$

$$\chi_{(1,2)}(q) = \frac{1}{2} q^{1/10} \{ \prod_{n=0}^{\infty} (1 + q^{5n+5/2})(1 + q^{5n+9/2})(1 + q^{5n+1/2}) + \prod_{n=0}^{\infty} (1 - q^{5n+5/2})(1 - q^{5n+9/2})(1 - q^{5n+1/2})\} \Big/ \prod_{n=0}^{\infty} (1 - q^{5n+1})(1 - q^{5n+4}) . \tag{2.5b}$$

Expanding these expressions in power series in q one can show[8] that $\dim(\hat{V}^{s+1}_{(1,1)}) = \dim(\hat{V}^{s}_{(1,2)}) + 1$ for $s = 1, 5, 7, 9, 11, 13$ and $\dim(\hat{V}^{s+1}_{(1,1)}) \leq \dim(\hat{V}^{s}_{(1,2)})$ in all other cases. So, this model possesses nontrivial IM (1.1) with spins 5,7,9,11,13. We conjecture that these are the first representatives of the infinite set of IM P_s with $\{s\}$ containing the numbers 1, 5, 7, 9, 11, 13, 17 repeated modulo 18.

$p = 6$

The characters $\chi_{(1,1)}(q)$ and $\chi_{(1,2)}(q)$ in (2.3) can be obtained from the formula[6]

$$\chi_{(n,m)}(q) = q^{\Delta(n,m)} \prod_{l=1}^{\infty} (1 - q^l)^{-1} \sum_{k \in \mathbb{Z}} [q^{\{(84k+7n+6m)^2-1\}/168} - q^{\{(84k+7n-6m)^2-1\}/168}], \tag{2.6}$$

where

$$\Delta_{(n,m)} = \{(7n - 6m)^2 - 1)/168 . \tag{2.7}$$

Comparing the dimensionalities of $\hat{V}^{s+1}_{(1,1)}$ and $\hat{V}^{s}_{(1,2)}$ computed from (2.3) and (2.6), one can prove the existence of the IM (1.1) with $s = 1, 5, 7, 11$. But these are not all IM one can discover in this model by the present method. The chiral algebra of the minimal CFT H_6 is larger than VIR. Since $\Delta_{(1,6)}$ takes the integer value 5, one can construct the local primary field $W_5(z)$ having the dimensions (5,0). Together with $V_{(1,1)}$ the space $V_{(1,6)}$ constitutes an extended chiral algebra of H_6 (more detailed study, which will be published elsewhere, shows that this is a particular case of a W-algebra associated with the root system E_6). The space $V_{(1,6)}$ can be taken as another source of local IM in the perturbed theory. The operator $\overline{D}$, defined as the contour integral in (2.1), can be applied to $V_{(1,6)}$; $\overline{D}$: $V_{(1,6)} \rightarrow V_{(1,5)}$ (The last property of $\overline{D}$ follows from the fusion rules of the CFT H_6). One can also define the factor spaces $\hat{V}_{(1,6)}$ and $\hat{V}_{(1,5)}$ and the operator $\hat{D} : \hat{V}_{(1,6)} \rightarrow \hat{V}_{(1,5)}$ and show that the new IM are related to the kernel of $\hat{D}$ in the same way. Again, the first few IM associated with $\hat{V}_{(1,1)}$ can be discovered by the simple dimensional counting. The operator $\hat{D}$ has the property $\hat{D} : \hat{V}^{s+1}_{(1,6)} \rightarrow \hat{V}^{s}_{(1,5)}$, where $\hat{V}^{s}$ are the spin components of the corresponding factor spaces. Using the character formulas (2.6) one can show that there is (at least) a one-dimensional kernel of $\hat{D}$ in $\hat{V}^{s+1}_{(1,6)}$ with $s = 4, 8$. So, the theory $\hat{H}_6$ possesses also the local IM (1.1) P_4 and P_8. There is an important difference between these new IM P_4 and P_8, associated with $V_{(1,6)}$, and the IM P_1, P_5, P_7, and P_{11}, which came from $V_{(1,1)}$. The CFT H_6, being the field theory of the tricritical point of the 3-state Potts model[9], possesses the global S_3 symmetry whose odd part $\mathbb{C}$ can be considered as "charge conjugation". The fields in $V_{(1,1)}$ ($V_{(1,6)}$) are $\mathbb{C}$ — even ($\mathbb{C}$ — odd) and the corresponding IM P_s in the perturbed theory inherit this symmetry property. The negative $\mathbb{C}$ — parity of IM P_4 and P_8 is manifested by the even values of their spins. (One can show that the spins of $\mathbb{C}$-even IM are always odd.)

3. Purely Elastic S-Matrices and Integrals of Motion

Here we will briefly discuss some general properties of PESM and in particular the limitations imposed by local IM to their "boootstrap" properties. We will restrict our attention to the "diagonal" S-matrices as only this sort of PESM will appear in our analysis of the models $\hat{H}_3$, $\hat{H}_4$ and $\hat{H}_6$.

Let us consider the purely elastic scattering theory which contains N

sorts of particles $A_a \in \{A_a\} = \{A_1, A_2, \ldots, A_N\}$ having the masses M_a. The charge conjugation $\mathbb{C}$ is the involution of $\{A_a\}$, i.e. $\mathbb{C}A_a = A_{\bar{a}} \in \{A_a\}$, so that the particle A_a is either neutral, $\mathbb{C}A_a = A_a$, or belongs to pair $(A_a, A_{\bar{a}})$, where $A_{\bar{a}} \in \{A_a\}$ is the corresponding antiparticle and $M_{\bar{a}} = M_a$. We will denote the n-particle asymptotic states as

$$|A_{a_1}(\theta_1)A_{a_2}(\theta_2)\ldots A_{a_n}(\theta_n)\rangle_{(\mathrm{in,\ out})} \tag{3.1}$$

where θ_i represents the rapidities of the corresponding particles A; the rapidity θ is related to the particle momentum

$$p = p^0 + P^1 = M\exp(\theta); \quad \overline{P} = p^0 - p^1 = M\exp(-\theta) . \tag{3.2}$$

In the "diagonal" case the purely elastic scattering is described by the equations

$$\begin{aligned} &|A_{a_1}(\theta_1)A_{a_2}(\theta_2)\ldots A_{a_n}(\theta_n)\rangle_{(\mathrm{in})} = \\ &S_{a_1,a_2,\ldots,a_n}(\theta_1,\theta_2,\ldots,\theta_n)|A_{a_1}(\theta_1)A_{a_2}(\theta_2)\ldots A_{a_n}(\theta_n)\rangle_{(\mathrm{out})} , \end{aligned} \tag{3.3}$$

where the n-particle S-matrix is the product of the two-particle ones

$$S_{a_1,a_2,\ldots,a_n}(\theta_1,\theta_2,\ldots,\theta_n) = \prod_{i<j} S_{a_i a_j}(\theta_{ij}) . \tag{3.4}$$

Here $S_{ab}(\theta)$ are the two particle scattering amplitudes

$$|A_{a_i}(\theta_i)A_{a_j}(\theta_j)\rangle_{(\mathrm{in})} = S_{a_i a_j}(\theta_{ij})|A_{a_i}(\theta_i)A_{a_j}(\theta_j)\rangle_{(\mathrm{out})} \tag{3.5}$$

and $\theta_{ij} = \theta_i - \theta_j$. We will imply the spatial reflection symmetry of the theory and so

$$S_{ab}(\theta) = S_{ba}(\theta) . \tag{3.6}$$

The amplitudes $S_{ab}(\theta)$ are meromorphic functions of θ, real at Re $\theta = 0$. They satisfy the crossing symmetry

$$S_{ab}(\theta) = S_{a\bar{b}}(i\pi - \theta) \tag{3.7}$$

and the unitarity condition

$$S_{ab}(\theta)S_{ab}(-\theta) = 1 . \tag{3.8}$$

Due to these properties the amplitudes $S_{ab}(\theta)$ are $2\pi i$ periodic functions of θ.

The simple poles of $S_{ab}(\theta)$ located in the "physical strip" $0 < \mathrm{Im}\theta < \pi$ at $\mathrm{Re}\theta = 0$ and having positive residues (if any) correspond to the "bound states" of the particles $A_a A_b$; the "bootstrap condition" requires these bound states to belong to the same set $\{A_a\}$. Let iu_{ab}^c be the position of the pole in $S_{ab}(\theta)$ corresponding to the "bound state" A_c. Then the relation

$$M_a^2 + M_b^2 - M_c^2 = -2M_a M_b \cos(u_{ab}^c) \ . \tag{3.9}$$

must hold. Geometrically, this condition means that $\overline{u}_{ab}^c = \pi - u_{ab}^c$ must be the internal angle of the Euclidean triangle with the sides m_a, m_b, m_c. The pole term

$$S_{ab}(\theta) \approx \frac{f(a,b;c)^2}{\theta - iu_{ab}^c} \tag{3.10}$$

corresponds to the diagram shown in Fig. 1, the constant $f(a,b;c)$ being associated with the three particle vertex shown in Fig. 2; it is evident that $f(a,b,c) = f(a,b;\overline{c})$ must be a symmetric function of a,b,c. This in particular means that if the particle A_c appears as the "bound state" of $A_a A_b$ then $A_{\overline{b}}\,(A_{\overline{a}})$ must appear as the "bound state" of $A_a A_{\overline{c}}\,(A_b A_{\overline{c}})$. The C symmetry implies $f(a,b,c) = f(\overline{a},\overline{b},\overline{c})$. As follows from (3.4), the amplitudes $S_{ab}(\theta)$ must satisfy the "bootstrap equations"[10]

$$S_{cd}(\theta) = S_{bd}(\theta - i\overline{u}_{bc}^a) S_{ad}(\theta + i\overline{u}_{ac}^b) \ . \tag{3.11}$$

This equation is illustrated by the diagram equality in Fig. 3, which is very similar in spirit to the Yang-Baxter triangle relation. The amplitude $S_{ab}(\theta)$ can also possess simple poles with negative residues in this region $0 < \mathrm{Im}\,\theta < \pi$, $\mathrm{Re}\,\theta = 0$; they correspond to the "bound states" of the cross channel $S_{a\overline{b}}(\theta)$. The multiple poles which are interpreted as the "anomalous thresholds"[11].

Let us now assume that the underlying field theory possesses a number of local IM P_s whose spins run over certain set $\{s\}$. The operators P_s are diagonalized by the states (3.1)

$$P_s|A_{a_1}(\theta_1)A_{a_2}(\theta_2)\dots A_{a_n}(\theta_n)\rangle_{(\mathrm{in,out})} =$$
$$\sum_{i=1}^{n} \alpha_{a_i}(s)\exp(s\theta_i)|A_{a_1}(\theta_1)A_{a_2}(\theta_2)\dots A_{a_n}(\theta_n)\rangle_{(\mathrm{in,out})} \tag{3.12}$$

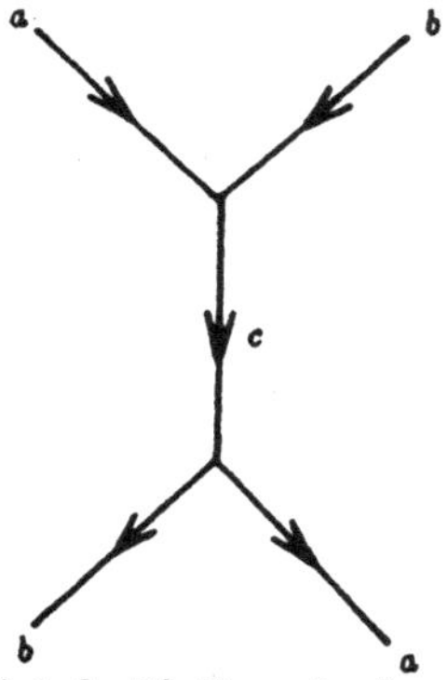

Fig. 1. The diagram associated with the pole singularity (3.10) in $S_{ab}(\theta)$.

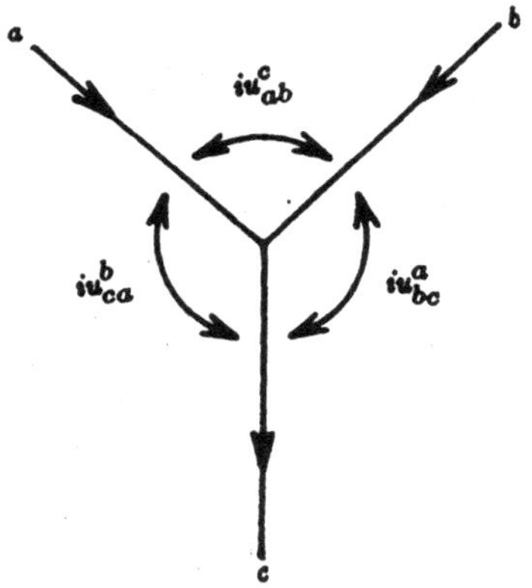

Fig. 2. The "three-particle vertex" associated with $f(a,b,c)$.

where $\alpha_a(s)$ are constants; $\alpha_a(1) = m_a$ since P_1 is the left light-cone component of momentum. It is possible to show that

$$\alpha_a(s) = (-1)^{s+1}\alpha_{\bar{a}}(s) \ . \tag{3.13}$$

which means that any IM P_s with even spin $s \in 2\,\mathbb{Z}$ annihilate any neutral particle $A_a : A_a \equiv A^{\bar{a}}$. In particular, if the theory contains only neutral particles, the IM with even spins are forbidden.

The bootstrap properties of the S-matrix must be consistent with these IM. This leads to the equations

$$\alpha_a(s)\exp(-is\overline{u}^b_{ac}) + \alpha_b(s)\exp(is\overline{u}^a_{bc}) = \alpha_c \tag{3.14}$$

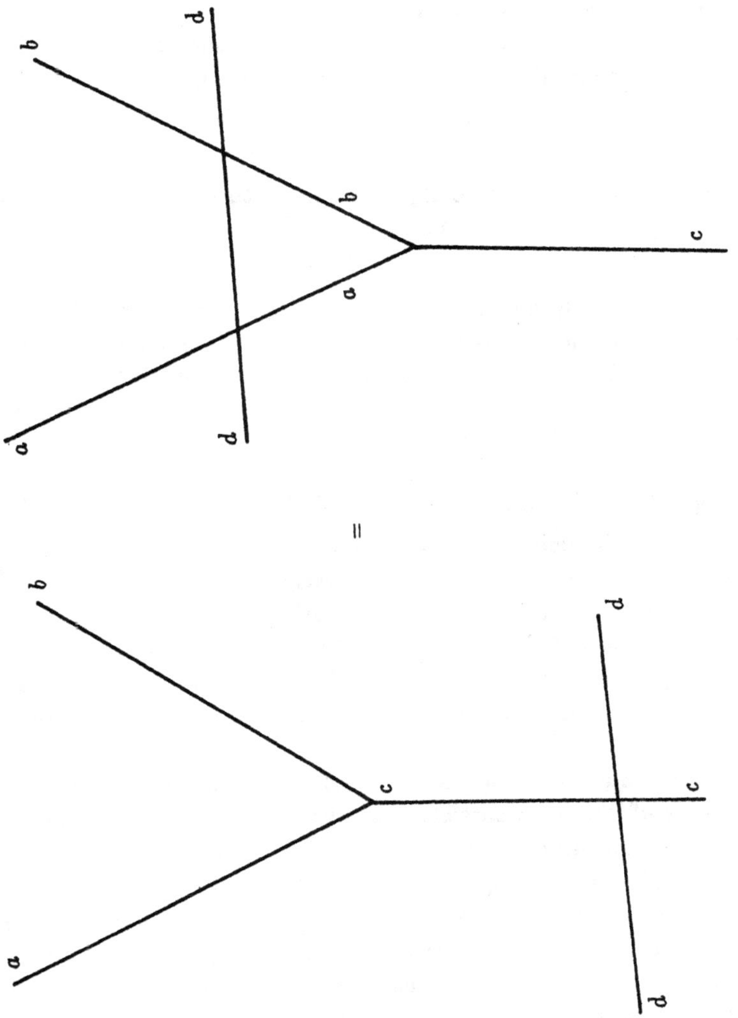

Fig. 3.

where $s \in \{s\}$ and a, b, c is any triple of particles such that $f(a,b,c) \neq 0$[5]. To see possible implications of these consistency conditions let us consider the PESM containing a neutral "fundamental" particle A_1 (i.e. all other particles of the theory can be obtained as the bound states of A_1) which has the "Φ^3 — property", i.e. A_1 itself appears as the "bound state" in the A_1A_1 scattering amplitude. Taking $a = b = c = 1$ in (3.14) one gets

$$2\cos(s\overline{u}_{11}^1) = 1; \quad s \in \{s\} , \tag{3.15}$$

$\alpha_1(s) \neq 0$ for all $s \in \{s\}$ since A_1 is assumed to be a "fundamental" particle) i.e.

$$u_{11}^1 = 2\pi/3 \tag{3.16}$$

and $\{s\}$ can include only numbers relatively prime to 6. From this simple calculation we can learn, for instance, that amongst the models (1.3), only $\hat{H}_3$, can be expected to possess a "fundamental" particle with the "Φ^3 — property" (see (1.4)). In the next three sections we will examine the sets $(1.4a, b, c)$ separately and conjecture the corresponding S- matrices.

4. Scaling Limit of the $T = T_c$ Ising Model with Magnetic Field

The CFT H_3 describes the critical point of the Ising model, the field $\sigma \equiv \Phi_{(1,2)}$ being the scaled local spin density[12]. Hence the model $\hat{H}_3$ appears to be the field theory of the scaled Ising model with $T = T_c$ but with nonzero magnetic field. This field theory certainly possesses finite correlation length $R_c \approx h^{-8/15}$ and is equivalent to the massive particle scattering theory which, according to the results of Sec. 2, must be purely elastic.

Let A_1 be the lightest particle of this scattering theory and let m_1 be its mass; $m_1 \approx h^{8/15}$. This particle is expected to be the "fundamental" one (i.e. if there are other particles in the theory, they can be obtained as "bound states" of an appropriate number of A_1). Also, since the Z_2 — symmetry of the Ising model is broken by the magnetic field, the particle A_1 is expected to possess the "Φ^3 — property", i.e. this particle itself should appear as the bound-state pole in the A_1A_1 scattering amplitude. This property of A_1 is consistent with the IM we have derived in Sec. 3 since the set (1.4a) does not contain numbers having 2 or 3 as divisors. But it does not explain why all the numbers divisible by 5 are also missing in the set (1.4a); for instance an IM with $s = 5$ would agree with the "bootstrap" properties of A_1 assumed above. Let us assume also that there is another stable particle, A_2, in this theory, which can be interpreted as the A_1A_1 "bound state" and, in addition, that A_1 appears as the "bound state" of A_2A_2. In other words, the "three particle couplings" f_{111}, f_{112} and f_{221} do not vanish. Substituting $a = b = 1$; $c = 2$ and $a = b = 2$; $c = 1$ in (3.14) one gets the equations

$$x_1^s + x_1^{-s} = \alpha_2(s)/\alpha_1(s) \ ; \quad x_2^s + x_2^{-s} = \alpha_1(s)/\alpha_2(s) \ , \tag{4.1}$$

where

$$x_1 = \exp(i\overline{u}_{12}^1) \ ; \quad x_2 = \exp(i\overline{u}_{21}^2) \ , \tag{4.2}$$

and s takes any value in the set (1.4a). Excluding the unknown constants α in the right-hand-side of (4.1) one obtains

$$(x_1^s + x_1^{-s})(x_2^s + x_2^{-s}) = 1 \ , \tag{4.3}$$

where $s = 1, 7, 11, 13, 17, 19$. Clearly, this system of algebraic equations is overdetermined. Nevertheless it admits the solution

$$x_1 = \exp(i\pi/5) \ ; \quad x_2 = \exp(2i\pi/5) \tag{4.4}$$

(of course one can interchange x_1 and x_2). Note that (4.4) satisfies (4.3) with any $s \neq 0(\text{mod } 5)$. Since $\alpha_0(1) = m_a$, this solution implies the mass ratio

$$M_2/M_1 = 2\cos(\pi/5) = 1.6180339\ldots \tag{4.5}$$

Thus we assume that the PEST of $\hat{H}_3$ contains at least two particles A_1 and A_2 with the mass ratio (4.5) and the bootstrap properties described above.

Let us construct the two particle S-matrix element $S_{11}(\theta)$ describing the $A_1A_1 \rightarrow A_1A_1$ scattering. This amplitude must possess poles at

$$\theta = iu_{11}^1 = 2\pi i/3 \ , \quad \theta = iu_{11}^2 = 2\pi i/5 \tag{4.6a}$$

having positive residues, and cross-channel poles at

$$\theta = i\overline{u}_{11}^1 = i\pi/3 \ , \quad \theta = i\overline{u}_{11}^2 = 3\pi i/5 \tag{4.6b}$$

with negative residues. Since A_1 appears as the bound state pole in $S_{11}(\theta)$, this amplitude must satisfy the "bootstrap" equation (3.11) with $a = b = c = d = 1$, i.e.

$$S_{11}(\theta) = S_{11}(\theta - i\pi/3)S_{11}(\theta + i\pi/3) \ . \tag{4.7}$$

It is impossible to satisfy this equation by a function with only the poles (4.6a) and (4.6b). It is necessary to add two poles at

$$\theta = iu_{11}^3 = i\pi/15\ , \quad \theta = i\overline{u}_{11}^3 = 14i\pi/15\ , \tag{4.8}$$

with positive and negative residues, respectively. Thus one obtains the expression

$$\begin{aligned} S_{11}(\theta) = {} & \mathrm{cth}(\theta/2 - i\pi/3)\mathrm{th}(\theta/2 + i\pi/3)\mathrm{cth}(\theta/2 - i\pi/5) \\ & \mathrm{th}(\theta/2 + i\pi/5)\mathrm{cth}(\theta/2 - i\pi/30)\mathrm{th}(\theta/2 + i\pi/30)\ . \end{aligned} \tag{4.9}$$

The additional pole at $\theta = iu_{11}^3$ represents a new particle A_3 having the mass

$$M_3/M_1 = 2\cos(\pi/30) = 1.9890437\ldots\ . \tag{4.10}$$

Now we can use Eq. (3.11) with $a = b = d = 1$, $c = 2$

$$S_{12}(\theta) = S_{11}(\theta - i\pi/5)S_{11}(\theta + i\pi/5) \tag{4.11}$$

to compute the amplitude

$$\begin{aligned} S_{12}(\theta) = {} & \mathrm{cth}(\theta/2 - 2\pi i/5)\mathrm{th}(\theta/2 + 2\pi i/5) \\ & \mathrm{cth}(\theta/2 - 3\pi i/10)\mathrm{th}(\theta/2 + 3\pi i/10)\mathrm{cth}(\theta/2 - 7\pi i/30) \\ & \mathrm{th}(\theta/2 + 7\pi i/30)\mathrm{cth}(\theta/2 - 4\pi i/30)\mathrm{th}(\theta/2 + 4\pi i/30) \end{aligned} \tag{4.12}$$

Note that besides the poles at $\theta = iu_{12}^1 = 4\pi/5$, $\theta = iu_{12}^2 = 3i\pi/5$ and $\theta = iu_{12}^3 = 7i\pi/15$, which represent the already known particles A_1, A_2 and A_3 respectively, this amplitude possesses a pole at $\theta = iu_{12}^4 = 4i\pi/15$ with positive residue; this pole corresponds to a new particle A_4 with the mass

$$M_4/M_1 = 4\cos(\pi/5)\cos(7\pi/30) = 2.4048671\ldots \tag{4.13}$$

The amplitude $S_{22}(\theta)$ computed from (3.11) with $a = b = 1$, $c = d = 2$

$$S_{22}(\theta) = S_{12}(\theta - i\pi/5)S_{12}(\theta + i\pi/5) \tag{4.14}$$

has the form

$$S_{22}(\theta) = S_{11}(\theta)S_{12}(\theta)\ . \tag{4.15}$$

Again, besides the poles $\theta = iu_{22}^1 = 4\pi i/5, \theta = iu_{22}^2 = 2\pi i/3, \theta = iu_{22}^4 = 7\pi i/15$ corresponding to the "familiar" particles A_1, A_2, A_4, this amplitude exhibits two positive-residue poles at $\theta = 4\pi i/15$ and $\theta = i\pi/15$ which represent the new particles A_5 and A_6 having the masses

$$\begin{aligned} M_5/M_1 &= 4\cos(\pi/5)\cos(2\pi/15) = 2.957295\ldots \\ M_6/M_1 &= 4\cos(\pi/5)\cos(\pi/30) = 3.2183404\ldots . \end{aligned} \tag{4.16}$$

In addition, the amplitude $S_{22}(\theta)$ possesses double poles at $\theta = 2\pi i/5$ and $\theta = 3\pi i/5$ which are interpreted as the anomalous thresholds[11].

Using the bootstrap equation (3.11) one can compute the two-particle amplitudes involving the particles A_3, A_4, A_5 and A_6. For example

$$\begin{aligned} &S_{13}(\theta) = \mathrm{cth}(\theta/2 - i\pi/60)\mathrm{th}(\theta/2 + i\pi/60) \\ &\mathrm{cth}(\theta/2 - 7\pi i/20)\mathrm{th}(\theta/2 + 7\pi i/20)\mathrm{cth}(\theta/2 - 13\pi i/60) \\ &\mathrm{th}(\theta/2 + 13\pi i/60)\mathrm{cth}(\theta/2 - i\pi/20)\mathrm{th}(\theta/2 + i\pi/20) \\ &[\mathrm{cth}(\theta/2 - 11\pi i/60)\mathrm{th}(\theta/2 + 11\pi i/60)]^2 \end{aligned} \tag{4.17a}$$

$$\begin{aligned} &S_{23}(\theta) = \mathrm{cth}(\theta/2 - i\pi/12)\mathrm{th}(\theta/2 + i\pi/12) \\ &\mathrm{cth}(\theta/2 - 19\pi i/60)\mathrm{th}(\theta/2 + 19\pi i/60)\mathrm{cth}(\theta/2 - 3\pi i/20) \\ &\mathrm{th}(\theta/2 + 3\pi i/20)[\mathrm{cth}(\theta/2 - 7\pi i/60)\mathrm{th}(\theta/2 + 7\pi i/60) \\ &\mathrm{cth}(\theta/2 - 13\pi i/60)\mathrm{th}(\theta/2 + 13\pi i/60)\mathrm{cth}(\theta/2 - i\pi/4)]^2 \end{aligned} \tag{4.17b}$$

$$\begin{aligned} &S_{33}(\theta) = \mathrm{cth}(\theta/2 - i\pi/6)\mathrm{th}(\theta/2 + i\pi/6) \\ &\mathrm{cth}(\theta/2 - i\pi/15)\mathrm{th}(\theta/2 + i\pi/15)\mathrm{th}(\theta/2 - i\pi/10) \\ &\mathrm{cth}(\theta/2 + i\pi/10)S_{11}(\theta)S_{22}(\theta) \end{aligned} \tag{4.17c}$$

$$\begin{aligned} &S_{44}(\theta) = \mathrm{cth}(\theta/2 - i\pi/15)\mathrm{th}(\theta/2 + i\pi/15) \\ &\mathrm{cth}(\theta/2 - 7\pi i/30)\mathrm{th}(\theta/2 + 7\pi i/30) \\ &[\mathrm{cth}(\theta/2 - i\pi/6)\mathrm{th}(\theta/2 + i\pi/6)]^2 S_{12}(\theta)S_{22}(\theta) \end{aligned} \tag{4.17d}$$

All simple positive-residue poles of these amplitudes correspond to the already known particles $A_1, \ldots, A_6$, the only exceptions being the pole

$\theta = 2\pi i/15$ exhibited by $S_{33}(\theta)$ and the pole $\theta = i\pi/15$ of $S_{44}(\theta)$; these poles represent additional particles A_7 and A_8 with the masses

$$
\begin{aligned}
M_7/M_1 &= 8\cos^2(\pi/5)\cos(7\pi/30) = 3.891156\ldots \\
M_8/M_1 &= 8\cos^2(\pi/5)\cos(2\pi/15) = 4.783386\ldots
\end{aligned}
\tag{4.18}
$$

We will not present here the rest of the two-particle S-matrix elements; they are computed using (3.11) straightforwardly. The bootstrap closes within the eight particles $A_1, \ldots, A_8$ already appearing above. We propose this PESM as the S-matrix for the field theory $\hat{H}_3$.

5. Thermal Shift Away From the Ising Tricritical Point

The Ising model exhibits the most generic type of criticality related to the spontaneous breakdown of Z_2 symmetry. By appropriate generalisation of the model (for instance, allowing nonvanishing probability of vacancies in the lattice of Ising spins) one can achieve the tricritical point which is described by the CFT $H_4(c = 7/10)$[9]. This CFT possesses two relevant spinless Z_2-symmetric operators $\varepsilon \equiv \Phi_{(1,2)}$ and $\rho \equiv \Phi_{(1,3)}$ having the dimensions (1/10, 1/10) and (3/5, 3/5), respectively. The most relevant of them, ε, is associated with thermal deformation. The renormalisation group flow in the vicinity of the fixed point H_4, generated by these two relevant operators, is shown schematically in Fig. 4. The RG trajectory connecting the fixed points H_4 and H_3 is generated by the operator ρ which preserves the Z_2 self-duality and supersymmetry (which seems to be related properties); this trajectory is discussed in Ref. 13 as an example of a spontaneous breakdown of the supersymmetry. The same operator ρ (added with another sign of the "coupling constant") generates another RG trajectory ($H_4 \to B$) which also goes along the self-duality line but rather leads to the massive theory with unbroken supersymmetry; this theory is integrable[2,5] but it will be discussed elsewhere. Here we are considering the RG trajectories ($H_4 \to A$) and ($H_4 \to A'$) in Fig. 4 generated by the "thermal" operator ε, i.e. the theory $\hat{H}_4$. As was already mentioned, the field $\varepsilon \equiv \Phi$ is Z_2-symmetric, but it changes sign under the duality transformation. Therefore the trajectories ($H_4 \to A$) and ($H_4 \to A'$) are interchanged by the duality transformation, i.e. the theory $\hat{H}_4$ does not essentially depend on the sign of λ in (1.3).

The results of Sec. 2 show that the scattering theory corresponding to the field theory $\hat{H}_4$ is purely elastic. Let us analyse the set of IM (1.4b).

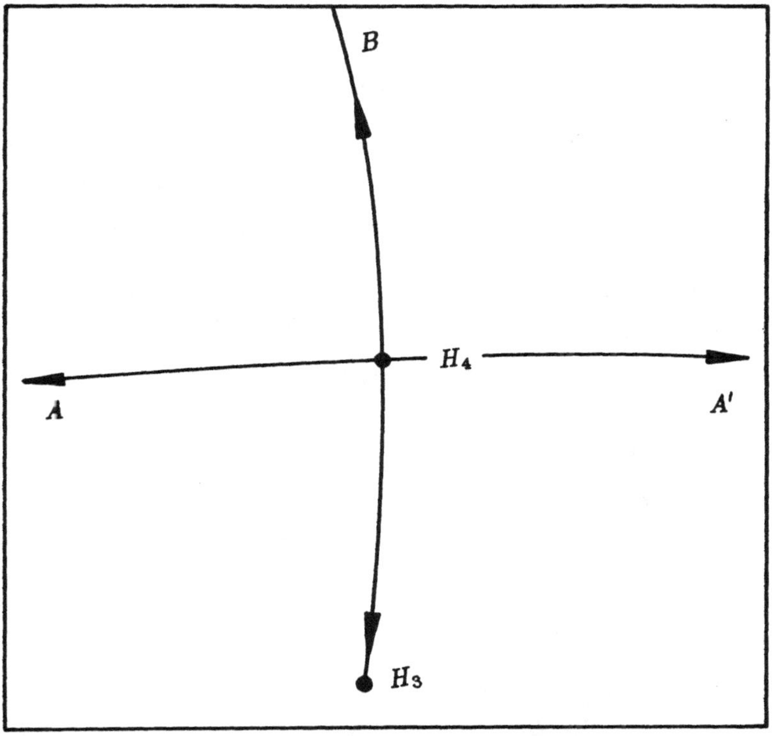

Fig. 4.

According to the analysis of Sec. 3, the "fundamental particle" of this PESM cannot possess the "Φ^3-property" since the set (1.4b) contains $s = 9$. This is quite reasonable. Since the Z_2-symmetry is not broken the particles are characterized by the Z_2 parity, and the "fundamental particle" must be Z_2-odd. However a Z_2-even particle with the " Φ^3-property" does not contradict the IM provided the operator P_9 annihilates this particle. Let us assume that the PESM indeed contains one or more Z_2-even particles with the "Φ^3-property" and denote the lightest of them as A_2 (saving the notation A_1 for the "fundamental particle" which is necessarily Z_2-odd). We have

$$P_9|A_2(\theta)\rangle = 0 \tag{5.1}$$

and $f_{222} \neq 0$. The particle A_2 can be considered as the "bound state" of two "fundamental" particles A_1. Since $\alpha_1(9) \neq 0$, (3.14) leads to the equation

$$\cos(9u_{11}^2/2) = 0\ . \tag{5.2}$$

Since $0 < u_{11}^2 < \pi$, there are four possibilities

$$\text{a.}\ u_{11}^2 = 7\pi/9 \quad \text{b.}\ u_{11}^2 = 5\pi/9 \quad \text{c.}\ u_{11}^2 = \pi/3 \quad \text{d.}\ u_{11}^2 = \pi/9 \tag{5.3}$$

The possibility (c) would correspond to an IM (1.1) having spin 3 which is absent in the set (1.4b). The detailed analysis of the possibility (a) (similar to that described below) shows that it contradicts the assumption that A_1 is the lightest of the Z_2-odd particles. As we will see, the choices (b) and (d) lead to the same "minimal" S-matrix and so we assume

$$u_{11}^2 = 5\pi/9 \tag{5.4}$$

i.e.

$$M_2/M_1 = 2\cos(5\pi/18) = 1.2855752\ldots\ . \tag{5.5}$$

This means that the amplitude $S_{11}(\theta)$ possesses a positive-residue pole at $\theta = 5\pi/9$ (and a negative-residue pole at $\theta = 4\pi i/9$), and that the amplitude $S_{12}(\theta)$ possesses a pole at $\theta = 13\pi i/18$, with positive residue, and the corresponding cross-channel pole at $\theta = 5\pi i/18$. These amplitudes are subject to the "bootstrap equations"

$$S_{12}(\theta) = S_{11}(\theta + 5\pi i/18)S_{11}(\theta - 5\pi i/18)\ , \tag{5.6a}$$

$$S_{11}(\theta) = S_{11}(\theta + 4\pi i/9)S_{12}(\theta - 5\pi i/18)\ . \tag{5.6b}$$

The "minimal" way to satisfy these equations is to introduce additional poles at $\theta = i\pi/9$ and $\theta = 8\pi i/9$ in the amplitude $S_{11}(\theta)$ and also add poles at $\theta = 7\pi i/18$ and $\theta = 11\pi i/18$ in $S_{12}(\theta)$. Thus we obtain

$$\begin{aligned} S_{11}(\theta) =&\,\mathrm{cth}(\theta/2 - 5\pi i/18)\mathrm{th}(\theta/2 + 5\pi i/18) \\ &\mathrm{cth}(\theta/2 - \pi i/18)\mathrm{th}(\theta/2 + \pi i/18) \end{aligned} \tag{5.7}$$

$$\begin{aligned} S_{12}(\theta) =&\,\mathrm{cth}(\theta/2 - 13\pi i/36)\mathrm{th}(\theta/2 + 13\pi i/36) \\ &\mathrm{cth}(\theta/2 - 7\pi i/36)\mathrm{th}(\theta/2 + 7\pi i/36)\ . \end{aligned} \tag{5.8}$$

The pole of $S_{11}(\theta)$ located at $\theta = i\pi/9$, which has positive residue, represents a new Z_2-even particle A_4; the pole of $S_{12}(\theta)$ at $\theta = 7\pi i/18$ corresponds to a new Z_2-odd particle A_3. The masses are

$$M_3/M_1 = 2\cos(\pi/9) = 1.879385\ldots \tag{5.9a}$$

$$M_4/M_1 = 2\cos(\pi/18) = 1.96961\ldots \tag{5.9b}$$

Using the equation

$$S_{22}(\theta) = S_{12}(\theta + 5\pi i/18) S_{12}(\theta - 5\pi i/18) \tag{5.10}$$

one obtains

$$\begin{aligned} S_{22}(\theta) =& \mathrm{cth}(\theta/2 - i\pi/3)\mathrm{th}(\theta/2 + i\pi/3)\mathrm{cth}(\theta/2 - 2\pi i/9) \\ & \mathrm{th}(\theta/2 + 2\pi i/9)\mathrm{cth}(\theta/2 - i\pi/18)\mathrm{th}(\theta/2 + i\pi/18)\ . \end{aligned} \tag{5.11}$$

Here the pole at $\theta = i\pi/9$ represents a new Z_2-even particle A_6 with the mass

$$M_6/M_1 = 4\cos(5\pi/18)\cos(\pi/18) = 2.532088\ldots \tag{5.12}$$

The amplitudes involving the "new" particles A_3, A_4 and A_6 are easily computed using (3.14). These new amplitudes contain poles corresponding to the two additional particles A_5 and A_7, which are Z_2-odd and Z_2 -even respectively and have the masses

$$\begin{aligned} M_5/M_1 &= 4\cos(\pi/9)\cos(2\pi/9) = 2.879385\ldots \\ M_7/M_1 &= 4\cos(\pi/18)\cos(\pi/9) = 3.701666\ldots \end{aligned} \tag{5.13}$$

It turns out that the bootstrap closes within these seven particles. We do not present here all elements of the two-particle S-matrix; the computation of them is straightforward. Let us only mention the amplitude

$$\begin{aligned} S_{33}(\theta) =& \mathrm{cth}(\theta/2 - 7\pi i/18)\mathrm{th}(\theta/2 + 7\pi i/18)\mathrm{cth}(\theta/2 - i\pi/18) \\ & \mathrm{th}(\theta/2 + i\pi/18)[\mathrm{cth}(\theta/2 - i\pi/3)\mathrm{th}(\theta/2 + i\pi/3) \\ & \mathrm{cth}(\theta/2 - 2\pi i/9)\mathrm{th}(\theta/2 + 2\pi i/9)]^2 \end{aligned} \tag{5.14}$$

which exhibits a pole at $\theta = 7\pi i/9$ representing the particle A_2 as the bound state A_3A_3; this pole corresponds to the choice (a) in (5.3). The particle A_3 can be taken as the "fundamental particle" of the theory instead of A_1 but it is not the lightest of the Z_2-odd particles.

We propose the above PESM as the S-matrix of the field theory $\hat{H}_4$.

6. Thermal Deformation of the Tricritical 3-State Potts Model

The CFT H_6 describes the tricritical point of the 3-state Potts model.[9] The model possesses the symmetry S_3 (which is the permutation group of three elements). The field $\varepsilon \equiv \Phi_{(1,2)}$ is the most relevant S_3-invariant operator of the theory and so we call the field theory $\hat{H}_6$ the "thermal deformation" of H_6.

The group S_3 is the semidirect product of two Abelian groups, Z_3 and Z_2. Let Ω and $\mathbb{C}$ be the generators of this group: $\Omega^3 = \mathbb{C}^2 = E, \mathbb{C}\Omega = \overline{\Omega}\mathbb{C}$. One can associate the operator $\mathbb{C}$ with the charge conjugation. It is easy to show that the IM P_s we derived in Sec. 2 obeys the commutation relations

$$P_s\Omega = \Omega P_s\ , \quad P_s\mathbb{C} = (-)^{s+1}\mathbb{C}P_s\ . \tag{6.1}$$

Let us first analyse the general properties of the PESM corresponding to $\hat{H}_3$. The particles involved in this scattering theory are characterised by their transformation properties with respect to S_3, i.e. they correspond to the irreducible representations of S_3. So, a particle A_a could be either singlet invariant with respect to Ω ($\mathbb{C}$-even or $\mathbb{C}$-odd) or belong to the particle-antiparticle doublet $(A, \overline{A})$ forming the basis of the two-dimensional representation of S_3. Obviously, the "fundamental particle" of this scattering theory should be the doublet $(A_1, \overline{A}_1)$ (since otherwise the whole space of states would be Ω-invariant) i.e.

$$\begin{aligned}\Omega|A_1(\theta)\rangle = \omega|A_1(\theta)\rangle\ ,\Omega|\overline{A}_1(\theta)\rangle = \overline{\omega}|\overline{A}_1(\theta)\rangle\ ,\\ \mathbb{C}|A_1(\theta)\rangle = |\overline{A}_1(\theta)\rangle\ ,\end{aligned} \tag{6.2}$$

where $\omega = \exp(2\pi i/3)$.

Due to nontrivial IM this theory possesses (see Sec. 2), the corresponding S-matrix is purely elastic and it factorises in terms of the two-particle scattering amplitudes. Since the particles A_1 and $\overline{A}_1$ have the same mass, one would expect the two-particle S-matrix to have the form

$$\begin{aligned}|A_1(\theta_1)A_1(\theta_2)\rangle_{(\text{in})} &= S_{11}(\theta)|A_1(\theta_1)A_1(\theta_2)\rangle_{(\text{out})}\ ;\\ |A_1(\theta_1)\overline{A}_1(\theta_2)\rangle_{(\text{in})} &= T_{1\overline{1}}(\theta)|A_1(\theta_1)\overline{A}_1(\theta_2)\rangle_{(\text{out})} +\\ &\quad R_{1\overline{1}}(\theta)|\overline{A}_1(\theta_1)A_1(\theta_2)\rangle_{(\text{out})}\end{aligned} \tag{6.3}$$

where $\theta = \theta_1 - \theta_2$ and $T_{1\overline{1}}(\theta)$ and $R_{1\overline{1}}(\theta)$ are the transition and reflection amplitudes of the $A_1\overline{A}_1$ scattering. It is possible to prove, however, that the reflection amplitude R in (6.3) vanishes and so the S-matrix of this theory is in fact diagonal. The constants $\alpha_1(s)$ and $\overline{\alpha}_1(s)$ defined as

$$\begin{aligned}P_s|A_1(\theta)\rangle &= \alpha_1(s)\exp(s\theta)|A_1(\theta)\rangle\\ P_s|\overline{A}_1(\theta)\rangle &= \overline{\alpha}_1(s)\exp(s\theta)|\overline{A}_1(\theta)\rangle\end{aligned} \tag{6.4}$$

obey the relation

$$\alpha_1(s) = (-)^{s+1}\overline{\alpha}_1(s) \tag{6.5}$$

which follows from (6.1). There are two $\mathbb{C}$-odd IM in the list (1.4c), namely P_4 and P_8. Applying, for instance, P_4 to (6.3) one obtains

$$R_{1\bar{1}}(\theta)\alpha_1(4)\mathrm{sh}(2\theta)|\overline{A}_1(\theta_1)A_1(\theta_2)\rangle_{(\mathrm{out})} = 0 \tag{6.6}$$

Since $\alpha_1(4) \neq 0$ (as A_1 is the "fundamental particle") and θ is arbitrary this equation implies

$$R_{1\bar{1}}(\theta) = 0 \tag{6.7}$$

Hence we can directly apply the results of Sec. 3 to these diagonal S-matrices.

The two-particle amplitudes $S_{11}(\theta)$ and $S_{1\bar{1}}(\theta) \equiv T_{1\bar{1}}(\theta)$, describing A_1A_1 and $A_1A_{\bar{1}}$ scatterings respectively, satisfy the crossing-symmetry relation

$$S_{11}(\theta) = S_{1\bar{1}}(i\pi - \theta) \tag{6.8}$$

and the unitarity conditions

$$S_{11}(\theta)S_{11}(-\theta) = 1\ ; \tag{6.9a}$$

$$S_{1\bar{1}}(\theta)S_{1\bar{1}}(-\theta) = 1 \tag{6.9b}$$

The S_3 symmetry allows the particle $\overline{A}_1$ to appear as the "bound state" pole in the A_1A_1 scattering amplitude $S_{11}(\theta)$. Moreover, this pole would agree with the IM (1.4c). Indeed, the corresponding equation (3.14)

$$2\alpha_1(s)\cos(\pi s/3) = \overline{\alpha}_1(s) \tag{6.10}$$

is satisfied for all s (1.4c) because of the relation (6.5). Therefore we assume that the amplitude $S_{11}(\theta)$ possesses the positive-residue pole at $\theta = u^{\bar{1}}_{11} = 2\pi i/3$ and satisfies the "bootstrap equation" (3.11), i.e.

$$S_{1\bar{1}}(\theta) = S_{11}(\theta + i\pi/3)S_{11}(\theta - i\pi/3) \tag{6.11a}$$

$$S_{11}(\theta) = S_{1\bar{1}}(\theta + i\pi/3)S_{1\bar{1}}(\theta - i\pi/3) \tag{6.11b}$$

If the amplitude would possess no other poles in the "physical strip" $0 < \mathrm{Im}\theta < \pi$, the scattering theory would be compatible with the IM (1.1) of spin $s = 2$ which is not present in the set (1.4c) (compare this set of IM with that analyzed in Ref. 4). Hence we have to assume that, besides the pole at $\theta = i\pi/3$, the amplitude $S_{1\bar{1}}(\theta)$ possesses other poles. The positive-residue poles of this amplitude could be interpreted as the $A_1\overline{A}_1$ bound states. Obviously, such a bound state would be an Ω-invariant neutral particle. Let A_2 be one of these bound states. Since $\mathbb{C}A_2 = A_2$ the commutation relations (6.1) imply

$$\alpha_2(4) = \alpha_2(8) = 0 . \tag{6.12}$$

Therefore Eq. (3.14) with $a = 1, b = \bar{1}$ and $s = 4, 8$ are reduced to

$$\sin(2u^2_{1\bar{1}}) = 0 ; \quad \sin(4u^2_{1\bar{1}}) = 0 . \tag{6.13}$$

The only solution to (6.13) belonging to the "physical strip" is $u^2_{1\bar{1}} = \pi/2$. So, the only $A_1\overline{A}_1$ "bound state" compatible with the IM (we call it A_2) has the mass

$$M_2/M_1 = 2\cos(\pi/4) . \tag{6.14}$$

The "minimal" solution of the bootstrap equations (6.11) possessing the pole described above is

$$S_{11}(\theta) = \frac{\mathrm{sh}(\theta/2 - i\pi/3)\mathrm{sh}(\theta/2 - i\pi/4)\mathrm{sh}(\theta/2 - i\pi/12)}{\mathrm{sh}(\theta/2 + i\pi/3)\mathrm{sh}(\theta/2 + i\pi/4)\mathrm{sh}(\theta/2 + i\pi/12)} . \tag{6.15}$$

Note the additional pole of $S_{11}(\theta)$ located at $\theta = iu^3_{11} = i\pi/6$ which represents the new particle $\overline{A}_3$ as the bound state A_1A_1. This particle is a component of the S_3 doublet $(A_3, \overline{A}_3)$ (A_3 is represented by the corresponding pole of the amplitude $S_{1\bar{1}}(\theta)$) having the mass

$$M_3/M_1 = 2\cos(\pi/12) = 1.93185\ldots \tag{6.16}$$

Using (3.11) one can compute the two-particle amplitudes

$$\begin{aligned} S_{12}(\theta) =& \mathrm{cth}(\theta/2 - 3\pi i/8)\mathrm{th}(\theta/2 + 3\pi i/8) \\ & \mathrm{cth}(\theta/2 - 5\pi i/24)\mathrm{th}(\theta/2 + 5\pi i/24) ; \end{aligned} \tag{6.17}$$

$$\begin{aligned} S_{22}(\theta) =& \mathrm{cth}(\theta/2 - i\pi/3)\mathrm{th}(\theta/2 + i\pi/3)\mathrm{cth}(\theta/2 - i\pi/12) \\ & \mathrm{th}(\theta/2 + i\pi/12)[\mathrm{cth}(\theta/2 - i\pi/4)]^2 . \end{aligned} \tag{6.18}$$

Note the pole of (6.18) at $\theta = i\pi/6$ which has positive residue and does not correspond to any of the already introduced particles; it represents the new neutral particle A_4 with the mass

$$M_4/M_1 = 4\cos(\pi/4)\cos(\pi/12) \tag{6.19}$$

By examining all other two-particle amplitudes one can check that the bootstrap closes within these six particles $(A_1, \overline{A}_1)$, A_2, $(A_3, \overline{A}_3)$ and A_4.

We conjecture that this PESM corresponds to the field theory $\hat{H}_6$.

7. Discussion

In the previous sections the sets of IM, the mass spectra and S-matrices for the models $\hat{H}_3, \hat{H}_4$ and $\hat{H}_6$ were proposed. As was mentioned in the Introduction the corresponding mass spectra are related to the root systems of E_8, E_7 and E_6, respectively. This relationship is inherited from hidden W-algebra structures of the CFT H_3, H_4 and H_6 which we will briefly discuss here.

Let $X_r(X = A, D, E)$ be a simply-laced Lie algebra of rank r. The associated W-algebra is generated by r analytic fields $W_{s+1}(z)$ whose spins s run over the set of exponents of X_r ($W_2(z)$ always coincides with the stress-energy tensor $T(z)$). The W-algebras admit the Feigin-Fuks realization in terms of an r-component free boson field ϕ; this realization for the W-algebras associated with the A and D series is constructed in Ref. 3. The generalisation of this construction for other X_r is straightforward.

For each of these W-algebras there is the discrete series of unitary "minimal" models of CFT with $c < r$;

$$c_p = r\left(1 - \frac{h(h+1)}{p(p+1)}\right) \ ; \quad p = h+1, h+2, \ldots \tag{7.1}$$

where h is the Coxeter number; we will call these models $\hat{X}_r^{(p)}$. There are finitely many primary (with respect to the W-algebra) fields in each of these "minimal" models. Each of these primary fields is characterized by the pair $(\mathbf{\Omega}, \mathbf{\Omega}')$ of dominant integral weights of X_r which satisfy the inequalities

$$(-\boldsymbol{\alpha}_0)\mathbf{\Omega} \leq p \, , \quad (-\boldsymbol{\alpha}_0)\mathbf{\Omega}' \leq p-1 \tag{7.2}$$

where $(-\boldsymbol{\alpha}_0)$ is the maximal root of X_r.

Here we will concentrate our attention on the special primary field associated with the weights $\mathbf{\Omega}' = \mathbf{\Omega}_0$ and $\mathbf{\Omega} = \mathbf{\Omega}_{ad}$, where $\mathbf{\Omega}_0$ and $\mathbf{\Omega}_{ad}$ correspond to the trivial and adjoint representations of X_r respectively; we will denote the corresponding local spinless primary field as Φ. The dimensions of this field are

$$\Delta = \overline{\Delta} = 1 - \frac{h}{p+1} \, . \tag{7.3}$$

Let us consider now the W-symmetric CFT with the central charge (7.3) perturbed by the operator $\lambda \int \Phi(z,\bar{z})d^2z$; we denote this field theory as $\hat{X}_r^{(p)}$. Using the general character formula for the "completely degenerate" representations of the W-algebras [3]

$$\chi(\Omega,\Omega') = [q^{1/24} \prod_{1=1}^{\infty} (1-q^1)]^{-r}$$

$$\Sigma_{\hat{s}\in w} \Sigma_{\lambda\in\Gamma\alpha} \det(\hat{s}) q^{[p\hat{s}\Omega-(p+1)\Omega+p(p+1)\lambda]^2/2p(p+1)} \tag{7.4}$$

where the sums run over the elements $\hat{s}$ of the Weyl group w and the root lattice $\Gamma\alpha$ of the Lie algebra X_r, one can show (in the way explained in Sec. 2) that each of these perturbed field theories possesses a number of nontrivial IM P_s, whose spins follow the set exponents of X_r modulo h. (A few nontrivial IM for $\hat{A}_n^{(p)}$ and $\hat{D}_n^{(p)}$ are given in Ref. 3; the analysis of the general case will be presented elsewhere.) We conjecture that in fact there are infinitely many IM with the spins running the "X_r-sequence", i.e. all the X_r exponents repeated modulo h.

The S-matrix analysis of Sec. 3 can be applied to the perturbed field theory $\hat{X}_r^{(p)}$ provided this model develops the finite correlation length. It is possible to show that this is indeed the case for the "basic" model $\hat{X}_r^{(h+1)}$ characterised by the minimal value of the central charge

$$c_r^{(h+1)} = 2r/(h+2) \; ; \tag{7.5}$$

in this case the dimension Δ of the field Φ is

$$\Delta = 2/(h+2) \, . \tag{7.6}$$

Note that for $X_r = E_8, E_7, E_6$ these models are precisely the field theories discussed in Secs. 4, 5 and 6 i.e. $\hat{E}_8^{(31)} = \hat{H}_3$, $\hat{E}_7^{(19)} = \hat{H}_4$, $\hat{E}_6^{(13)} = \hat{H}_6$.

As we have seen in Sec. 3 the nontrivial IM imposes hard limitations on the mass spectrum of the theory; the PESM have to satisfy Eq. (3.14) and the "bootstrap equations" (3.11). In the case where the set $\{s\}$ is the "X_r-sequence", the following simple reasoning allows one to determine the "minimal" particle spectrum consistent with these IM. The "X_r-sequence" of IM is characteristic for the completely integrable field theory known as the 2D Toda system associated with X_r and described by the Lagrangian density

$$L = 1/2(\partial\phi)^2 + \sum_{i=0}^{r} \exp(\beta\alpha_i\phi) \tag{7.7}$$

where α_i are positive simple roots of X_r, $(-\alpha_0)$ is the maximal root and β is a "coupling constant".[b] Evidently, the mass spectrum of this theory, which is related to the matrix (1.6) (the squares of the masses are proportional to the eigenvalues of this matrix), should be consistent with the "X_r-sequence" of the IM. The Toda mass spectrum for $X_r = E_8, E_7$ and E_6 is given by (1.5a), (1.5b) and (1.5c) respectively. For $X_r = A_{n-1}$ and D_n the Toda spectra have the form

$$A_{n-1} : M_k = M_1 \frac{\sin(\pi k/n)}{\sin(\pi/n)} \; ; k = 1, 2, \ldots, n-1 \, . \tag{7.8a}$$

$$D_n : M_n = M_{n-1} = M . \, M_k = 2M \sin(\pi k/2n-2) \; ; \; k = 1, \ldots, n-2 \tag{7.8b}$$

Our main conjecture is that for general simply-laced X_r the mass spectrum of the model $\hat{X}_r^{(h+1)}$ is precisely this Toda mass spectrum, and that the corresponding PESM is the "minimal" solution of the bootstrap equations with this spectrum.

One readily checks this conjecture for the models $\hat{D}_n^{(2n-1)}$ which are equivalent to the Sine-Gordon model

$$L = \frac{1}{2}(\partial\phi)^2 + \lambda \cos(\beta\phi) \tag{7.9}$$

with the special values of the coupling constant

$$\beta_n = \sqrt{8\pi/n} \; ; \; n = 3, 4, \ldots \tag{7.10}$$

The Sine-Gordon mass spectrum is well-known; for the points (7.10) this spectrum coincides with (7.8b). Note that (7.10) are precisely the points where the soliton-antisoliton reflection amplitude vanishes;[14] at these points the Sine-Gordon S-matrix indeed reduces to the "minimal" solution of the bootstrap equations consistent with the "D_n-sequence" of the IM P_s

$$s = 1, 3, 5, \ldots, 2n-3 \; ; \; n-1 \, \mathrm{mod}(2n-2) \, . \tag{7.11}$$

Appearance of the extra IM of spins $n-1$ mod $(2n-2)$, which are absent at arbitrary β, distinguishes the points (7.10). In fact the vanishing of the

---

[b]It is possible to show that in the Feigin-Fuks free-field realisation the IM of the models $\hat{X}_r^{(p)}$ commute with the Hamiltonian corresponding to (7.7) with appropriate (imaginary) value of the coupling constant β.

reflection amplitude at $\beta = \beta_n$ is the reflection of the hidden D_n structure of the models (7.9) and (7.10).

References

1. A.B. Zamolodchikov and Al. B. Zamolodchikov, *Ann. Phys.* **120**(1979) 235.
2. A.B. Zamolodchikov, *JETP Lett.* **46** (1987) 160.
3. S. Lukyanov and V.A. Fateev, Lectures given in II Spring School "Contemporary Problems in Theoretical Physics", Kiev 1988, Preprints ITF–88–74–76P, Kiev 1988 (in Russian).
4. A.B. Zamolodchikov, *Int.J.Mod. Phys.* **A3** (1988) 746.
5. A.B. Zamolodchikov, *Proceedings of Taniguchi Symposium on Integrable Models in Field Theory and Statistical Mechanics.*
6. *Conformal Invariance and Applications to Statistical Mechanics*, eds. *C. Itzykson, H. Saleur and J.B. Zuber* (World Scientific, 1988).
7. A. Rocha-Caridi, in *Vertex Operators in Mathematics and Physics*, eds. *J. Lepowsky, S. Mandelstam and I.M. Singer* (Springer 1984).
8. S.K. Yang, private communication.
9. D. Friedan, Z. Qiu and S. Shenker, *Phys. Rev. Lett.* **52** (1984) 1575.
10. B. Schroer, T.T. Truong and P. Weiss, *Phys. Lett.* **B63** (1976) 422.
11. S. Coleman and H.-J. Thun, *Commun. Math. Phys.* **61** (1978) 31.
12. A.A. Belavin, A.M. Polyakov and A.B. Zamolodchikov, *Nucl. Phys.* **B241** (1984) 333.
13. D. Kastor, E. Martinec and S. Shenker, EFI preprint 88–31.
14. L.D. Faddeev and V.E. Korepin, *Teor. Mat. Fiz.* **25** (1975) 147.

Note Added in Proof

When this work was finished we received the recent preprints[15–18] which have considerable overlap with this paper.

Added References

15. T. Eguchi and S. -K. Yang, Kyoto University preprint RIFP-797 (1989).
16. H. W. Braden, E. Corrigan, P. E. Dorey and R. Sasaki, Durham University preprint UDCPT-89-23 (1989).
17. P. Christe and G. Mussardo, Santa Bharbara Preprint UCSBTH-89-19.
18. G. Sotkov and S. J. Zhu, SISA Preprint 7689 EP (1989).

REPRESENTATIONS OF AFFINE KAC-MOODY ALGEBRAS AND BOSONIZATION

Boris L. Feigin
117218, Profsoyuznaya Street, 13/12, Apt 91, Moscow, USSR

and

Edward V. Frenkel
140410, Suvorova Street, 100, Apt 27, Kolomna, Moscow district, USSR

In memory of Vadim Knizhnik

We study a new class of representations of affine Kac-Moody algebras, so-called Wakimoto modules, introduced in our previous work[9]. These modules have boson realization, endowed with special objects-chains and intertwining operators. We give explicit constructions of these objects, using composition vertex operators and point out the connection with Virasoro algebra and W-algebras. Our results yield explicit description of primary fields in WZW-models and allow to give integral representation of correlation functions in these models. We also use Wakimoto modules for investigation of the structure of highest weight modules on the singular hyperplane $k = -\overline{g}$.

1. Introduction

In Ref. 9 we introduced the new class of modules with heighest weight over affine Lie algebras, which we called Wakimoto modules. These modules in general may be characterized by homological properties. In particular, their composition series quotients coincide with those of the corresponding Verma modules. In our forthcoming work,[10] we will explain the geometrical meaning of Wakimoto modules by studying of the semi-infinite flag manifolds and their Schubert stratifications. We also will establish in Ref. 10 analogues of Bernstein-Gelfand-Gelfand resolution of the irreducible module with dominant integral highest weight over arbitrary affine Kac-Moody algebra, consisting of Wakimoto modules (in spirit of Ref. 36).

In this work we deal with the structure of Wakimoto modules. It turned out that these modules have remarkable boson realizations. These realizations enable one to give a new bosonization procedure for Wess-Zumino-Witten (WZW) models. We have explicit formulas of these realisations for affine algebras of type $Lsl_n^\wedge$. Note that the formulas for the simplest affine algebra $Lsl_2^\wedge$ first appeared in M. Wakimoto's work[33]. Explicit realizations for other affine algebras may be obtained in the same way but they seem to be much more cumbersome. That is why in this work we will be working mainly with affine Lie algebras of type $Lsl_n^\wedge$ however our results may be directly generalized to arbitrary affine Lie algebras.

We pay special attention to the algebra $Lsl_2^\wedge$. It turns out that Wakimoto modules over $Lsl_2^\wedge$ are of great resemblance to semi-infinite forms (SIF) modules over Virasoro algebra, introduced by the first author and D.B. Fuchs.[11,12] This resemblance has deep foundations. There is a functor between (derived) categories of highest weight representations of $Lsl_2^\wedge$ and Virasoro algebra. This functor translates general irreducible representation of $Lsl_2^\wedge$ to irreducible representation of Virasoro algebra and it always translates a Wakimoto module to a SIF-module.

Also, there is a functional correspondence between highest weight modules over arbitrary simply-laced affine algebras and over Fateev-Zamolodchikov-Lukyanov algebras (or W-algebras)[25], so that Wakimoto modules over affine algebras correspond to boson representations of W-algebras, investigated in Refs. 1-2, 6 and 7. This correspondence has place also for non-simply-laced algebras, but it is more subtle. We hope to explain these questions in detail in the next work. In this work we will see a number of examples of appearance of this correspondence, but we will only mention them when necessary without detailed explanations. Note also that this functorial correspondence is closely related to the new approach to quantum gravity developed by Knishnik, Polyakov and Zamolodchikov[23,30] (correspondence between quantum gravity and WZW models).

The special objects, appear in the representation theory of affine algebras. They are chains and composition vertex operators, introduced in Ref. 12. While the chains over Virasoro algebra were modules of tensor fields on the circle, the chains over the affine algebra $Lg^\wedge$ are modules of currents on the circle to highest weight representation of the finite-dimensional algebra g. Composition vertex operators yield intertwining operators between Wakimoto modules as well as in Ref. 12. The functor, mentioned above, translates these objects to each other.

Technique of vertex operators, developed in this work, may be used for the contruction of integral representations of correlation functions of WZW models in the same way as Feigin-Fuchs integral representations.[5] These correlation functions seem to be connected with Feigin-Fuchs integrals in 2D conformal field theory and maybe they are expressed via Feigin-Fuchs integrals in many important cases. In this setting the chains are primary fields,[24,32] intertwining operators are screening operators and so on.

Our investigation also allows us to describe completely the composition series structure of Wakimoto modules over $Lsl_2^\wedge$ and to connect this structure outside the singular line $k = -2$ with one of SIF-modules over Virasoro algebra.

In the exceptional case $k = -2$ the structure of highest weight modules over $Lsl_2^\wedge$ essentially changes. This also happens with all affine Lie algebras at the singular hyperplane $k = -\bar{g}$ (where $\bar{g}$ is dual Coxeter number[17]). It is explained by the appearance of a great number of singular vectors of imaginary degrees in Verma modules with highest weights, which belong to the singular hyperplane. These vectors may be obtained by the application of Segal-Sugawara operators to vacuum vector of a Verma module. The quotient of a Verma module by its submodule, generated by these vectors, is irreducible in a general point of the singular hyperplane[26,16,34] and it coincides with restricted Wakimoto module. It turns out that the composition series structure of these quotients resemble that of Verma modules over the corresponding finite-dimensional algebra. It is possible to use (restricted) Wakimoto modules to investigate this structure. We obtain complete results for $Lsl_2^\wedge$ and particular results for the general affine algebras. We believe, that our geometrical concept, developed in Ref. 10 may be used to obtain final results.

Briefly the contents of the paper are as follows:

Section 2 is devoted to the main definitions and notations. We introduce Wakimoto modules and consider their properties.

In Sec. 3 we introduce the main objects related to Wakimoto modules: chains and intertwining operators and give explicit formulae for them via vertex operators , following the scheme of Ref. 12. Because that work is not open to general use, we give all the necessary definitions (in particular, of composition vertex operators) and account for the basic ideas and results to emphasize the correspondence with Virasoro algebra mentioned above. We beign with $Lsl_2^\wedge$, after which we study $Lsl_n^\wedge$, and finally we show that these results can be directly generalized to all simply-laced affine algebras and with some changes to non-simply-laced algebras.

Section 4 is devoted to $Lsl_2^\wedge$. We formulate the theorem about the structure of Wakimoto modules outside the singular line. The proof is yielded by Sec. 3 and the considerations in Refs. 12 and 27. Then we describe the structure of restricted Wakimoto modules as well as Verma modules on the singular line.

In Sec. 5 we consider the restricted Wakimoto modules over arbitrary affine algebra on the singular hyperplane and generalize the results of Sec. 4 and formulate some conjectures.

In the Appendix we formulate Wick theorem, which is often used in Sec. 3.

In the conclusion, we note that our results have super-analogues. There exist super-Wakimoto modules over superconformal current algebras[21] with all the additional structures, studied in this work, including the correspondence with Neveu-Schwarts and Ramond algebras.[29] We will treat it in the next work.

2. Preliminaries

2.1. Let g be complex semi-simple Lie algebra of rank r, $n_- \oplus h \oplus n_+$ – its Cartan decomposition, Δ-root system, $\alpha_1, \ldots, \alpha_r$ – the set of simple roots, Λ-root lattice. We have $\Delta = \Delta_+ \cup \Delta_-$, where $\Delta_+(\Delta_-)$ is the set of positive (negative) roots, $n_\pm = \oplus_{\alpha \in \Delta \pm} g_\pm$. Let $E_i, H_i, F_i, i = 1, \ldots, r$ be canonical generators of g.[17]

The affine algebra $Lg^\wedge$ is the unique central extension of the current algebra $Lg = g \otimes \mathbb{C}((t))$ (see Ref. 17 for details). Commutation relations in $Lg^\wedge$ are as follows. Denote by $A(n)$ all elements $A \otimes t^n$ of $Lg^\wedge$, where $A \in g$. Then we have

$$[A(n), B(m)] = [A, B](n+m) + \langle A, B\rangle \cdot n \cdot K\ , \tag{2.1}$$

where $\langle , \rangle$ is (normalized) killing form on g and K is the central element.

Cartan decomposition of $Lg^\wedge$ is given by $Lg^\wedge = \hat{n}_- \oplus \hat{h} \oplus \hat{n}_+$ where $\hat{n}_- = n_- \otimes 1 \oplus g \otimes t^{-1}\mathbb{C}[t^{-1}], \hat{h} = h \otimes 1 \oplus \mathbb{C} \cdot K, \hat{n}_+ = n_+ \otimes 1 \oplus g \otimes t\mathbb{C}[[t]]$. Let $\hat{\Delta}$ be the root system of $Lg^\wedge, \hat{\Delta} = \hat{\Delta}_+ \cup \hat{\Delta}_-$, where $\hat{\Delta}_+(\hat{\Delta}_-)$ is the set of positive (negative) roots. We have $\hat{n}_\pm = \oplus_{\hat{\alpha} \in \hat{\Delta}_\pm} Lg^\wedge_{\hat{\alpha}}$. Let $\alpha_0, \alpha_1, \ldots, \alpha_r$ be the set of simple roots. Let $(,)$ be invariant scalar product on $\hat{h}^*$. We assume that $(\bar{\alpha}_i, \bar{\alpha}_j) = \langle H_i, H_j\rangle$ where $\bar{\alpha}_i = \dfrac{2\alpha_i}{(\alpha_i, \alpha_i)}$. The root $\hat{\alpha} \in \hat{\Delta}$ is real,if $(\hat{\alpha}, \hat{\alpha}) > 0$ and imaginary, if $(\hat{\alpha}, \hat{\alpha}) = 0$. It is well-known that the imaginary roots of $Lg^\wedge$ are of the form $l \cdot \delta$, where $l \in \mathbb{Z}, \delta = \alpha_0 + \alpha_{\max}, \alpha_{\max}$ being the maximal root of g. Any real root may be written as follows: $\hat{\alpha} = l \cdot \delta + \alpha$

or $\hat{\alpha} = l \cdot \delta - \alpha$, where $\alpha \in \Delta_+$. If $\hat{\alpha}$ is real we define the reflection $S_{\hat{\alpha}}$ on $\hat{h}^*$ as follows: $S_{\hat{\alpha}}(\hat{\chi}) = \hat{\chi} - \frac{2(\hat{\alpha},\hat{\chi})}{(\hat{\alpha},\hat{\alpha})\hat{\alpha}}$. Elements $S_{\alpha_i}, i = 0, 1, \ldots, r$ generate the affine Weyl group. Put

$$S^{\rho}_{\hat{\alpha}}(\hat{\chi}) = S_{\hat{\alpha}}(\hat{\chi} + \rho) - \rho \tag{2.2}$$

where $\rho \in \hat{h}^*$, $(\rho, \alpha_i) = 1, i = 0, 1, \ldots, r$.

Each character $\hat{\chi}$ of $\hat{h}^*$ determines the character of $\hat{b} = \hat{n}_+ \oplus \hat{h}$: $\hat{b} \to \hat{h} \xrightarrow{\hat{\chi}} \mathbb{C}$. We have $\hat{h}^* = h^* \oplus (\mathbb{C}K)^*$, hence $\hat{\chi} \in \hat{h}^*$ is the pair (χ, k) where $\chi \in h^*$ and $k \in \mathbb{C}$. Let $M_{\chi,k}$ be the Verma module over $Lg^\wedge$ with highest weight (χ, k), that is $M_{\chi,k} = U(Lg)^\wedge \otimes_{U((\hat{b}))} \mathbb{C}_{\chi,k}$, where $\mathbb{C}_{\chi,k}$ is the one-dimensional representation of $\hat{b}$, determined by the charater (χ, k). The vector $1 \otimes 1_{\chi,k}$ is called vacuum vector of Verma module $M_{\chi,k}$. It is annihilated by $\hat{n}_+$. Let $M^*_{\chi,k}$ be contragradient module to $M_{\chi,k}$. These modules are characterized by the following properties:

a. They belong to the category $\mathcal{O}$ of representations of $Lg^\wedge$;[3,4]

b. $H_0(\hat{n}_-, M_{\chi,k})^* \cong H^0(\hat{n}_+, M^*_{\chi,k}) \cong \mathbb{C}^*_{\chi,k}$;
$H_i(\hat{n}_-, M_{\chi,k}) = 0$, $H^i(\hat{n}_+, M^*_{\chi,k}) = 0$, $i \neq 0$.

The modules $M_{\chi,k}$ and $M^*_{\chi,k}$ are naturally graded.

The character of $M_{\chi,k}$ is given by ch $M_{\chi,k} = e^{(\chi,k)} \cdot \prod_{\hat{\alpha} \in \hat{\Delta}_+} (1 - e^{-\hat{\alpha}})^{-\dim Lg^{\wedge}_{\hat{\alpha}}}$. The character of $M^*_{\chi,k}$ coincides with the one of $M_{\chi,k}$.

We also need Verma and contragradient Verma modules over the finite-dimensional algebra g. It is convenient to consider that these modules are of lowest weight, not of highest weight. So we put $M_\chi = U(g) \otimes_{U(b_-)} \mathbb{C}_\chi$, where $b_- = n_- \oplus h$ and $\mathbb{C}_\chi$ is one-dimensional representation of b_- : $b_- \to h \xrightarrow{\chi} \mathbb{C}$.

The problem of reducibility of $M_{\chi,k}$ was solved in Ref. 19. Evidently, $M_{\chi,k}$ is reducible if and only if there is a singular vector w in $M_{\chi,k}$, such that $\hat{n}_+ w = 0$ and $w \neq v$. We can consider that w is homogeneous. According to Ref. 19, the highest weights of reducible Verma modules lie on the union of a countable number of hyperplanes. Any hyperplane is determined by the pair $(m, \hat{\alpha})$, where $\hat{\alpha} \in \hat{\Delta}_+, m = 1, 2, \ldots$. The corresponding equation is

$$2(\hat{\chi} + \rho, \hat{\alpha}) = m(\hat{\alpha}, \hat{\alpha}) \ . \tag{2.3}$$

If (1.3) is satisfied then $M_{\chi,k}$ contains a singular vector of degree $-m \cdot \hat{\alpha}$. This equation is called the Kac-Kazhdan equation.

If $\hat{\alpha}$ is real, that is $\hat{\alpha} = l \cdot \delta \pm \alpha$, where $\alpha \in \Delta_+$, then we can rewrite (2.3) as

$$S^{\rho}_{\hat{\alpha}}(\hat{\chi}) - \hat{\chi} = m \cdot \hat{\alpha} \ . \tag{2.4}$$

We can also rewrite it as

$$l(k + \overline{g}) \pm (\chi + \rho, \alpha) = m(\alpha, \alpha) \ , \tag{2.5}$$

where $\overline{g}$ is the dual Coexter number, $\overline{g} = \rho(\delta)$.[17]

If $\hat{\alpha}$ is an imaginary root, then all equations (2.3) coincide and we have

$$k + \overline{g} = 0 \ . \tag{2.6}$$

It is the equation of the singular hyperplane. Verma modules has singular vectors of (any) imaginary degree if and only if its highest weight belongs to singular hyperplane. In Refs. 15 and 20 it was noticed that if $k = -\overline{g}$ then the central charge of the Virasoro algebra (which consists of the so-called Segal-Sugawara operators), lying in the completing of the universal enveloping algebra of $Lg^\wedge$, is infinite, that is the Virasoro algebra degenerates to commutative algebra, commuting with $Lg^\wedge$. The idea of Ref. 28 to use Segal-Sugawara operators for the construction of all singular vectors was realised in Refs. 26 and 16. It turned out that to each central element Δ_j of $U(g)(j = 1, \ldots, r)$ (for example, Casimir element) we can attach a countable set of operators $T_i^{(j)}(i \in \mathbb{Z}\ )$ (for example, Segal-Sugawara operators). When $k \neq -\overline{g}$ these operators compose of W-algebra (for example, Virasoro algebra). But when $k = -\overline{g}$ they compose of a commutative algebra, which also commute with $Lg^\wedge$. Applying these operators to the vacuum vector we obtain a great number of singular vectors of imaginary degrees. In Refs. 26 and 16 it was shown that in a general point of the singular hyperplane $k = -\overline{g}$ there are no other singular vectors in the Verma module and so the quotient $\overline{M}_{\chi,-\overline{g}}$ of $\overline{M}_{\chi,-\overline{g}}$ by its submodule, generated by these singular vectors, is irreducible. We also have $M_{\chi,-\overline{g}} = \mathbb{C}[T_i^{(j)}]_{\substack{j=l,\ldots,r \\ i\in \ ,i<0}} \otimes \overline{M}_{\chi,-\overline{g}}$ as linear spaces. Hence the character of $\overline{M}_{\chi,-\overline{g}}$ is equal to ch $\overline{M}_{\chi,-\overline{g}} = e^{(\chi,-\overline{g})} \prod_{\hat{\alpha}\in\hat{\Delta}^{re}_+} (1 - e^{-\hat{\alpha}})^{-1}$ where $\hat{\Delta}^{re}_+$ is the set of positive real roots. We call $\overline{M}_{\chi,-\overline{g}}$ restricted Verma modules.

We also have to introduce other $Lg^\wedge$ modules. Let N be a g-module. Then $LN = N \otimes \mathbb{C}((t))$ is $Lg^\wedge$-module, with natural action of $Lg^\wedge$ (in particular, $k = 0$). If we put $C(m) = C \otimes t^m$ where $C \in N$, then we have

$A(n).C(m) = (A \cdot C)(m+m)$, where $A \in g$. If we put $N = M^*_\mu$, then we can obtain the family of $Lg^\wedge$-modules LM^*_μ, parametrized by $\mu \in h^*$, which we need in Sec. 3.

2.2. Now we pass to Wakimoto modules. These modules were constructed in Ref. 9 (in the simplest case of algebra $L\hat{sl}_2$ it was done in Ref. 33). We discuss details in Ref. 10. Here we briefly review these questions.

Let $a = n_- \otimes \mathbb{C}((t)) \oplus h \otimes t^{-1}\mathbb{C}[t^{-1}]$. We have the decomposition $a = a_+ \oplus a_-$, where $a_\pm = \hat{n}_\pm \cap a$.

Wakimoto modules $W_{\chi,k}$ are characterized by the following properties, which are analogous to the properties a) and b) of 2.1:

a') $W_{\chi,k}$ belongs to category $\mathcal{O}$;

b') $H_{\infty/2+i}(a, W_{\chi,k}) = 0$, $i \neq 0$; $H_{\infty/2}(a, W_{\chi,k}) \cong \mathbb{C}_{\chi,k}$ where $H_{\infty/2+i}(a, W_{\chi,k})$ is the semi-infinite homology of a with respect to decomposition $a = a_+ \oplus a_-$.[8]

The property b') means, in particular, that $W_{\chi,k}$ is free over a_- and "co-free" over a_+ (that is dual to $W_{\chi,k}$ is free over a_+). Thus, $W_{\chi,k}$ are intermediate between $M_{\chi,k}$ and $M^*_{\chi,k}$.

The character of $W_{\chi,k}$ coincides with the character of the Vermal module: $\operatorname{ch} W_{\chi,k} = e^{(\chi,k)} \prod_{\hat{\alpha}\in\hat{\Delta}_+} (1-e^{-\hat{\alpha}})^{-\dim Lg^\wedge_{\hat{\alpha}}}$, and if the Verma module is irreducible then it is isomorphic to $W_{\chi,k}$.

We mark out the case $k = -\check{g}$ (singular hyperplane). Consider the subalgebra $\bar{a} = n_- \otimes \mathbb{C}((t))$ of g. Let $\overline{W}_{\chi,-\check{g}}$ be the $Lg^\wedge$ module, which satisfies the condition a') and the following modification of the condition b"):

b") $H_{\infty/2+i}(\bar{a}, \overline{W}_{\chi,-\check{g}}) = 0$; $i \neq 0$; $H_{\infty/2}(\bar{a}, \overline{W}_{\chi,-\check{g}}) \cong \mathbb{C}_{\chi,k}$

We call $\overline{W}_{\chi,-\check{g}}$ the restricted Wakimoto module. Its character coincides with the character of $\overline{M}_{\chi,-\check{g}}$: $\operatorname{ch} \overline{W}_{\chi,-\check{g}} = e^{(\chi,-\check{g})} \prod_{\hat{\alpha}\in\hat{\Delta}^{re}} (1-e^{-\hat{\alpha}})^{-1}$.

Certainly, if $M_{\chi,-\check{g}}$ is irreducible, it is isomorphic to $\overline{W}_{\chi,-\check{g}}$.

The following lemma is purely technical.

Lemma

1. $M_{\chi,k}$ is reducible and contains a singular vector of degree less than γ if and only if $W_{\chi,k}$ or $W^*_{\chi,k}$ contains a singular vector of degree less than γ.

2. $\overline{M}_{\chi,-\check{g}}$ is reducible and contains a singular vector of degree less than γ if and only if $\overline{W}_{\chi,-\check{g}}$ or $\overline{W}_{\chi,-\check{g}}$ contains a singular vector of degree

less than γ.

The existence of Wakimoto modules and restricted Wakimoto modules was proven in Ref. 9. Namely, they were constructed algebro-geometrically in the same way as Verma or contragradient Verma modules (as certain modules over the algebra of vector fields on the semi-infinite flag mainfold, where $Lg^\wedge$ imbeds, for details see Ref. 10). We will not repeat this construction and only write down the explicit formulae for $\mathrm{Lsl}_n^\wedge$, obtained in Ref. 9.

Remark.

In Ref. 10 we give a more general definition of Wakimoto modules $W_{\hat{\chi},p}$ where p is the parabolic subalgebra of g. Namely, let r be the nilpotent radical of p and $p = r \oplus S$. Consider the subalgebra $a_p = r \otimes \mathbb{C}((t)) \oplus S \otimes t^{-1}\mathbb{C}[t^{-1}]$. $W_{\hat{\chi},p}$ are characterized by the property a) and b) $H_{\infty/2+i}(a_p, W_{\hat{\chi},p}) = 0, i \neq 0; H_{\infty/2}(a_p, W_{\hat{\chi},p}) \cong \mathbb{C}_{\hat{\chi},k}$:

In particular, $W_{\hat{\chi},b_-} = W_{\hat{\chi}}$ and $W_{\hat{\chi},g} = M_{\hat{\chi}}$. In this work we deal only with $W_{\hat{\chi},b_-}$.

2.3. To write down our formulae we need some preparations.

The algebra sl_{n+1} is of rank $n, g = n+1$. Let $E_i, H_i, F_i, i = 1, \ldots, n$ be its generators.

Let $C_{ij} = 2\delta_{ij} - \delta_{i,j-1} - \delta_{ij+1}$ be Cartan matrix of sl_{n+1}. We put $(\alpha_i, \alpha_j) = \langle H_i, H_j \rangle = C_{ij}, \langle E_i, F_j \rangle = \delta_{ij}$. So we have the following commutation relations which determine uniquely the structure of Lsl_{n+1}:

$$
\begin{aligned}
[E_i(m), F_j(l)] &= \delta_{ij} H_i(m+1) + \delta_{ij}\delta_{m-l} m \cdot K \ , \\
[H_i(m), E_j(l)] &= C_{ij} F_j(m+1) \ , \\
[H_i(m), F_j(l)] &= -C_{ij} F_j(m+l) \ , \\
[H_i(m), H_j(l)] &= C_{ij}\delta_{m,-l} m \cdot K \ .
\end{aligned} \tag{2.7}
$$

(We omitted the relations of second order.) Evidently, $E_i(m), H_i(m), F_i(m)$ generate $\mathrm{Lsl}_{n+1}^\wedge$.

Let Γ be the Heisenberg algebra with generators $a_{ij}(m), a_{ij}^*(m), b_i(m)$, $i = 1, \ldots, n; j = i, \ldots, n; m \in \mathbb{Z}$ and commutation relations:

$$
\begin{aligned}
[a_{ij}(m), a_{pq}(l)] &= [a_{ij}^*(m), a_{pq}^*(l)] \\
= [a_{ij}(m), b_k(l)] &= [a_{ij}^*(m), b_k(l)] = 0 \ , \\
[a_{ij}(m), a_{pq}^*(l)] &= \delta_{ip}\delta_{jq}\delta_{m,-l} \ , \\
[b_i(m), b_j(l)] &= c_{ij}\delta_{m,-l} \cdot m \ .
\end{aligned} \tag{2.8}
$$

Let M be the irreducible representation of Γ with vacuum vector v, such that $a_{ij}(m)v = 0, m > 0\ ; a^*_{ij}(l)v = 0\ , b_i(l)v = 0\ , l \geq 0$.

Introduce normal ordering as usual (see Appendix) for any countable set of operators $A(n)$ by putting $A \equiv A(z) = \sum_{n\in\mathbb{Z}} A(n)z^n$ (generating function) and $\dot{A} \equiv \dot{A}(z) = \sum_{n\in\mathbb{Z}} nA(n)z^n$. Using generating functions we may express $E_i(z), H_i(z), F_i(z)$ via $a_{ij}(z), a^*_{ij}(z), b_j(z)$ and so imbed $\mathrm{Lsl}_{\hat{n}}$ into Γ. This embedding depends on $(n+1)$ parameters $\chi_1, \ldots, \chi_{n,\nu}$ and so M becomes $(n+1)$ -parameter family of $\mathrm{Lsl}^{\wedge}_{n+1}$ modules. This family is the family of Wakimoto modules $W_{\chi,k}$, where $\chi \in h^*$ and $\chi_i = \chi(H(0)), k = \nu^2 - (n+1)$. In the following we denote them by $W_{\chi,\nu}$. We put

$$E_i(z) =: a^*_{ii}\Big(\sum_{j=1}^{i-1} a_{j,i-1}a^*_{j,i-1} - \sum_{j=1}^{i} a_{ji}a^*_{ji}\Big) : -\nu a^*_{ii}b_i - \chi_i a^*_{ii}$$

$$+ \sum_{j=i+1}^{n} a_{i+1,j}a^*_{ij} - \sum_{j=1}^{i-1} a_{j,i-1}a^*_{ji} + (i+1-\nu^2)\dot{a}^*_{ii}\ ,$$

$$H_i(z) = 2 : a_{ii}a^*_{ii} : + \sum_{j=1}^{i-1}(: a_{ji}a^*_{ji} : - : a_{j,i-1}a^*_{j,i-1} :)$$

$$+ \sum_{j=i+1}^{n} (: a_{ij}a^*_{ij} : - : a_{i+1,j}a^*_{i+1,j} :) + \nu b_i + \chi_i\ ,$$

$$F_i(z) = a_{ii} + \sum_{j=i+1}^{n} a_{ij}a^*_{i+1,j}\ , k = \nu^2 - (n+1)\ . \tag{2.9}$$

Everywhere a_{ij}, a^*_{ij} and b_i means $a_{ij}(z), a^*_{ij}(z), b_i(z)$. We also want to give the formulae for $\mathrm{Lsl}^{\wedge}_2$. They are especially simple. We put $a(m) = a_{11}(m), a^*(m) = a^*_{11}(m)$ and $b(m) = \frac{1}{\sqrt{2}}b_1(m)$.

$$E(z) = - : a(z)a^*(z)a^*(z) : -(\nu b(z) + \chi)a^*(z) + (2 - \frac{\nu^2}{2})\overset{\circ}{a}{}^*(z)$$

$$H(z) = 2 : a(z)a^*(z) : +\nu b(z) + \chi\ ,$$

$$F(z) = a(z)\ , k = \frac{\nu^2}{2} - 2\ . \tag{2.10}$$

We see that if $\nu = 0, k = -(n+1) = -\bar{g}$ and we pass to the singular hyperplane while all terms which include $b_i(z)$ disappear.

That is why we can consider the restricted Heisenberg algebra $\overline{\Gamma}$, generated by $a_{ij}(m), a^*_{ij}(m)$ and its irreducible representation $\overline{M}$ with vacuum vector V, annihilated by $a_{ij}(m), m > 0, a^*_{ij}(m), m \geq 0$.

So we have an n-parameter embedding of $\mathrm{Lg}^\wedge$ into $\overline{\Gamma}$ and hence $\overline{M}$ becomes an n-parameter family of $\mathrm{Lg}^\wedge$ modules. These modules are restricted Wakimoto modules $\overline{W}_{\chi,-\bar{g}}$ (which we denote by $\overline{W}_{\chi,0}$). We will denote the Cartan generators, acting on $\overline{W}_{\chi,0}$ by small letters: $e_i(m), h_i(m), f_i(m)$.

We will explain briefly how such formulae are obtained.

Recall that the contragradient Verma module M over a finite-dimensional algebra, in particular, over sl_{n+1} may be realized as the space of the sections of one-dimensional fibering, determined by the character χ, over the big cell of the corresponding flag manifold. This cell can be identified with Lie group N_-, corresponding to Lie algebra n_-. We choose coordinates on $N_- x_{ij}, i = 1, \ldots, n; j = i, \ldots, n$, corresponding to matrix elements. The fibering over the big cell is identified with the trivial one and so the Lie algebra sl_{n+1} imbeds into the algebra of differential operators of order less than 2 or the big cell, that is on N_-. So, we can express $E_i, H_i, F_i, i = 1, \ldots, n$ as differential operators of order less than 2 on x_{ij}.

If we put $a_{ij} = \dfrac{\partial}{\partial x_{ij}}, a^*_{ij} = x_{ij}$, then a_{ij}, a^*_{ij} form a finite-dimensional Heisenberg algebra with relations

$$[a_{ij}, a_{pq}] = [a^*_{ij}, a^*_{pq}] = 0 \ , [a_{ij}, a^*_{pq}] = \delta_{ip}\delta_{jq}$$

and generators of sl_{n+1} are expressed via a_{ij}, a_{ij} as polynomials of degree 1 on a_{ij}. These polynomials coincide with (2.9) if we put $\nu = 0$ and remove the last term in $E_i(z)$.

After that, to pass to $\mathrm{Lsl}^\wedge_{n+1}$ we have to pass to generating functions and change a_{ij}, a^*_{ij} by $a_{ij}(z), a^*_{ij}(z)$. We also must introduce normal ordering to give meaning to our expressions. The commutation relations between them can be computed using the Wick theorem (see Appendix). As our expressions are polynomials of degree 1 on a_{ij}, we may obtain only ordinary and binary pairings between them. We know from Proposition A. 2 that only binary pairings change the right answer. But, fortunately, we can compensate this change, adding the term $(-\nu^2 + i + 1)\dot{a}^*_{ii}(z)$ to the expression of $E_i(z)$, so that the result differs from the right answer only by the central charge $-\bar{g} = -(n+1)$. This important fact was proved in Ref. 9 by homological methods. It is true also for an arbitrary affine algebra.

The formulae, obtained in this way, give the module $\overline{W}_{\chi,0}$. To obtain the formulae for $W_{\chi,\nu}$ with arbitrary ν one should add $b_i(z)$. It is done without any problems.

Using this scheme, one can in principle construct the explicit boson realizations for other affine algebras. One needs an infinite-dimensional Heisenberg algebra with generators $a_\alpha(m), a^*_\alpha(m),\ b_i(m), \alpha \in \Delta_+$, $i = 1, \dots, r, +m \in \mathbb{Z}$, and commutation relations

$$\begin{aligned}
[a_\alpha(m), a_\beta(l)] &= [a^*_\alpha(m), a^*_\beta(l)] = \\
[a_\alpha(m), b_i(l)] &= [a^*_\alpha(m), b_i(l)] = 0 \ , \\
[a_\alpha(m), a^*_\beta(l)] &= \delta_{\alpha,\beta}\delta_{m,-l} \ , \\
[b_i(m), b_j(l)] &= m\delta_{m,-l}(\bar{\alpha}_i, \bar{\alpha}_j) \ .
\end{aligned} \tag{2.11}$$

One has to consider irreducible representation of this Heisenberg algebra with vacuum vector v, annihilated by $a_\alpha(m), m > 0, a^*_\alpha(m), b_i(m), m \geq 0$.

The affine algebra acts on this representation. The operators $H_i(z)$ are always of the form

$$H_i(z) = \sum_{\beta \in \Delta_+} (\bar{\alpha}_i, \beta) \ : a^{(z)}_\beta a^*_\beta(z) ; + \nu b_i(z) + \chi_i \ .$$

The operators $F_\alpha(z)$ are of the form

$$F_\alpha(z) = a_\alpha(z) + Q(a_\beta(z), a^*_\beta(z)) \ .$$

The operators $E_\alpha(z)$ are very cumbersome in general cases.

The central charge is $k = -\bar{g} + \nu^2$ by definition of $-\bar{g}$.

Note also that $E_\alpha(m), H_i(m), F_\alpha(m)$ can be considered as differential operators on infinitely many variables.

Let

$$a_\alpha(m) = \begin{cases} x_\alpha(m), & m \leq 0 \\ \frac{\partial}{\partial x_\alpha(m)}, & m > 0 \ , \end{cases} \qquad a^*_\alpha(m) = \begin{cases} x_\alpha(-m), & m < 0 \\ \frac{\partial}{\partial x_\alpha(-m)}, & m \geq 0 \end{cases}$$

$$b_i(m) = \begin{cases} m e_i \cdot \frac{\partial}{\partial y(m)}, & m > 0 \\ 0, & m = 0 \\ e_i \cdot y(-m), & m < 0 \end{cases}$$

where

$$\begin{aligned} y(m) &= (y_1(m), \ldots, y_r(m)) \\ \frac{\partial}{\partial y(m)} &= \left(\frac{\partial}{\partial y_1(m)}, \ldots, \frac{\partial}{\partial y_r(m)}\right), \end{aligned}$$

and e_i are such vectors in r-dimensional space, that $e_i \cdot e_j = (\alpha_i, \alpha_j)$ in the usual orthonomal basis.

Then

$$M = \mathbb{C}[x_\alpha(n), y_i(m)]_{\substack{\alpha\in\Delta_+; n\in\mathbb{Z} \\ i=1,\ldots,r; m\in\mathbb{Z}, m>0}}$$

$$\overline{M} = \mathbb{C}[x_\alpha(n)]_{\alpha\in\Delta_+; n\in\mathbb{Z}}$$

So $Lg^\wedge$ acts on the space of polynomial on infinitely many variables.

2.4. In this subsection we introduce some objects concerning Virasoro algebra.

Virasoro algebra $\mathcal{L}^\wedge$ is the unique central extension of the algebra of vector fields on the circle $\mathcal{L}$.[14] We can choose the basis $L(n), n \in \mathbb{Z}$ and C with the following commutation relations

$$\begin{aligned} &[L(n), L(m)] = (n-m)L(n+m) + \frac{1}{12}(n^3 - n)\cdot C \\ &[L(n), C] = 0 \end{aligned}$$

"Cartan decomposition" of $\mathcal{L}^\wedge$ is given by $\mathcal{L}^\wedge = \mathcal{L}_- \oplus \mathcal{L}_0 \oplus \mathcal{L}_+$, where $\mathcal{L}_- = \oplus_{i<0}\mathbb{C}L(i)$, $\mathcal{L}_0 = \mathbb{C}L(0) \oplus \mathbb{C}C$, $\mathcal{L}_+ = \oplus_{i>0}\mathbb{C}L(i)$.

We define the Verma module $M_{h,c}$ over the Virasoro algebra as $U(\mathcal{L}^\wedge)\ S \otimes_{U(\mathcal{L}_0\oplus\mathcal{L}_+)} \mathbb{C}_{h,c}$ where $\mathbb{C}_{h,c}$ is one-dimensional representation of $\mathcal{L}_0 \oplus \mathcal{L}_+$, whose restriction to $\mathcal{L}_+$ is trivial, and restriction to $\mathcal{L}_0$ is the character χ, such that $\chi(L(0)) = h, \chi(C) = c$.

Besides Verma modules we will also treat two remarkable two-parameter families of representations of Virasoro algebra. They are tensor fields on the circle $\mathcal{F}_{\lambda,\mu}$ and semi-infinite forms (SIF) modules $\mathcal{H}_{\xi,\eta}$.

The modules $\mathcal{F}^{14}_{\lambda,\mu}$ have the basis $f_j, j \in \mathbb{Z}$ and $\mathcal{L}^\wedge$ acts as follows

$$L(i)\cdot f_j = (\mu - j + (i-1)\lambda) f_{i+j}\ . \tag{2.13}$$

Note that this formula coincides with the transformation law of the primary fields in the conformal field theory.

The modules $\mathcal{H}_{\xi,\eta}$ may be realized in the Fock representation of infinite-dimensional Heisenberg algebra with basis $b(m), m \in \mathbb{Z}$ and commutation relations $[b(m), b(l)] = m\delta_{m,-l}$. The Fock representation $\mathcal{H}_\eta$ is its irreducible representation with vacuum vector v_η, such that $b(m)V_\eta = 0, m > 0, b(0)V_\eta = \eta V_\eta$. The Virasoro operators (their generating functions) can be expressed via $b(m)$ as follows:

$$L(z) = \frac{1}{2} : b(z)^2 : + \xi \overset{\circ}{b}(z) - \frac{\xi^2}{2}$$
$$c = 1 - 12\xi^2 \ . \tag{2.14}$$

Hence $\mathcal{H}_\eta$ becomes a two-parameter family of $\mathcal{L}^\wedge$ modules.

The characters of these modules are equal to the character of the Verma module and if $\mathcal{H}_{\xi,\eta}$ is irreducible, then $\mathcal{H}_{\xi,\eta}$ is isomorphic to $M_{h(\xi,\eta),c(\xi)}$, where $h(\xi,\eta) = \frac{1}{2}(\eta^2 - \xi^2), c = 1 - 12\xi^2$.

Note that these modules allow fermion description as semi-infinite forms.[11] This explains the origin of their title. SIF-modules play the same role in the representation theory of Virasoro algebra as Wakimoto modules in representation theory of affine algebras. We will see in the next section some corroborations of this fact.

3. Chains and Intertwining Operators

In this section we will study Wakimoto modules outside the singular hyperplane. We mentioned earlier that there is a deep analogy between Wakimoto modules over affine algebras and boson representations of W-algebras. The appearance of chains and intertwining operators in homomorphisms of Wakimoto modules is the important example of this analogy. B.L. Feigin and D.B. Fuchs in Ref. 12 and V.A. Fateev, A.B. Zamolodchikov and S.L. Lukyanov in Refs. 6 and 7 (see also Refs. 1-2) constructed explicitly such objects in boson representations of W-algebras. In Refs. 5, 6, 7, 1-2 these objects were used in computing correlation functions in conformal field theory with $\mathbb{Z}_N$-symmetry. Our results may be used in the same way as in WZW models.

3.1. *Virasoro algebra*

We begin with recalling the key definitions and results of Ref. 12, which we will use for the study of Wakimoto modules over $\mathrm{Lsl}_2^\wedge$.

Let M_1 and M_2 be two highest weight modules over Virasoro algebra. The chain of the type (λ, μ) is the submodule of $\mathrm{Hom}_{\mathbb{C}}(M_1, M_2)$, isomorphic to $\mathcal{F}_{\lambda,\mu}$. It turns out that if M_1 and M_2 are SIF modules, then it is possible to construct explicitly the chains using vertex operators.

Vertex operators $B_j(\beta), \beta \in \mathbb{C}, j \in \mathbb{Z}$ act on SIF modules $\mathcal{H}_{\xi,\eta}$. Their generating function is

$$B(\beta, z) = \sum_{j\in\mathbb{Z}} B_j(\beta)z^j = \exp\Big(-\sum_{i<0} \frac{\beta b(i)z^i}{i}\Big) \exp\Big(-\sum_{i>0} \frac{\beta b(i)z^i}{i}\Big)T_\beta\,, \tag{3.1}$$

where $T_\beta : \mathcal{H}_{\xi,\eta-\beta} \rightarrow \mathcal{H}_{\xi,\eta}$ is the linear operator, which commutes with $b(i), i \neq 0$, and maps $V_{\xi,\eta-\beta}$ to $V_{\xi,\eta}$.

So $B_j(\beta)$ are well-defined operators acting from $\mathcal{H}_{\xi,\eta-\beta}$ to $\mathcal{H}_{\xi,\eta}$.

The key property of vertex operators, which characterizes them uniquely, is their commutation relations with Heisenberg algebra:

$$[b(i), B_j(\beta)] = \beta B_{i+j}(\beta)\ . \tag{3.2}$$

To compute the commutation relations of $B_j(\beta)$ with Virasoro operators $L(i)$ (2.14) one should use Wick theorem (see Appendix) and take into consideration that

$$\underbrace{b(z_1)B(\beta, z_2)} = \frac{\beta z_1}{z_2 - z_1} B(\beta, z_2)\,, |z_1| < |z_2|\,;$$
$$\dot{B}(\beta, z) = \beta\ : b(z)B(z)\ :$$

We have:

$$[L(i), B_j(\beta)] = \Big(-j - i + \frac{\beta^2}{2}(i-1) + \beta(\eta + \xi i)\Big) B_{i+j}(\beta)\,, \tag{3.3}$$

where we put $B_j(\beta) : \mathcal{H}_{\xi,\eta-\beta} \rightarrow \mathcal{H}_{\xi,\eta}$.

The meaning of (3.3) in conformal field theory is that $B(\beta, z)$ is the primary field of conformal weight $\frac{\beta^2}{2}$. On the other hand, we see that $B(\beta, z)$ gives the embedding

$$\mathcal{F}_{\beta^2/2+\beta\xi-1,\beta(\xi+\eta)-1} \rightarrow \mathrm{Hom}_{\mathbb{C}}(\mathcal{H}_{\xi,\eta-\beta}, \mathcal{H}_{\xi,\eta})\,,$$

that is $B(\beta, z)$ is the chain of the type $(\beta^2/2 + \beta\xi - 1, \beta(\xi + \eta) - 1)$ in $\mathrm{Hom}_{\mathbb{C}}(\mathcal{H}_{\xi,\eta-\beta}, \mathcal{H}_{\xi,\eta})$.

The case $\lambda = \beta^2/2 + \beta\xi - 1 = 0$ (or, equally, $\xi = \frac{1}{\beta} - \frac{\beta}{2}$) is of great importance. If $\lambda = 0$ and $\mu = l \in \mathbb{Z}$ then $B_l(\beta)$ commutes with the Virasoro algebra ($\mathcal{F}_{o,l}$ is the space of the functions on the circle, and F_l is constant). So, $B_l(\beta)$ in this case yields the intertwining operator $\mathcal{H}_{\xi,\eta-\beta} \to \mathcal{H}_{\xi,\eta}$. In particular, if $l > 0$, then this mapping at general point of the corresponding curve in $C^2 = (\xi, \eta) : \xi = \frac{1}{\beta} - \frac{\beta}{2}, \eta = \frac{l}{\beta} + \frac{\beta}{2}$ is the projection onto irreducible quotient; if $l < 0$, then this mapping at a general point is the embedding of the irreducible submodule and $B_l(\beta)V_{\xi,\eta-\beta}$ is a singular vector in $\mathcal{H}_{\xi,\eta}$ of the degree l. At other points this mapping has a kernel and a cokernel.

Thus, we see that for every pair of SIF modules $\mathcal{H}_{\xi,\eta-\beta}$ and $\mathcal{H}_{\xi,\eta}$ there is the chain of the type (λ, μ) with appropriate (λ, μ) and if $(\lambda, \mu) = (0, l)$, where l is an integer, then this chain contains intertwining operator of degree l.

Further generalization of this construction is based on the following trick. Suppose that we have a chain of vertex operators of the type $(0, \mu_i)$ in $\mathrm{Hom}_{\mathbb{C}}(\mathcal{H}_{\xi,\eta_1}, \mathcal{H}_{\xi,\eta_2})$ and μ_1 is an arbitrary complex number. Let us imagine that there is an operator (of the degree μ_1), $B_{\mu_1}(\eta_2 - \eta - 1)$ which intertwines modules $\mathcal{H}_{\xi,\eta_1}$ and $\mathcal{H}_{\xi,\eta_2}$. Such an operator does not exist, but suppose that there are also chains of vertex operators of the types $(0, \mu_i)$ in $\mathrm{Hom}_{\mathbb{C}}(\mathcal{H}_{\xi,\eta_i}, \mathcal{H}_{\xi,\eta_{i+1}}), i = 1, \ldots, m$, where μ_i are complex numbers, but their sum $\sum_{i=1}^{m} \mu_i$ is an integer. So, the "composition" of these unreal intertwining operators $B_{\mu_i}(\eta_{i+1} - \eta_i)$ of complex degrees has integer degree, and we can suppose that this composition exists; however, its components do not exist, and hence we can obtain real intertwining operator between $\mathcal{H}_{\xi,\eta_1}$ and $\mathcal{H}_{\xi,\eta_{m+1}}$. It turns out that it is possible to give meaning to this procedure, that is to define composition vertex operators as $B_{\mu_m}(\eta_{m+1} - \eta_m) \cdot \ldots \cdot B_{\mu_1}(\eta_2 - \eta_1)$, where μ_i are not necessary integers but their sum is an integer.

Let us define, following Ref. 12 that the family of composition vertex operators with parameters $(\beta; \gamma) = (\beta_1, \ldots, \beta_m; \gamma_1, \ldots, \gamma_m)$ where $\sum_{i=1}^{m} \gamma_i = 0$ as the family of operators $B_{l_1,\ldots,l_m}(\beta; \gamma) : \mathcal{H}_{\xi,\eta-\beta_1-\cdots-\beta_m} \to \mathcal{H}_{\xi,\eta}, l_i \in \mathbb{Z}$, commuting with Heisenberg algebra and Virasoro algebra as

$$[b(i), B_{l_1,\ldots,l_m}(\beta;\gamma)] = \sum_{s=1}^{m} \beta_s B_{l_1,\ldots,l_s+i,\ldots,l_m}(\beta;\gamma) \ , \tag{3.4}$$

$$[L(i), B_{l_1,\dots,l_m}(\beta;\gamma)] = \sum_{s=1}^{m}[\mu_s - (l_s + \gamma_s) + (l-1)\lambda_s] \qquad (3.5)$$

$\times B_{l_1,\dots,l_s+i,\dots,l_m}(\beta;\gamma)$,where

$$\lambda_s = \frac{\beta s^2}{2} + \beta_s\xi - 1\,, \mu_s = \beta_s(\xi + \eta - \beta_{s+1} - \dots - \beta_m) - 1\,. \qquad (3.6)$$

So we see that $B_{l_1,\dots,l_m}(\beta;\gamma)$ behave like $B_{l_1+\gamma_1}(\beta_1)\cdot B_{l_2+\gamma_2}(\beta_2)\cdot\dots\cdot B_{l_m+\gamma_m}(\beta_m)$ and Virasoro algebra acts on these operators as on $\mathcal{F}_{\lambda_1,\mu_1-\gamma_1}\otimes\mathcal{F}_{\lambda_2,\mu_2-\gamma_2}\otimes\dots\otimes\mathcal{F}_{\lambda_m,\mu_m-\lambda_m}$.

Taking (2.14) into consideration one obtains the system of equations from (3.4) and (3.5), which depends, generally, on ξ,η,β,γ. But it is amazing that this system really does not depend on ξ and η and always has solutions, as shown in Ref. 12. Moreover, the dimension of the space of the solutions is not always less than $(m-1)!$. It is proved in Ref. 12 by homological methods.

In Refs. 12 and 31 explicit formulae for composition vertex operators are proposed. They are based on contour integrals of the type

$$\int_\Gamma dz_1\dots dz_m z_1^{-l_1-1}\dots z_m^{-l_m-1} : B(\beta_1,z_1)\dots B(\beta_m,z_m) :$$
$$\prod_{1\le i<j\le m}(1-\frac{z_i}{z_j})^{\beta_i\beta_j}\prod_{1\le i\le m} z_i^{-\gamma_i}\,. \qquad (3.7)$$

The function in (3.7) is multi-valued, and to compute the integral we should choose the element Γ of the m-th homology group of the manifold $M_m = \{(z_i,\dots,z_m)\in(\mathbb{C}^*)^m\,; z_i\ne z_j\}$ with coefficients in the local system, which is determined by the monodromy of this holomorphic function. The number $(m-1)!$ is nothing but the lower bound of the dimension of this homology group. Integrals (3.7) (Feigin-Fuchs integrals) are used in 2D conformal theory for computing correlation funtions.[5]

Evidently, if $\gamma_1,\dots,\gamma_m = 0$, then operators $B_{l_1}(\beta_1)\cdot\dots\cdot B_{l_m}(\beta_m)$ satisfy (3.4) and (3.5).

Note that (3.4) does not depend on γ_i, but γ_i give the contribution $-\sum_{s=1}^{m}\gamma_s B_{l_1,\dots l_s+i,\dots,l_m}(\beta;\gamma)$ to (3.5). So we could imagine $B_{l_1,\dots,l_m}(\beta;\gamma)$ also as composition $B_{l_1+\gamma_1}(\beta_1)\cdot\dots\cdot B_{l_m}(\beta_m;\gamma_m)$, where

$$[b(i), B_{l_j}(\beta_j,\gamma_j)] = \beta_j B_{l_j+i}(\beta_j,\gamma_j)\,, \qquad (3.8)$$

$$\beta_j : b(z)B(\beta_j,\gamma_j,z) : = \overset{\circ}{B}(\beta_j,\gamma_j,z) + \gamma_j B(\beta_j,\gamma_j,z)\,. \qquad (3.9)$$

The last term in (3.9) gives the required contribution to (3.5).

If γ_j are not all integers, then operators $B_{l_j}(\beta_j, \gamma_j)$ do not exist, but we can use this model to clarify the meaning of composition vertex operators.

Now we apply composition vertex operators to produce new intertwining operators and chains. We noticed that $B_{l_1,\dots,l_m}(\beta;\gamma)$ form the module $\mathcal{F}_{\lambda_1,\mu_1-\gamma_1} \otimes \dots \otimes \mathcal{F}_{\lambda_m,\mu_m\gamma_m}$ over Virasoro algebra. So, if we put all $\lambda_j = 0$ and $\mu_i - \gamma_i$ to be integers l_i, then $B_{l_1,\dots,l_m}(\beta;\gamma)$ appears to be intertwining operator of degree $l_1 + \dots + l_m$ (due to $\sum_{s=1}^{m} \gamma_s = 0$). In particular, if we are interested in singular vectors in $\mathcal{H}_{\xi,\eta}$, we may put $l_1 = \dots = l_m = -l$ where l is positive integer. Then $B_{-l,\dots,-l}(\beta;\gamma)V_{\xi,\eta-\beta_1-\dots-\beta_m}$ gives singular vector of the degree $-l \cdot m$ in $\mathcal{H}_{\xi,\eta}$ if and only if β_i are equal roots of the quadratic equation

$$\frac{\beta^2}{2} + \beta\xi - 1 = 0(\beta_\pm = -\xi \pm \sqrt{\xi^2 + 2}) \quad \text{and}$$
$$\gamma_i = \beta(\xi + \eta_m + (i - m)\beta) + l - 1 .$$

The equation $\sum_{i=1}^{m} \gamma_i = 0$ gives

$$\xi = -\frac{\beta}{2} + \frac{1}{\beta}\,, \eta = \frac{m \cdot \beta}{2} - \frac{l}{\beta} \tag{3.9}$$

where m and l are positive integers (these integers are the labels of the Kac equation for reducible Verma modules — see Ref. 18).

In Refs. 12 and 31 it is proved that this vector is nonzero.

If we want to obtain a composition chain, we should claim $\lambda_i = 0$ for all i besides S and $\lambda_s = \lambda$. Again, $\mu_i = \gamma_i, i \neq S$ are necessary integers, for example, they are equal to 0. So, $B_{\underbrace{0,\dots,0}_{(s-1)},\, l,\, \underbrace{0,\dots,0}_{(m-s)}}(\beta;\gamma)$ is the chain in $\mathrm{Hom}_{\mathbb{C}}(\mathcal{H}_{\xi,\eta-\beta_1-\dots-\beta_m}; \mathcal{H}_{\xi,\eta})$ if the following conditions are valid:

$$\beta_i = \beta_+, i = 1, \dots, s-1; \beta_s = \beta : \beta_i = \beta_-, i = s+1, \dots, m$$
$$\gamma_i = \beta_i(\xi + \eta - \beta_{i+1} - \dots - \beta_m) - 1, i \neq s; \gamma_s = -\sum_{i\neq s} \gamma_i$$

The type of this chain is $\lambda = \dfrac{\beta^2{}_s}{2} + \beta_s - 1, \mu = (s-1)\beta_+ + (m-s)\beta_- + \beta$.

These formulae were obtained in Ref. 12.

Now we pass to $L\hat{sl_2}$ and apply similar arguments to construct affine analogues of chains and intertwining operators.

3.2. $L\hat{sl_2}$

It is natural to give the following definition.

The chain of the type μ is the submodule in $\mathrm{Hom}_{\mathbb{C}}(M_1, M_2)$ (where M_1, M_2 are $L\hat{sl_2}$-modules), isomorphic to the module LM_μ^* of currents on the circle to the contragradient module M_μ^* of Verma module over sl_2 with *lowest* weight μ (see 2.1).

We are in a position to give the first example of the chain in $\mathrm{Hom}_{\mathbb{C}}(W_{\chi_1,\nu}, W_{\chi_2,\nu})$. To do it we put that vertex operators $B_j(\beta)$ to act from $W_{\chi-\beta\nu,\nu}$ to $W_{\chi,\nu}$(according to (3.1)).

Proposition 3.1

Homogeneous components of $C_l(\beta, z) = a^*(z)^l B(\beta, z)$ form the chain of the type $\mu = \beta\nu$ in $\mathrm{Hom}_{\mathbb{C}}(W_{\chi-\beta\nu,\nu}, W_{\chi,\nu})$.

Proof

It is easy to compute the commutation relations of $L\hat{sl_2}$ with $C_l(\beta, z)$ using Wick theorem. We obtain

$$\begin{aligned} [F(i), C_j(\beta, z)] &= z^{-i} j C_{j-1}(\beta, z)\,, \\ [H(i), C_j(\beta, z)] &= z^{-i}(\beta\nu + 2j) C_j(\beta, z)\,, \\ [E(i), C_j(\beta, z)] &= z^{-i}(-\beta\nu - j) C_{j+1}(\beta, z) \end{aligned} \tag{3.10}$$

and Proposition 3.1. follows.

We also produce another proof, which can be easily generalized to arbitrary affine algebra.

Recall that M_0^* may be realized as the space of polynomials on the group N_-, or, equally, polynomials on one variable $x = a^*$. If we put $\frac{\partial}{\partial x} = a$, then the action of sl_2 in this space is

$$\begin{aligned} E &= -a^* a^* a \\ H &= 2a^* a \\ F &= a \end{aligned}$$

When we pass to the currents we change a and a^* by $a(z)$ and $a^*(z)$. So, $L\hat{sl_2}$ acts on homogeneous components of $a^*(z)^l$ as on LM_0^* . To prove that $Lsl_2^\wedge$ acts on $a^*(z)^l B(\beta, z)$ as on $LM^*_{\beta\nu}$ one should verify that

$$[E(i), B(\beta, z)] = -\beta\nu z^{-i} B(\beta, z) a^*(z) ,$$
$$[H(i), B(\beta, z)] = \beta\nu z^{-i} B(\beta, z) , \qquad (3.11)$$

and it can be easily done. Proposition 3.1 is proved.

Proposition 3.1 gives the simplest example of a chain. In 3.1 we have found the modules $\mathcal{F}_{0,\mu}$ over Virasoro algebra, which contain elements $f\mu$, being annihilated by Virasoro algebra, and used them to construct intertwining operators and composition chains.

In the affine case there are also such modules with similar property. These modules are deformations of the LM_0-module of currents on the circle to the Verma module M_0 with highest weight 0 over sl_2. It is evident that elements $V \otimes t^i \in LM_0, i \in \mathbb{Z}$, where V is vacuum vector of M_0 are annihilated by $F(j)$ and $H(j)$. We consider deformation $LM_{0,[i]}$ which contains the element $v \otimes t^i$ with fixed $i \in \mathbb{Z}$, being annihilated also by $E(j)$.

It is the analogue of $\mathcal{F}_{0,\mu}$.

Put $D(\beta, z) = a(z) B(\beta, z)$.

Proposition 3.2

The homogeneous components of $D(\frac{2}{\nu}, z), C_l(\frac{2}{\nu}, z), l = 0, 1, \ldots$ form the $Lsl_2^\wedge$-module in Hom $(W_{\chi-2,\nu}, W_{\chi,\nu})$ which is isomorphic to LM_0 or $LM_{0,[i]}$.

Proof

The proposition follows from the commutation relations:

$$[F(i), D(\frac{2}{\nu}, z)] = 0 ,$$
$$[H(i), D(\frac{2}{\nu}, z)] = 0 ,$$
$$[E(i), D(\frac{2}{\nu}, z)] = z^{-i}[(\chi + i\frac{\nu^2}{2} B(\frac{2}{\nu}, z) - \frac{\nu^2}{2} \dot{B}(\frac{2}{\nu}, z)] \qquad (3.12)$$

Corollary 3.3

If $\chi - l\frac{\nu^2}{2} = 0$ is satisfied, where l is an integer, then $D_l(\frac{2}{\nu})$ is the intertwining operator between $W_{\chi-2,\nu}$ and $W_{\chi,\nu}$. If l is negative, then

$D_l(\frac{2}{\nu})V_{\chi-2,\nu}$ is a singular vector in $W_{\chi,\nu}$ of the degree $(l-1)\alpha_1 + l\alpha_0$ and $\chi - l\frac{\nu^2}{2} = 0$ is the Kac-Kazhan equation. ($V_{\chi,\nu}$ is a vacuum vector in $W_{\chi,\nu}$.)

It follows from Proposition 3.2.

Operators $D_L(\frac{2}{\nu})$ enable us to construct new intertwining operators and chains over $\mathrm{Lsl}_2^\wedge$ in the same way as over Virasoro algebra.

As above, we should take the composition of $D_l(\frac{2}{\nu})$ where $l = \frac{2\chi}{\nu^2}$ and only the sum of the degrees of components is an integer. So we need operators $D_{l_1,\dots,l_m}(\frac{2}{\nu},\dots,\frac{2}{\nu};\gamma_1,\dots,\gamma_m)$ with $\sum_{i=1}^{m}\gamma_i = 0$ which may be considered as composition $D_{l_1}(\frac{2}{\nu},\gamma_1)\dots D_{l_m}(\frac{2}{\nu},\gamma_m)$ where $D_l(\frac{2}{\nu},\gamma)$ commute with $\mathrm{Lsl}_2^\wedge$ as follows

$$
\begin{aligned}
[F(i), D_l(\tfrac{2}{\nu},\gamma)] &= 0\\
[H(i), D_l(\tfrac{2}{\nu},\gamma)] &= 0\\
[E(i), D_l(\tfrac{2}{\nu},\gamma)] &= (\chi - (l+\gamma)\tfrac{\nu^2}{2})B_{l+i}(\tfrac{2}{\nu},\gamma)\ . \qquad (3.13)
\end{aligned}
$$

It is clear that $\sum_{j\in\mathbb{Z}} B_{l-j}(\frac{2}{\nu},\gamma)a(j)$ satisfy (3.13), where $B_j(\frac{2}{\nu},\gamma)$ are operators introduced in 3.1. This shows that one should put

$$D_{l_1,\dots,l_m}(\frac{2}{\nu},\dots,\frac{2}{\nu};\gamma_1,\dots,\gamma_m) = \sum_{j_1,\dots,j_m\in\mathbb{Z}} a(j_1)\dots a(j_m)$$

$\times B_{l_1-j_1,\dots,l_m-j_m}(\frac{2}{\nu},\dots,\frac{2}{\nu};\gamma_1,\dots,\gamma_m)$, where $B_{l_1,\dots,l_m}(\frac{2}{\nu},\dots,\frac{2}{\nu};\gamma_1,\dots,\gamma_m)$ are composition vertex operators introduced in 3.1.

We have, as required:

$$
\begin{aligned}
&[F(i), D_{l_1,\dots,l_m}(\tfrac{2}{\nu},\dots,\tfrac{2}{\nu};\gamma_1,\dots,\gamma_m)] = 0\ ,\\
&[H(i), D_{l_1,\dots,l_m}(\tfrac{2}{\nu},\dots,\tfrac{2}{\nu};\gamma_1,\dots,\gamma_m)] = 0\ ,\\
&[E(i), D_{l_1,\dots,l_m}(\tfrac{2}{\nu},\dots,\tfrac{2}{\nu};\gamma_1,\dots,\gamma_m)] = \sum_{s=1}^{m}(\chi_s - (l_s+\gamma)\tfrac{\nu^2}{2}) \sum_{j_1,\dots,j_m\in\mathbb{Z}}\\
&a(j_1)\dots a(\hat{j}_s)\dots a(j_m)B_{l_1-j_1,\dots,l_s-j_s+i,\dots,l_m-j_m}(\tfrac{2}{\nu},\dots,\gamma_1,\dots,\gamma_m)\ , \qquad (3.14)
\end{aligned}
$$

where $\chi_s = \chi - 2(s-1)$.

Operators act from $W_{\chi-2m,\nu}$ to $W_{\chi,\nu}$.

These relations may be proved using explicit definition of composition vertex operators as contour integrals (3.7). It is possible also to consider the corresponding system of equations by virtue of Ref. 5.

Now we are in a position to construct new intertwining operators. We should take the module $LM_{0,[l_1]} \otimes \ldots \otimes Lm_{0,[l_m]}$ and its vector $(v \otimes t^{l_1}) \otimes \ldots \otimes (v \otimes t^{l_m})$

Theorem 3.4

Let $\gamma_j = \dfrac{2\chi - 4(j-1)}{\nu^2} - l_j, j = 1, \ldots, m$ and $\sum_{j=1}^{m} \gamma_j = 0$. Then $D_{l_1,\ldots,l_m}(\frac{2}{\nu}, \ldots, \frac{2}{\nu}; \gamma_1, \ldots, \gamma_m)$ is intertwining operator between $W_{\chi-2m,\nu}$ and $W_{\chi,\nu}$. In particular, $D_{l,\ldots,l}(\frac{2}{\nu}, \ldots, \frac{2}{\nu}; \gamma_1, \ldots, \gamma_m)V_{\chi-2m,\nu}$ (where l is a negative integer) is a singular vector in $W_{\chi,\nu}$ of the degree $m((l-1)\alpha_1 + l\alpha_0)$, the equation $\sum_{i=1}^{m} \gamma_i = 0$ being the Kac-Kazhdan equation $\chi - l\frac{\nu^2}{2} - (m-1) = 0$.

The proof of this is the direct computation using (3.14). Nonvanishing theorem follows from Refs. 12 and 31 immediately.

Analogous to 3.1 we also construct chains. We take $LM_{0,[0]} \otimes \ldots \otimes LM_{0,[0]} \otimes LM^*_\mu$ and its submodule $(V \otimes 1) \otimes \ldots \otimes (v \otimes 1)LM_\mu$.

Theorem 3.5

Let $\gamma_j = \dfrac{2\chi - 4(j-1)}{\nu^2}, j = 1, \ldots, m; \gamma_{m+1} = -\sum_{j=1}^{m} \gamma_j$. The homogeneous components of $\overline{D}_j(\beta, z)_m = \sum_{j \in \mathbb{Z}} \sum_{j_1,\ldots,j_m \in \mathbb{Z}} a(j_1) \ldots a(jm)$ $B_{-j_1,\ldots,-jm,j}(\frac{2}{\nu}, \ldots, \frac{2}{\nu}, \beta; \gamma_1, \ldots \gamma_{m+1})z^j$ and of $a^*(z)^l \overline{D}(\beta, z)_m$ form the chain of the type $\mu = \beta\nu$ in $\mathrm{Hom}_{\mathbb{C}}(W_{\chi-2m-\beta\nu,\nu}; W_{\chi,\nu})$.

Proof is direct computation.

So we constructed for any pair of weights χ_1, χ_2 such that $\chi_2 - \chi_1 = \mu + 2m, m \in \mathbb{Z}$ the chain of the type μ in $\mathrm{Hom}_{\mathbb{C}}(W_{\chi_1,\nu}, W_{\chi_2,\nu})$. It turns out that for a general pair of weights χ_1, χ_2 there is a unique chain of the type μ if and only if $\chi_2 - \chi_1 - \mu \in 2\mathbb{Z}$ (it may be proved independently). So we have constructed explicitly all chains in honomorphisms of general Wakimoto (and Verma) modules.

Note also that Theorem 3.4 leads to many interesting combinatorial identities if one changes the set $(l_1, \ldots, l_m)$ by $(l'_1, \ldots, l'_m)$, where $\sum_{s=1}^{m} l_s = \sum_{s=1}^{m} l'_s$. These identities may be viewed as identities of hypergeometric functions.

3.3 $\mathrm{Lsl}_n^\wedge$

Now we pass to affine Lie algebras of the type $\mathrm{Lsl}_n^\wedge$. At first we want to account briefly the "Virasoro-counterpart" of our constructions.

Consider the algebra $W_n^{25,6,1,2}$ which corresponds to $\mathrm{Lsl}_n^\wedge$ in the same manner in which the Virasoro algebra (W_2) corresponds to $\mathrm{Lsl}\,_2^\wedge$. This algebra is $\mathbb{Z}$-graded and contains $(n-1)$ countable sets of operators $L_i^{(1)}, \ldots, L_i^{(n-1)}$. Operators $L_i^{(1)}$ form the Virasoro algebra and this algebra acts on $L_i^{(j)}$ as on module $\mathcal{F}_j$. The relations between other $L_{i_1}^{(j_1)}$ and $L_{i_2}^{(j_2)}$ are quadratic.

We can define the analogue of the chain over algebra W_n. However it is not possible to construct a W_n module with the required properties. We consider the set of operators which form the modules over the Virasoro algebra, isomorphic to $\mathcal{F}_{\lambda,\mu} \oplus \mathcal{F}_{\lambda+1,\mu} \oplus \ldots \oplus \mathcal{F}_{\lambda+n-2,\mu}$ and define the quadratic relations of these operators with $L_i^{(j)}$.

These relations connect $\mathcal{F}_i \otimes \mathcal{F}_{\lambda,\mu} (i = 2, \ldots, n-1)$ with $\mathcal{F}_{\lambda+i-1,\mu}$ by the invariant differential operator (as well-known, there exists an invariant differential operator $\mathcal{F}_{\lambda_1,\mu_1} \otimes \mathcal{F}_{\lambda_2,\mu_2} \to \mathcal{F}_{\lambda_1+\lambda_2-1,\mu_1+\mu_2}$, determined by $(f(z)dz^{-\lambda_1}, g(z)dz^{-\lambda_2} \to [\lambda_2 f'(z)g(z) - \lambda_1 f(z)g'(z)]dz^{-\lambda_1-\lambda_2+1})$. This operator is unique up to a constant and so we have $(n-2)$ relations determined up to $(n-2)$ parameters.

Other relations are uniquely determined by these and the structure of the algebra W_n. They are also quadratic. Thus, the object we constructed depend on n parameters (including λ, μ). We call it the chain over algebra W_n. These chains appear in boson realizations of W algebras.[6,7,1-2]. In particular, their "top" part $\mathcal{F}_{\lambda,\mu}$ may be expressed via vertex operators. They give intertwining operators between boson representations. In 2D conformal theory with symmetry $\mathbb{Z}_n$ they play the role of screening operators. The chains over $\mathrm{Lsl}_n^\wedge$, discussed below, are connected with chains over the algebra W_n.

As in 3.2 we define the chain of the type μ over $\mathrm{Lsl}_{n+1}^\wedge$ where μ is the character of the Cartan subalgebra of sl_{n+1} as the module LM_μ^* of currents on the circle to M_μ^*-contragradient Verma module over sl_{n+1} with the lowest weight μ.

We need direct generalization of vertex operators:

$$B(\beta^1,\dots,\beta^m;z)=B(\beta;z)=\sum_{l\in\mathbb{Z}}B_l(\beta)z^l=\exp\Big(-\sum_{i<0}\frac{\beta\cdot b(i)z^i}{i}\Big)$$

$\exp\big(-\sum_{i>0}\frac{\beta\cdot b(i)z^i}{i}\big)$, where $b_j(i)$ form Heisenberg algebra of 2.3 with the relations

$$[b_l(i),b_m(j)]=ic_{l,m}\delta_{i,-j}\ ,$$

where $c_{l,m}=\langle H_l,H_m\rangle=(\bar\alpha_l,\bar\alpha_m)$, and $b(i)=(b_1(i),\dots,b_n(i))$. Vertex operators act in $\mathrm{Hom}_{\mathbb{C}}(W_{\chi-\sum_{s=1}^{n}\beta_s\alpha_s\nu;\nu};W_{\chi,\nu})$.

We obtain the direct generalization of (3.2):

$$[b_l(i),B_j(\beta)]=\sum_{s=1}^{n}c_{is}\beta_sB_{j+l}(\beta)\ . \tag{3.15}$$

The first example of the chain is given by the following Proposition which is a direct generalization of Proposition 3.1.

Proposition 3.6

Homogeneous components of $\prod_{\substack{1\le j\le n\\ i\le j\le n}}a^*_{ij}(z)^{d_{ij}}B(\beta,z),d_{ij}=0,1\dots$ form the chain of the type $\mu=\nu\sum_{s=1}^{n}\beta_s\alpha_s$ in $\mathrm{Hom}_{\mathbb{C}}(W_{\chi-\mu,\nu};W_{\chi,\nu})$.

Proof is analogous to the proof of Proposition 3.1 and uses realization of M^*_μ as the space of polynomials on the group N_- and explicit formulae (2.9).

The next step, as above, consists of the construction of modules which could contain an intertwining operator. Let L_0 be trivial one-dimensional module over sl_{n+1}. As is well-known, $\mathrm{Ext}^1(L_0,M_{\alpha_i})=\mathbb{C}$. So there is non-trivial extension $M_0^{(i)}$.

The modules we need are $LM_0^{(i)}$ and $LM_{0,[l]}^{(i)}$, where $i=1,\dots,n;l\in\mathbb{Z}$. Module $M_{0,[l]}^{(i)}$ contains element $v\otimes t^l$, which is annihilated by $L\mathrm{sl}^\wedge_{n+1}$ (note that $v\otimes t^s\in LM_0^{(i)}$ are annihilated by $F_j(m),H_j(m)$ and $E_j(m),j\ne i$).

To construct $LM_0^{(i)}$ (or $LM_{0,[l]}^{(i)}$), explicitly we need operators

$$\overline{F}_i(z)=-a_{ii}(z)-\sum_{j=1}^{i-1}a_{ji}(z)a^*_{j,i-1}(z)\ . \tag{3.16}$$

These operators are interesting due to their remarkable commutation relations with generators of $\mathrm{Lsl}_{n+1}^{\wedge}$.

Proposition 3.7

$\overline{F}_i(m)$ generate algebra Ln_- and commute with generators of $\mathrm{Lsl}_{n+1}^{\wedge}$ as follows:

$$\begin{aligned}
[F_i(z), \overline{F}_j(m)] &= 0 \\
[H_i(z), \overline{F}_j(m)] &= -c_{ij} z^{-m} \overline{F}_j(z) \\
[E_i(z), \overline{F}_j(m)] &= -c_{ij} z^{-m} a_{ii}^*(z) \overline{F}_j(z) - \delta_{ij} \cdot \nu b_i(z) z^{-m} \\
&\quad - \delta_{ij} m(\chi_i + \nu^2 - 2) z^{-m} .
\end{aligned} \tag{3.17}$$

Proof

Return to our realization of the contragradient Verma module over sl_{n+1} as the space of the polynomials on the group N_-. We have the left action of the group SL_{n+1} on the flag manifold, and so the left infinitesimal action of the algebra sl_{n+1} on the group N_- as on the big cell of the flag manifold. But the group N_- acts on itself also on the right side, this right action of N_- commuting with left action of this group. So we obtain the right infinitesimal action of n_- on N_-, commuting with its left action. If we write down the corresponding vector fields in coordinates $\mathcal{X}_{ij}$ (so, as polynomials on $a_{ij} = \dfrac{\partial}{\partial x_{ij}}$ and $a_{ij}^* = x_{ij}$) and pass to currents, then we obtain the action of the algebra Ln_- on Wakimoto module. To compute (3.17) one should compute the corresponding relations between left and right vector fields on the finite-dimensional flag manifold and also take into consideration and binary pairings. This can be done easily. The proposition follows.

Now we introduce the operators $D^{(j)}(\beta, z) = \overline{F}_j(z) B(\beta, z)$. They are analogues of $D(\beta, z)$ of 3.2.

Let $\varepsilon_i = (0, \ldots, 1_i, \ldots, 0)$. The following proposition is a direct generalization of the Proposition 3.2.

Proposition 3.8

Homogeneous components of $D^{(l)}(\frac{\varepsilon_l}{\nu}, z)$ and $\prod\limits_{\substack{1 \le i \le n \\ i \le j \le n}} a_{ij}^*(z)^{d_{ij}} B(\varepsilon_l/\nu, z)$, $d_{ij} = 0, 1, \ldots$ form in $\mathrm{Hom}_{\mathbb{C}}(W_{\chi-\alpha_l,\nu} W_{\chi,\nu})$ the module, isomorphic to LM_0^l for general values (χ, ν).

Proof follows from the commutation relations

$$
\begin{aligned}
[F_i(m), D_l^{(j)}(\frac{\varepsilon_j}{\nu})] &= 0 \\
[H_i(m), D_l^{(j)}(\frac{\varepsilon_j}{\nu})] &= 0 \\
[E_i(m), D_l^{(j)}(\frac{\varepsilon_j}{\nu})] &= \delta_{ij}(\chi_j - l\nu^2)B_{l+i}(\frac{\varepsilon_j}{\nu})\,.
\end{aligned}
\tag{3.18}
$$

Corollary 3.9

If $\chi_j - l\nu^2 = 0$ is satisfied, where l is an integer, then $D_l^{(j)}(\frac{\varepsilon_j}{\nu})$ is an intertwining operator between $W_{\chi-\alpha_j,\nu}$ and $W_{\chi,\nu}$. If l is a negative integer, then $D_l^{(j)}(\frac{\varepsilon_j}{\nu})V_{\chi,\alpha_j,\nu}$ is a singular vector in $W_{\chi,\nu}$ of the degree $l\delta - \alpha_j$, the equation $\chi_j - l\nu^2 = 0$ being Kac-Kazhdan equation.

At the next step we introduce the composition vertex operators $B_{l_1,\dots,l_m}(\beta_1,\dots,\beta_m;\gamma_1,\dots,\gamma_m)(\sum_{i=1}^{m}\gamma_i = 0)$ in the same way as in 3.1. After that we put

$$
\begin{aligned}
D_{l_1,\dots,l_m}^{j_1,\dots,j_m}(\gamma_1,\dots,\gamma_m) &= \sum_{i_1,\dots,i_m\in \mathbb{Z}} \overline{F}_{j_1}(i_1)\cdot\ldots\cdot\overline{F}_{jm}(i_m) \\
&\times B_{l_1-i_1,\dots,l_m-i_m}(\frac{\varepsilon_{j_1}}{\nu},\dots,\frac{\varepsilon_{jm}}{\nu},\gamma_1,\dots,\gamma_m) \quad \text{and} \\
D_{l_1,\dots,l_m,l}^{(j_1,\dots,j_m)}(\beta;\gamma_1,\dots,\gamma_{m+1}) &= \sum_{i_1,\dots,i_m\in \mathbb{Z}} \overline{F}_{j_1}(i_1)\cdot\ldots\cdot\overline{F}_{jm}(i_m) \\
&\times B_{l_1-i_1,\dots,l_m-i_m,l}(\frac{\varepsilon_{j_1}}{\nu},\dots,\frac{\varepsilon_{jm}}{\nu},\beta;\gamma_1,\dots,\gamma_{m+1})\,.
\end{aligned}
$$

We have the following generalizations of the theorems 3.4 and 3.5.

Theorem 3.10

Let $\gamma_i = \frac{1}{\nu^2}(\chi - \sum_{s=1}^{i-1}\alpha_{j_s},\alpha_{ji}) - l_i$ and $\sum_{i=1}^{m}\gamma_i = 0$. Then $D_{l_1,\dots,l_m}^{(j_1,\dots,j_m)}$ $(\gamma_1,\dots,\gamma_m)$ is intertwining operator between $W_{\chi-\sum_{s=i}^{m}\alpha_{j_s},\nu}$ and $W_{\chi,\nu}$.

In particular,

$$
D_{l,\dots,l;0,\dots,0,\dots,0\dots,0}^{(i+1,\dots,i+j;i+1,\dots,i+j;\dots;i+1,\dots,i+j)}
(\gamma_1^{(1)},\dots,\gamma_j^{(1)};\gamma_j^{(2)},\dots,\gamma_{ji}^{(2)};\dots;\gamma_1^{(m)},\dots,\gamma_j^{(m)})
$$

$V_{\chi-m}(\alpha_{i+1}+\ldots+\alpha_{i+j}),\nu$ is singular vector in $W_{\chi,\nu}$ of the degree $m(l\delta-\sum_{s=i+1}^{i+j}\alpha_s)$, if the Kac-Kashdan equation is satisfied:

$$\sum_{s=i+1}^{i+j}\chi_s - l\nu^2 = m-j$$

Theorem 3.11

Let $\gamma_i = \frac{1}{\nu^2}(\chi - \sum_{s=1}^{i-1}\alpha_{j_s},\alpha_{j_i}), i=1,\ldots,m, \gamma_{m+1} = -\sum_{i=1}^{m}\gamma_i$. Then $\overline{D}^{(j_1,\ldots,j_m)}(\beta,z) = \sum_{l\in\mathbb{Z}} D^{(j_1,\ldots,j_m)}_{0,\ldots,0,l}(\beta;\gamma)z^l$ and $\prod_{\substack{1\leq i\leq n\\ i\leq j\leq n}} a^*_{ij}(z)^{d_{ij}}$ $\overline{D}^{(j_1,\ldots,j_m)}(\beta,z)$ form the chain of the type $\gamma\sum_{s=1}^{n}\beta_s\cdot\alpha_s = \mu$ in $\mathrm{Hom}_{\mathbb{C}}$ $(W_{\chi-\alpha_{j_1}-\ldots-\alpha_{j_m}-\mu,\nu};W_{\chi,\nu})$.

These theorems are proved directly using (3.18). The nonvanishing theorem may be obtained in the same way.

Let χ_1 and χ_2 be a general pair of the highest weights. One can prove that there is the chain of the type μ in $\mathrm{Hom}_{\mathbb{C}}(W_{\chi_1,\nu};W_{\chi_2,\nu})$ if and only if $\chi_2-\chi_1=\mu+\gamma$, where γ is a positive element of the root lattice Λ of sl_{n+1}, the dimension of the linear space of such chains being equal to the Kostant function $n(\lambda)$: $\sum_{\lambda\in\Lambda_+} n(\lambda)e^\lambda = \prod_{\alpha\in\Delta_+}(1-e^{-\alpha})^{-1}$. Theorem 3.11 gives exactly the same dimension of the space of the chains (it follows from the existence of the relations between $D^{(i)}(\frac{\varepsilon_i}{\nu},z)$ and $D^{(i+1)}(\frac{\varepsilon_{i+1}}{\nu},z)$ of the third order depending on ν). So, in general, we have constructed all chains explicitly.

We also point out two interesting corollaries of the theorems 3.10. In many cases the intertwining operator is unique and does not depend on the choice of the ordered sets $(j_1,\ldots,j_m)$ and $(l_1,\ldots,l_m)$. So, we obtain a great number of combinatorial identities.

Theorem 3.10 also gives explicit formulae for singular vectors of the degrees $-m\alpha$, where α is the root of $\mathrm{sl}_{n+1},\alpha\in\Delta_+$. So, we obtain new expressions of these vectors in addition to Refs. 28 and 35; they depend only on the coefficients of some vertex operators.

3.4. General case

In this subsection we discuss how to generalise our results to other affine Lie algebras.

In the case of simply-laced algebras the generalization is direct. We should write down formulae for boson realization of Wakimoto modules as in 2.3. After that, we introduce vertex operators $B(\beta,z) = \exp\left(-\sum_{i<0}\frac{\beta\cdot b(i)z^i}{i}\right)\exp\left(-\sum_{i>0}\frac{\beta\cdot b(i)z^i}{i}\right)$, where $b_j(m)$ generates the Heisenberg algebra with the commutation relations:

$$[b_i(m), b_j(l)] = (\alpha_i,\alpha_j)m\delta_{m,-l}, \text{ and } \quad \beta = (\beta^1,\ldots,\beta^r),$$

We have: $[b_i(m), B(\beta,z)] = \sum_{s=1}^r(\alpha_i,\alpha_s)\beta_s B(\beta,z)\cdot z^{-m}$, if we say that $B(\beta,z)$ act from $W_{\chi-\beta\nu,\nu}$ to $W_{\chi,\nu}$.

The chains are modules LM_ν^*. The homogeneous components of $\prod_{\alpha\in\Delta_+} a_\alpha^*(z)^{d_\alpha}B(\beta,z), d_\alpha = 0,1,\ldots$ form the chain of the type $\beta\nu$ in $\mathrm{Hom}_{\mathbb{C}}$ $(W_{\chi-\beta\nu,\nu}; W_{\chi,\nu})$.

After that, one should introduce the operators $\overline{F}_i(z)$ as operators of right action of Ln_-. Then $D^{(i)}(\frac{\varepsilon_i}{\nu},z) = \overline{F}_i(z)B(\frac{\varepsilon_i}{\nu},z)$, where ε_i is fundamental weight, gives the intertwining operators between the appropriate Wakimoto modules. These operators may also be used as building blocks of composition operators, and one may construct chains and intertwining operators, in the general case one obtains all chains. Theorems 3.10 and 3.11 pass without changes. Note that to write down explicit formulae for chains and intertwining operators it is sufficient to dispose explicit expressions of $\overline{F}_i(z)$. One can use it for the computation of correlation functions in WZW-models.

The situation with non-simply-laced algebras is more subtle. For example,

$$[H_i(m), D^{(j)}(\frac{\varepsilon_j}{\nu},z)] = z^{-m}\left[(\check{\alpha}_i,\check{\alpha}_j) - \frac{2(\alpha_i,\alpha_j)}{(\alpha_i,\alpha_i)}\right]B(\frac{\varepsilon_j}{\nu},z)\ .$$

When the algebra is not simply-laced, this does not vanish.

So, one should, as always, add fermions: one to each short root. It means that one should pass to the corresponding super-conformal algebra.[21] After that, all operators can easily be constructed. In the conclusion of this section we note that in conformal field theory chains are nothing but primary fields, intertwining operators are screening operators. Explicit formulae obtained in this section allow one not only to compute correlation functions but also to investigate the action of monodromy group (for example, braid group) on these functions, in virtue of Ref. 32. In particular, it seems to give new information about hypergeometric functions.

4. The Structure of Wakimoto Modules Over $Lsl_2^\wedge$

This section is devoted to the study of Wakimoto modules over $Lsl_2^\wedge$. In 4.1 we will discuss some known results concerning the structure of Verma modules over $Lsl_2^\wedge$ outside the singular line $k = -2$. In 4.2 we will describe the structure of Wakimoto modules outside the singular line and we will see a great resemblance with the structure of restricted Wakimoto and Verma modules on the singular line and we will see that it is similar to the structure of Verma modules over sl_2.

4.1 *Verma modules*

Consider the Verma modules $M_{\chi,k}$ over $Lsl_2^\wedge (k \neq -2)$. This module is reducible if and only if Kac-Kazhdan equation (2.3) is satisfied. More exactly, $M_{\chi,k}$ contains the singular vector of the degree $-m(l\delta \pm \alpha_1)$ (where $\delta = \alpha_0 + \alpha_1, m \geq 1 l \geq 0 \text{ or } 1$) if and only if

$$S^{\rho}_{l\delta\pm\alpha_1}(\chi, k) - (\chi, k) = n(\pm 2, 0) \tag{4.1}$$

where $S^{\rho}_{l\delta\pm\alpha_1}(\chi, k)$ is the ρ-centered reflection relative to the root $l\delta \pm \alpha_1$ defined in (2.2).

Consider the line $k = \text{const}$ in the space $(\chi, k) = \hat{h}^*$ and two different $(k \neq -2)$ points ρ_1 and ρ_0 of it: $\chi = -1$ and $\chi = k + 1$ (Fig. 1).

Affine Weyl group S acts in $\hat{h}^*$, stabilizing the lines $k = \text{const}$. S is generated by two simple reflections: $S_1 = S^{\rho}_{\alpha_1}$ and $S_0 = S^{\rho}_{\alpha_0}$; S_0, S_1 acting as follows: S_i reflects the line $k = \text{const}$ relative to the point ρ_i. Further

$$S^{\rho}_{l\delta-\alpha_1} = S_0(S_1S_0)^{l-1}\,, S^{\rho}_{l\delta+\alpha_1} = S_1(S_0S_1)^{l}$$

To determine whether $M_{\chi,k}$ is reducible or not one should consider the orbit of our highest weight (χ, k) under the action of S. If there are points $(\chi_1, k), \ldots, (\chi_a, k)$, which belong to this orbit and $S^{\rho}_{l\delta\pm\alpha_1}(\chi_i, k) = (\chi_{i+1}, k)$ and $(\chi_{i+1}, k) - (\chi_i, k) = m_i(\pm 2, 0), i = 0, 1, \ldots, a - 1 (\chi_0 = \chi)$, then one concludes that there is a sequence of inclusions $M_{\chi,k} \supset M_{\chi_1,k} \supset \ldots \supset M_{\chi_a,k}$ so that $M_{\chi,k}$ contains a singular vector of the degree $\sum_{i=1}^{a-1} m_i(l_i\delta \pm \alpha_1)$.

We have several possibilities. At first, let the S-orbit of $q_0 = (\chi, k)$ be of general position. Then it is the disjoint union of two sets: $q_i^{(1)} =$

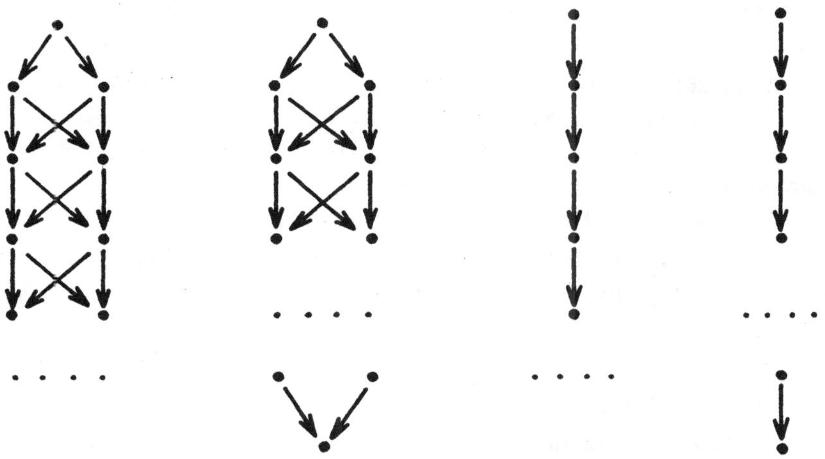

III_- III_+ III°_- III°_+

Fig. 1.

$S_{1(\text{or }0)}\dots S_1S_0S_1q_0$ and $q_i^{(0)} = S_{1(\text{or }0)}\dots S_0S_1S_0q_0$. If $q_{2i+1}^{(1)} - q_0^{(1)} \neq 2m$ and $q_{2i+1}^{(0)} - q_0^{(0)} \neq 2m$ for all integer $i \geq 0$ and $m \geq 1$, then $M_{\chi,k}$ is irreducible (case I). If only one of these equations is satisfied, then $M_{\chi,k}$ contains a unique singular vector of the corresponding degree and it generates the irreducible submodule of $M_{\chi,k}$, the quotient also being irreducible (case II). If there are more than one equations being satisfied, then $M_{\chi,k}$ may contain finitely many (case III_+) or infinitely many singular vectors (case III_-). The case III_+ happens if $k < -2$ and case III_- if $k > -2$ (note also that if $M_{\chi_1,k} \subset M_{\chi_2,k}$ then $M_{-2-\chi_2,-4-k} \subset M_{-2-\chi_1,-4-k}$).

Now let the S-orbit of q_0 be not of general position. It means that there are i and j such that $q_i^{(1)} = q_j^{(0)}$, so there is such $s \in S$ that $sq_0 = q_0$. We know that $(S_0S_1)(\chi,k) = (\chi + 2(k+2), k)$ so it happens if $\chi = -1 + i(k+2), i \in \mathbb{Z}$.

In this case some singular vectors in $M_{\chi,k}$ are "glued" to each other and again if $k < -2, M_{\chi,k}$ contains finitely many singular vectors (case III°_+) if $k > -2$, infinitely many singular vectors (case III°_-).

Now we are in position to formulate the final result (it is proved by F. Malikov[27]).

Theorem 4.1

In the case I, the Verma module $M_{\chi,k}$ is irreducible.

In the case II $M_{\chi,k}$ contains a unique singular vector and the quotient of $M_{\chi,k}$ by its submodule generated by this singular vector is irreducible.

The structure of Verma modules in cases $\mathrm{III}_{\pm}$, $\mathrm{III}^{\circ}_{\pm}$ is shown in Fig. 1. The points denote singular vectors. An arrow, or a chain of arrows, from one point to another means that the second vector belongs to the submodule generated by the first vector.

4.2 *Wakimoto modules outside the singular line*

We formulate the theorem which describes the structure of Wakimoto modules over $\mathrm{Lsl}_2^\wedge$ outside the singular line. Note that $W_{\chi,\nu}$ is isomorphic to $W_{\chi,-\nu}$.

Theorem 4.2 In the case I Wakimoto module $W_{\chi,\nu}$ is irreducible and isomorphic to $M_{\chi,\frac{\nu^2}{2}-2}$.

In the case II Wakimoto module $W_{\chi,\nu}$ is isomorphic to $M_{\chi,\frac{\nu^2}{2}-2}$ if the unique singular vector of $M_{\chi,\frac{\nu^2}{2}-2}$ has the degree $-m(l\delta+\alpha_1)$ and it is isomorphic to $M^*_{\chi,\frac{\nu^2}{2}-2}$ if this vector has a degree $-m(l\delta-\alpha_1)$.

The cases III_{+}, $\mathrm{III}^{\circ}_{\pm}$ are divided into two subcases $\mathrm{III}_{+}(\pm)$ and $\mathrm{III}^{\circ}_{\pm}(\pm)$. If the degree of the highest singular vector in M_{χ,ν^2-2} is equal to $-m(l\delta+\alpha_1)$, then we have the subcase $(+)$, if it is equal to $-m(l\delta-\alpha_1)$, then we have the subcase $(-)$.

The structure of the Wakimoto modules $W_{\chi,\nu}$ in these cases is shown in Fig. 2.

The points on the picture denote singular vectors in $M_{\chi,\frac{\nu^2}{2}-2}$ or in its quotient, which correspond to the singular vectors in $M_{\chi,\frac{\nu^2}{2}-2}$. Interrelations between these vectors are expressed by the arrows. The chain of arrows leads from one vector to another if and only if for any choice of vectors projecting under some factorizations into singular vectors the second vector is contained in the submodule generated by the first vector.

It is easy to see that the structure of Wakimoto modules coincides with the structure of SIF modules. This is due to functorial correspondence which we mentioned above. Namely, the functor translates $W_{\chi,\nu}$ to $\mathcal{H}_{\xi,\eta}$, where

$$\xi=-\frac{1}{\nu}+\frac{\nu}{2}$$
$$\eta=\frac{\chi+1}{\nu}\,. \tag{4.2}$$

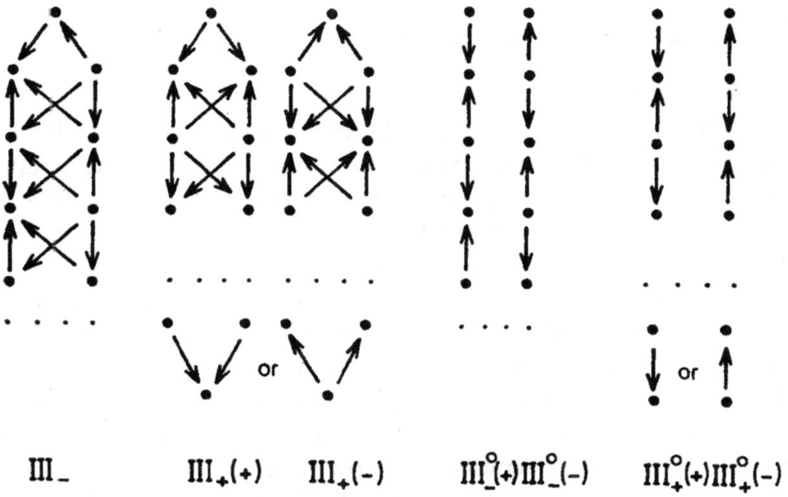

Fig. 2.

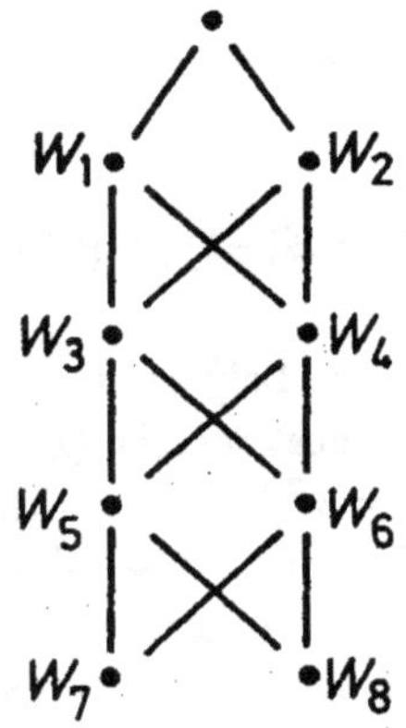

Fig. 3.

It is clear that the Kac-Kazhdan equations

$$\chi + l\frac{\nu^2}{2} - (m-1) = 0 \tag{4.3}$$

are translated to the Kac equations[18] (3.9):

$$\xi = -\frac{\beta}{2} + \frac{1}{\beta}$$

$\eta = \frac{m\beta}{2} - \frac{l}{\beta}$ where $\beta = \frac{2}{\nu}$, $l, m = 1, 2, \ldots$.

Proof of Theorem 4.2

To prove Theorem 4.2 one should apply the arguments of Ref. 12 (theorem 1.10). In Ref. 12 two Jantzen filtrations (corresponding to maps $M_{h,c} \to \mathcal{H}_{\xi,\eta}$ and $\mathcal{H}_{\xi,\nu} \to M^*_{h,c}$) were used and information about singular vectors and cosingular vectors (that is singular vectors in contragradient module) in $\mathcal{H}_{\xi,\eta}$ which was yielded by vertex operators.

We can apply these arguments directly and refer to Ref. 12 for details. For example, we show how to use Theorem 3.4 in order to obtain information about singular and cosingular vectors in the most complicated case III_-. The structure of the Verma module in this case is shown in Fig. 4. Theorem 3.4 shows that vectors w_{4n+1} are singular in $W_{\chi,\nu}$ (may be obtained using vertex operators). To prove Theorem 4.2 we also have to show that vectors w_{4n+2} are cosingular in $W_{\chi,\nu}$ or that they are singular in $W_{\chi,\nu}$. To do it we have to take into consideration the following statement.

There is an automorphism of $\mathrm{L}\hat{\mathrm{sl}}_2$ which transforms its Cartan generators as follows:

$$E(0) \leftrightarrow F(1)\,, F(0) \leftrightarrow E(-1)\,, H(0) \leftrightarrow K - H(0)\,.$$

Under this transformation the $\mathrm{L}\hat{\mathrm{sl}}_2$ module with highest weight (χ, k) becomes a $\mathrm{L}\hat{\mathrm{sl}}_2$ module with highest weight $(k - \chi, k)$. It turns out that $W_{\chi,\nu}$ becomes $W_{k-\chi,\nu}$ where $k = \nu^2/2 - 2$, which may be verified directly using explicit formulae. By Theorem 3.4 we then obtain that W_{4n+2} are singular vectors in $W_{k-\chi,\nu}$ and so are cosingular vectors in $W_{k-\chi,\nu}$.

Then we obtain the structure of $W_{\chi,\nu}$ in the case III_-. Other cases may be considered analogously. Theorem 4.2 is proved.

4.3 *Singular line*

Now we pass to restricted Wakimoto and Verma modules on the singular line $k = -2$ (see 2.1 and 2.2).

We have the following result.

Theorem 4.3

If $\chi = 0, 1, 2 \ldots$, then $\overline{M}_{\chi,-2}$ is isomorphic to $\overline{W}_{\chi,0}$ and it contains the unique singular vector of the degree $(\chi + 1)\alpha_1$, the quotient by the submodule generated by this singular vector being irreducible.

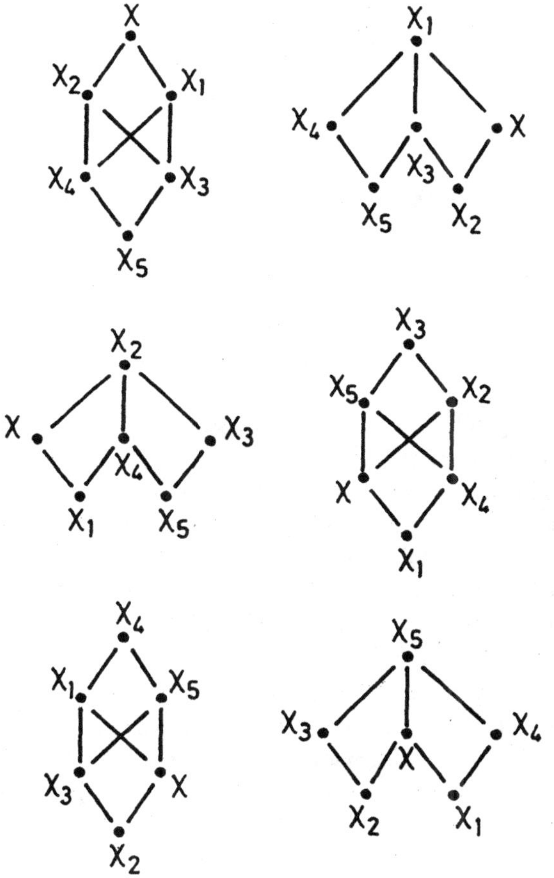

$$\chi_1 = S_1(\chi + \rho) - \rho$$
$$\chi_2 = S_2(\chi + \rho) - \rho$$
$$\chi_3 = S_2 S_1(\chi + \rho) - \rho$$
$$\chi_4 = S_1 S_2(\chi + \rho) - \rho$$
$$\chi_5 = S_1 S_2 S_1(\chi + \rho) - \rho$$

Fig. 4.

If $\chi = -2, -3, \dots$, then $\overline{M}_{\chi,-2}$ is isomorphic to $\overline{W}^*_{\chi,0}$ and it contains the unique singular vector of degree $(-1-\chi)\alpha_0$, the quotient by the submodule generated by this singular vector being irreducible.

If $\chi \neq 0, 1, 2, \dots$ or $-2, -3, \dots$, then $\overline{M}_{\chi,-2}$ is irreducible and isomorphic to $\overline{W}_{\chi,0}$

Proof

We will prove that theorem using explicit realizations of the modules $\overline{W}_{\chi,0}$ in the space of polynomials on infinite number of variables $x(i), i \in \mathbb{Z}$.

Let $\overline{W}_{\chi,0}$ contain a singular vector. Because $f(i) = \dfrac{\partial}{\partial x(i)}$ this vector is polynomial only on half of the variables: $x(i), i \leq 0$ or, equally, the polynomial on $f(i), i \leq 0$. We will show that only $f(0)^m$ may be the singular vector.

Let $W = P(f(i))v$ be a singular vector on $\overline{W}_{\chi,0}$. Let j be minimal negative integer, such that variables $f(j)$ is contained in P. So $P = f(j)^{k_1} P_1(f(0), \dots, f(j+1)) + Q_1(f(0), \dots, f(j))$, where the degree of Q_1 on $f(j)$ is less than k_1. Singular vector is annihilated by $h(1)$: $h(1)W = [f(j)^{k_1} \tilde{P}(f(0), \dots, f(j+1)) - 2k_1 f(j)^{k_1-1} P_1(f(0), \dots, f(j+1)) + \tilde{Q}(f(0), \dots, f(j))]v$ where $\tilde{P}(f(0), \dots, f(j+1))v = h(1)P_1(f(0), \dots, f(j+1))v$. The only term in the expression of the degree k_1 on $f(j)$ is the first term, so we conclude that $h(1)P_1(f(0), \dots, f(j+1))v = 0$. If we rewrite $P_1(f(0), \dots, f(j+1))$ as above $P_1(f(0), \dots, f(j+1)) = f(j+1)^{k_2} P_2(f(0), \dots, f(j+2)) + Q_2(f(0), \dots, f(j+1))$ where the degree of Q_2 on $f(j+1)$ is less than k_2, then we analogously conclude that $h(1)P_2(f(0), \dots, f(j+2))v = 0$. Proceeding we obtain that $f(i)^{k_{i-j+1}} f(0)^m v$ is annihilated by $h(1)$ where i is the greatest negative integer, such that $f(i)$ is contained in $P, P_1, \dots, P_{i-j}$, but it is impossible. Hence, a singular vector is of the form $f(0)^m v$ and it is really singular if and only if $\chi = m - 1 = 0, 1, 2, \dots$.

On the other hand, analogously, a singular vector in $\overline{W}^*_{\chi,0}$ is of the form $\chi(1)^m$ and it is really singular if and only if $\chi = -m-1 = -2, -3, \dots$.

Now let $\chi = 0, 1, \dots$ Then $\overline{W}_{\chi,0}$ contains the unique singular vector and does not contain cosingular vectors. Hence, the natural mapping $\overline{W}^*_{\chi,0} \to \overline{M}^*_{\chi,-2}$ has no kernel and $\overline{W}_{\chi,0}$ is isomorphic to $\overline{M}_{\chi,-2}$. Thus, singular vectors in these modules coincide and so $\overline{M}_{\chi,-2}$ has the unique singular vector of the degree $(\chi+1)\alpha_1$.

Analogously, if $\chi = -2, -3 \dots$, then $\overline{W}^*_{\chi,0}$ is isomorphic to $\overline{M}_{\chi,-2}$ and the latter module contains the unique singular vector of the degree $(-\chi-1)\alpha_0$.

After all, if $\chi \neq 0,1,\ldots$ and $-2,-3\ldots$, then $\overline{W}_{\chi,0}$ does not contain singular and cosingular vectors. So $\overline{W}_{\chi,0}$ is isomorphic to $\overline{M}_{\chi,-2}$ and both modules are irreducible. The theorem is proved. Note that F. Malikov has proved it by other means.[27]

Theorem 4.3 shows that the structure of the restricted Verma modules over $Lsl_2^\wedge$ is similar to the structure of the Verma modules over sl_2. There seems to be equivalence between certain categories of representations of sl_2 and $Lsl_2^\wedge$ on the singular line.

In the next section we will generalize these results to an arbitrary affine algebra.

5. The Structure of Modules on the Singular Hyperplane

In this section we deal with restricted Verma and Wakimoto modules (see 2.1 and 2.2) over $Lg^\wedge$ on the singular hyperplane $k = -\check{g}$.

We will formulate some results and conjectures which generalize Theorem 4.3.

We know from Refs. 26 and 16 that the restricted Verma modules are irreducible at a general point of the singular hyperplane, that is at such points which does not belong to another Kac-Kazhdan hyperplane (corresponding to a real root).

Kac-Kazhdan equation for the pair $(m, l\delta \pm \alpha)$ reduces on the singular hyperplane to

$$\pm 2(\chi + \rho, \alpha) = m(\alpha, \alpha)\ . \tag{5.0}$$

These equations do not depend on l. That is why one could expect the appearance of the great number of the singular vectors in $\overline{M}_{\chi,-\check{g}}$ when (5.0) is satisfied. However, we are sure that it is not so. Otherwise, there appears only one singular vector of the degree $m\alpha$ or $m(\delta - \alpha)$ if (5.0) is satisfied. Singular vectors of the degrees $m(l\delta + \alpha), l = 1,2,\ldots$ and $m(l\delta - \alpha), l = 2,3,\ldots$ also appear in $M_{\chi,-\check{g}}$, but they are contained in the submodule generated by Segal-Sugawara operators and do not appear in $\overline{M}_{\chi,-\check{g}}$. We proved it for $sl_2^\wedge$ in 4.3. In this section we use the realization of $\overline{W}_{\chi,0}$ in the space of polynomials on infinitely many variables. But we manage to prove the theorem about the structure of these modules and corresponding restricted Verma modules $\overline{M}_{\chi,-\check{g}}$ only in the case when $\overline{W}_{\chi,0}$ does not contain cosingular vectors and is isomorphic to $\overline{M}_{\chi,-\check{g}}$. It happens *if and only if the corresponding Verma module M_χ over finite-dimensional* Lie algebra g (with *highest* weight χ) is not contained in another Verma module as a proper submodule. In other words M_χ is projective in category $\mathcal{O}$ of representations of a finite-dimensional algebra and we call the weight χ projective. In particular, a dominant weight is projective, and a general weight is projective.

If χ is projective, then χ may belong only to hyperplanes

$$2(\chi+\rho,\alpha)=m(\alpha,\alpha) \tag{5.1}$$

and does not belong to hyperplanes

$$-2(\chi+\rho,\alpha)=m(\alpha,\alpha)\ . \tag{5.2}$$

It is clear that (5.1) is the Kac-Kazhdan equation for singular vector of the degree $m\alpha$ in M_χ. These singular vectors may be considered as singular vectors in $\overline{M}_{\chi,-g}$ if we identify g with $g\otimes 1\subset Lg^\wedge$. The following theorem show that there are no other singular vectors if χ is projective.

Theorem 5.1

If the weight χ is projective, then the singular vectors of $\overline{M}_{\chi,-g}$ over Lg^ coincides with the singular vectors of M_χ over g. In particular, if χ is general, then $\overline{M}_{\chi,-g}$ is irreducible.

Proof

As in 4.3 we should use the trick of reducing to half of the variables.

Consider the realization of $\overline{W}_{\chi,0}$ in the space of polynomials on variables $\chi_\alpha(m),\alpha\in\Delta_+,m\in\mathbb{Z}$. There are two commuting algebras Ln_-, acting on $\overline{W}_{\chi,0}$.

The first is generated by $f_\alpha(m),\alpha\in\Delta_+,m\in\mathbb{Z}$ and the second is generated by $\bar{f}_\alpha(m),\alpha\in\Delta_+,m\in\mathbb{Z}$ (see 3.3 and 3.4). They are of the form $f_\alpha(m)=a_\alpha(m)+P_\alpha(a_\beta(i),a^*_\beta(i)),\bar{f}_\alpha(m)=-a_\alpha(m)+Q_\alpha(a_\beta(i),a^*_\beta(i))$, where

$$a_\alpha(m)=\begin{cases}x_\alpha(m), & m\le 0\\ \frac{\partial}{\partial x_\alpha(m)}, & m>0\end{cases}\qquad a^*_\alpha(m)=\begin{cases}-\frac{\partial}{\partial x_\alpha(m)}, & m\ge 0\\ x_\alpha(-m), & m<0\end{cases}$$

It is clear that $\overline{W}_{\chi,0}$ is generated by $\bar{f}_\alpha(m),m\le 0$ and $f^*_\alpha(m),m>0$, where $f^*_\alpha(m)$ is obtained from $f_\alpha(m)$ by changing $x_\alpha(i)$ by $\dfrac{\partial}{\partial x_\alpha(i)}$ and vice versa.

A singular vector in $\overline{W}_{\chi,0}$ does not contain $f^*_\alpha(m),m>0$ that is why it is polynomial on $\bar{f}_\alpha(m),m\le 0$.

Let us introduce a basis in the universal enveloping algebra of $\bar{f}_\alpha(m)$, $m \leq 0$. Let $ht(\alpha)$ be a linear function of Δ_+, such that $ht(\alpha_i) = 1, i = 1, \ldots, r$. By Poincare-Birkhoff-Witt we can choose the monomial basis which consists of the monomials of the form $\bar{f}_{\beta_1}^{m_1}(n_1)\bar{f}_{\beta_2}^{m_2}(n_2)\ldots\bar{f}_{\gamma_1}^{l_1}(0)\ldots\bar{f}_{\gamma_p}^{l_p}(0)$, where $ht(\beta_1) < ht(\beta_2) < \ldots$ and if $\gamma_i = \gamma_{i+j}$, then $\gamma_{i+s} = \gamma_i, s = 1, \ldots, j$.

Let w be a singular vector. So it is a linear combination of our monomials. We have $w = P(\bar{f}_\alpha(m))v$, where $P(\bar{f}_\alpha(m)) = \bar{f}_{\delta_1}^{m_1}(n_1)P_1(\bar{f}_\alpha(m)) + Q_1(\bar{f}_\alpha(m))$, and where $ht(\delta_1)$ is minimal, and n_1 is minimal among $\bar{f}_{\delta_1}(n)$, provided that $P(\bar{f}_\alpha(m))\,; P_1(\bar{f}_\alpha(m))$ does not include $\bar{f}_{\delta_1}(n_1)$ and the degree of $Q_1(\bar{f}_\alpha(m))$ on $\bar{f}_{\delta_1}(n_1)$ is less than m_1.

Now we act on w by $h : (1), i = 1, \ldots, r$. Recall that $h_i(j)$ commute with $\bar{f}_\alpha(m)$ as with $f_\alpha(m)$.

We have $h_i(1)w = 0, i = 1, \ldots, r$. Now we can apply arguments of 4.3. We see that $P_1(\bar{f}_\alpha(m))$ is annihilated by $h_i(1), i = 1, \ldots, r$.

Proceeding with this procedure we arrive at the vector of the form $\bar{f}_\delta^m(n)R(\bar{f}_\alpha(0))v$, where $n < 0$, which must be annihilated by $h_i(1), i = 1, \ldots, r$. But it is impossible. Hence $w = R(\bar{f}_\alpha(0))v = R'(x_\alpha(0)v = R''(f_\alpha(0))v$ and we conclude that it is a singular vector of the finite-dimensional algebra g. Thus singular vectors in $W_{\chi,0}$ coincide with singular vectors of M_χ over g.

Applying these arguments to $\overline{W}^*_{\chi,0}$ we obtain that singular vectors in $\overline{W}^*_{\chi,0}$ are of the form $S(\bar{f}^*_\alpha(1))$, and so their degrees are of the form $-\sum_{i=1}^{l} m_i(\delta - \beta_i)$ (they correspond to singular vectors in $M_{-2\rho-\chi}$ over g).

Let χ be projective. We should show that in this case there are no singular vectors of the degree $-\sum_{i=1}^{l} m_i(\delta - \beta_i)$ on $\overline{M}_{\chi,-g}$. It may be proved by induction. We say that the singular vector w in $\overline{M}_{\chi,-g}$ has the depth $k (k \geq 1)$ if there is the sequence of singular vectors $V \to W_1 \to \ldots \to W_{k-1} \to W$ such that W_i generates W_{i+1} and there are no other sequences of the length greater than k. Then it is clear that there are no singular vectors of depth 1 and of the degree $-\sum m_i(\delta - \beta_i)$ if χ is projective. Let the singular vector of depth 2 have the degree $-\sum m_i(\delta - \beta_i)$. Then we have the sequence $V \to W_1 \to W$ where W_1 is of the degree $-m\beta$ and

$$2(\chi + \rho, \beta) = m(\beta, \beta)$$

Simple analysis of Kac-Kazhdan equations shows that the degree of

W in the Verma submodule generated by W_1 may only be $-m_1(l_1\delta + \beta')$ or $-m(l_2\delta - \beta)$. So we have: $\sum m_i(\delta - \beta_i) = m\beta + m_1(l_1\delta + \beta')$ or $\sum m_i(\delta - \beta_i) = m\beta + m(l_1\delta - \beta)$, but it is impossible. This shows that there are no singular vectors of degree $\sum m_i(\delta - \beta_i)$ and of depth 2 in $\overline{M}_{\chi,-\mathfrak{g}}$ where χ is projective. Proceeding by induction to the depth of singular vector one can show that there are no such vectors in $\overline{M}_{\chi,-\mathfrak{g}}$ at all.

Hence, if χ is projective, then $\overline{W}^*_{\chi,0}$ does not contain singular vectors. So $\overline{W}_{\chi,0}$ is isomorphic to $\bar{M}_{\chi,-\mathfrak{g}}$ and their singular vectors coincide. But we have shown above that these vectors are singular vectors of finite-dimensional algebra, and the theorem follows.

We also have the conjecture about nonprojective highest modules.

Let $\bar{M}_{\chi,-\mathfrak{g}}$ be restricted Verma module with highest weight χ which is not projective. In particular, it means that $\bar{M}_{\chi,-\mathfrak{g}}$ is reducible. Consider the orbit of χ under the action of the finite Weyl group. Choose weights $\chi_1, \dots, \chi_l$ which belong to this orbit, such that for each χ_i there is a sequence $\chi \to \chi_{j_1} \to \dots \to \chi_i$, where $\chi_{i_j} - \chi_{i_{j+1}} = k_j\beta_j$ and $2(\chi_{i_j} + \rho, \beta_j) = m_i(\beta_j, \beta_j)$, where m_i is an arbitrary integer, not necessarily positive and $\beta_j \in \Delta_+$. Let $\chi_{\max}$ be the highest weight among $\chi_1, \dots, \chi_l$. In other words, $M_{\chi_{\max}}$ over g is projective in corresponding finite-dimensional category σ and $M_{\chi_{\max}} \supset M_\chi$. The weight $\chi_{\max}$ is projective. We have the following conjecture.

Conjecture 5.2

The number of singular vectors of $\overline{M}_{\chi,-\mathfrak{g}}$ coincides with the number of singular vectors of $\bar{M}_{\chi,-\mathfrak{g}}$ (which are described by Theorem 5.1). In particular, if $\chi = S(\lambda + \rho) - \rho$, where S is an element of finite Weyl group and λ is dominant weight, then the number of singular vectors in $\bar{M}_{\chi,-\mathfrak{g}}$ is equal to the order of this Weyl group.

Note that we proved it in 4.3 for $L\hat{sl}_2$.

Using Wakimoto modules one can describe the structure of $\bar{M}_{\chi,-3}$ over $L\hat{sl}_3$ with the highest weight χ which belongs to the orbit of dominant weight under the action of the Weyl group (see Fig. 4).

We also have the conjecture about the structure of the modules $\bar{M}_{\chi,-\mathfrak{g}}$, where χ belongs to the orbit of dominant weight for arbitrary affine algebra.

In Ref. 10 we prove the two-sided Bernstein-Gelfand-Gelfand resolution of Wakimoto modules. Let $\hat{\chi}$ be integral dominant weight. Then there is the resolution $\overline{R}_*(\hat{\chi})$, such that $\overline{R}_i(\chi) = \bigoplus_{lt(s)=i} W_{s(\hat{\chi}+\rho)-\rho}$, whose homologies are concentrated on the 0th dimension and are isomorphic to the irreducible representation $L(\hat{\chi})$.

The Affine Weyl group S acts on h^* such that the root lattice acts trivially. We define the modified length of the element s of S as in Ref. 10 Let $\varphi_s^{\pm} = \langle\hat{\gamma} \in \hat{\Delta}_+ ; S^{-1}\hat{\gamma} < 0 ; \hat{\gamma} = l\delta \pm \gamma\}$ and let the modified length be $lt(s) = \#\varphi_s^+ - \#\varphi_s^-$. (The usual length is $l(s) = \#\varphi_s^+ + \#\varphi_s^-$.)

Let χ be the dominant weight. We conjecture the resolution consisting of modules $\bar{M}_{s(\chi+\rho)-\rho,-\bar{g}}$, where $s \in S$

Conjecture 5.3

There is acyclic resolution $R_*(\chi)$ of Lg^− modules, such that $R_n(\chi)$ $= \bigoplus_{lt(s)=n} \bar{M}_{s(\chi+\rho)-\rho,-\bar{g}}$.

The structure of the resolution over $Ls\hat{l}_2$ and $Ls\hat{l}_3$ is shown in Fig. 5. The points denote restricted Verma modules and the arrows denote the action of differential.

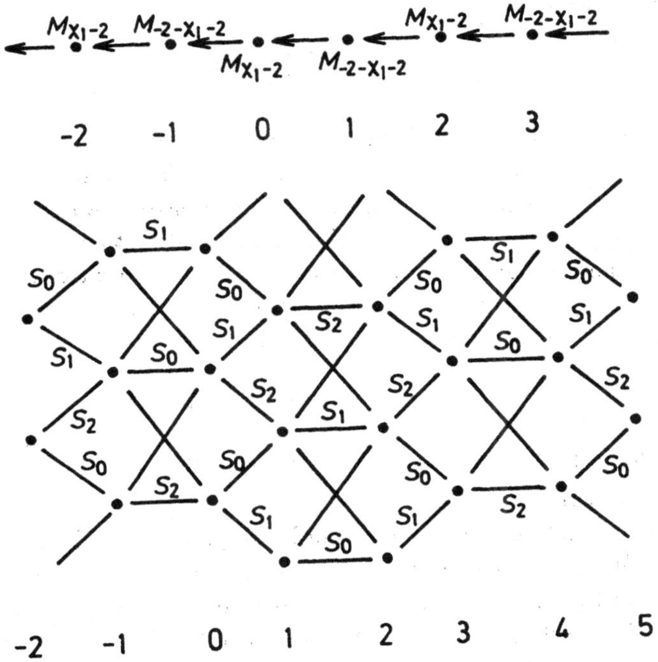

Fig. 5.

One can determine the structure of singular vectors of $\bar{M}_{S(\chi+\rho)-\rho,-\bar{g}}$ using this resolution. To do it one should take the "elementary part" of the resolution with the "top" $S(\chi+\rho)-\rho$ (see Fig. 4 for $L\mathfrak{sl}_3^{\wedge}$) . For example, for $\bar{M}_{\chi,-\bar{g}}$ one obtains the structure of M_χ over g.

Our resolutions appear as the Cousin-Grothendieck resolutions of the invertible sheaf ξ_χ on the semi-infinite flag manifold[10] in the same way as in the finite-dimensional situation.[22] It consists of the local "semi-infinite" cohomology with support on the Schubert cell. We have: $H_{C_s}^{\dim C_s}(\xi_\chi) \cong \bar{M}_{S(\chi+\rho)-\rho,-\bar{g}}$, where C_s is Schubert cell and dim C_S is its relative semi-infinite dimension. See Ref. 10 for details.

Note in conclusion that there is a correspondence between categories of representations of finite-dimensional Lie algebras and affine algebras on the singular hyperplane. This correspondence seems to be realized via the correspondence between certain categories of g-modules on a finite-dimensional flag manifold and semi-infinite flag manifold.

Appendix. Wick Theorem

Wick theorem yields a very effective method for computations with boson operators. In particular, it is the main technical tool in the proof of commutation relations in Sec. 3.

Here we will account this method following Ref. 13.

Let a_1 and a_2 be two elements of the Heisenberg algebra. If the representation of this algebra is given, then one can introduce normal ordering : $:\underbrace{a_1 a_2}:$ of these operators as the product ordered so that the right operator is an annihilation operator and the left is a creation operator. Now put $a_1 a_2 = a_1 a_2 -\, :a_1 a_2:$. This scalar operator is called the pairing. We define normally ordered product with pairings as:

$$:a_1 \ldots \underbrace{a_i \ldots a_j} \ldots a_k := \underbrace{a_i a_j} : a_1 \ldots \hat{a}_i \ldots, \hat{a}_j \ldots a_k :$$

We give the formulation of Wick theorem from Ref. 13.

Theorem A.1

The product of two normally ordered products of linear operators $(:a_1 \ldots a_k:)(:b_1 \ldots b_l:)$ is equal to the sum of all normally ordered products of these operators: $:a_1 \ldots a_k b_1 \ldots b_l:$ with all possible pairings between elements of two sets $(a_1, \ldots, a_k)$ and $(b_1, \ldots, b_l)$ including normal product without pairings.

This theorem is very suitable for computing operators whose generating functions are expressed via generating functions of the elements of an infinite-dimensional Heisenberg algebra.

Let $a_1, \ldots, a_n; a_1^*, \ldots, a_n^*$ be generators of a finite-dimensional algebra with commutation relations

$$[a_i, a_j] = [a_i^*, a_j^*] = 0, [a_i, a_j^*] = \delta_{i,j} .$$

Consider the infinite-dimensional Heisenberg algebra with generators $a_i(k), a_i^*(k), i = 1, \ldots, n; k \in \mathbb{Z}$ and commutation relations

$$[a_i(k), a_j(l)] = [a_i^*(k), a_j^*(l)] = 0, [a_i(k), a_j^*(l)] = \delta_{i,j}\delta_{k,-l} .$$

We put $a_i(z) = \sum_{k\in\mathbb{Z}} a_i(k), z^k, a_i^*(z), = \sum_{k\in\mathbb{Z}} a_i^*(k)z^k$. Let $a_i(k), k > 0$ and $a_j^*(l), l \geq 0$ be annihilation operators.

We have $a_i(z_1)a_j(z_2) = a_i^*(z_1)a_j^*(z_2) = 0$,

$$\begin{aligned} \underbrace{a_i(z_1)a_j^*(z_2)} &= \frac{\delta_{i,j}z_1}{z_2 - z_1}, |z_1| < |z_2|, \\ \underbrace{a_j^*(z_2)a_i(z_1)} &= \frac{\delta_{i,j}z_1}{z_2 - z_1}, |z_2| < |z_1| . \end{aligned} \tag{A.1}$$

Let $P(a_1, \ldots, a_n; a_1^*, \ldots, a_n^*) = P(a_i, a_i^*)$ and $Q(a_i, a_i^*)$ be the polynomials on $a_i, a_i^*, i = 1, \ldots, n$. And let $P(z) = P(a_i(z), a_i^*(z)), Q(z) = Q(a_i(z), a_j^*(z))$ be the corresponding generating functions.

So $P(z) = \sum_{k\in\mathbb{Z}} P(k)z^k, Q(z) = \sum_{k\in\mathbb{Z}} Q(k)z^k$, where $P(k)$ and $Q(k)$ may be expressed via $a_i(l), a_j^*(l), l \in \mathbb{Z}, P(k) = \int_C P(z)z^{-k}\frac{dz}{z}$, where C is the contour around the origin (we omit the factor $(2\pi i)^{-1}$).

We want to compute the commutation relations $[P(z), Q(k)]$. To do this we should use Cauchy residue formula. $[P(z_0), Q(k)] = \int_{C_R} P(z_0)Q(z) z^{-k-1}dz - \int_{C_r} Q(z)P(z_0)z^{-k-1}dz$ where C_R and C_r are contours around the origin and $r < R$ (see Fig. 6).

We have by Wick theorem

$$\begin{aligned} P(z_0)Q(z) =&: P(z_0)Q(z) : + \sum_{i=1}^{n} \frac{z_0}{z - z_0} : P_{i,0}^{(i)}(z_0)Q_{1,0}^{(i)}(z) : \\ &+ \sum_{i=1}^{n} \frac{z}{z_0 - z} : P_{0,1}^{(i)}(z_0)Q_{0,1}^{(i)}(z) : + \sum_{i,j=1}^{n} \frac{z_0^2}{(z - z_0)^2} : P_{2,0}^{(i,j)}(z_0)Q_{2,0}^{(i,j)}(z) : \\ &+ \sum_{i,j=1}^{n} \frac{z_0 z}{(z - z_0)^2} : P_{1,1}^{(i,j)}(z_0)Q_{1,1}^{(i,j)}(z) : + \sum_{i,j=1}^{n} \frac{z^2}{z - z_0} : P_{0,2}^{(i,j)}(z_0)Q_{0,2}^{(i,j)}(z) : \end{aligned}$$

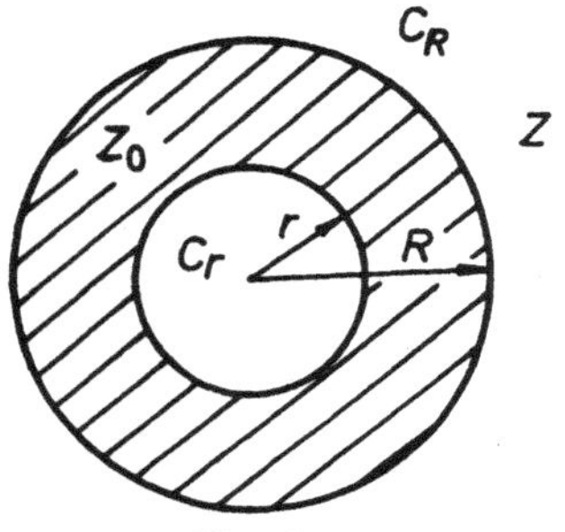

Fig. 6.

$+$ (terms with 3 contractions) $+\ldots$ if $|z_0| < |z|$.

Here $: P^{(i)}_{1,0}(z_0) Q^{(i)}_{1,0}(z) :$ is an ordered product with pairing $a_i(z) a_i^*(z)$ etc.

Analogously,

$$Q(z) \cdot P(z_0) =:. P(z_0) Q(z) : + \sum_{i=1}^{n} \frac{z_0}{z_0 - z} : P^{(i)}_{1,0}(z_0) Q^{(i)}_{1,0}(z) :$$
$$+ \sum_{i=1}^{n} \frac{z}{z - z_0} : P^{(i)}_{0,1}(z_0) Q^{(i)}_{0,1}(z) \ : + \ldots, \text{if} |z| < |z_0| \ .$$

So we obtain $[P(z_0), Q(k)] = \int_{C_R/C_r} \frac{dz}{z^{k+1}} (\sum_{i=1}^{n} \frac{z_0}{z - z_0} : P^{(i)}_{1,0}(z_0) Q^{(i)}_{1,0}(z) : - \sum_{i=1}^{n} \frac{z}{z - z_0} : P^{(i)}_{0,1}(z_0) Q^{(i)}_{0,1}(z) : + \ldots)$. Applying Cauchy residue formula we obtain $[P(z_0), Q(k)] = \sum_{i=1}^{n} (: P^{(i)}_{1,0}(z_0) \cdot Q^{(i)}_{1,0}(z_0) : - : P^{(i)}_{0,1}(z_0) Q^{(i)}_{0,1}(z_0) :) z_0^{-k} + \ldots$. Other terms are of the form

$$\int_{C_R/C_r} \frac{z_0^l z^m}{(z_0 - z)^{c+m}} : P_{l,m}(z_0) Q_{l,m}(z) : \frac{dz}{z^{k+1}}$$
$$= \frac{z_0^l}{(m + l - 1)!} \frac{d^{m+l-1}}{d_z^{m+l-1}} (z^{m-k-1} : P_{l,m}(z_0) Q_{l,m}(z) :) z = z_0$$

by Cauchy residue formula. In this way one can compute all terms of the commutation relations. The first two terms are of great importance.

Proposition A.2

Let the infinite-dimensional Heisenberg algebra $[P(a_i, a_i^*), Q(a_i, a_i^*)] = R(a_i, a_i^*)$ and let $S_k(z_0) = \sum_{i=1}^{n} z_0^{-k}(: P_{1,0}^{(i)}(z_0)Q_{1,0}^{(i)}(z_0) : - : P_{0,1}^{(i)}(z_0)Q_{0,1}^{(i)}(z_0) :)$ is the sum of the first order terms in $[P(z_0), Q(k)]$. Then $S_k(z_0) = z_0^{-k}R(z_0)$.

Proof is straightforward.

This proposition shows that when one computes commutation relations in an infinite-dimensional Heisenberg algebra, one takes into consideration only ordinary pairings (first order terms), the terms of higher order (Shwinger terms) appear in the expansion only in the infinite-dimensional algebra because of normal ordering.

Consider now the Heisenberg algebra with generators $b_i(k), i = 1, \dots, r; k \in \mathbb{Z}$ and commutation relations: $[b_i(k), b_j(l)] = \delta_{i,j}\delta_{k,-l} \cdot k$. Wicks theorem can be applied for computing commutation relation in this algebra as well. If we put $b_i(k), k \geq 0$ to be annihilation operators, then $\underbrace{b_i(z_1)b_j}(z_2) = \frac{z_1 z_2}{(z_1 - z_2)^2}, |z_1| < |z_2|$ and we can use the Cauchy residue formula as above.

In conclusion we give an example of the computations. We compute the commutation relations of $E(m), H(l)$ in $\mathrm{Lsl}_2^{\wedge}$, see (2.10).

$$
\begin{aligned}
E(z) &= - : a(z)a^*(z)a^*(z) : -(\nu b(z) + \chi)a^*(z) + (2 - \frac{\nu^2}{2})a^*(z) \\
H(z) &= 2 : a(z)a^*(z) : +\nu b(z) + \chi
\end{aligned}
$$

Note that $\underbrace{a(z_1)a^*}(z_2) = -\frac{z_1 z_2}{(z_1 - z_2)^2}, |z_1| < |z_2|$.

We have:

$$
\begin{aligned}
H(z_0)E(z) &= -4\frac{z_0}{z - z_0} : a(z)a^*(z)a^*(z_0) : +\frac{2z}{z - z_0} : a(z_0)a^*(z)a^*(z) : \\
&+ 4\frac{z_0 z}{(z - z_0)^2}a^*(z) - 2\frac{z_0}{z - z_0}(\nu b(z) + \mu)a^*(z_0) \\
&- 2(2 - \frac{\nu^2}{2})\frac{z_0 z}{(z - z_0)^2}a^*(z_0) - \nu^2\frac{z_1 z_0}{(z_1 - z_0)^2}a^*(z). \\
\Big[H(z_0), E(k)\Big] &= \int_{C_R/C_r} \frac{dz}{z^{k+1}}\Big(- 4\frac{z_0}{z - z_0} : a(z)a^*(z)a^*(z_0) : \\
&+ 2\frac{z}{z - z_0} : a(z_0)a^*(z)a^*(z) : -2\frac{z_0}{z - z_0}(\nu b(z) + \mu)a^*(z_0)
\end{aligned}
$$

$$+2\left(2-\frac{\nu^2}{2}\right)\frac{z_0 z}{(z-z_0)^2}(a^*(z)-a^*(z_0)) = \Big[-2 : a(z_0)a^*(z_0)a^*(z) :$$
$$-2(\nu b(z_0)+\mu)a^*(z_0)\Big]z_0^{-k}$$
$$+2\left(2-\frac{\nu^2}{2}\right)z_0\frac{d}{dz}(a^*(z)z^{-k}-a^*(z_0)z^{-k})_{z=z_0}\Big]$$
$$= 2z_0^{-k}E(z_0)\ ,$$

as required.

Acknowledgment

We would like to thank D.B. Fuchs and A.A. Beilinson for helpful discussions.

Note added in proof

When this work was finished, we learned that V.Fateev and V. Dotsenko have applied our results for the computation of correlation functions in WZW modules. We were also informed that Alekseev, Gerasimov, Marshakov, Morozov, Olshanetski and Shatashvili obtained results similar to Sec. 3, in physical point of view. A. Zamolodchikov independently obtained explicit formulae for $\mathrm{Lsl}_2^\wedge$ and $\mathrm{Lsl}_3^\wedge$, which coincide with our formulae (Sec. 2).

References

1. F. Bais, P. Bouwknegt, M. Surridge and K. Schoutens, *Nucl. Phys.* **B304** (1988) 348.
2. F. Bais, P. Bouwknegt, M. Surridge and K. Schoutens, *Nucl. Phys.* **B304** (1988) 371.
3. J. Bernstein, I. Gelfand and S. Gelfand, *Funkt. Anal. Prilozhen.* **10** (1976) 1–8 (in Russian).
4. V. Deodhar, O. Gabber and V. Kac, *Adv. Math.* **45** (1982) 92–116.
5. Vl.S. Dotsenko and V.A. Fateev, *Nucl. Phys.* **B240 [FS12]** (1984) 312–348.
6. V.A. Fateev and S.L. Lukyanov, *Int. J. Mod. Phys.* **A3** (1988) 507–520.
7. V.A. Fateev and A.B. Zamolodchikov, *Nucl. Phys.* **B280 [FS18]** (1987) 644–660.

8. B.L. Feigin, *Usp. Mat. Nauk (Russian Math. Surv.)* **39** (1984) 195–196 (in Russian).
9. B.L. Feigin and E.V. Frenkel, *Usp. Mat. Nauk* (Russian Math. Surv.), **43** (1988) 227–228 (in Russian).
10. B.L. Feigin and E.V. Frenkel, "Representations of affine Kac-Moody algebras and semi-infinite flag manifolds" to appear in *Commun. Math. Phys.*
11. B.L. Feigin and D.B. Fuchs, *Funkt. Anal. Prilozhen.* **16** (1982) 47–63 (in Russian).
12. B.L. Feigin and D.B. Fuchs, "Representations of the Virasoro algebra", in *Seminar on Supermanifolds 5*, ed. D. Leites, Reports of Dept. Math. Stockholm University N 25 (1986). To be published in *Representation of Infinite-Dimensional Lie Groups and Lie Algebras*, (Gordon and Breach, New York).
13. I. Frenkel, in *Lect. Appl. Math.* **21** (1985) 325–353.
14. D.B. Fuchs, *Cohomology of Infinite-Dimensional Lie Algebras*, (Plenum Press, 1986).
15. B. Goodman and N. Wallach, *J. Reine Arioew. Math.* **347** (1984) 69–133.
16. T. Hayashi, *Invent. Math.* **94** (1988) 13–52.
17. V. Kac, Infinite-Dimensional Lie Algebras (Birkhäuser, 1983).
18. V. Kac, in *Lect. Notes Math.* **94** (1979) 441–445.
19. V. Kac and D. Kazhdan, *Adv. Math.* **34** (1979) 97–108.
20. V. Kac and D. Peterson, *Adv. Math.* **50** (1984) 165–257.
21. V. Kac and I. Todorov, *Commun. Math. Phys.* **102** (1985) 337–347.
22. G. Kempf, *Adv. Math.* **29** (1978) 310–396.
23. V. Knizhnik, A. Polyakov and A. Zamolodchikov, *Mod. Phys. Lett.* **A3** (1988) 819–826.
24. V. Knizhnik and A. Zamolodchikov, *Nucl. Phys.* **B247** (1984) 83–103.
25. S. Lukyanov, *Funkt. Anal. Prilozhen.* **22** (1988) 1–10 (in Russian).
26. F. Malikov, *Funkt. Anal. Prilozhen.* **23** (1989) 76–77 (in Russian).
27. F. Malikov, "Verma modules over affine rank 2 algebras", to appear in *Algebra and Analysis* (1989) (in Russian).
28. F.G. Malikov, B.L. Feigin and D.B. Fuchs, *Funkt. Anal. Prilozhen.* **20** (1986) 25–37 (in Russian).
29. A. Meurman and A. Rocha-Caridi, *Commun. Math. Phys.* **107** (1986) 263–294.
30. A. Polyakov, *Mod. Phys. Lett.* **A2** (1987) 893–898.
31. A. Tsuchiya and Y. Kanie, *Publ. RIMS,* Kyoto Univ., *22 (1986) 259–327.*
32. A. Tsuchiya and Y. Kanie, *Adv. Stud. Pure Math.* **16** (1988) 297–372.

33. M. Wakimoto, *Commun. Math. Phys.* **104** (1986) 604–609.

34. N. Wallach, *Math. Z.* **196** (1987) 303–313.

35. D. Zhelobenko, "Introduction to the theory of S-algebras over reductive Lie algebras", to be published in *Representations of Infinite Dimensional Lie Groups and Lie Algebras*, (Gordon and Breach, New York).

36. G. Felder, Nucl. Phys. **B317** (1989) 215.

TWO-DIMENSIONAL CONFORMAL TRANSFORMATIONS REPRESENTED BY QUANTUM FIELDS IN MINKOWSKI SPACE-TIME *

R. Jackiw

Center for Theoretical Physics
Laboratory for Nuclear Science
and Department of Physics
Massachusetts Institute of Technology
Cambridge, Massachusetts 02139 U.S.A.

* This work is supported in part by funds provided by the U. S. Department of Energy (D.O.E.) under contract #DE-AC02-76ER03069.

ABSTRACT

Aspects of the conformal group in two-dimensional Minkowski space-time are reviewed. Representations of the algebra and of the group, *i.e.* of infinitesimal generators and of finite group elements, are given within a field theoretic Schrödinger representation for bosons and fermions. The role of self-dual fields in representing conformal transformations is stressed, their kinematics and dynamics are shown to obey the Hill-Feynman boson-fermion duality.

I. INTRODUCTION

Although I never met V. Knizhnik, I admire his work and am saddened by his death at an early age. This paper is offered to honor his memory, with the hope that he would have found my observations on conformal transformations interesting. This is one of the topics to which Knizhnik contributed significantly, but within a string-motivated framework, somewhat different from the one employed here.

In Minkowski space-time, the conformal group transforms the coordinate x^μ into $\tilde{x}^\mu(x)$ according to the following rules: $\tilde{x}^\mu = x^\mu + a^\mu$ [translation]; $\tilde{x}^\mu = \Lambda^\mu_\nu x^\nu$, $\Lambda^\mu_\alpha \Lambda^\nu_\beta \eta_{\mu\nu} = \eta_{\alpha\beta}$ [Lorentz]; $\tilde{x}^\mu = \alpha x^\mu$ [dilation]; $\tilde{x}^\mu = \frac{x^\mu - c^\mu x^2}{1-2c\cdot x+c^2x^2}$ [special conformal]. The last is closely related to a coordinate inversion $x^\mu \to x^\mu/x^2$, as is seen by taking the large-c limit; indeed the special conformal transformation is just a translation in an inverted coordinate. Infinitesimal forms of the above, $\tilde{x}(x) = x^\mu + \delta x^\mu(x)$, involve the vector fields $f^\mu(x) \equiv -\delta x^\mu$, $f^\mu(x) = a^\mu$ [translation]; $f^\mu(x) = \omega^\mu_\nu x^\nu, \omega_{\mu\nu} = -\omega_{\nu\mu}$ [Lorentz]; $f^\mu(x) = \alpha x^\mu$ [dilation]; $f^\mu(x) = 2x^\mu x\cdot c - c^\mu x^2$ [special conformal]. [Here the parameters are understood to be the infinitesimal portions of the previous.] The composition law for these transformations is succinctly presented in terms of the Lie algebra satisfied by the [Hermitian] generators: P^μ [translation], $M^{\mu\nu}$ [Lorentz], D [dilation], K^μ [special conformal]. Their commutators form an $SO(n,2)$ algebra in n-dimensional space-time; the non vanishing ones are

$$
\begin{aligned}
&i[M^{\mu\nu}, M^{\alpha\beta}] = g^{\mu\alpha}M^{\nu\beta} - g^{\nu\alpha}M^{\mu\beta} + g^{\mu\beta}M^{\alpha\nu} - g^{\nu\beta}M^{\alpha\mu} \quad ; \\
&i[M^{\mu\nu}, P^\alpha] = g^{\mu\alpha}P^\nu - g^{\nu\alpha}P^\mu \quad ; \quad i[M^{\mu\nu}, K^\alpha] = g^{\mu\alpha}K^\nu - g^{\nu\alpha}K^\mu \\
&i[D, P^\mu] = P^\mu \quad ; \quad i[D, K^\mu] = -K^\mu \quad ; \\
&i[P^\mu, K^\nu] = -2g^{\mu\nu}D + 2M^{\mu\nu}
\end{aligned}
\tag{1}
$$

A dynamical field theoretic model possesses conformal symmetries if a conserved [translation invariance], symmetric [Lorentz invariance] and traceless [conformal invariance] energy-momentum tensor $T^{\mu\nu}$ can be constructed. This is demonstrated by forming the *Bessel–Hagen* current.

$$J_f^\mu = T_\nu^\mu f^\nu \tag{2}$$

Conservation and symmetry of $T^{\mu\nu}$ show that the divergence of J_f^μ reads $\partial_\mu J_f^\mu = T^{\mu\nu}(\partial_\mu f_\nu + \partial_\nu f_\mu)$ and the current is conserved when f^μ if a *Killing vector*, *i.e.* satisfying.

$$\partial_\mu f_\nu + \partial_\nu f_\mu = 0 \tag{3}$$

However when the energy-momentum tensor is also traceless, J_f^μ is conserved even when the left-hand side of (3) is non-vanishing, but proportional to the metric tensor [here assumed flat]: $\partial_\mu f_\nu + \partial_\nu f_\mu \propto \eta_{\mu\nu}$. The proportionality factor is evaluated by taking the trace, and f^μ is recognized to be a *conformal Killing vector*, which by definition satisfies

$$\partial_\mu f_\nu + \partial_\nu f_\mu = \frac{2}{n}\eta_{\mu\nu}\partial_\alpha f^\alpha \tag{4}$$

The only solutions to (4) in flat space-times with $d \geq 3$ are those given above.

[In curved space-time described by the metric tensor $g_{\mu\nu}$, the derivatives in (3) and (4) are covariant and the expression on the left-hand side reproduces the Lie derivative of the metric.

$$D_\mu f_\nu + D_\nu f_\mu = f^\alpha \partial_\alpha g_{\mu\nu} + \partial_\mu f^\alpha g_{\alpha\nu} + \partial_\nu f^\alpha g_{\mu\alpha} \tag{5}$$

Hence the Killing equation in curved space-time requires that the metric tensor be invariant against the [infinitesimal] coordinate transformation f^μ, while its conformal generalization requires that the response of the metric tensor be proportional to itself. The Bessel–Hagen construction then produces a covariantly conserved current. If the energy-momentum tensor is viewed as the functional derivative of a matter action with respect to $g_{\mu\nu}$, the couplings to $g_{\mu\nu}$ must be conformally invariant. In general this involves non-minimal interactions with $g_{\mu\nu}$ and gives rise to a "new improved" energy momentum tensor.[1]]

Invariance of a [classical] field theory against conformal transformations was first established for the free Maxwell theory of electrodynamics by H. Bateman and E. Cunningham at the beginning of this century.[2] In subsequent years the subject was pursued sporadically,[3] receiving a widely noticed impetus from the theoretical work of G. Mack and A. Salam,[4] who showed that many other classical field theories, which upon quantization are of interest to particle physicists, are either conformally invariant or fail to be so only because of non-vanishing mass terms.

The field transformation law that is consistent with the Lie algebra (1) and leads to invariant equations is the usual one for translations and Lorentz transformations [Poincaré group], while for dilations a linear formula is most frequently adopted.

$$\delta\Phi = i[D,\Phi] = x^\alpha \partial_\alpha \Phi + d\Phi \tag{6}$$

Here Φ is a multiplet of fields and d is the *scale dimension* matrix, which can be taken diagonal. The conformal transformation reads

$$\delta\Phi = i\,[K^\mu,\Phi] = \left(2x^\mu x^\alpha - g^{\mu\alpha}x^2\right)\partial_\alpha\Phi + 2x_\alpha\left(g^{\alpha\mu}d - \Sigma^{\alpha\mu}\right)\Phi \tag{7}$$

where Σ is the spin matrix, defined by the Lorentz transformation. However, other transformation rules are also possible; see below.

[It is worth remarking here that, contrary to statements made occasionally, scale (dilation) invariance does *not* imply conformal invariance, although the converse is true as is seen from (1). Scale invariant dynamics, arising from a Lagrangian density $\mathcal{L}$, will also be conformally invariant if the *field virial* V^μ is a total derivative.[1]

$$V^\mu \equiv \frac{\partial \mathcal{L}}{\partial \partial_\alpha \Phi} (g^{\alpha\mu} d - \Sigma^{\alpha\mu}) \Phi = \partial_\alpha \sigma^{\mu\alpha} \tag{8}$$

(The nomenclature derives from classical mechanics.) For all popular and sensible models (8) is true, but it is not a mathematical identity and (classical) scale invariant models can be constructed that are not conformally invariant.]

The theoretical speculations about the conformal group acquired practical significance when MIT–SLAC experiments on deep inelastic [high energy] electron-nucleon scattering provided evidence that scale and conformal symmetries become operative at sufficiently high energies, where masses can be ignored.[5]

In spite of the ubiquity of classical field theories that are conformally invariant or whose invariance is weakly broken by mass terms, quantized field theories rarely exhibit this symmetry, because generically it is beset by symmetry violating anomalies: the quantized energy-momentum tensor possess a non-vanishing anomalous trace even in the absence of masses.[6] Technically the anomaly-producing mechanism is very similar to that responsible for breaking the chiral symmetry of massless fermions: in both cases the regularization procedure introduces cut-off mass-scales, which survive renormalization and break the symmetry.[7] [Chiral anomalies can also be understood in more immediate physical terms: defining the second quantized vacuum by filling the fermionic negative-energy sea violates fermion chirality.[7] Such a "physical" explanation of scale and conformal anomalies would be most welcome, but none is available at present.] Because of the scale anomalies, the renormalization group equations,[8] rather than those of the conformal group, govern high energy behavior, and conformal symmetry is inapplicable except at a possible non-trivial zero of the Gell–Mann–Low function.

Owing to the above circumstance, attention of conformal theorists shifted to models in two-dimensional space-time. These are simpler, and it is easier than in four dimensions to expose their conformal and other properties. Moreover, since the degree of divergence is milder than in four dimensions, anomalies are not so severe; conformal symmetry *can* survive renormalization, but parameters describing the representation [*e.g.* dimensions of fields, *etc.*] may acquire dynamical [anomalous] corrections, like in the two-dimensional Thirring model where K. Wilson realized his notion of anomalous dimensions.[9] Furthermore, two-dimensional field theories are physically relevant as continuum limits of various models in statistical/condensed physics models, whose critical parameters can be derived algebraically with conformal symmetry, if it is present.[10] Finally conformally invariant two-dimensional field theories are at the heart of string theory. Statistical physics and string theory lead to Euclidean two-dimensional field theories, and there is wide interest at the present time in such conformally invariant models.[11] Here I shall remain in Minkowski space-time.

II. TWO-DIMENSIONAL CONFORMAL SYMMETRY

The conformal Killing equation (4) in two dimensions, $n = 2$, possesses solutions beyond the finite set available with $n \geq 3$, because any function f^μ with $f^+ = f^+(x^+)$, $f^- = f^-(x^-)$ solves (4), where the $\pm$ components are defined by $\pm = \frac{1}{\sqrt{2}}(0 \pm 1)$.

This enlargement of conformal symmetry in two dimensions was appreciated in the early days of string theory.[12] There the formalism was not that of quantum field theory in two-dimensional Minkowski space-time, rather the "field theory" resided on a compact Riemannian manifold. Nevertheless, the relationship between the two was established by the invention of radial quantizations.[13] In this scheme, the dilation operator takes the role of the generator of dynamical evolution: D is viewed as the "Hamiltonian." Within this approach, one can construct for Euclidean fields a unitary quantum theory in Hilbert space, and "string" results can be understood in field-theoretic terms. Specifically, the Virasoro algebra — the Lie algebra of the infinite-dimensional conformal group together with its central extension— was recognized as the equal time commutator algebra of energy-momentum tensor components with the Schwinger term, necessarily present as a consequence of a positive Hilbert space inner product,[14] giving rise to the center.[13,15]

Thus a quantum field theoretic representation of the [extended] Lie algebra for the infinite two-dimensional conformal group is given by the anomalous commutators of energy-momentum tensor components. These have been computed in many ways and are well-understood and well-known.[11] What is not so very well-known is how *finite* conformal group elements are represented in the field theory, and in this essay I wish to publicize these results. I shall explicitly construct unitary, field theoretic operators that implement finite conformal transformations; I shall show that after regularization and renormalization these operators satisfy a composition law that follows the group composition law, up to a phase called at *2-cocycle i.e.* the infinite conformal group is represented projectively. Of course the infinitesimal form of the unitary operator defines the Lie algebra generator, while the infinitesimal form of the projective phase [2-cocycle] gives the center in the Lie algebra.

Our construction gives yet another determination of the extension in the Virasoro algebra. But our presentation differs in one interesting respect from previous ones: we never choose a Fock vacuum, nor do we normal order with respect to a preselected vacuum.

The traditional way of constructing field theoretic representations for a transformation algebra is to select a dynamical model that admits these transformations as symmetries, construct the symmetry generators, normal order with respect to the Fock vacuum of the model [or regulate in some other way specific to the dynamical model under consideration] and compute the commutator algebra. On the other hand, while a specific model certainly may be invariant against conformal transformations, and therefore carry a representation of the algebra and the group, these mathematical structures are independent of the specific physical/dynamical model and transcend it. Consequently, it is satisfying to develop representation theory independent of dynamics, just as it is done in quantum mechanics. This is what we offer, not only *for the finite group elements, but also for the infinitesimal* generators.

We use a fixed time, Schrödinger representation and picture, both for bosons and fermions,[16] and we construct a representation for the group of reparametrizations of the [infinite] line, whose coordinate x transforms infinitesimally by an arbitrary function f.

$$\delta_f x = -f(x) \tag{9}$$

This is what the two-dimensional conformal transformations reduce to at fixed time. The Lie algebra encapsulates the infinitesimal composition law.

$$\begin{gathered}[\delta_f, \delta_g]\, x = -\delta_{(f,g)} x \\ (f,g) = fg' - gf'\end{gathered} \tag{10}$$

[The dash ($'$) denotes differentiation with respect to the argument x.] A transformation law for a general field $\Theta(x)$, consistent with (10), may be taken as

$$\delta_f \Theta = (f\Theta)' + (d-1)f'\Theta + \frac{\alpha}{\sqrt{2\pi}} f^{(d+1)} \tag{11}$$

Because we are in the fixed-time Schrödinger picture, all operators are considered at the same time, hence the time argument is suppressed; only the $\{x\}$ [position]-dependence of Θ is indicated. In (11) d is the scale dimension of Θ and the last inhomogenous term, involving an arbitrary constant α and $d+1$ derivatives of f, may be present for $d = 0, 1, 2$. One verifies that the composition law for two transformations f and g of the form (11), which generalizes (6) and (7), follows the defining composition law (10).

A unitary representation for the finite transformation and a Hermitian representation for the generator will be found in terms of specific quantized field operators that transfrom according to specific realizations of (11) [special values of d and α].

We expect that the algebra of the field theoretic generators follows the Lie algebra of the group (10), but with an extension dictated by positivity.

$$[:Q_f:, :Q_g:] = i : Q_{(f,g)} : -\frac{i}{48\pi} c \int dx \Big(f(x) g'''(x) - g(x) f'''(x) \Big) \tag{12}$$

Here the double dots (:) indicate that the generator has been regulated and renormalized, though not necessarily by normal ordering. The extension is measured by the constant c which is positive.

III. CONSTRUCTION OF REPRESENTATIONS

A. To begin, we obtain first a representation in terms of a quantized Hermitian boson field which transforms according to (11) with $d = 1$ and $\alpha = 0$. To this end we consider a field operator $\chi(x)$ that satisfies the [equal time] commutation relation consistent with scale dimensionality 1 — the Abelian Kač–Moody algebra.

$$[\chi(x), \chi(y)] = i\delta'(x-y) \equiv k(x,y) = \int \frac{dp}{2\pi} e^{-ip(x-y)} p \tag{13}$$

[One may think of χ as $\frac{1}{\sqrt{2}}(\Pi + \Phi')$ where Π and Φ are canonically conjugate, but this is not necessary.]

The formal generator of the transformation

$$Q_f = \frac{1}{2} \int dx \chi(x) f(x) \chi(x) \tag{14}$$

transforms the field operator χ as

$$\delta_f \chi = i [Q_f, \chi] = (f\chi)' \tag{15}$$

which coincides with (11) for $d = 1$ and $\alpha = 0$. The generators follow the Lie algebra (10), $[Q_f, Q_g] = iQ_{(f,g)}$, when the commutator is evaluated formally, without care about the product of χ with itself at the same point. But the coincident-point product of operators in (14) renders Q_f ill-defined, and the evaluation of its commutator is unreliable; indeed absence of a center violates the positivity requirements mentioned earlier.

To overcome the problem of an ill-defined Q_f, that quantity must be regulated, so that it is well-defined, and renormalized, so that when the regulators are removed a finite expression remains.

To regulate the generator, we promote f to a bilocal function $F(x, y)$ and define the regulated generator by

$$Q_F = \frac{1}{2} \int dx\, dy\, \chi(x) F(x, y) \chi(y) \tag{16}$$

while removing the regulator consists of passing to the local limit.

$$F(x, y) \to \frac{1}{2} (f(x) + f(x))\, \delta(x - y) \tag{17}$$

$F(x, y)$ is taken to be real and symmetric in (x, y) and sufficiently well-behaved near $x \approx y$ to permit all formal manipulations.

Commutators of the regulated generators may be readily evaluated since now there are no singularities associated with operators at coincident points.

$$[Q_F, Q_G] = iQ_{(F,G)} \tag{18}$$

$$i(F, G) \equiv FkG - GkF \tag{19}$$

[A self-evident matrix/kernel notation is used: *e.g.* FkG denotes the kernel

$$(FkG)(x, y) = \int dz\, dz' F(x, z) k(z, z') G(z'y) = i \int dz\, F(x, z) \frac{d}{dz} G(z, y) \quad .$$

In the local limit, when the regulators are removed according to (17), the Hermitian kernel F becomes diagonal in position space.] Of course Q_F no longer generates conformal transformations, rather general linear canonical transformations.

To renormalize the generator, in this simple low-dimensional setting where divergences are not very severe, it is sufficient to perform a c-number subtraction q_F which removes that portion of Q_F that becomes singular in the absence of regulators. By construction, this defect does not afflict the difference, $Q_F - q_F$, and regulators can be removed, leaving the renormalized generator.

$$:Q_f: \equiv \lim_{F \to f} (Q_F - q_F) \tag{20}$$

It is clear that such a c-number modification will not affect (15), but the non-linear expression in the generators (18) acquires the extra contribution $iq_{(F,G)}$, which can survive as the regulators are removed and can give rise to a center in the Lie algebra.

It remains to determine q_F, or what is equivalent, to isolate the potentially singular portion of Q_F. Conventionally, this is achieved by normal ordering with respect to some preselected Fock vacuum, *i.e.* q_F is the expectation value of Q_F in the chosen Fock vacuum. We prefer another method: rather than selecting a Fock space, and thereby tying the discussion to a particular dynamical model or class of models, we use bosonic field states of the Schrödinger representation.

In the field theoretic Schrödinger representation and picture,[16] states [kets] are described by wave functionals $\Psi(\varphi)$ of a c-number field $\varphi(x)$ at fixed time.

$$|\Psi\rangle \longleftrightarrow \Psi(\varphi) \tag{21a}$$

Dual states [bras] are the complex confugated functionals,

$$\langle\Psi| \longleftrightarrow \Psi^*(\varphi) \tag{21b}$$

and the inner product is defined by functional integration.

$$\langle\Psi_1|\Psi_2\rangle = \int \mathcal{D}\varphi \Psi_1^*(\varphi)\Psi_2(\varphi) = \langle\Psi_2|\Psi_1\rangle^* \tag{22}$$

Operators $\mathcal{O}$ are represented by functional kernels, which act by functional integration.

$$\mathcal{O} \longleftrightarrow \mathcal{O}(\varphi, \tilde{\varphi}) \tag{23a}$$

$$\mathcal{O}|\Psi\rangle \longleftrightarrow \int \mathcal{D}\tilde{\varphi}\mathcal{O}(\varphi,\tilde{\varphi})\Psi(\tilde{\varphi}) \tag{23b}$$

For a canonical field operator at fixed time, $\Phi(x)$, we use a diagonal kernel $\Phi(x) \longleftrightarrow \varphi(x)\delta(\varphi-\tilde{\varphi})$, the canonical commutation relations determine the canonical momentum kernel as $\Pi(x) \longleftrightarrow \frac{1}{i}\frac{\delta}{\delta\varphi(x)}\delta(\varphi-\tilde{\varphi})$. Both kernels involve a functional δ-function. Evidently Φ acts by multiplication on functionals of φ, while Π acts by functional differentiation. In this way, the action of any operator constructed from Π and Φ is

$$\mathcal{O}(\Pi,\Phi)|\Psi\rangle \longleftrightarrow \mathcal{O}\left(\frac{1}{i}\frac{\delta}{\delta\varphi},\varphi\right)\Psi(\varphi) \tag{24}$$

One may also view these formulas as arising when eigenstates of the field operator $|\varphi\rangle$ are used.

$$\Phi(x)|\varphi\rangle = \varphi(x)|\varphi\rangle \tag{25a}$$

$$\langle\varphi|\tilde{\varphi}\rangle = \delta(\varphi-\tilde{\varphi}) \tag{25b}$$

$$\Psi(\varphi) \equiv \langle\varphi|\Psi\rangle \tag{25c}$$

$$\mathcal{O}(\varphi,\tilde{\varphi}) \equiv \langle\varphi|\mathcal{O}|\tilde{\varphi}\rangle \tag{25d}$$

$$\langle\varphi|\Phi(x)|\tilde{\varphi}\rangle = \varphi(x)\delta(\varphi-\tilde{\varphi}) \tag{25e}$$

$$\langle\varphi|\Pi(x)|\tilde{\varphi}\rangle = \frac{1}{i}\frac{\delta}{\delta\varphi(x)}\delta(\varphi-\tilde{\varphi}) \qquad (25f$$

[However, this interpretation is not available in the Schrödionger representation for fermion fields, which is discussed below.]

We recognize that all this is just ordinary quantum mechanics in the Schrödinger representation, but extended to the infinite number of degrees of freedom that comprise a field. Manipulation within this formalism is aptly described as "analysis on infinite-dimensional manifolds."

We use the above scheme to give an explicit functional expression to Q_F. This requires a functional formula for the operator χ, which according to (13) is not a canonical field. However, it is clear that the commutator (13) is realized by the following representation for χ.

$$\chi(x) \longleftrightarrow \frac{1}{\sqrt{2}}\left(\frac{1}{i}\,\frac{\delta}{\delta\varphi(x)} + \varphi'(x)\right)\delta(\varphi - \tilde{\varphi}) \equiv \langle\varphi|\chi(x)|\tilde{\varphi}\rangle \tag{26a}$$

$$\chi(x)|\Psi\rangle \longleftrightarrow \frac{1}{\sqrt{2}}\left(\frac{1}{i}\,\frac{\delta}{\delta\varphi(x)} + \varphi'(x)\right)\Psi(\varphi) \tag{26b}$$

The regulated generator (16) is represented in this formalism by

$$Q_F \longleftrightarrow \frac{1}{4}\int dx\,dy\left(\frac{1}{i}\,\frac{\delta}{\delta\varphi(x)} + \varphi'(x)\right)F(x,y)\left(\frac{1}{i}\,\frac{\delta}{\delta\varphi(y)} + \varphi'(y)\right)\delta(\varphi - \tilde{\varphi})$$
$$\equiv \langle\varphi|Q_F|\tilde{\varphi}\rangle \tag{27}$$

However, owing the functional delta function, this expression is not suited for a study of the singularities in Q_F as F becomes diagonal.

We can however consider the operator that is the phase exponential of Q_F.

$$U_F = e^{-iQ_F} \tag{28a}$$

In the Schrödinger representation it is better behaved than $\langle\varphi|Q_F|\tilde{\varphi}\rangle$, because

$$U_F(\varphi,\tilde{\varphi}) \equiv \langle\varphi|U_F|\tilde{\varphi}\rangle \tag{28b}$$

is an ordinary functional of φ and $\tilde{\varphi}$, not a functional distribution involving a δ-function like $\langle\varphi|Q_F|\tilde{\varphi}\rangle$. Consequently the singularities of $U_F(\varphi,\tilde{\varphi})$ when $F \to f$ can be explicitly examined. [Here analogy is drawn to ordinary quantum mechanics. The operator $\frac{1}{2}\mathbf{p}^2$ (in d dimensions) is represented by a distribution: $\langle\mathbf{r}|\frac{1}{2}\mathbf{p}^2|\mathbf{r}'\rangle = -\frac{1}{2}\nabla^2\delta(\mathbf{r}-\mathbf{r}')$; however, the representation of its phase exponential $U = e^{-it\frac{1}{2}\mathbf{p}^2}$ is an ordinary function: $\langle\mathbf{r}|U|\mathbf{r}'\rangle = (2\pi i t)^{-d/2}\exp\frac{i}{2t}(\mathbf{r}-\mathbf{r}')^2$.]

Also U_F is of interest in its own right. It is the unitary operator that implements the finite regulated conformal transformation and composes according to the rule,

$$U_F U_G = U_{F\circ G} \tag{29a}$$

$$F \circ G = F + G + \frac{1}{2}(F,G) + \dots \tag{29b}$$

which in the Schrödinger representation is a statement about the functional integral of two U's.

$$\int \mathcal{D}\varphi U_F(\varphi_1,\varphi)U_G(\varphi,\varphi_2) = U_{F\circ G}(\varphi_1,\varphi_2) \tag{29c}$$

In order to calculate U_F, a one parameter path for passing from the identity $U_{F=0}$ to U_F is selected: we consider $e^{-i\tau Q_F}$ and the corresponding functional $U_{\tau F}(\varphi,\tilde{\varphi})$ satisfied a Schrödinger-like equation that follows from (26), (27)

$$i\frac{\partial}{\partial\tau}U_{\tau F}(\varphi,\tilde{\varphi}) = \frac{1}{4}\int dx\,dy\left(\frac{1}{i}\frac{\delta}{\delta\varphi(x)} + \varphi'(x)\right)F(x,y)\left(\frac{1}{i}\frac{\delta}{\delta\varphi(y)} + \varphi'(y)\right)U_{\tau F}(\varphi,\tilde{\varphi}) \tag{30a}$$

together with the initial condition

$$U_{\tau F}(\varphi,\tilde{\varphi})\big|_{\tau=0} = \delta(\varphi-\tilde{\varphi}) \tag{30b}$$

Equations (30) comprise the infinitesimal version of the composition law (29).

It should be remarked that for ordinary, finite-dimensional Lie groups one knows that any finite group element can be reached by exponentiating generators along a one-parameter path from the identity. No such assurance can be given for infinite-dimensional groups, and therefore we cannot assert that every finite group element is represented by a functional of the type that we are calculating. This example of differences between analysis on finite-dimensional and infinite-dimensional manifolds is one of many,[16] but it does not appear to negate the results that we draw from our construction.

Equation (30) involves only quadratic operators; it may be solved. The result is a Gaussian functional, times a normalization factor N_F.

$$U_F(\varphi_1,\varphi_2) = N_F \exp - \int \varphi_1 k \varphi_2 \exp \frac{i}{2} \int (\varphi_1 - \varphi_2) K_F (\varphi_1 - \varphi_2) \tag{31}$$

$$N_F = \det{}^{1/2} F^{1/2} \left(\frac{2\pi i}{\mathcal{F}} \sin \frac{\mathcal{F}}{2}\right) F^{1/2} \tag{32}$$

$$K_F = F^{-1/2} \left(\mathcal{F} \mathrm{ctn} \frac{\mathcal{F}}{2}\right) F^{-1/2} \tag{33}$$

$$\mathcal{F} \equiv F^{1/2} k F^{1/2} \tag{34}$$

Constant factors in N_F have been fixed by the initial condition (30b).

The representation is clearly unitary. That the composition law (29) is indeed satisfied may be verified with the help of trigonometric identities when F is proportional to G, the general case is checked by expanding $F \circ G$.

With the explicit formula for $U_F(\varphi, \tilde{\varphi})$ in hand, we can explicitly examine its behavior in the local limit, as the regularization is removed according to (17). Of course, at issue is the behavior of K_F and N_F. One finds that K_F attains a well-defined expression.

$$\begin{aligned} K_F(x,y) \to K_f(x,y) &= \frac{1}{f(x)}\left\{\int \frac{d\lambda}{2\pi}\left(\lambda \mathrm{ctn}\frac{1}{2}\lambda\right)\exp\left(-i\lambda\int_y^x \frac{dz}{f(z)}\right)\right\}\frac{1}{f(y)} \\ &= -i\pi\frac{1}{f(x)}P\csc^2\left\{\pi\int_y^x \frac{dz}{f(z)}\right\}\frac{1}{f(y)} \end{aligned} \tag{35}$$

[P means principle value.] However, the normalization constant, N_F, diverges. The divergence resides in an unimportant constant factor Z, which may be removed by redefining the measure of functional integration, and in a phase e^{-iq_F}, which is determined in imaginary $\tau[F \to -iF]$; this continuation, rather than $F \to iF$, is appropriate when the energy spectrum is bound below.

$$q_F = \frac{1}{4}\operatorname{tr} F\omega \tag{36}$$

$$\omega(x,y) \equiv |k|(x,y) = \int \frac{dp}{2\pi}e^{-ip(x-y)}|p| = -P\frac{1}{\pi(x-y)^2} \tag{37}$$

[We use the notation $|\ldots|$ on a kernel to represent the absolute value kernel, defined through its spectral representation by taking the absolute value of the eigenvalues, as in (13) and (37).] It follows that the redefined operator

$$Z^{-1}e^{-i(Q_F-q_F)} \longleftrightarrow Z^{-1}e^{iq_F}U_F(\varphi_1,\varphi_2) \tag{38}$$

possesses a well-defined local limit. Thus we are led to define the renormalized generator by

$$:Q_f: \equiv \lim_{F\to f}\left(Q_F - \frac{1}{4}\operatorname{tr} F\omega\right) \tag{39}$$

Since q_F is a numerical quantity, independent of φ, it is a c-number.

Because U_F has been redefined in (38) by a phase, the composition law for the redefined quantities differs from (29): a projective phase — the 2-cocycle — arises.

$$\begin{aligned}\int \mathcal{D}(Z\varphi)\,(Z^{-1}e^{iq_F}U_F(\varphi_1,\varphi))\,(Z^{-1}e^{iq_G}U_G(\varphi,\varphi_2))\\ = e^{-2\pi i\omega_2(F,G)}Z^{-1}e^{iq_{F\circ G}}U_{F\circ G}(\varphi_1,\varphi_2)\end{aligned} \tag{40}$$

The 2-cocycle is

$$2\pi\omega_2(F,G) = q_{F\circ G} - q_F - q_G = \frac{1}{4}\,\mathrm{tr}\,(F\circ G - F - G) \tag{41}$$

When F and G are the regulated, bilocal expressions, ω_2 is "trivial" in the sense that it can be removed by redefining phases; indeed it arose because an additional phase was multiplied into U_F in (38). But in the local limit, when the regularization is removed and the bilocal functions become local, ω_2 is non-trivial — no phase redefinitions removes it.

The infinitesimal portion of ω_2 [in the local limit] gives the center of the conformal Lie algebra. One finds, *e.g.* from (18) and (39), formula (12) with $c = 1$. This of course is the conventional result, here derived without normal ordering or choosing a Fock vacuum. Rather a kinematical analysis is given, which also produces expressions for the finite group elements.

Although no Fock space is assumed in the derivation, it should be noted that ultimately our representation selects that Fock space within which a Gaussian with covariance K_f of (35) is well-defined.

Our representation functional allows computing how generic states transform under conformal transformations.

$$\Psi(\varphi) \xrightarrow[F]{} \Psi_F(\varphi) = \int \mathcal{D}(Z\tilde{\varphi})\,(Z^{-1}e^{iq_F}U_F(\varphi,\tilde{\varphi}))\,\Psi(\tilde{\varphi}) \tag{42}$$

In particular, for a Gaussian functional with covariance $\Omega(x,y)=\Omega(y,x)$

$$\Psi_\Omega(\varphi)\equiv \det{}^{-1/4}\frac{\Omega}{\pi}\exp-\frac{1}{2}\int\varphi\Omega\varphi \tag{43}$$

the transformed state is again Gaussian with transformed covariance.

$$\Omega_F=\Omega-(\Omega+k)(\Omega-iK_F)^{-1}(\Omega-k) \tag{44a}$$

Also the transformed state acquires an additional phase θ_F.

$$\theta_F=q_F+\operatorname{Im}\left(\ln N_F-\frac{1}{2}\operatorname{tr}\ln(\Omega-iK_F)\right) \tag{44b}$$

From (44a) one learns that the covariance Ω provides a representation for the algebra without center, the latter resides in the representation provided by the phase θ_F in (44b).

B. It is possible to obtain inequivalent Hermitian representations of the conformal Lie algebra with centers $c>1$ in (12). For these we employ the transformation law (11) with $d=1$, but $\alpha\neq 0$. The inhomogenous transformation law can be realized on our field χ by "improving" the generator (14) with the addition $\frac{\alpha}{\sqrt{2\pi}}\int dx\, f'(x)\chi(x)$. The modified generators[17]

$$Q_f^\alpha=\frac{1}{2}\int dx\,\chi(x)f(x)\chi(x)+\frac{\alpha}{\sqrt{2\pi}}\int dx\, f'(x)\chi(x) \tag{45}$$

effect the transformation

$$i\left[Q_f^\alpha,\chi\right]=(f\chi)'+\frac{\alpha}{\sqrt{2\pi}}f'' \tag{46}$$

and they possess an extension already on the classical level: when their commutator is evaluated with Poisson brackets one finds $c=12\alpha^2$. In the quantum theory, after proper definition of the bilinear, the quantum extension adds the additional 1, so that $: Q_f^\alpha :$ satisfies (12) with $c = 1 + 12\alpha^2$.

$$[: Q_f^\alpha :, : Q_g^\alpha :] = i : Q_{(f,g)}^\alpha : -\frac{i}{48\pi}(1 + 12\alpha^2) \int dx \Big((f(x) g'''(x) - g(x) f'''(x) \Big) \tag{47}$$

The finite group element corresponding to the improved generator is the previous expression, (31) appropriate to $\alpha = 0$, times the additional factor

$$\exp -i\frac{\alpha}{\sqrt{2\pi}} \int dx \Big(\ln f(x) \Big)' \Big(\varphi_1(x) - \varphi_2(x) - \frac{\alpha}{\sqrt{16\pi}} f'(x) \Big) \quad .$$

C. Fields that transform with $d = 2$ and inhomogenously are provided by $: \chi^2 :$. This is seen from (47) with $\alpha = 0$, which upon functional differentiation with respect to the arbitrary function $g(x)$ yields

$$i\, [: Q_f :, : \chi^2 :] = f : \chi^2 :' + 2f : \chi^2 : - \frac{1}{12\pi} f''' \tag{48}$$

Here, unlike (46), the inhomogeneous term is a quantum effect, not present classically.

Fields with $d = 0$ are constructed from our χ by

$$\theta(x) = \frac{1}{2} \int dy\, \epsilon(x - y) \chi(y) \tag{49}$$

When χ is transformed inhomogenously as in (46), θ transforms as in (11) with $d = 0$. Note that by virtue of (13) θ satisfies commutation relations appropriate to $d = 0$.

$$[\theta(x), \theta(y)] = -\frac{i}{2} \epsilon(x - y) \tag{50}$$

These fields will be of interest to us below.

D. Conformal transformations may also be represented by fermionic variables. We choose the simplest fermionic entity: a Weyl–Majorana fermion defined [at fixed time] on the line. This is a Hermitian one-component object that satisfies the equal-time anti-commutation relation, which may be viewed as an infinite-dimensional Clifford algebra.

$$\{\psi(x),\psi(y)\} = \delta(x-y) = \int \frac{dp}{2\pi} e^{-ip(x-y)} \tag{51}$$

$$\psi^\dagger = \psi \tag{52}$$

The conformal transformation law (11) is taken with $d = 1/2$, $\alpha = 0$ in keeping with (51).

$$\delta_f \psi = i\,[Q_f, \psi] = (f\psi)' - \frac{1}{2} f' \psi \tag{53}$$

The formal generator for this transformation is

$$Q_f = \frac{i}{4} \int dx \Big(\psi(x) f(x) \psi'(x) - \psi'(x) f(x) \psi(x) \Big) \tag{54}$$

whose commutator formally follows the Lie algebra (10), but Q_f suffers from coincident-point operator product singularities. For a well-defined regularized generator we take

$$Q_F = \frac{1}{2} \int dx\, \psi(x) F(x,y) \psi(y) \tag{55}$$

and removal of regulators entails replacing the bilocal, anti-symmetric Hermitian kernel $F(x,y)$ by a local expression.

$$F(x,y) \to \frac{i}{2} (f(x) + f(y)) \delta'(x-y) = \frac{1}{2} (f(x) + f(y))\, k(x,y) \tag{56}$$

The regulated generators satisfy the algebra (18) except that $i(F, G)$ differs from (19).

$$i(F, G) = [F, G] \tag{57}$$

Notice that (51), (55), (56) and (57) arising in the fermionic development are analogs of the corresponding bosonic expressions (13), (16), (17) and (19), with the interchange of $k \leftrightarrow \delta$; see also below.

In order to remove the regulator, the above expressions must be renormalized [subtracted]; in order to fix the renormalization procedure [subtraction] without normal ordering with respect to a preselected Fock vacuum, we need a Schrödinger-like representation for fermions. The challenge here is that "field state", analogous to the bosonic ones used in (25), can no longer be defined, because fermionic operators do not commute. Nevertheless, a functional space *can* be given with a well-defined action of the operator ψ on the functionals.[16]

The functional space consists of functionals $\Psi(u)$ of anti-commuting Grassmann fields at fixed time.

$$\{u(x), u(y)\} = 0 \tag{58}$$

In particular fermionic states [kets] are represented by functionals of u.

$$|\Psi\rangle \longleftrightarrow \Psi(u) \tag{59a}$$

We also need a rule for representing the dual states [bras] $\langle\Psi|$. In the bosonic case this is simple: duals are obtained by complex conjugation, see (21b). Here such a rule is inapplicable, in particular we have no room for the complex conjugate of the

Grassmann variable u. Nevertheless we use the asterisk to denote the functional corresponding to the bra.

$$\langle\Psi| \longleftrightarrow \Psi^*(u) \tag{59b}$$

It remains to give the rule for constructing $\Psi^*(u)$ from $\Psi(u)$. That rule is somewhat involved; we present it first and then motivate by an example.

To form the dual functional $\Psi^*(u)$, we first introduce auxiliary Grassmann variables $\bar{u}(x)$ and the auxiliary dual functional $\bar{\Psi}(\bar{u})$. $\bar{\Psi}(\bar{u})$ is constructed from $\Psi(u)$ by 1) reversing the order of Grassmann variables in $\Psi(u)$ and replacing u by $\bar{u}$, 2) complex conjugating all c numbers in $\Psi(u)$; *e.g.* $\Psi(u) = au(x)u(y)$, $\bar{\Psi}(\bar{u}) = a^*\bar{u}(y)\bar{u}(x)$. Then $\Psi^*(u)$ is obtained from $\bar{\Psi}(\bar{u})$ by a Grassmann functional integral.

$$\Psi^*(u) = \int \mathcal{D}\bar{u}\, e^{\int dx\, \bar{u}(x)u(x)}\bar{\Psi}(\bar{u}) \tag{60}$$

Moreover, the inner product is defined in the obvious way, again by functional Grassmann integration.

$$\begin{aligned}\langle\Psi_1|\Psi_2\rangle &= \int \mathcal{D}u\Psi_1^*(u)\Psi_2(u)\\ &= \int \mathcal{D}u\,\mathcal{D}\bar{u}\, e^{\int dx\, \bar{u}(x)u(x)}\bar{\Psi}_1(\bar{u})\Psi_2(u) = \langle\Psi_2|\Psi_1\rangle^*\end{aligned} \tag{61}$$

To motivate this construction, and gain some understanding of it, let us apply it to the same problem but on a space $\{x\}$ consisting of two points $i = 1, 2$ on which two fermion operators $\psi(i)$ are defined, satisfying the two-dimensional Clifford algebra.

$$\{\psi(i), \psi(j)\} = \delta_{ij} \tag{62}$$

We seek to represent this in terms of Grassmann variables $u(i)$.

$$\{u(i), u(j)\} = 0 \tag{63}$$

A specific state $|\Psi_f\rangle$ is represented by a Grassmann function of $u(i)$ that can be expanded in a four-dimensional basis,

$$|\Psi_f\rangle \leftrightarrow \Psi_f(u) = f_0 + f_1 u(1) + f_2 u(2) + f_{1\,2} u(1)u(2) \tag{64}$$

where the f's are c-numbers. The inner product with another state $|\Psi_g\rangle$

$$|\Psi\rangle_g \longleftrightarrow \Psi_g(u) = g_0 + g_1 u(1) + g_2 u(2) + g_{12} u(1)u(2) \tag{65}$$

is defined in the natural way.

$$\begin{aligned}\langle\Psi_g|\Psi_f\rangle &= g_0^* f_0 + g_1^* f_1 + g_2^* f_2 + g_{1\,2}^* f_{1\,2} \\ &= \langle\Psi_f|\Psi_g\rangle^*\end{aligned} \tag{66}$$

This can be expressed as

$$\langle\Psi_g|\Psi_f\rangle = \int d^2u \Psi_g^*(u)\Psi_f(u) \tag{67}$$

provided the dual [bra] $\langle\Psi_g|$ is represented by

$$\langle\Psi_g| \longleftrightarrow \Psi_g^*(u) = g_{1\,2}^* + g_2^* u(1) - g_1^* u(2) + g_0^* u(1)u(2) \tag{68}$$

Equation (66) follows from (64), (67) and (68) since only the following Grassmann integral is non-vanishing.

$$\int d^2u\, u(1)u(2) = 1 \tag{69}$$

Formula (68) for the dual does indeed emerge if the rules of our construction are followed. The auxiliary Grassmann variables are $\bar{u}(i)$ and an intermediate dual to $\Psi_g(u)$ is

$$\bar{\Psi}_g(\bar{u}) = g_0^* + g_1^* \bar{u}(1) + g_2^* \bar{u}(2) + g_{1\,2}^* \bar{u}(2)\bar{u}(1) \tag{70}$$

Multiplying (70) by $e^{\sum_i \bar{u}(i)u(i)} = 1 + \sum_i \bar{u}(i)u(i) + \bar{u}(1)u(1)\bar{u}(2)u(2)$ and, as indicated by (60), integrating with the help of the analog of (69)

$$\int d^2\bar{u}\, \bar{u}(2)\bar{u}(1) = 1 \tag{71}$$

results in (68).

The duality transformation that takes a ket (65) into a bra (68) is that of differential forms in two dimensions. There are four basis forms: the zero-form, two one-forms dx^μ, $\mu = 1,2$ and one area two-form $\frac{1}{2}\epsilon_{\mu\nu}dx^\mu dx^\nu$. These are the analogs of the four basis elements in (65), while the duals of these forms are the basis elements in the dual formula (68). While the field theoretic generalization of this would use infinite-dimensional differential forms and their duals, the analytic formulation (60) circumvents the need to define and deal with such difficult concepts, replacing them by manipulations with functional Grassmann integrals.

Note that in the finite-dimensional example our representation for the Clifford algebra (62) in terms of Grassmann variables $u(i)$ satisfying (63) is reducible: we require four elements to represent (64), (65) or (68), hence the representation is four-dimensional. But any two Pauli matrices also reproduce (62) two-dimensionally. [Reducible representations of infinite-dimensional Clifford algebras have been considered within mathematics.[18]]

We return now to our infinite-dimensional field theoretic problem, and observe that action on functionals of u of the operator ψ satisfying (51) can be represented by

$$\psi(x)|\Psi\rangle \longleftrightarrow \frac{1}{\sqrt{2}}\left(u(x) + \frac{\delta}{\delta u(x)}\right)\Psi(u) \tag{72}$$

Moreover, the above dualization rules have the consequence that the adjoint of $u(x)$ is $\frac{\delta}{\delta u(x)}$ and $\frac{1}{\sqrt{2}}\left(u(x)+\frac{\delta}{\delta u(x)}\right)$ is Hermitian, as it must be if is to represent the Hermitian operator $\psi(x)$ of (52).

With the above functional Schrödinger representation for fermions, we can represent the [regulated] conformal generators (55); actually we first represent the finite group elements, *i.e.* we seek unitary functionals $U_F(u,\tilde{u})$ that represent e^{-iQ_F} and compose according to

$$\int \mathcal{D}u\, U_F(u_1,u)U_G(u,u_2) = U_{F\circ G}(u_1,u_2) \tag{73a}$$

$$F\circ G = F+G+\frac{1}{2}(F,G)+\ldots \tag{73b}$$

By considering one-parameter exponentials $e^{-i\tau Q_F}$, we see that $U_{\tau F}(u_1,u_2)$ satisfies the Schrödinger-like equation

$$\begin{aligned} & i\frac{\partial}{\partial\tau}U_{\tau F}(u_1,u_2) \\ & = \frac{1}{4}\int dx\,dy\left(u_1(x)+\frac{\delta}{\delta u_1(x)}\right)F(x,y)\left(u_1(y)+\frac{\delta}{\delta u_1(y)}\right)U_{\tau F}(u_1,u_2) \end{aligned} \tag{74a}$$

with an initial condition at $\tau=0$, involving a Grassmann functional δ-function.

$$U_{\tau F}(u_1,u_2)\big|_{\tau=0} = \delta(u_1-u_2) \tag{74b}$$

Equations (74) are the infinitesimal statement of (73) and their solution in a Gaussian Grassmann functional, times a normalization factor N_F.

$$U_F(u_1,u_2) = N_F \exp -\int u_1u_2 \exp\frac{i}{2}\int (u_1-u_2)K_F(u_1-u_2) \tag{75}$$

$$N_F = \det{}^{1/2} i\sin\frac{F}{2} \tag{76}$$

$$K_F = \operatorname{ctn}\frac{F}{2} \tag{77}$$

Constants are adjusted in N_F so that (74b) holds.

That U_F is indeed the correct transformation kernel, satisfying the composition law (73) can be verified explicitly when F is proportional to G; the more general case can be checked by expanding $F \circ G$. The definition of the inner product on our space determines the form of the adjoint kernel.

$$U_F^\dagger(u_1, u_2) = N_F^* \exp - \int u_1 u_2 \exp - \frac{i}{2} \int (u_1 - u_2) K_F^\dagger (u_1 - u_2) \tag{78}$$

Since F and K_F are Hermitian, the above is just $U_{-F}(u_1, u_2)$; hence, the representation is unitary.

The argument now proceeds just as in the bosonic case; we shall be brief in summarizing the completely analogous statements.

In the local limit, as the regularization is removed according to (56), K_F attains a well-defined expression.

$$\begin{aligned} K_F(x,y) \to K_f(x,y) &= \frac{1}{\sqrt{f(x)}} \left\{ \int \frac{d\lambda}{2\pi} \left(\operatorname{ctn} \frac{\lambda}{2} \right) \exp \left(-i\lambda \int_y^x \frac{dz}{f(z)} \right) \right\} \frac{1}{\sqrt{f(y)}} \\ &= \frac{1}{\sqrt{f(x)}} \left\{ P \operatorname{ctn} \pi \int_y^x \frac{dz}{f(z)} \right\} \frac{1}{\sqrt{f(y)}} \end{aligned} \tag{79}$$

The normalization factor N_F diverges. The analysis, as in the bosonic case, is performed for imaginary $\tau[F \to -iF]$, where N_F becomes $e^{\frac{1}{2} \operatorname{tr} \ln \sinh \frac{1}{2}|F|}$ and the anti-symmetric kernel F is replaced by the absolute value kernel. The result, after returning to real τ, is that apart from an infinite constant Z, $e^{iq_F} N_F$ attains a limit, provided

$$q_F = -\frac{1}{4} \operatorname{tr} F \frac{k}{|k|} \tag{80}$$

To renormalize, we absorb the constant divergence Z in the functional integration measure, and remove the divergent phase. Thus

$$Z^{-1}e^{-i(Q_F-q_F)} \longleftrightarrow Z^{-1}e^{iq_F}U_F(u_1,u_2) \tag{81}$$

possesses a finite limit, but the composition law (73) acquires a projective phase given by a trivial 2-cocycle,

$$2\pi\omega_2(F,G) = q_{F\circ G} - q_F - q_G = -\frac{1}{4}\,\mathrm{tr}\,(F\circ G - F - G)\frac{k}{|k|} \tag{82}$$

which becomes non-trivial in the local limit. This implies that the renormalized charges

$$:Q_f :\equiv \lim_{F\to f}\left(Q_F + \frac{1}{4}\,\mathrm{tr}\,F\frac{k}{|k|}\right) \tag{83}$$

satisfy (12), with $c = 1/2$, in agreement with a general theorem.[19]

The action of the transformation on a generic state $\Psi(u)$ is given by

$$\Psi(u) \underset{F}{\longrightarrow} \Psi_F(u) = \int \mathcal{D}(Zu)\left(Z^{-1}e^{iq_F}U_F(u,\tilde{u})\right)\Psi(\tilde{u}) \tag{84}$$

In particular, the transform of a Grassmann Gaussian functional, with covariance $\Omega(x,y) = -\Omega(y,x)$,

$$\Psi_\Omega(u) = \det{}^{-1/4}\Omega \exp -\frac{1}{2}\int u\Omega u \tag{85}$$

is again a Gaussian with transformed covariance,

$$\Omega_F = \Omega + (I-\Omega)(\Omega + iK_F)^{-1}(I+\Omega) \tag{86a}$$

and an additional phase θ_F.

$$\theta_F = q_F + \mathrm{Im}\left(\ln N_F + \frac{1}{2}\,\mathrm{tr}\,\ln(\Omega + iK_F)\right) \tag{86b}$$

As in the bosonic case, Ω is a representation for the conformal algebra without center; the latter resides in the representation provided by θ_F.

E. We remark on a duality between the bosonic and fermionic structures. The analogy between the corresponding operators, when k and δ are interchanged, has already been noted below Eq. (57). Let us now observe that the bosonic and fermionic representations of the finite group element, U_F in (31) – (34) and in (75) – (77), are also analogous when the commutator of the bosonic fields, k, is replaced by δ the anti-commutator of fermionic fields: we appreciate that the first Gaussian factor with mixed fields is the same in the two cases, as are the two K_F once k in $\mathcal{F}$ of (34) is replaced by δ, *i.e.* $\mathcal{F}$ is replaced by F. Of course, differences in the Jacobian factor between bosons and fermions have to be taken into account in comparing the normalization factors, N_F.

If we view U_F as a propagation kernel for dynamics governed by Q_F, we can say that in the bosonic case the symplectic structure [commutator] is given by k, see (13), and the dynamics by δ, modulated by f in the sense of (16) and (17). For fermions there is an exchange: symplectic structure [anti-commutator] is δ, see (51), and dynamics involves k, modulated by f as in (55) and (56).

Rather than treating the boson and fermion fields separately and observing the duality between them, they may be combined supersymmetrically and described by a superfield. Within this framework one may represent the superconformal group. The compact and elegant expressions that emerge in this unified treatment have been given recently by R. Floreanini and L. Vinet.[20]

The above field theoretic duality and supersymmetric unification is the generalization to an infinite number of degrees of freedom of the boson-fermion duality observed by C. Hill and R. Feynman in quantum mechanics.[21] With a pair of operators, the non-Hermitian a and its adjoint $a^\dagger$, form their commutator and anti-commutator.

$$[a, a^\dagger] = C_- \tag{87a}$$

$$\{a, a^\dagger\} = C_+ \tag{87b}$$

Assume now that one of $C_\pm$ is a c-number, which by choice of normalization can be set to 1, and the other describes dynamics. If $C_- = 1$, then C_+ is [proportional to] the Hamiltonian and we are dealing with ordinary bosonic quantum mechanics of a harmonic oscillator. The dual situation arises when $C_+ = 1$ and C_- is [proportional to] the Hamiltonian — but this is a fermionic two-level system.

E. Fermionic representations with $c = 1$ are constructed simply by doubling the degrees of freedom, and the two Majorana–Weyl fields ψ_i $i = 1, 2$ can be combined into one charged Weyl field $\psi = \frac{1}{\sqrt{2}}(\psi_1 + i\psi_2)$. The functional Schrödinger representation is simply built as the direct product of the separate representation.

Once there are available charged fermions, one may also obtain $c > 1$ by using the improved generator

$$: Q_f^\alpha :\equiv: Q_f : +\alpha \int dx\, f'(x)\rho(x) \tag{88}$$

where ρ is the charge density $: \psi^\dagger \psi :$. Owing to the regularization and renormalization needed to define the product, ρ satisfies the [Abelian Kač–Moody] algebra,

$$[\rho(x), \rho(y)] = \frac{i}{2\pi}\delta'(x - y) \tag{89}$$

transforms under Q_f with dimension $d = 1$, and generates a $U(1)$ gauge transformation on ψ. Consequently, $:Q_f^\alpha:$ generates a conformal transformation on ψ with complex d,

$$i\left[:Q_f^\alpha:\psi\right] = (f\psi)' - \left(\frac{1}{2} + i\alpha\right) f'\psi \tag{90}$$

and satisfies (47) — the conformal algebra (12) with $c = 1 + 12\alpha^2$.

Comparison with (13), (45) and (47) shows that χ may be identified with $\sqrt{2\pi}\,\rho$ and the improved fermionic generator (88) coincides with the bosonic one (45) — this being an example of the Sugawara–Sommerfield construction, known to hold for fermions in two-dimensional quantum field theory.[22]

Note that the improvement of the fermionic generator, which lifts the center c by an arbitrary amount above 1, is unavailable for Majorana fermions, because no charge density operator can be constructed. For these c is fixed at 1/2 and cannot be continuously deformed by this improvement device.

Indeed, a general theorem states that c, when below 1, is restricted to discrete values.[19]

$$c = 1 - \frac{6}{m(m+1)}\ , \quad m = 3, 4, \ldots \tag{91}$$

[Then also allowed dimensions of fields are fixed.] This result has been derived in the Euclidean framework, *not* in Minkowski space-time quantum field theory, though presumably it follows here as well from positivity/unitarity constraints. It would be interesting to know where precisely these constraints are encoded. [The result $c > 0$ is encoded in the positive expectation value for Q_f^2.]

IV. CONFORMAL DYNAMICS IN MINKOWSKI SPACE-TIME — SOME SIMPLE EXAMPLES

Representations of the conformal group and algebra have here been constructed without reference to dynamics. Let me conclude by recalling that the conformal group is a symmetry group for the free, massless, two-dimensional boson theory and for the Liouville theory [exponential interaction]. The action is invariant when the field transforms with $d = 0$, and in the free case any α is allowed, while in the Liouville model α must be the coupling constant. Then χ is identified with $\frac{1}{\sqrt{2}}(\Pi + \Phi')$ and the generator is constructed from the energy-momentum tensor, appropriately improved for $\alpha \neq 0$.[17]

An even simpler conformally invariant theory is that of a self-dual field governed by a non-local Lagrangian.

$$L = \frac{1}{4}\int dx\, dy\, \chi(t,x)\epsilon(x-y)\dot{\chi}(t,y) - \frac{1}{2}\int dx\, \chi^2(t,x) \tag{92}$$

[The dot signifies time differentiation.] The equation of motion satisfied by the field χ is the self-dual one,

$$\dot{\chi} = \chi' \tag{93}$$

and the Lagrangian changes by the total derivative $\frac{d}{dt}\left(-\frac{\alpha}{\sqrt{8\pi}}f'\chi\right)$ when the field is transformed as

$$\delta\chi(t,x) = (f(t+x)\chi(t,x))' + \frac{\alpha}{\sqrt{2\pi}}f''(t+x) \tag{94}$$

so that the conserved generator is Q_f^α, and $\frac{1}{2}\int dx \chi^2(x)$ is the Hamiltonian which we write as

$$H = \frac{1}{2}\int dx\, dy\, \chi(x)\delta(x-y)\chi(y) \tag{95}$$

Canonical quantization of this first order theory leads to the Kač–Moody commutator (13) for χ. [The quantization is carried out by modern symplectic methods; L of (91) does *not* define a constrained system and Dirac's quantization method is not needed.[23]] The Hamiltonian equations of motion reproduce (93). We see that our bosonic realization of the conformal group coincides with the fixed-time ($t = 0$) quantization of the self-dual boson field, described by the above model.

Observe also that θ, introduced in (49), is a self-dual field, when χ is. The dynamics for θ may be deduced from (49) and (91). The Lagrangian is local.

$$L = \frac{1}{2}\int dx \left(\theta'(x,t)\dot{\theta}(t,x) - (\theta'(t,x))^2\right) \tag{96}$$

The non-locality, hidden in (49), appears only in the commutator (50).

Conformally invariant fermionic theories include the free, massless Majorana–Weyl model, whose fields enter in the $c = 1/2$ representation of the conformal algebra. The Lagrangian

$$L = \frac{i}{2}\int dx\, \psi(t,x)\left(\dot{\psi}(t,x) - \psi'(t,x)\right) \tag{97}$$

gives rise to self-dual equations and to the Hamiltonian $H = \frac{i}{2}\int dx\, \psi(x)\psi'(x)$, which reproduces them upon commutation with ψ, when the canonical anticommutator (51) is used. Upon writing the Hamiltonian as

$$H = \frac{1}{2}\int dx\, dx\, \psi(x)k(x,y)\psi(y) \tag{98}$$

one recognizes once again the Hill–Feynman duality between bosons and fermions, when symplectic structure and dynamics are interchanged: $k \leftrightarrow \delta$.

The doubled, charged version of the above, $\psi = \frac{1}{\sqrt{2}}(\psi_1 + i\psi_2)$, describes free massless charged Weyl fermions. The bosonization of the latter produces the self-dual boson theory governed by (92), (95) or (96). The identifications are[23]

$$\psi = \sqrt{\frac{m}{2\pi}} : \exp -i\sqrt{2\pi}\theta : \tag{99a}$$

$$\rho = \sqrt{2\pi}\,\chi \tag{99b}$$

where m provides infrared regularization. The fermionic Lagrangian is

$$L = i \int dx\, \bar{\psi}\gamma^\mu \partial_\mu \psi \tag{100}$$

and the Hamiltonian $i\int dx\, \psi^\dagger(x)\psi'(x)$ coincides after bosonization with that of self-dual scalar field χ, (95).

A conformally invariant, interacting fermionic theory is the Thirring model;[9] there the conformal algebra is realized[13,15] with $c = 1$.

As mentioned already, no explicit dynamical models in Minkowski space-time that represent the conformal algebra by using the discrete series in the interval $\frac{1}{2} < c < 1$ are known today, and finding them remains an interesting open problem. Because radial quantization provides a bridge between models in Minkowski space-time and those in Euclidean space, it should be possible to learn more about the discrete series within space-time physics.

With radial quantization, evolution is not in time but in the radial coordinate — the dilation operatgor replaces the energy as the "Hamiltonian." The conformal group suggests another alternative evolution operator: one can take the "Hamiltonian" to be $P^0 + K^0$, which generates a compact subgroup. While some investigations have made use of this idea,[24] such applications of the conformal group await future developments.

REFERENCES

1. C. Callan, S. Coleman and R. Jackiw, "A New Improved Energy-Momentum Tensor," *Ann. Phys.* (NY) **59**, 42 (1970).

2. E. Cummingham, "The Principle of Relativity in Electrodynamics and an Extension Thereof," *Proc. London Math. Soc.* **8**, 77 (1910); H. Bateman, "The Transformation of the Electrodynamical Equations," *Proc. London Math. Soc.* **8**, 223 (1910).

3. E. Bessel–Hagen, "Über die Erhaltungssätze der Elektrodynamik," *Math. Ann.* **84**, 258 (1921); P. Dirac, "Wave Equations in Conformal Space," *Ann. Math.* **37**, 429 (1936)' F. Gursey, "On a Conform-Invariant Spinor Wave Equation," *Nuovo Cim.* **3**, 988 (1956); H. Dürr, W. Heisenberg, W. Mitter, S. Schlieder and K. Yamazaki, "Zur theorie der Elementarteilchen," *Zeit. Naturforsch* **14a**, 441 (1959); J. Wess, "The Conformal Invariance in Quantum Field Theory," *Nuovo Cim.* **18**, 1086 (1960); H. Kastrup, "Zur physikalischen Deutung und darstellungstheoretischen Analyse der konformen Transformationen von Raum and Zeit," *Ann. Physik* **9**, 338 (1962); T. Fulton, F. Rohrlich and L. Witten, "Conformal Invariance in Physics," *Rev. Mod. Phys.* **34**, 442 (1962).

4. G. Mack and A. Salam, "Finite-Component Field Representations of the Conformal Group," *Ann. Phys.* (NY) **53**, 174 (1969).

5. R. Jackiw, "Introducing Scale Symmetry," *Physics Today* **25**, No. 1, 23 (1972).

6. S. Coleman and R. Jackiw, "Why Dilatation Generators Don't Generate Dilatations," *Ann. Phys.* (NY) **67**, 552 (1971).

7. Anomalies and other topological effects in quantum field theory are reviewed in *Current Albegra and Anomalies*, S. Treiman, R. Jackiw, B. Zumino and E. Witten, (Princeton University Press/World Scientific, Princeton NJ/Singapore, 1985).

8. E. Stuckelberg and A. Petermann, "La Normalisation des Consantes dans la Théorie des Quanta," *Helv. Phys. Acta* **26**, 449 (1953); M. Gell-Mann and F. Low, "Quantum Electrodynamics at Small Distances," *Phys. Rev.* **95**, 1300 (1954); *Introduction to the Theory of Quantized Fields*, N. Bogoliubov and D. Shirkov, (Interscience, New York, NY, 1959).

9. K. Wilson, "The Renormalization Group and Critical Phenomena," *Rev. Mod. Phys.* **55**, 583 (1983).

10. A. Belavin, A. Polyakov and A. Zamolodchikov, "Infinite Conformal Symmetry in Two-Dimensional Quantum Field Theory," *Nucl. Phys.* **B241**, 333 (1984); D. Friedan, Z. Qiu and S. Shenker, "Conformal Invariance, Unitarity and Two-Dimensional Critical Exponents," in *Vertex Operators in Mathematics and Physics*, J. Lepowsky, S. Mandelstam and I. Singer, eds. (Springer, Berlin, 1985); J. Cardy, "Conformal Invariance," in *Phase Transitions*, vol. 11; C. Domb and J. Lebowitz, eds., (Academic Press, New York, NY, 1987).

11. For a review of modern results see P. Ginsparg, "Applied Conformal Field Theory," in *Champs, Cordes et Phénomènes Critiques/Fields, Strings and Critical Phenomena*, E. Brézin and J. Zinn-Justin, eds. (Elsevier, Amsterdam, 1989).

12. For a review of early work, see *Currents in Hadron Physics*, V. de Alfaro, S. Fubini, G. Furlan and C. Rossetti, (Elsevier, Amsterdam, 1973); *Dual Theory*, M. Jacob, ed. (North-Holland, Amsterdam, 1974).

13. S. Fubini, A. Hanson and R. Jackiw, "New Approach to Field Theory," *Phys. Rev. D* **7**, 1732 (1972).

14. D. Boulware and S. Deser, "Stress Tensor Commutators and Schwinger Terms," *J. Math. Phys.* **8**, 1468 (1967).

15. S. Ferrara, R. Gatto and A. Grillo, "Conformal Algebra in Two Space-Time Dimensions and the Thirring Model," *Nuovo Cim.* **12A**, 959 (1972).

16. For a recent review of applications of the Schrödinger representation to quantum field theory, see R. Jackiw, "Analysis on Infinite-Dimensional Manifolds — Schrödinger Representation for Quantized Fields," to be published in *Séminaire de Mathématiques Supérieures* (Montréal, Canada, 1988), MIT preprint CTP#1632 (1988).

17. E. D'Hoker and R. Jackiw, "Classical and Quantal Liouville Field Theory," *Phys. Rev. D* **26**, 3517 (1982).

18. L. Gross, "On the Formula of Mathews and Salam,"' *J. Funct. Anal.* **25**, 162 (1977).

19. Friedan *et al.*, Ref. [10].

20. R. Floreanini and L. Vinet, "Schrödinger Picture for Self-Dual Superfields," Trieste/Montréal preprint (1989).

21. C. Hill and R. Feynman, unpublished conversation.

22. S. Coleman, D. Gross and R. Jackiw, "Fermion Avatars of the Sugawara Model," *Phys. Rev.* **180**, 1459 (1969).

23. R. Floreanini and R. Jackiw, "Self-Dual Fields as Charge-Density Solitons," *Phys. Rev. Lett.* **59**, 1873 (1987); L. Faddeev and R. Jackiw, "Hamiltonian Reduction of Unconstrained and Constrained Systems," *Phys. Rev. Lett.* **60**, 1692 (1988).

24. V. de Alfaro, S. Fubini and G. Furlan, "Conformal Invariance in Quantum Mechanics," *Nuovo Cim.* **34A**, 569 (1976); R. Jackiw, "Dynamical Symmetry of the Magnetic Monopole," *Ann. Phys.* (NY) **129**, 183 (1980).

Riemann surfaces, operator fields, strings. Analogues of the Fourier-Laurent bases

I.M. KRICHEVER, S.P. NOVIKOV

Institut des Hautes Etudes Scientifiques
35, route de Chartres
91440 Bures-sur-Yvette (France)

Avril 1989

In the previous works [1-3], started in 1986, the grounds of the operator theory of the interacting closed bosonic strings were constructed. Our theory is the direct development of the classical algebraic operator theory of non-interacting bosonic string of Virasoro. Mandelstam et. al. [4-5]. This theory corresponds to the case of the Riemann surface of the genus $g = 0$. The approach which we proposed unifies the ideas of the operator formalism with the ideas of the geometrical approach of Polyakov and others [6-8].

This paper is a brief review and an extension of the works [1-3].

§1. Fourier-Laurent-type bases on the Riemann surfaces. Almost graded analogues of the Heizenberg and Virasoro algebras.

The classical definition of the "Heizenberg algebra" (the algebra of the creation and annihilation operators for the non-interacting bosonic string) uses the Fourier expansions of the co-ordinates $X^{\mu}(\sigma)$ and momentums $P^{\mu}(\sigma)$, $0 < \sigma \leq 2\pi$, $\mu = 1, \ldots \mathcal{D}$. It turns out that this definition can be easily generalized for the case of the interacting string in which the world sheet can be a surface of an arbitrary genus. To obtain this generalization it is sufficient to introduce the special Fourier-Laurent-type bases, which are defined by the surface itself.

Let's consider a nonsingular Riemann surface Γ of a genus g with punctures P_i and with fixed real numbers p_i such that

$$\sum_{i=1}^{m} p_i = 0 \tag{1.1}$$

The set of data (Γ, P_i, p_i) will be called the "multistring diagram". For any such diagram there exists a unique differential dk which satisfies the following properties : a) it is holomorphic on Γ outside the punctures P_i ; b) at every point P_i it has a simple pole with the residue equal to p_i ; c) the periods of dk over an arbitrary closed cycle γ on Γ are pure imaginary, i.e.

$$\mathrm{Re} \int_{\gamma} dk = 0 \qquad (1.2)$$

From the last condition it follows that the real part of the multivalued function $k(z)$ is single-valued. This function $\tau(z) = \mathrm{Re}\, k(z)$ is called the "euclidian time". We denote the curves $\tau(z) = \mathrm{const} = \tau$ by C_τ and the domains $\tau_1 \le \tau(z) \le \tau_2$ by $C_{\tau_1 \tau_2}$ (Riemann annulus). The curves c_τ for $\tau \to \pm\infty$ tend to small circles around the points P_α for which $p_\alpha < 0$ and $p_\alpha > 0$, respectively. The transformation $p_\alpha \to ap_\alpha$ corresponds to the transformation $\tau \to a\tau$, which preserves the "chronological ordering" in the case $a > 0$. That's why in the simplest case of "one-string" diagram $(\Gamma, P_\pm)$ we can always assume that $p_+ = 1$, $p_- = -1$.

For any integer $\lambda \neq 0,1$ and "one-string" diagram $(\Gamma, P_\pm)$ in the general position, $g \neq 1$, for any integer $n + g/2$ there exists the unique up to a constant factor tensor $f_n^\lambda = f_n^\lambda(z)$ of the weight λ with the following analytical properties : a) it is holomorphic on Γ outside the points $P_\pm$; b) it has the form

$$f_n^\lambda = \varphi_{n,\lambda}^{\pm}\, z_\pm^{\pm n - S}\, (1 + O(z_\pm))(dz_\pm)^\lambda, \; S = g/2 - \lambda(g-1), \qquad (1.3)$$

near the points $P_\pm$. Here $z_\pm = z_\pm(Q)$ are local co-ordinates in the neighbourhoods of the points $P_\pm$ respectively.

For the exceptional cases ($g > 1$, $\lambda = 0,1$; $g = 1$, λ-arbitrary integer) the tensor fields f_n^λ are defined by the conditions (1.3) for all except the finite number of n (see [1,2]). For other $n \in Z - g/2$ the definition of f_n^λ is slightly more complicated. We don't present them here because their exact form is not essential for us now.

The bases f_n^λ and $f_{1-\lambda}^m$ is dual to each other

$$\frac{i}{2\pi i} \cdot \int_{C_\tau} f_n^\lambda \cdot f_{1-\lambda}^m = \delta_{n,-m} \,. \tag{1.4}$$

Theorem 1.1 [1] Let C_τ be non-singular, then for any smooth tensor $f^\lambda(\sigma)$ of the weight λ on C_τ the expansion (Fourier-type)

$$f^\lambda(\sigma) = \sum_n f_n^\lambda(\sigma) \left(\frac{1}{2\pi i} \int_{C_\tau} f^\lambda(\sigma') \, f_{-n}^{1-\lambda}(\sigma')\right) \tag{1.5}$$

is valid. The same expansion is valid for the tensors $f^\lambda(z)$ which are holomorphic in the Riemann annulus $C_{\tau_1 \tau_2}$ (Laurent-type expansion). (The theorem is valid for singular contours C_τ but smoothness conditions in this cases are slightly more rigourous. The theorem is valid independently of whether the C_τ contour is connected or not).

The proof of this theorem is based on the connection of the bases f_n^λ with the "well-known discrete Baker-Akhiezer functions" in the soliton theory (about the details see [3] and [9]).

Remark. Recently the construction of the Fourier-Laurent-type bases for general multi-string diagram and the proof of theorem 1.1 for them was obtained by one of the authors with the help of the discrete Baker-Akhiezer vertor-functions. These functions appeared in the theory of the commutative difference operators with the matrix coefficients. ([10]). We shall give in the appendix the definition of the correspondence bases in the case $p_i = \pm 1$, $i = 1, \dots 2m$.

The important properties of our bases (which immediately follow from the definition) are their almost-graded structure in respect to the multiplication

$$f_n^\lambda f_m^\mu = \sum_{|k| \le g/2} Q_{n,m}^{\lambda,\mu,k} f_{n+m-k}^{\lambda+\mu} \tag{1.6}$$

and in respect to the action of the vector-fields

$$[e_n, f_m^\lambda] = \sum_{|k| \leq g_0} R_{n,m}^{\lambda,k} f_{n+m-k}^\lambda, \quad g_0 = {}^{3g}/_2 \tag{1.7}$$

(Here and below we shall use the special notations for bases for the $\lambda = -1, 0, 1/2, 1,2$:

$$e_n = f_n^{-1}, A_n = f_n^0, f_n^{1/2} = \varphi_n, f_n^1 = d\omega_{-n}, f_n^2 = d^2\Omega_{-n}.) \tag{1.8}$$

For exceptional cases $\lambda, \mu = 0,1$, $|n|$ or $|m| \leq {}^g/_2$ the sums in (1.6, 1.7) must include the additional terms. But in any case the sums include the terms only for $|k| \leq N$, where N depends on g, λ, μ and doesn't depend on the n, m. This property leads us to the definition of the almost-graded algebras and modules.

Definition . The almost-graded (N-graded) algebra L (or module M over L) is the algebra (or module) which can be expanded into the direct sum of the subspaces

$$L = \sum_i L_i \; , \; M = \sum_i M_i \tag{1.9}$$

so that

$$L_i L_j \subset \sum_{|k| \leq N} L_{i+j-k} \; ; \; L_i M_j \subset \sum_{|k| \leq N_1} M_{i+j-k} \tag{1.10}$$

The basis $A_n = f_n^0$, as it follows from (1.6) with $\lambda = \mu = 0$, defines the almost-graded structure in the commutative algebra A^Γ of the meromorphy on Γ functions which are holomorphic everywhere except the points $P_\pm$.

The functions A_n with $\pm n > {}^g/_2$, as it follows from (1.3), are holomorphic in the neighbourhood of the points $P_\pm$ respectively and have at least the simple zero in these points. The choice of the base functions A_n for $|n| \leq {}^g/_2$ is not canonical. But in any case we can do it in such a way that

$$A_{-g/2} = 1 \tag{1.11}$$

and the other functions A_n, $-{}^g/_2 < n \le {}^g/_2$ have the poles in both points $P_\pm$.

The analogue of the Heizenberg algebra is the algebra which is generated by the elements α_n, t with the following commutative relations

$$[\alpha_n, \alpha_m] = \gamma_{nm} \cdot t, \gamma_{nm} = \frac{1}{2\pi i} \int_{C_\tau} A_m \, d A_n, [\alpha_n, t] = 0 \tag{1.12}$$

As it follows from (1.3), we have for $|n| > {}^g/_2$, $|m| > {}^g/_2$ that

$$\gamma_{nm} = 0, \ |n+m| > g, \tag{1.13}$$

For all values of n, m the slightly weak condition is held : $\gamma_{nm} = 0$ if $|n+m| > 2g$. In the particular case : $g = 0$, $P_+ = 0$, $P_- = \infty$ the definition (1.12) leads to the ordinary Heizenberg algebra of the creation and annihilation operators of the free closed bosonic string.

In the case $\lambda = -1$ we have the almost-graded algebra $\mathcal{L}^\Gamma$, $(e_n = f_n^{-1})$

$$[e_n, e_m] = \sum_{k=-g_0}^{g_0} C_{nm}^k e_{n+m-k} \tag{1.14}$$

The analogue of the Virasoro-type algebra is the central extended algebra $\hat{\mathcal{L}}^\Gamma$ with the basis $e_n \cdot t$ with the commutative relations :

$$[e_n, e_m] = \sum_{k=-g_0}^{g_0} c_{nm}^k e_{n+m-k} + \chi_{nm} \cdot t, [e_n, t] = 0 \tag{1.15}$$

where the cocycle $\chi_{nm} = \chi(e_n, e_m)$ is defined by the formula

$$\chi(f, g) = \frac{1}{48\pi i} \int\limits_{C_\tau} (f'''g - g'''f) - 2(f'g - g'f) \times R) \, dz \qquad (1.16)$$

Here $f = f(z)\, \partial/\partial z$, $g = g(z)\, \partial/\partial z$ vector-fields, $R(z)$-projective connection which is holomorphic on Γ except the points $P_\pm$. (The projective connection is the value $R(z)$ which is transformed by the following way

$$R(w) = R(z)\,(w')^2 + \left(\frac{w'''}{w'} - \frac{3}{2}\left(\frac{w''}{w'}\right)^2\right), \quad w' = \frac{dw}{dz} \qquad (1.17)$$

in respect to the transformation of local system of the co-ordinate : $w = w(z)$. If the projective connection is holomorphic in the points $P_\pm$ also then

$$\chi_{nm} = 0, \quad \text{if } |n + m| > 3g \qquad (1.18)$$

<u>Remark</u>. The central extensions of the algebras A^Γ and $\mathfrak{L}^\Gamma$ can be defined for any closed contour γ on $\Gamma(P_+ \cup P_-)$ by the previous formula if one changes in them the contours C_τ for γ.

<u>Conjecture</u>.

$$H^2(\mathfrak{L}^\Gamma, R) = H_1(\Gamma \setminus (P_+ \cup P_-), R) \qquad (1.19)$$

<u>Theorem</u>. ([1,2]) The cohomology class of contours $[C_\tau]$ is one and only one cohomology class of contours γ such that corresponding central extensions of algebras A^Γ and $\mathfrak{L}^\Gamma$ are also almost graduated.

§2. <u>Riemann analogues of Heizenberg and Virasoro algebras in string theory.</u>

Let $X^\mu(Q)$ and $P^\nu(Q)$, $Q \in \Gamma$, be the operator-valued scalar function and 1-form in respect to the variable Q on Γ. They commute with each other in different moments of time (i.e. $\tau(Q) \neq \tau(Q')$) and satisfy the commutative relations

$$\left[X^{\mu}(Q), P^{\nu}(Q')\right] = - i\, \eta^{\mu\nu} \Delta_{\tau}(Q,Q'), \text{ if } \tau(Q) = \tau(Q') = \tau \tag{2.1}$$

Here $\Delta_{\tau}(Q,Q')$ is the "δ-function" on the contour C_{τ} (it is function in respect to Q and 1-form in repect to Q' ; for any smooth function $f(Q)$ we have by definition

$$f(Q) = \frac{1}{2\pi i} \int_{C_{\tau}} f(Q')\, \Delta_{\tau}(Q,Q')).$$

$\eta^{\mu\nu} = \eta^{\mu} \delta^{\nu}_{\mu}$ the metric of the physical space (Minkovskii or Euclidean), $\mu = 1, \ldots, \mathcal{D}$.

In the work [2] it was proved that the coefficients of the expansion

$$\mathfrak{g}^{\mu}(\sigma) = \partial_{\sigma} X^{\mu}\, d\sigma + \pi P^{\mu}(\sigma) = \sum \alpha^{\mu}_{n}\, d\omega_{n}(\sigma) \tag{2.2}$$

$(d\omega_{n} = f^{1}_{-n})$ satisfy the commutative relations of the Heizenberg-type algebra

$$\left[\alpha^{\mu}_{n}, \alpha^{\nu}_{m}\right] = \eta^{\mu\nu} \gamma_{nm} \tag{2.3}$$

The coefficients $\overline{\alpha}^{\mu}_{n}$ of the expansion

$$\overline{\mathfrak{g}}^{\mu}(\sigma) = X^{\mu}_{\sigma}\, d\sigma - \pi P^{\mu} = \sum_{n} \overline{\alpha}^{\mu}_{n}\, \overline{d\omega}_{n}(\sigma) \tag{2.4}$$

satisfy the commutative relations of the conjugate Heizenberg-type algebra

$$\left[\overline{\alpha}^{\mu}_{n}, \overline{\alpha}^{\nu}_{m}\right] = \overline{\gamma}_{nm}\, \eta^{\mu\nu} \tag{2.5}$$

At the same time ([2])

$$[\alpha_n^\mu, \overline{\alpha}_m^\nu] = 0 \tag{2.6}$$

That's why we can restrict ourselves and can consider only the holomorphic part of the theory. The full Fock space would be as usual a tensor product of the holomorphic and antiholomorphic parts.

The holomorphic parts of the "vacuum-sectors" of in - and out - Fock spaces are defined as the spaces, which are generated by the left and right actions of the operators α_n^μ from the vacuum-vectors $|0>$ and $<0|$

$$\alpha_n^\mu |0> = 0 \ , \ n > g/2 \ , \ n = -g/2$$
$$<0| \alpha_n^\mu = 0 \ , \ n \leq -g/2 \tag{2.7}$$

According to the previous definitions this means that the vacuum-vectors are annihilated by the operators, corresponding to the basic functions A_n which are holomorphic in the neighbourhoods of the points P_+ (for $|0>$) and P_- (for $<0|$).

It can be shown that such defined spaces $H_\Gamma^\pm$ (which of course depend on $P_\pm$ also) are isomorphic to the vacuum sectors of the ordinary Fock spaces for small circles around $P_\pm$. More precisely, let $a_{N,\pm}^\mu$, $N \in Z$, be the operators with the ordinary commutative relations

$$[a_{N\pm}^\mu, a_{M\pm}^\nu] = \eta^{\mu\nu} N \delta_{N,-M} \tag{2.8}$$

Then it can be easily checked that the operators α_n^μ which are defined for $|n| > g/2$ by both of the formulae (2.9) and in the slightly different form for $|n| \leq g/2$ satisfy the commutator relations (2.5).

$$\alpha_n^{\mu} = \sum_{s=0}^{\infty} \xi^{+}_{s}(n)\, a^{\mu,+}_{n+s-g/2} = \sum_{s=0}^{\infty} \xi^{-}_{s}(n)\, a^{\mu,-}_{s-n-g/2} \tag{2.9}$$

where

$$A_n = z_{\pm}^{\pm n-g/2} \sum_{s=0}^{\infty} \xi^{\pm}_{s}(n)\, z_{\pm}^{s}$$

The conditions (2.7) do coincide with the ordinary one

$$a^{\mu}_{N,+} | 0 > = 0 \ , \ < 0 | a^{\mu}_{N,-} = 0 \ , \ N \geq 0 \tag{2.10}$$

which define the "local" Fock spaces.

Remark. As it was mentioned by A. Polyakov (private talk) the global definition of the basic function A_n and the corresponding possibility to express them in terms of both local systems of co-ordinates $z_{\pm}$ near the points $P_{\pm}$ must contain in some sense the information about the scattering process of the string exitations.

Of course, one can define using (2.9) the formal Bogoluobov transformation

$$a^{\mu}_{N,-} = \sum_{n} (U_{-})^{-1}_{N,n} (U_{+})_{nM}\, a^{\mu}_{M,+} \tag{2.11}$$

where matrix elements $(U_{\pm})_{nM}$ are given by the right hand side of the equalities (2.9). But it looks as if even for each N, M the corresponding series in r·h·s of (2.11) diverge. Up to now the question of the regularization of (2.11) is not exactly clear .

In the classical case the densities of the Hamiltonian and momentum are equal to the sum and difference of the values

$$T(z) = \frac{1}{2}\, \mathcal{J}^2(z) \ , \ \overline{T}(z) = \frac{1}{2}\, \overline{\mathcal{J}}^2(z) \tag{2.12}$$

The definitions of the corresponding quantum operators require, as usual, the definition of the "normal ordering", of the products $\alpha_n^\mu \alpha_m^\nu$.

There exists the arbitrariness in this definition because operators α_n^μ, α_m^ν don't commute for $|n|, |m| \leq g/2$.

Let's dissect the integer (or half-integer) plane of pairs (n,m) into two parts $\sum^\pm$ such that $\sum^+$ differs from the integer half-plane $m \leq n$ only in the finite number of points. The definition of the normal ordering depends on the choice of $\sum^\pm$

$$:\alpha_n \alpha_m: = \alpha_n \alpha_m , \ (n, m) \in \textstyle\sum^+ ; \ :\alpha_n \alpha_m: = \alpha_m \alpha_n , \ (n, m) \in \textstyle\sum^- \quad (2.13)$$

As it follows from (2.7) the operator

$$T(Q) = \frac{1}{2} : \mathfrak{z}^2(Q) : = \frac{1}{2} \sum_{n,m} : \alpha_n \alpha_m : d\omega_n (Q) \, d\omega_m (Q) \qquad (2.14)$$

is correctly defined. It is quadratic differential on Γ. That's why it can be expanded in the form

$$T(Q) = \sum_k L_k \, d^2 \Omega_k (Q) \ , \ d^2 \Omega_n = f_{-n}^2 \, . \qquad (2.15)$$

$$L_k = \frac{1}{2} \sum_{n,m} l_{nm}^k : \alpha_n \alpha_m : \ , \ l_{nm}^k = \frac{1}{2\pi i} \int_{C_\tau} e_k \, d\omega_n \, d\omega_m \qquad (2.16)$$

If $|k| > g_0$ then

$$l_{nm}^k = 0 \ , \ |n + m - k| > g/2$$

For $|k| \leq g_0$ the width of the strip in the plane of the pairs (n, m), such that l_{nm}^k may differ from zero, is slightly larger. But in any case it remains finite.

<u>Theorem 2.1</u>. [2] The operators $e_k = - L_k$, where L_k are given by the formula (3.13), (3.15), satisfy the commutator relations (1.15) of the Riemann analogues of the Virasoro algebra with the central charge $t = \mathcal{D}$. The cocycle $\chi_{nm} = \chi^{\Sigma}_{nm}$ depends on the choice of the normal ordering but his cohomology class does not depend on this choice.

We shall call the normal ordering admissible if it corresponds to the choice of $\sum^{\pm}$ such that $\sum^{+}$ differs from the half plane $n \leq m$ only in the strip $|n + m| \leq g - 2$. The corresponding projective connection R_{Σ} (which defines χ^{Σ}_{nm}) is holomorphic on Γ everywhere (including the points $P_{\pm}$). For admissible normal ordering the following important conditions of the regularity of vacuum

$$L_k | 0 > = 0 , \ k \geq g_o - 1 , \ < 0 | L_k = 0 , \ k \leq -g_o + 1 \tag{2.17}$$

are fulfilled (details see [3]).

<u>Remark</u>. The quadratic expressions of the form (2.16) are well-known in the case of genus $g = 0$. They are the special case of the Sugavara-construction of the Virasoro generators through the generators of the Kac-Moody algebras. The Riemann analogues of the untwisted Kac-Moody algebras was proposed in [1]. The generalization of the Sugavara construction for the case of the Riemann surfaces of the genus $g > 0$ was proposed in [12].

The projective connection R_{Σ} corresponding to the co-cycle χ^{Σ}_{nm} depends, as well as the tensor $T(z) = T_{\Sigma}(z)$, on the choice of the normal ordering. But the operator-valued projective connection

$$\check{T}(z) = T_{\Sigma}(z) + \frac{\mathcal{D}}{2} R_{\Sigma} \tag{2.18}$$

doesn't depend on this choice and is a *canonically defined* pseudo-tensor of the energy-momentum. The last statement follows from the theorem.

Theorem 2.2. ([3]) The chronological product $\mathcal{J}(z)\,\mathcal{J}(w) = \mathcal{J}^{\mu}(z)\,\eta^{\mu\nu}\,\mathcal{J}^{\nu}(w)$, where $\tau(z) > \tau(w)$ is correctly defined. For $z \to w$ the following expansion is valid

$$\mathcal{J}(z)\,\mathcal{J}(w) = \mathcal{D}\,\frac{dz\,dw}{(z-w)^2} + 2\mathcal{T}(z) + \mathcal{O}(z-w) \tag{2.19}$$

For the pseudo-tensors of the energy-momentum on the arbitrary Riemann surfaces the ordinary operator expansions are fulfilled.

Theorem 2.3. ([3]) The chronological product $\mathcal{T}(z)\,\mathcal{T}(w)$, $\tau(z) > \tau(w)$ is correctly defined. For $z \to w$ we have

$$\mathcal{T}(z)\,\mathcal{T}(w) = \frac{\mathcal{D}}{2(z-w)^4} + \frac{2\mathcal{T}(z)}{(z-w)^2} + \frac{\mathcal{T}_z(z)}{z-w} + O(1). \tag{2.20}$$

The definition of the in - and out - Fock spaces, the construction of the globally defined operator fields on Γ, $\mathcal{J}(z)$, $\mathcal{T}(z)$ is only the first part of our program. The following important step is the construction of the bilinear product (coupling) between the spaces $H_{\Gamma}^{\pm}$. The conformal invariance requires that the operators L_k must be formally self-adjoint in respect to this coupling, i.e.

$$<\phi| L_k\,\psi> = <\phi\, L_k|\psi> = <\phi|L_k|\psi>,\ \phi \in H_{\Gamma}^{-},\ \psi \in H_{\Gamma}^{+} \tag{2.21}$$

It is clear that these conditions would be fulfilled in the case when

$$<\phi|\alpha_n^{\mu}\,\psi> = <\phi\,\alpha_n^{\mu}|\psi> = <\phi|\alpha_n^{\mu}|\psi>. \tag{2.22}$$

In the case $g = 0$ the conditions (2.22) and the condition $<0|0> = 1$ uniquely define the product $<\phi|\psi>$ of any elements of the vacuum sectors of the in - and out - Fock spaces. In the general case $g > 0$, these conditions are insufficient because the operators α_n^{μ} with $-g/2 < n \le g/2$ don't annihilate neither in - nor out - vacuum vectors.

§3. <u>Bloch's half-differentials and "fermionization formulae".</u>

Let's fix the contour σ on Γ, connecting the points $P_\pm$, and the unitary representation of the fundamental group $\rho : \pi_1(\Gamma) \to S^1 = \{ z \in C , |z| = 1\}$. If we fix on Γ the canonical basis of cycles a_i, b_j with the intersection matrix $a_i \bullet a_j = b_i \bullet b_j = 0$, $a_i \bullet b_j = \delta_{ij}$, then this representation can be defined by the real numbers u_j, $j = 1, \ldots, g$: $\rho(a_j) = \exp(2\pi i\, u_j)$, $\rho(b_j) = \exp(2\pi i\, u_{g+j})$.

<u>Lemma 3.1</u>. For the representation ρ in the general position and for the fixed number p there exists the unique half-differential $\phi_\nu(z,\rho)$, $\nu - p \in Z+1/2$, such that it is holomorphic on Γ, cutting along the cycles a_i, b_j and the contour σ, everywhere except the points $P_\pm$. In the neighbourhoods of these points it has the form

$$\phi_\nu(z,\rho) = \varphi^{\pm}_{\nu 1/2}\, z_\pm^{\pm\nu-1/2}\,(1 + O(z_\pm))(dz_\pm)^{1/2}, \qquad (3.1)$$

$\varphi^{+}_{\nu 1/2} \equiv 1$. It is multiplied by $\rho(\gamma)$ when one goes around the cycle γ. Its boundary values $\phi_{\nu\pm}$ on the contour σ are connected by the relation

$$\phi_{\nu+}(t,\rho) = \exp(2\pi i p)\, \phi_{\nu-}(t,\rho)\ ,\ t \in \sigma. \qquad (3.2)$$

It can be mentioned that the representations ρ such that $\rho(\gamma) = \pm 1$ for any $\gamma \subset \Gamma$ correspond to the spinor structures on Γ. The spinor structures are in the general position (in the sense of the statement of lemma) only if it is an even spinor structure.

Let's consider the ordinary Dirac fermionic operators ψ_ν, ψ^+_ν, $\nu - 1/2 \in Z$, with the anti-commutator relations

$$[\psi_\nu, \psi_\mu]_+ = [\psi^+_\nu, \psi^+_\mu]_+ = 0\ ,\ [\psi_\nu, \psi^+_\mu]_+ = \delta_{\nu+\mu,0} \qquad (3.3)$$

The Fock spaces $\mathcal{H}^{\pm}$ of the Dirac fermions are generated by the operators ψ_v, ψ^+_μ from the vacuum vectors $|0_F>, <0_F|$

$$\psi_v|0_F> = \psi^+_v|0_F> = 0\,, \quad v > 0$$
$$<0_F|\psi_v = <0_F|\psi^+_v = 0\,, \quad v < 0 \tag{3.4}$$

Let's introduce the "fermionic" fields

$$\psi(z,\rho) = \sum_v \psi_v \phi_{-v}(z,\rho)\,, \quad v - {}^1/_2 \in Z$$
$$\psi^+(z,\rho) = \sum_v \psi^+_v \phi^+_{-v}(z,\rho)\,, \quad \phi^+_v(z,\rho) = \phi_v(z,\rho^{-1})$$

Theorem 3.1. The chronological product $\psi(z,\rho)\,\psi^+(w,\rho)$, $\tau(z) > \tau(w)$ is correctly defined. For $z \to w$

$$\psi(z,\rho)\,\psi^+(w,\rho) = \frac{\sqrt{dz\,dw}}{z-w} + \mathcal{J}(z,\rho) + \mathcal{O}(z-w) \tag{3.5}$$

The coefficients of the expansion

$$\mathcal{J}(z,\rho) = \sum \alpha_n(\rho)\, dw_n(z)\,, \quad n - {}^g/_2 \in Z \tag{3.6}$$

satisfy the commutator relations (1.12) (with $t = 1$), i.e.

$$[\alpha_n(\rho), \alpha_m(\rho)] = \gamma_{nm} \tag{3.7}$$

of the Riemann analogue of the Heizenberg algebra.

Below we shall define the set of "Szego"-type kernels $S_p(z, w, \rho)$ which are holomorphic $^1/_2$-differential in respect to the variables z, w on Γ, cutting along the cycles a_i, b_j and contour σ, everywhere except the points $P_\pm$ and the diagonal $z = w$. For the fixed w (resp. z) it is multiplied by $\rho(\gamma)$ (resp $\rho^{-1}(\gamma)$) when one goes around the cycle γ. The boundary values S_p on σ satisfy the relation

$$S_p^+(t, w, \rho) = e^{2\pi i p} S^-(t, w, \rho), \; S^+(z, t, \rho) = e^{-2\pi i p}(z, t, \rho), \; t \in \sigma. \quad (3.8)$$

In the neighbourhoods of the points $P_\pm$ we have

$$S_p(z_\pm, w, \rho) = z_\pm^{\mp p} \mathcal{O}(1)(dz_\pm)^{1/2},$$
$$S_p(z, w_\pm, \rho) = w_\pm^{\pm p} \mathcal{O}(1)(dw_\pm)^{1/2} \quad (3.9)$$

The last condition, which uniquely defines the kernel S_p, requires that near the diagonal it has the form

$$S_p(z, w, \rho) = \frac{\sqrt{dz\,dw}}{z - w} + ds_p(z, \rho) + \mathcal{O}(z - w). \quad (3.10)$$

It can be shown that $ds_p(z, \rho)$ is the single-valued holomorphic differential except the points $P_\pm$ where it has the simple poles with the residues $\pm p$.

<u>Corollary</u>. The following formula ("ferminization")

$$\alpha_n(\rho) = \sum_{\nu,\mu} a^n_{\nu\mu} : \psi_\nu \psi^+_{-\mu} : + a_n \quad (3.11)$$

gives the representation of the Heizenberg-type algebra in the spaces $\mathcal{H}^\pm$. Here the coefficients

$$a^n_{\nu\mu} = \frac{1}{2\pi i} \int_{C_\tau} \phi_{-\nu} \phi^+_\mu A_n, \quad a_n = \frac{1}{2\pi i} \int_{C_\tau} A_n \, ds_o \quad (3.12)$$

depend on the representation ρ. From (1.3 , 3.1), it follows that

$$a^n_{\nu\mu} = 0, \quad |n| > g/2, \quad |n - \nu - \mu| > g/2 \quad (3.13)$$

In the case of $|n| \leq g/2$ the "strip" in the (ν, μ)-plane, outside $a^n_{\nu\mu} = 0$, becomes slightly bigger, but it is less than $|n - \nu - \mu| \leq g$.

The differential $ds_0(z,\rho)$ is holomorphic. Hence

$$a_n = a_n(\rho) = 0 \ , \ |n| > g/2 \ , \ n = - g/2. \tag{3.14}$$

The "ferminization formulae" give the possibility to introduce the "coupling" between $H^{\pm}_{\Gamma}$ using the natural coupling between the fermionic Fock spaces $\mathcal{H}^{\pm}$. The latter is uniquely defined by the requirements

$$< \phi|\psi \Psi > = < \phi \psi|\Psi > = < \phi|\psi|\Psi > , \ \phi \in \mathcal{H}^- , \ \Psi \in \mathcal{H}^+ \tag{3.15}$$

$$< 0_F|0_F > = 1 \tag{3.16}$$

Let the vectors $< 0_F | , | 0_F >$ have the charge zero and the operator ψ and ψ^+ have the charge $+ 1$ and $- 1$ respectively. Then

$$\mathcal{H}^{\pm} = \sum_p \mathcal{H}^{\pm}_p \ , \ p \in Z \ \text{- charge} \tag{3.17}$$

For any set of the representations ρ_μ, $\mu = 1, \ldots, \mathcal{D}$ in general positions, the correspondence

$$\alpha^{\mu}_n \to 1 \otimes \ldots \otimes \sqrt{\eta^{\mu}} \ \alpha_n \ (\rho_\mu) \otimes 1 \otimes \ldots \otimes 1 \tag{3.18}$$

$$| 0 > = \otimes | 0_{\rho_\mu} > \ , \ < 0 | = \otimes < 0_{\rho_\mu} |$$

where

$$|0_\rho > = Z_+(\rho) \, | 0_F > \ , \ < 0_\rho | = Z_-(\rho) < 0_F | \tag{3.19}$$

defines the isomorphism

$$H_T^{\pm} \equiv (\mathcal{H}_0^{\pm})^{\otimes \mathcal{S}}. \tag{3.20}$$

The product of the constants $Z_{\pm}(\rho)$ which is the norm of the vacuum vectors

$$<0_\rho \mid 0_\rho> = Z_+(\rho)\, Z_-(\rho) < 0_F \mid 0_F > = Z(\rho) \tag{3.21}$$

can be naturally considered as the density of the partition function.

In the next paragraph we shall return to the problem of their definition. The normalized expectation values, i.e. the values of the form

$$< H >_\rho = \frac{< 0_\rho | H | 0_\rho >}{< 0_\rho | 0_\rho >} = \frac{< 0_\rho | H | 0_\rho >}{Z(\rho)} \tag{3.22}$$

don't depend on the constant $Z(\rho)$. Hence, we can consider them just now.

In the previous work ([3]) it was proved that

$$< \psi(z, \rho)\, \psi^+(w, \rho) >_\rho = S_0(z, w, \rho) \tag{3.23}$$

In the modern physical literature the formula (3.23) is playing the role of the definition of the propagator of the free fermionic fields (in the case $\rho(\gamma) = \pm 1$) without any constructions of the proper fields .

The calculations of the normalized expectation values of the products of the operators can be easily done using the standard Vick-formula, the correctness of which in our theory was proved in [3].

For example,

$$< \mathfrak{J}(z)\, \mathfrak{J}(w) >_\rho = - S_0(z, w, \rho)\, S_0(w, z, \rho) \tag{3.24}$$

When $z \to w$ the r.h.s of this equality has the form

$$- S_o(z, w, \rho)\, S_o(w, z, \rho) = \frac{dz\, dw}{(z - w)^2} + 2R_\rho(z) + \mathcal{O}(z - w), \tag{3.25}$$

where $R_\rho(z)$ is the Szego projective connection. The comparison of (3.25, 3.24) and (2.19) gives immediately

$$< T(z) >_{\vec{\rho}} = \sum_{\mu=1}^{\wp} R_{\rho_\mu}(z) \qquad \vec{\rho} = (\rho_1, \ldots, \rho_{\wp}) \tag{3.26}$$

<u>Remark</u>. It must be specially mentioned that the projective connection $R_\rho(z)$ for any g does not depend on the punctures $P_\pm$.

<u>Example. g = 1</u>. Consider the elliptic curve Γ with the periods $2\omega = 1$, $2\omega' = \tau$. The Szego-kernel, corresponding to the representation $\rho(1) = \exp(2\pi i\, u_1)$, $\rho(\tau) = \exp(2\pi i\, u_2)$ has the form

$$S_o(z, w, \rho) = \frac{\sigma(z - w + x)}{\sigma(z - w)\, \sigma(x)}\, e^{y(z-w)} \sqrt{dz\, dw} \tag{3.27}$$

where $\quad x = x_1 + ix_2\,, \; y = y_1 + iy_2$

$$y_1 = -2\eta\, x_1\,, \quad y_2 = -2\frac{\mathcal{I}m\, \eta'}{\mathcal{I}m\, \tau}\, x_2$$

$$u_1 = \frac{x_2}{\mathcal{I}m\, \tau}, \quad u_2 = -x_1 + \frac{x_2}{2\pi}\left(\mathrm{Re}\, \eta' - \mathrm{Re}\, \tau \cdot \frac{\mathcal{I}m\, \eta'}{\mathcal{I}m\, \tau}\right) \tag{3.28}$$

Here and below - $\sigma, \zeta, \wp$-Weierstrass elliptic functions $\eta = \zeta(1/2)$, $\eta' = \zeta(\tau/2)$.

From the substitution of (3.27) into the (3.25) we obtain that in the global plane system of co-ordinate on Γ

$$R_\rho(z) = \frac{1}{2}\, \wp(x)\, (dz)^2. \tag{3.29}$$

There are three even spinor structures on Γ. They correspond to the following values of $x = \omega_\alpha$, ($\omega_1 = 1/2$, $\omega_2 = \tau/2$, $\omega_3 = \frac{\tau + 1}{2}$). In these cases the formula (3.29) coincides with the results of [13], which were obtained with the help of the conformal Ward identities. This coincidence was briefly mentioned in [3]; it was not stressed sufficiently that (3.29) coincides with the mean value of energy-momentum tensor, which was obtained in [13] for the free fermionic field on Γ but not for the scalar field.

It is well-known that in the $c = 1$ case there are many different conformal field theories. The most important question : is there the possibility to introduce the coupling between $H_{\Gamma}^{\pm}$ which corresponds to the scalar part of the bosonic string theory ?

The calculations which was made recently by one of the authors [10] makes very reliable the following conjecture.

Conjecture. The normalized vacuum expectation values $< H >_0$ for the scalar field theory are equal to the averaging values $< H >_\rho$ in respect to all representations $\rho : \pi_1(\Gamma) \to S^1 \subset C^2$, i.e.

$$<H>_0 = \int_0^1 \dots \int_0^1 \prod_{i=1}^{2g} du_i < H >_\rho ,$$

$$\rho(a_j)^* = \exp(2\pi i\, u_j) \,,\ \rho(b_j) = \exp(2\pi i\, u_{i+g}). \tag{3.30}$$

We cite here only one example : $H = \tilde{T}(z)$ and $g = 1$. In that case

$$< \tilde{T}(z) >_0 = \frac{\mathcal{D}}{2} \int_0^1 \int_0^1 du_1\, du_2\, \wp(x) = \frac{\mathcal{D}}{2} (4\eta - \frac{2\pi}{\mathrm{Im}\,\tau}) .$$

where x is given by the formula (3.28). The r.h.s is exactly the same which was obtained from the Ward identities in [13] for the scalar fields.

§4. The semi-infinite forms and "the principle of the normalization"

The formulae (2.15, 3.11) define the structure of the Verma modules over the algebra $\hat{\mathcal{L}}^{\Gamma}$ in the spaces $\mathcal{H}^{\pm}_{p}$. These modules are the particular cases of the general Verma modules which were introduced in [1,2]. The geometrical realization of these modules, which was proposed there, is the following (in the case $g = 0$ it was done in the work [14]).

Let's fix the number p and the contour σ, connecting the points $P_{\pm}$. In the general position, for any n, $n - p - g/2 \in Z$, there exists the unique up to the constant factor tensor f^{λ}_{n} of the weight $\lambda \in Z$ (if λ is half-integer then it is necessary to fix also the representation $\rho : \pi_1(\Gamma) \to S^1$) such that a) it is holomorphic on Γ except the points $P_{\pm}$ and the contour σ ; b) in the neighbourhoods of the points $P_{\pm}$ it has the form (1.3) ; c) its boundary values on σ are connected by the following relation

$$f^{\lambda}_{n,+} = e^{2\pi i p} f^{\lambda}_{n,-} \tag{4.1}$$

Consider the right and left semi-infinite forms - exterior products of the form

$$\begin{gathered} f^{\lambda}_{S+p+n_0} \wedge f^{\lambda}_{S+p+n_1} \wedge \dots f^{\lambda}_{S+p+n_k} \wedge \dots \text{ right-form} \\ \dots \wedge f^{\lambda}_{-S+p+m_{-k}} \wedge \dots \wedge f^{\lambda}_{-S+p+m_{-1}} \wedge f^{\lambda}_{-S+p+m_0} \text{ -left-form} \end{gathered} \tag{4.2}$$

$S = S(\lambda, g) = g/2 - \lambda(g - 1)$, such that the sequences $n_0 < n_1 < \dots$, become stable from number on. This means for some k_0 we have $n_k = k$ if $k > k_0$. The spaces of the finite linear combinations of the basic vectors of the form (4.2) would be denoted by $M^{\pm}_{p\lambda} = M^{\pm}_{p,\lambda\sigma}$ (the sign + and - for right and left semi-infinite forms respectively). In [1,2] it was proved that the spaces $M^{\pm}_{p,\lambda}$ are the Verma-modules over $\hat{\mathfrak{L}}^{\Gamma}$. The generating (singular) vectors $|\overset{+}{\psi}_{\lambda p}>$ and $<\overset{-}{\psi}_{\lambda p}|$ have the form (4.2) where the corresponding sequences of indices are $n_i = i$, $i = 0, 1, \ldots$. They satisfy the relations

$$L_n |\overset{+}{\psi}_{\lambda p}> = 0\,,\;\; n > g_o = {}^{3g}/_{2}\,,\;\; L_{g_o} |\overset{+}{\psi}_{\lambda p}> = \overset{+}{h}_{\lambda p} |\overset{+}{\psi}_{\lambda p}>$$

$$<\overset{-}{\psi}_{\lambda p}| L_n = 0\,,\;\; n < -\, g_o\,,\;\; <\overset{-}{\psi}_{\lambda p}| L_{-g_o} = \overset{-}{h}_{\lambda p} <\overset{-}{\psi}_{\lambda p}| \tag{4.3}$$

The central charge of the corresponding representation is given by the well-known formula

$$t = -\,12\,\lambda^2 + 12\,\lambda - 2$$

The highest weights equal $\overset{\pm}{h}_{\lambda p} = \frac{1}{2}\, p(2\lambda + p - 1)$.

There exists the natural "coupling" between the spaces of all right and left semi-infinite form

$$W^{\pm}_{\lambda} = \bigoplus_{p} W^{\pm}_{\lambda p}$$

(may-be it will be better to write the direct integral of these spaces). For the basic forms $f \in W^{+}_{\lambda}$, $g \in W^{-}_{\lambda}$ let's consider the product $f \wedge g$. If this infinite (in both directions) form coincides after the permutation with the standard form (the exterior product of all basic tensors :

$$\bigwedge_{n} f^{\lambda}_{n}\,,\; n - p - {}^{g}/_{2} \in Z)$$

then we define

$$< f|g > = (-1)^{\varepsilon},$$

where ε is the sign of the corresponding permutation. In other cases the product $< f|g > = 0$ would be equal to zero by the definition. The scalar products between basic elements define the scalar products of any elements $f \in W_{\lambda}^{+}$, $g \in W_{\lambda}^{-}$ by the linearity. It is non-trivial between subspaces $W_{\lambda,p_{\pm}}^{\pm}$ when

$$S + p_{+} + (S + p_{-}) = 1 \tag{4.7}$$

__Lemma__. ([1]) The operators L_n acting in the spaces of right and left semi-infinite forms are self-ajoint in the respect to the "coupling" which was defined above (i.e. the equalities (2.21) are fulfilled).

In the works [3,9] we have discussed in details the "regularity conditions" for the vacuum vectors. They require that such vectors $|0_{\lambda}>$, $<0_{\lambda}|$ have satisfied the equalities

$$L_n | 0_{\lambda} > = 0\, , \ n \geq g_o - 1\, ; < 0_{\lambda} | L_n = 0\, , \ n \leq - g_o + 1 \tag{4.8}$$

It was proved that they are fulfilled for the vectors $|\overset{+}{\psi}_{\lambda o}>$ and $<\overset{-}{\psi}_{\lambda o}|$.

Hence these vectors are proportional to $|0_{\lambda}>$ and $<0_{\lambda}|$:

$$|\overset{+}{\psi}_{\lambda o}> = (\overset{+}{Z}_{\lambda})^{-1} | 0_{\lambda} >\, , < \overset{-}{\psi}_{\lambda o} | = (\overset{-}{Z}_{\lambda})^{-1} < 0_{\lambda} |. \tag{4.9}$$

Again as in the previous paragraph it arises the problem of the definition of these normalizing constant.

We assume that the vacuum vectors, as in the case $g = 0$, are equal to the exterior product of all non-negative powers of the local parameters. Then

$$|0_\lambda> = 1 \wedge z_+ \wedge z_+^2 \wedge \dots \quad \text{-"in-vacuum"}$$

$$< 0_\lambda | = \dots \wedge z_-^2 \wedge z_- \wedge 1 \quad \text{-"out-vacuum"} \tag{4.10}$$

The basic tensor-fields f_n^λ are defined by the conditions (1.3) up to the factor. There are two different types of their normalization. In the case of in-normalization when we fix $\overset{+}{\varphi}_{n,\lambda} \equiv 1$

$$Z_\lambda^+ = 1\,,\ (Z_\lambda^-)^{-1} = \prod_{n \le -S(\lambda)} (\overset{-}{\varphi}_{n,\lambda}) \tag{4.11}$$

and

$$Z_\lambda = < 0_\lambda | 0_\lambda > = Z_\lambda^+ Z_\lambda^- < \overset{-}{\Psi}_{\lambda o} | \overset{+}{\Psi}_{\lambda o} > \tag{4.12}$$

(In the case of out-normalization when $\overset{-}{\varphi}_{n,\lambda} \equiv 1$ we have $Z_\lambda^- = 1$, but as it can be seen below the product $Z_\lambda^+ Z_\lambda^-$ does not depend on the choice of in - or out - normalization. Below we shall always fix the in-normalization.)

From (4.7) it follows that

$$< 0_\lambda | 0_\lambda > = 0\,, \quad \text{if } \lambda \ne {}^1/_2 \tag{4.13}$$

In the case $\lambda = 1/2$ (the only case when $2S(\lambda,g) = 1$ all the quantities depend on the representation. For the brevity we shall denote vectors $| 0_{1/2,\rho} >$, $< 0_{1/2,\rho} |$ by $|0_\rho >$ and $< 0_\rho |$ respectively. From (4.11, 4.12) it follows that

$$< 0_\rho | 0_\rho > = Z(\rho) = \prod_{n \le -1/2} (\overset{-}{\varphi}_{n,1/2}(\rho))^{-1} \tag{4.14}$$

Remark. Here and below (till the end of this paragraph) the infinite products of the form (4.11) are considered formally. Later we shall argue the problem of their regularization.

The operators $\psi_\nu, \overset{+}{\psi}_\nu, \nu - {}^1/_2 \in Z$ have the representation in the spaces $W^{\pm}_{k,1/2}$, $k \in Z$

$$\psi_\nu \to \phi_\nu \wedge \tag{4.15}$$

(the exterior multiplication of the semi-infinite form by ϕ_ν)

$$\overset{+}{\psi}_\nu \to {}^{\partial}/_{\partial\phi_{-\nu}} \cdot \tag{4.16}$$

(the differentiation of the semi-infinite form in respect to $\phi_{-\nu}$). This correspondence gives the isomorphism of $W^{\pm}_{1/2}$ and $\mathcal{H}^{\pm}$ which is consistent with the coupling between the right and left spaces.

<u>A few remarks about the ghost sector</u>.

The Polyakov-Faddeew-Popov ghost-fields in the string theory have the tensor weights -1 and 2 and are fermionic. We shall define them by

$$b(z) = \Sigma\, b_n\, d^2\, \Omega_n\ ,\ c(z) = \Sigma\, c_n\, e_n(z) \tag{4.17}$$

The coefficients b_n, c_n have the ordinary anti-commutators

$$[\, b_n, b_m]_+ = [c_n, c_m]_+ = 0\ ,\ [\, b_n, c_m] = \delta_{n,m} \tag{4.18}$$

As it was shown in [15], the definitions of the stress-energy operator of the ghost-fields and the operator of the BRST-sharge can be easily generalized for the case of the Riemann surfaces of the genus $g > 0$ with the help of the bases which were introduced in §1.

The full Fock space includes the tensor product of the "physical" and "ghost" sectors. In particular, the vacuum vector has to be the tensor product of the "physical" and "ghost" vacuum vectors. The regularity conditions of the ghost vacuum have the form (see [3,9])

$$\begin{gathered} b_n \,|\, 0_{gh} > = 0 \quad n \geq g_o\text{-}1,\ \ c_n \,|\, 0_{gh} > = 0,\ \ n < g_o\text{-}1 \\ < 0_{gh} \,|\, b_n = 0,\ \ n \leq -\, g_o\text{+}1,\ \ < 0_{gh} \,|\, c_n = 0,\ \ n > -\, g_o\text{+}1. \end{gathered} \tag{4.19}$$

Let's define the action of the operators b_n, c_n in the spaces $W^{\pm}_{\lambda=2}$ with the help of multiplication and differentiation of the semi-infinite forms

$$b_n \to f_n^{\lambda=2} \wedge \ldots,\ c_n \to \partial/\partial f_n\ \lambda = 2 \qquad (4.20)$$

This representation with the correspondence $|0_{gh}> = |0_2>$, $<0_2| = <0_{gh}|$ defines the isomorphisms between $W^{\pm}_2$ and $H^{\pm}_{\Gamma,gh}$. Hence the "coupling" between $W^{\pm}_2$ defines the "coupling" between the in - and out - ghost's Fock spaces. From (4.13) it follows that $<0_{gh}|0_{gh}> = <0_2|0_2> = 0$. The most simple non-zero expressions can be obtained only in the presence of the insertions. For example

$$g = 0,\ <0_2|c_{-1}c_0c_1|0_2> = 1$$

$$g = 1,\ <0_2|b_{1/2}c_{-4/2}|0_2>$$

$$= \prod_{n=-\infty}^{-g_0-1} (\dot{\varphi}_{n,2})^{-1} \neq 0 \qquad (4.21)$$

$$g > 1,\ <0_2|b_{-g_0+2}\ldots b_{-g_0-2}|0_2>$$

The operators b_n for $|n| \leq g_0-2$, $g > 1$ correspond to the holomorphic quadratic differentials which are the basis of the co-tangent bundle over the modular space of the genus g surfaces. That's why the square of the modulus of the value (4.21) defines the measure on the modular space. The connection of this measure with Polyakove-Belavin-Knighnik measure is under consideration now.

<u>Remarks about the states with non-zero momentum.</u>

The ground states with the momentums $p_{\pm} = (p^{\mu}_{\pm})$ at the points $P_{\pm}$ are defined by the following conditions

$$\alpha^{\mu}_n|\vec{p}^{\mu}_n> = 0,\ n > g/2,\ \alpha^{\mu}_{-g/2}|\vec{p}_+> = p^{\mu}_+|\vec{p}_+>,$$

$$<\vec{p}_-|\alpha^{\mu}_n = 0,\ n < -g/2,\ <\vec{p}_-|\alpha^{\mu}_{-g/2} = p^{\mu}_-<\vec{p}_-|. \qquad (4.22)$$

The corresponding spaces would be denoted by $H^{\pm}_{\Gamma}$. Let's introduce for the fixed real number p the spaces $\mathcal{H}^{\pm}(p)$ which are generated by the operators $\psi_\nu, \psi^{+}_{-\mu}, \nu,\mu \in Z_{\pm p+1/2}$. The generating vectors are defined by the conditions

$$\psi_\nu | p_F > = 0, \nu > p \ , \ \psi^{+}_{-\mu} | p_F > = 0 \ , \ \mu < p$$

$$< p_F | \psi_\nu = 0 \ , \nu < -p \ \ < p_F | \psi^{+}_{-\mu} = 0 \ , -\mu > -p \qquad (4.23)$$

If the vectors $| p_F >$ and $| p_F |$ have the charges p and the operators $\psi_\nu \psi^{+}_\nu$, as before, have charges ± 1 respectively then

$$\mathcal{H}^{\pm}(p) = \sum_{\kappa} \mathcal{H}^{\pm}_{\kappa}(p) \quad , \kappa\text{-}p \in Z \qquad (4.24)$$

The spaces $\mathcal{H}^{\pm}_{\kappa}(p)$ and $\mathcal{H}^{\pm}_{\kappa}(p')$ are naturally isomorphic to each other if $p\text{-}p' \in Z$. Hence, we can use the notation $\mathcal{H}^{\pm}_{\kappa}$.

Let's define the operators $\alpha_n(p)$ in the spaces $\mathcal{H}^{\pm}_{p}$

$$\alpha_{n,p}(\rho) = \sum a^{n}_{\nu,\mu} : \psi_\nu \psi^{+}_{-\mu} :_p + \alpha_{n,p} \ , \ \nu,\mu \in Z\text{-}p\text{-}1/2 \qquad (4.25)$$

where the coefficients $a^{n}_{\nu,\mu}$ again are given by the first of the formulae (3.12), and

$$\alpha_{n,p} = \frac{1}{2\pi i} \int_{C_\tau} A_n ds_p \qquad (4.26)$$

The symbol $: :_p$ means the normal ordering in respect to the vectors $|p>$, $<-p|$.

<u>Theorem 4.1</u> ([3]). The operators $\alpha_{n,p}(\rho)$ satisfy the commutator relations (3.7). Their actions on $\mathcal{H}^{\pm}_{\kappa}(p)$, $\mathcal{H}^{\pm}_{\kappa}(p')$, $p\text{-}p' \in Z$ are the same

as

$$\alpha_{n,p}(\rho) = \alpha_{n,p'}(\rho) \tag{4.27}$$

Their actions on the vectors $|p_F\rangle$, $\langle p_F|$ satisfy the conditions (4.22).

The vectors $|p_F\rangle$ and $\langle p_F|$ are annihilated by the operators L_n for $n > g_0$ and $n < -g_0$, respectively and are the eigenvectors for L_{g_0} and L_{-g_0} with the eigenvalues corresponding to the conformal weights $p^2/2$.

Corollary. The spaces $(\mathcal{H}^{\pm}_{\rho\mu})^{\otimes\mathcal{B}}$ and $H^{\pm}_{\Gamma}(\vec{p})$ are isomorphic. This isomorphism can be defined using the representations (4.25) for the set of characters $\rho_1,\ldots,\rho_{\mathcal{B}}$ and the correspondence

$$|\vec{p}\rangle \rightarrow \otimes |p^{\mu}_{\rho}\rangle , \quad \langle\vec{p}| = \otimes \langle p^{\mu}_{\rho}|$$

$$|p_\rho\rangle = Z^{+}_{p}(\rho)\,|\,p_F\rangle , \quad \langle p_\rho| = Z^{-}_{p}(\rho) \langle\, p_F| \tag{4.28}$$

The constants $Z^{\pm}_{p}(\rho)$ can be defined from the nomralization principle. We obtain (for in-normalization) that $Z^{+} \equiv 1$ and

$$Z_p(\rho) = Z^{-}_{p}(\rho) = \prod_{n\leq S+p} (\varphi^{-}_{n,1/2}(\rho))^{-1} \tag{4.29}$$

Hence, we obtain the following formula

$$A(p,\Gamma,P_{\pm}) = \frac{\langle -p_\rho|p_\rho\rangle}{0_\rho|0_\rho\rangle} = \frac{1}{Z(\rho)} \prod_{n\leq -S+p} (\varphi^{-}_{n,1/2}(\rho))^{-1}$$

(4.30)

For general p *the quantity* (4.29) *must be regularized,* as well as (4.11). For the integer p all factors in the ratio (4.30) except the finite number are cancelled. Hence, for $p > 0$-integer

$$A_\rho(\pm p,\Gamma,P_\pm) = \prod_{\nu=\pm 1/2}^{\pm p \pm 1/2} (\bar{\varphi}_{\nu,1/2}(\rho))^{\pm 1} \quad , \ p > 0 \tag{4.31}$$

From the definition of $\bar{\varphi}_{\nu,1/2}$ it follows that the quantity (4.31) depends on $P_\pm$ as the tensor of the weight $p^2/2$.

From the results of the soliton theory the formula

$$\bar{\varphi}_{\nu,1/2} = \frac{\theta[\rho]((\nu-1/2)(A(P_+)-A(P_-)))}{\theta[\rho]((\nu+1/2)(A(P_+)-A(P_-)))} E^{-2\nu}(P_+,P_-) \tag{4.32}$$

follows, where $\theta[\rho]$-theta-function with characteristic corresponding to the representation ρ ; $E(P_+,P_-)$ - Prym-form

$$E^{-2}(P_+,P_-) = \frac{\theta^2[m](A(P_+)-A(P_-))}{(\sum_i \omega_i(P_+)\theta_i[m])(\sum_i \omega_i(P_-)\theta_i[m])} \ .$$

Let's consider now the infinite product (4.14). Using the ζ-funcitonal regularization for the product of the factors $E^{-2\nu}(P_+,P_-)$ we obtain the ordinary conformal anomaly which cancels the corresponding conformal anomaly for the ghost sector (it appears after the regularization of the (4.21)). Hence, we can introduce (without loss of generality) the local co-ordinates $z_\pm$ near $P_\pm$ such that $E^2(P_+,P_-) = 1$. In the product (4.14) the numerator of each factor cancels the denominator of the next factor. That's why it naturally regularizes the infinite product (4.14) so that

$$\langle 0_\rho | 0_\rho \rangle = \prod_{\nu \le -1/2} (\bar{\varphi}_{\nu,1/2})^{-1} = \theta[\rho](0) \tag{4.33}$$

The same regularization for the quantity (4.29) gives

$$\langle -p_\rho | p_\rho \rangle = \theta[\rho] \ (p(A(P_+)-A(P_-))) \tag{4.34}$$

For integer p the ratio of (4.34) and 4.33) coincides with (4.31).

The regularization, which was proposed above, using the exact theta-functional formulae looks naturally but not the usual one. Their exist two ways for obtaining this regularization without using the exact formulae. First of all one can compute the logarithmic derivative in respect to the deformation along the modular space. In [10] these computations were done and it was proved that after the usual regularization the logarithmic derivatives of (4.14) satisfy the Ward identities.

The second way of the regularization, which is also discussed in [10], is based on the computation of the logarithmic derivatives in respect to the changing of the representation ρ . It must be specially emphasized that according to the variables u_i (which define ρ) the values $\varphi^{-}_{n,1/2}$ satisfy the equations of the hierarchy of the two-dimensional Toda lattice. It gives us the hope that the quantity of the form (4.14) can be expressed in terms of the spectral theory of the two-dimensional Schrödinger operators 16]. This possibility is under considerations now.

Appendix

Let's consider the multi-string diagram of the following form $(\Gamma, P_{\pm\alpha, p_{\pm\alpha}}$ where $p_{\pm\alpha} = \pm 1$ respectively, $\alpha = 1,\dots,\ell$, in the general position.

Lemma. For any $\lambda \neq 0,1$, $g \neq 1$ (or $g = 1$ and λ any integer) and any integer n their exists the unique tensor $f^{\lambda}_{n,\alpha}(z)$ of the weight λ with the following analytical properties : a) it is holomorphic on Γ outside the points $P_{\pm\alpha}$; b) it has the form

$$f^{\lambda}_{n,\alpha} = \varphi^{\pm\beta}_{n,\lambda,\alpha} z_{\pm\beta}^{\pm n - S_{\pm\beta}(\lambda,g)}(1+O(z_{\pm\beta}))(dz_{\pm\beta})^{\lambda} \tag{A.1}$$

near the points $P_{\pm\beta}$, $\alpha,\beta = 1,\dots,\ell$. Hear $z_{\pm\beta} = z_{\pm\beta}(Q)$ are local co-ordinates in neighbourhoods of the points $P_{\pm\beta}$; $S_{\pm\beta}(\lambda,g)$ are an arbitrary fixed set of integers such that

$$\sum_{\beta} S_{\beta}(\lambda,g)+S_{-\beta}(\lambda,g) = g-2\lambda(g-1)+\ell-1 \tag{A.2}$$

One can choose them in such a way that

$$S_{\pm\beta}(\lambda,g) + S_{\pm\beta}(1-\lambda,g) = -1 \tag{A.3}$$

In this case as it follows from (A.1) we have

$$\frac{1}{2\pi i}\int_{C_\tau} f^{\lambda}_{n,\alpha}\, f^{1-\lambda}_{m,\beta} = \delta_{\alpha\beta}\cdot \delta_{n,-m} \tag{A.4}$$

Theorem. Let C_τ be non-singular, then for any smooth tensor $f^{\lambda}(\sigma)$ on C_τ the expansion (fourier-type)

$$f^{\lambda}(\sigma) = \sum_{n\in Z}\sum_{\alpha=1}^{\ell} f^{\lambda}_{n,\alpha}(\sigma)\cdot \frac{1}{2\pi i}\int_{C_\tau} f^{\lambda}(\sigma)\, f^{1-\lambda}_{-n,\alpha}(\sigma')))$$

is valid. The same expansion is valid for the tensors $f^{\lambda}(z)$ which is holomorphic in the Riemann annulus C_{τ_1,τ_2} (Laurent-type).

Recall that here the contours C_{τ} are defined with the help of the differential dk corresponding to the diagram $(\Gamma, P_{\pm\alpha}, p_{\pm\alpha} = \pm 1)$.

For the exceptional cases the definition of $f^{\lambda}_{n,\alpha}$ is given by the conditions (A.1) for all n except the finite number. For finite number of n the definition $f^{\lambda}_{n,\alpha}$ has to be slightly changed. It can be done so that the statement of theorem would be valid.

As for one-string diagram the bases $f^{\lambda}_{n,\alpha}$ satisfy the important properties of the almost-graded structure in respect to the action of the basic vector-field $e_{n,\alpha} = f^{-1}_{n,\alpha}$. The only difference is that now the number N (which is introduced in the definition of the N-graded structure) is linear, depending on λ, g and ℓ . The constant term in his representation depends on the choice of constant $S_{\pm\beta}(\lambda, g)$. But N does not depend on n,m .

Again, as in the case of one-string diagram, we obtain the almost-graded algebras of functions and vector-fields which are holomorphic outside the punctures $P_{\pm\alpha}$.

These algebras have the unique central extensions - the analogues of Heizenberg and Virasoro algebras.

References

1. Krichever I.M., Novikov S.P., Virasoro-type algebras, Riemann surfaces and structures of the soliton theory, functional analys i pril., 1987, V.21, N2, 46-63.

2. Krichever I.M., Novikov S.P., Virasoro-type algebras, Riemann surfaces and string in Minkovsky space, Funct. analys i priloz., 1987, v.21, N4, 47-61.

3. Krichever I.M., Novikov S.P., Virasoro-type algebras, pseudo-tensor of energy-momentum and operator expansions on the Riemann surfaces, Funct. analys i pril., 1989, v.23, N1.

4. Mandelstam, Dual resonance models, Physics reports, 1974, 13C, 259.

5. Venetziano S., An introduction to dual models, Physics reports, 1974, 9C, 199.

6. Polyakov A.M., Phys. Lett., 1981, 103B, 207, 211.

7. Friedan D., Introduction to Polyakov's string theory, Elsevier (Amsterdam), 1984.

8. Belavin A., Knizhnik V. Phys. Lett., B 168, 1986, 201.

9. I.M. Krichever, S.P. Novikov, Virasoro-Gel'fand-Fuks-type algebras, Riemann surfaces, operator's theory of closed string. Journal of Geometry and Physics, 1989 (to appear).

10. Krichever I.M., Virasoro-type algebras and Ward identities, Funct. analiz i pril., 1989, V.23 (to appear).

11. Bonora L., Matone M., Rinaldi M., preprint SISSA 119/88/EP September 1988.

12. Bonora L., Rinaldi M., Russo J., Wu K., The Sugavara construction on genus g Riemann surfaces, Phys. Lett. B, 1988, v.208, N3,4, 440.

13. Eguchi T., Ooguri H., Conformal and current algebras on general Riemann surface, Nuclear Phys. B. 282, 1987, 308.

14. Feigin B.L., Fuks D.B., Func. analys i pril., 1982, v.16, N2, 47-63.

15. Bonora L., Bregola M., Cotlo-Ruminisino, Martellini M., Virasoro-type algebras and BRST operators on Riemann surfaces, Phys. Lett. B 205 (1988), 53.

16. Krichever I.M., Spectral theory of two-dimensional periodic operators and its applications, Uspechi Mat. Nauk, 1989, v.44, N2, 121-183.

Criticality, Catastrophes, and Compactifications

Emil J. Martinec[1]

Ecole Normale Superieur and Univ. Pierre et Marie Curie
Paris Cedex 05 France

and

Enrico Fermi Institute and Department of Physics
5640 S. Ellis Ave., Chicago IL 60637 USA

——*Dedicated to the memory of V. G. Knizhnik*——

1. Motivations

Our understanding of string theory will be greatly advanced when we find a heuristic scheme which frames its fundamental concepts. An important clue in this regard is the observation by a number of groups[1] that the classical string equation of motion is the condition of conformal invariance of the 2d quantum field theory (qft) describing string propagation in a static background. Conformal Ward identities guarantee the unitarity of the string scattering amplitudes. The weak coupling expansion – string propagation in large smooth spacetimes – yields the Einstein equations coupled to matter. It was stressed particularly by Friedan and Shenker and by Lovelace that conformal symmetry is the key property, so that one has the right to call any combination of conformal field theories a 'spacetime'. Thus the focus shifted to understanding conformal field theory, leaving a geometrical spacetime interpretation aside for the moment.

[1] Work supported in part by DOE grant DE-AC02-80ER10587, an NSF Presidential Young Investigator Award, the Alfred P. Sloan Foundation, and CNRS.

A general framework for string dynamics along these lines was proposed by T. Banks and the author[2]. The nonlinear sigma model (NLSM) suggests one consider the space of metrics and matter fields as a configuration space of string backgrounds. Generalizing to the neighborhood of an arbitrary renormalization group (RG) fixed point suggests that the space of couplings g to scaling fields play this role in general, the scaling fields being simply a choice of basis for background perturbations that diagonalizes the spacetime 'Laplacian', which is the scaling operator[3]. Regarding the *space of 2d qfts* as a configuration space we have a framework in which it is possible that nonperturbative dynamics, exploring the whole configuration space, could choose among the wealth of critical points describing different string vacua.

Thus it behooves us to study (and to be honest, define) this space of 2d qfts. According to Zamolodchikov[4], the action principal which governs dynamics is the Schwinger term $c(g)$ in the algebra of stress tensors. Thus semiclassically we may restrict ourselves to the space of small c qft's as a sort of model in which to explore structure, symmetry, and dynamics[4][5][6]. Among other things one finds that the phase structure in the simplest arena $c < 1$ can be characterized by a Landau-Ginsburg (LG) effective potential model, especially when considering topological properties like the index $\mathrm{tr}(-)^F$ in the supersymmetric case[6]. In particular there are only a finite number of relevant operators about any fixed point, and the space of theories is effectively finite dimensional and amenable to study.

In the above context we are moved to ask, what kind of spacetime geometry is described by these models? For large radius/weak coupling the spectrum of scaling fields reproduces the eigenfunctions of the Laplacian. Equivalently the weak coupling measure is concentrated on pointlike strings, so we reproduce the algebra of functions on a manifold. As we move towards strong coupling, this algebra becomes stringy in a way that bears a strong resemblance to the structure of quantum groups[7][8]. The microscopic 2d dynamics of the discrete series of $c < 1$ models is a lattice spin system[9][10][11] which renormalizes to a LG-type effective scalar field theory. The field in the microscopic theory is constrained to take a discrete set of values; macroscopically, renormalization spreads out the measure; the sum of delta functions smooths out to a polynomial effective potential. The measure is merely peaked at the potential minima, with the amount of peaking governed by the coupling. Quantum effects 'smear out' the spacetime geometry. Simple examples consist of dynamics on a set of heights labelled by an A_n dynkin diagram, where the effective potential is simply $V = \phi^{2n-2}$, and the order parameters are the powers of the field ϕ^p,

$p = 0, ..., 2n - 4$ of dimensions approximately $\frac{p^2}{4n}$. That is, the order parameters are precisely the relevant operators. Their algebra is just that of Wick contractions in structure (or equivalently spherical harmonics Y_{l0}):

$$\phi^p \cdot \phi^q \sim \sum_{r=|p-q|}^{p+q} \phi^r .$$

In the limit of large n, we might expect the theory to become roughly classical (in a sense similar to that of [8]); the quantum fields would be the algebra of eigenfunctions of the Laplacian on some one-dimensional manifold (because c accumulates at 1), with quadratic spectrum. The microscopic theory being a map from a discrete world sheet to a discrete spacetime consisting of a finite chain of points, we might expect that the continuum smearing turns the dynamics into some kind of string theory in a box with walls. Indeed, the potential wells of the effective potential become numerous and dense in some interval of size $\sim n\sqrt{\alpha'}$ (since there are $O(n)$ relevant operators). Spacetime is effectively compact because the probability measure is very small at large field values, even though the configuration space is noncompact. The models based on D_n diagrams are then qft on the $\phi \rightarrow -\phi$ orbifold of the interval. The renormalized kinetic term $f(\phi)(\partial\phi)^2$ determines the metric on the interval; of course in strong coupling it is not really legitimate to separate $f(\phi)$ from $(\partial\phi)^2$ due to composite operator renormalizations. Spacetime is continuous in the sense that we have at our disposal operators to probe at any point along this 'continuous' interval down to a resolution of order $\sqrt{\alpha'}$. I hesitate to use the word continuous because it really has no operational significance in the present context.

Two dimensional renormalization washes out features of the spacetime background on scales smaller than the inverse string tension $\sqrt{\alpha'}$. To see this, let us compare a spacetime which is a d-dimensional lattice, lattice spacing a, to one which is a continuum. Strings moving on the former can be described by a 2d field theory with potential $V = g\prod_{i=1}^{d}(1 - cos(2\pi X_i/a))$. V is an external potential from the point of view of string theory, forcing the field to particular points on the spacetime manifold. When g is very large the energetics dictates that the field be restricted to the lattice points. However, the dimension of the operator V is roughly α'/a^2, so when a is smaller than the string size the potential is an irrelevant operator and the continuum dynamics is described by free field theory. The classical background geometry is insensitive to perturbations whose wavelength is shorter than the typical size of a string. One cannot distinguish between discrete spacetime and continuous spacetime, or more generally probe any features of the geometry on scales smaller than the string scale. On the other hand, the dynamics is unstable to the addition of perturbations of longer wavelength; we gain information by studying the bulk behavior and correlations of relevant operators (*i.e.* the order parameters of the statistical system). Roughly speaking, perturbation by relevant operators causes flow to the trivial fixed point (unless we engage in fine tuning), isolating a particular 'point' in field space – one of the minima of the potential and the small fluctuations about it – by application of an external force. Further evidence is provided by the spin systems on affine Dynkin diagrams[11][12], whose continuum limits are $c = 1$ conformal field theories. Here, the $\hat{A}_n$ Dynkin diagrams are quantum field theory on Z_n, discrete versions of qft on a circle. The order parameters are Z_n Fourier transforms of the position on the ring, which renormalize to plane waves in the continuum. There are only as many order parameters, only as many 'points' to examine, as the size of the circle in units of $\sqrt{\alpha'}$. The $\hat{D}_n$ models, which are discrete versions of qft on the Z_2 orbifold of the circle, also fit the pattern.

If one accepts the premise that Landau theory is capable of describing some sort of generalized, 'stringy' or 'quantum' spacetime geometry (stringy because for small n in the above example the geometry is not very smooth – geometry on a finite number of fuzzy or smeared out points), then in turn it becomes an interesting question to consider what class of geometries are so described. Conversely, one would like to know the extent to which the vicinity of a generic 2d fixed point is well-described by Landau theory. Part of our interest in [6] was to develop an approximate language to characterize the neighborhood of fixed points. I originally hoped that there would be a LG description for any 2d fixed point, however it has become clear that such a description generally only works well in N=2 supersymmetric models, for reasons I will try to explain below. As a natural part of the program outlined above, the LG analysis for N=2 supersymmetric 2d qfts was considered in [6]. We noted particularly that

(i) *Conjectured nonrenormalization theorems implied that the superpotential W (almost) completely characterizes the fixed point because (a) the classical potential determines the form of the full quantum effective potential, and (b) the kinetic terms are irrelevant perturbations and therefore determined at the fixed point by the superpotential (conversely the location of the fixed point is governed by the wavefunction renormalization since the potential is marginal at the fixed point). In particular naively marginal perturbations of W*

are exactly marginal and therefore constitute moduli of the conformal field theory at the fixed point. Moreover, nonrenormalization implies additivity of scaling dimensions, so the algebra of chiral fields behaves like an algebra of functions.
(ii) *Gepner[13] conjectured the equivalence of an orbifold of tensor products of N=2 discrete series with integer c on the one hand, and the NLSM on algebraic varieties with vanishing first Chern class on the other. In the handful of examples where the corresponding algebraic variety was then known[13],* $W = 0$ *considered as an equation in projective space was the defining equation of the variety. The marginal perturbations of* W *are from (i) the algebraic moduli of the variety.*

Thus one was led to believe that N=2 supersymmetric order parameters could be regarded as *algebraic variables* describing what one might call 'quantum algebraic geometry'.

Point (i) suggests that a large subset of N=2 fixed points can be characterized by elementary functional analysis. Point (ii) seems at variance with the kind of geometrical picture of the LG theories I have been describing. We will see below that both are valid, being spacetimes belonging to different universality classes characterized by the same superpotential. Basically the naive LG theory corresponds to qft on a finite patch as above, and is related to the description of algebraic singularities (catastrophes). To obtain the NLSM of (ii) requires an orbifold projection; I will show that the orbifold theory describes another universality class of theories with superpotential W.

Following these two observations, it remained to classify effective potentials and to find a uniform description of the algebraic varieties corresponding to the tensor product models and their generalizations. These two tasks were completed recently by the author[14], as well as by Vafa and Warner[15] regarding the classification problem and Greene, Vafa, and Warner[16] regarding the second issue (these authors offer in addition an explanation of (ii)). In the present article I will present the details of the results announced in [14]. Section 2 describes N=2 scalar field Lagrangians and the structure of the RG flows and fixed points they determine. We provide a rigorous proof of formulas in [14][15] determining the central charge in terms of the effective potential W. Section 3 examines some common features of the spectrum of cft's at low c. Section 4 presents an introduction to singularity (catastrophe) theory; the discussion is rather technical and can probably be skipped on a first reading. The precise connection between LG models and singularity theory remains somewhat mysterious. Not all singularities correspond to conformal field theories; not all N=2 cft's are obviously described by LG effective lagrangians. Section 5 lays out the connection between N=2 LG theory and projective algebraic varieties. We improve on the GVW explanation of why an orbifold of the naive LG theory describes the NLSM. An exhaustive analysis of K3 constructions is carried out (for threefolds the investigator becomes exhausted first), not because of immense relevance for physics but to illustrate some basic features of the LG–algebraic geometry connection. A major point will be that the twisted states in Gepner's orbifold projection come directly from the resolution of singularities of the corresponding algebraic variety. The theme throughout will be that the LG theory encodes the topological and algebraic features of the qft and corresponding spacetime geometry in much the same way that the defining equations which embed an algebraic variety in projective space carry the data about its topological and algebraic structure. While the presentation will throughout stress the structure of Landau-Ginsburg models, much of the analysis is completely general and applies to any N=2 theory.

2. N=2 effective Lagrangians[6][14][15]

A complex supersymmetric order parameter is a function on N=2 superspace, parametrised by $z, \theta, \theta^*; \bar{z}, \bar{\theta}, \bar{\theta}^*$. The natural covariant derivatives in a local coordinate patch are

$$\begin{aligned} D^* &= \frac{\partial}{\partial \theta^*} + \theta \partial_z \\ D &= \frac{\partial}{\partial \theta} + \theta^* \partial_z \end{aligned} \qquad D^2 = D^{*2} = 0\ , \qquad \{D, D^*\} = 2\partial_z\ . \tag{2.1}$$

The representation theory is quite analogous to N=1 in 4d. Besides vector superfields depending on all the coordinates, the simplest superfields are

$$\begin{aligned} D^* X &= \overline{D}^* X = 0 && \text{chiral} \quad i.e.\ \ X = X(z + \theta\theta^*, \theta; \overline{z + \theta\theta^*}, \bar{\theta}) \\ D X^* &= \overline{D} X^* = 0 && \text{antichiral} \\ D^* \phi &= \overline{D} \phi = 0 && \text{twisted chiral} \\ D \phi^* &= \overline{D}^* \phi^* = 0 && \text{twisted antichiral}\ . \end{aligned} \tag{2.2}$$

The component expansion of the (anti)chiral field $X(X^*)$ is

$$\begin{aligned} X &= x(z, \bar{z}) + \theta \psi(z, \bar{z}) + \bar{\theta} \bar{\psi}(z, \bar{z}) + \theta \bar{\theta} F(z, \bar{z}) \\ X^* &= x^* + \theta^* \psi^* + \bar{\theta}^* \bar{\psi}^* + \theta^* \bar{\theta}^* F^* \end{aligned}$$

up to total derivatives. Note that it is important to distinguish field conjugation $X \leftrightarrow X^*$, *i.e.* hermitian conjugation in field space, from complex conjugation in 2d parameter space

$z \leftrightarrow \bar{z}$. I will use the word 'holomorphic' to denote 2d complex analytic objects, whereas 'chiral' will denote analyticity with respect to the complex structure in field space.

The possible terms in the 2d Lagrangian are the nonchiral kinetic terms (sometimes called D-terms)

$$\int d^2z d^2\theta d^2\theta^* \, K(X_i, X_i^*; \phi_a, \phi_a^*) \tag{2.3}$$

(with $d^2\theta = d\theta d\bar{\theta}$) which in weak coupling can be thought of as the Kahler potential of the spacetime manifold, *e.g.* for flat spacetime

$$\int XX^* = \int [\partial x \bar{\partial} x^* + h.c. + \psi \bar{\partial} \psi^* + \bar{\psi} \partial \bar{\psi}^* + FF^*] \; ; \tag{2.4}$$

free N=2 field theory consists of a complex boson and complex fermion, and we will normalize the central charge $c = 1$ for this system – one third of the usual normalization. In addition we may allow chiral superpotential terms (sometimes called (twisted) F-terms)

$$\int d^2z (d^2\theta \, W(X_i) + h.c.) + \int d^2z (d\theta d\bar{\theta}^* U(\phi_a) + h.c.) \, . \tag{2.5}$$

We ignore dynamical gauge fields since they are massive in 2d and cannot appear in the effective Lagrangian. The compatibility of this general Lagrangian with N=2 supersymmetry turns out to require *two* commuting complex structures in field space[17], which implies the existence of an additional U(1) current algebra structure on the world sheet. Therefore the class of Lagrangians depending only on the chiral fields,

$$S = \int d^2z d^2\theta d^2\theta^* \, K(X_i, X_i^*) + \int d^2z (d^2\theta \, W(X_i) + h.c.) \, , \tag{2.6}$$

is actually more general since fewer symmetry requirements are imposed on the spacetime geometry. Although they will not usually appear in the Lagrangian, twisted chiral fields can arise as twist fields, in particular in the orbifold constructions of Gepner[13]. They are thus like magnetic operators in the Gaussian model; they occur in the theory as dual fields and therefore cannot appear in the Lagrangian except under certain conditions. This is because the product $\phi \cdot X$ has the holomorphic structure of a chiral field but the antiholomorphic structure of a nonchiral field. Since these have entirely different properties under renormalization it is unlikely that a theory with both in the action could be consistent unless either (a) the operator product coefficients involving chiral and twisted chiral fields vanish, or (b) these fields operator product to a holomorphic *i.e.* $h = (n, 0)$ field. Often it is possible to transform twisted chiral fields into chiral fields by a so-called duality transformation[17]; we will find nontrivial examples of such duality transformations in some K3 constructions below.

As was observed in [6], the full superspace measure $d^2z d^2\theta d^2\theta^*$ is scale invariant; since any scaling operator in a unitary quantum field theory has positive dimension, the kinetic terms are irrelevant. However the chiral measure $d^2z d^2\theta$ has dimension $-\frac{1}{2}$, so any chiral superfield of dimension $h < \frac{1}{2}$ is a relevant operator, and $h = \frac{1}{2}$ fields are marginal.[2] Thus the kinetic term is stable under renormalization, and as one flows to the infrared the theory will reach a fixed point (possibly trivial) where the couplings in the kinetic terms are fixed in terms of those in the potential[6][14][15]. However, a priori the space of kinetic terms could divide into several basins of attraction of the flows for a given W, in which case there would be a discrete number of possible fixed points for a given superpotential labelled by the values of the kinetic couplings. For the $c < 1$ models one may eliminate this possibility, since the models are exactly soluble and unique for a given superpotential, but generically we will have this discrete ambiguity. We will call the universality class of the quadratic kinetic term (2.4) the naive LG model. Thus, specification of the superpotential W specifies a discrete set of universality classes, and the fixed point values of the kinetic term couplings are then determined in terms of those in the superpotential. This special feature of N=2 in 2d makes it feasible to classify a vast array of N=2 fixed points by classifying the allowed superpotentials[14][15] – modulo field redefinitions, since these can have no physical effect.

One might ask whether effective Lagrangian (mean field) analysis is generally useful or even applicable at a generic 2d fixed point. After all, the N=0,1 A_n discrete series models appear to be described rather well by effective Lagrangians[10][4][5][6]. However, starting with a generic classical potential, the renormalizations of the various terms X, X^2, and so on are independent; by the time one reaches the vicinity of a fixed point, the field content, the structure of the operator algebra, and the form of the potential may have very little to do with the initial classical theory; to study the fluctuations near the fixed point one should probably introduce a separate source for each order parameter. The correlations of X^2 might bear no relation to those of X. Long range fluctuations, very strong in 2d, might obstruct the use of a simple mean field Lagrangian for the description of the dynamics,

[2] The absence of a superpotential does not mean that there are no points on the geometry in the above sense, but rather that such scaling operators are not the highest components of superfields and break supersymmetry when added to the action.

or even to reliably describe the phase structure. The effective action is typically quite nonlocal. So although universality classes might be characterized by various potentials, one doesn't know which ones go with which fixed points, when two potentials are in the same universality class, which are the relevant perturbations, etc., and the program is probably less useful. This is not to say that the models are not in the universality class of some scalar field theory; just that the fixed point is so far from the semiclassical, weak coupling region that renormalization washes away all the information content in such a description. The effective Lagrangian must include all possible terms consistent with the low-energy symmetries that we impose. Generally renormalization induces all of these with strength of order one, and we learn very little about fixed point physics by writing down a general Lagrangian. On the other hand, by imposing some low-energy symmetries we may learn something (as we will see, by imposing N=2 supersymmetry we learn a lot). In general the central charge c and the spectrum of scaling dimensions h are not determined by the effective potential, but they are in N=2. One good but not infallible guide to the applicability of Landau theory is the degree to which the theory is weakly coupled. In weak coupling the effective Lagrangian is well approximated by the semiclassical theory. A rough measure of the strength of coupling is the difference between the value of the fixed point central charge and its free field value, the number of LG fields. The A_n models accumulate at $c = 1$, and are well described by a single LG field (basically because there are few choices for the effective Lagranigian of the simplest n-fold multicritical point), whereas the D and E theories must be described by a two-variable potential[11][6], so that c is very far from its free field value. Indeed it seems rather difficult to come up with a coherent LG description of these latter models. The models on the lower end of the A_n series are also strongly coupled, but there one may reliably identify the fundamental field and its powers because the possible terms in the effective Lagrangian are extremely limited. Theories for which one may write down a class of effective Lagrangians include the Z_3 coset models $\frac{A_2(k)\times A_2(1)}{A_2(k+1)}$ and $\frac{A_1(k)\times A_1(4)}{A_1(k+4)}$ (spin 4/3 parafermion models), which belong to the class of effective potentials

$$V(\varphi,\varphi^*) = f(\varphi\varphi^*, \varphi^3 + \varphi^{*3}) \,. \tag{2.7}$$

For instance the 3-state Potts model is $V = g(\varphi^3 + \varphi^{*3}) + (\varphi\varphi^*)^2$, and the tricritical 3-state Potts model is $V = g(\varphi^3 + \varphi^{*3}) + (\varphi\varphi^*)^3$. Note that these series both accumulate at $c = 2$, the value for a complex scalar field. Also the Z_n parafermion models have potentials of the same form as (2.7), except with the power in the second argument n instead of 3. It is difficult in general to determine the precise form of the potential from the operator algebra because of the manifold possibilities.

Returning to our development of N=2 models, in a given universality class the only renormalization of the coefficients in the superpotential is through wavefunction renormalization[18] [19] – the coefficients in the kinetic D-terms. The original graphical argument of [18] has a refined version[19] that holds at the nonperturbative level given mild assumptions. This means in particular that the resummation of perturbation theory necessary to make the superpotential marginal (it is dimensionless in the loop expansion) does not affect our conclusions, nor does the massless limit provided the fields in the effective action are the same as in the microscopic theory. Consequently, once we set the superpotential at the classical level it is essentially not altered by quantum effects; the field X^n scales like n powers of X, and quasiclassical methods like mean field theory are valid. One need only introduce a source for X to probe the effective dynamics, rather than an independent source for each order parameter. This is why we might expect an effective Lagrangian approach to be useful: the fixed point is not only characterized by the effective potential, the effective potential is determined *classically*. Here we are of course assuming that the wavefunction renormalization does not result in a drastic change of field variables, *e.g.* $X \rightarrow \sin(X)$; we assume that we are in the universality class of the fixed point we are trying to describe, with for instance the same supersymmetry index and other characteristic features.

At an N=2 fixed point, the algebra of superderivatives (2.1) embeds in the larger superconformal algebra generated by the stress tensor

$$\mathcal{T}(z,\theta,\theta^*) = J(z) + \theta G(z) + \theta^* G^*(z) + \theta\theta^* T(z) \ .$$

The coordinate derivatives correspond to the modes

$$D = G_{-\frac{1}{2}} \ , \quad D^* = G^*_{-\frac{1}{2}} \ , \quad \partial = L_{-1} \ . \tag{2.8}$$

The holomorphic U(1) current J rotates the two supercurrents G, G^* by opposite phases; the zero mode generates what is sometimes called an R-symmetry. This symmetry only appears at the fixed point; otherwise it is anomalous (a discrete subgroup may be preserved). The chiral fields X_i have charges q_i under this symmetry, normalized so that the coordinate variables $\theta(\theta^*)$ carry charge +1(−1) (cf. (2.8)). The fixed point Lagrangian must be invariant under this R-symmetry, so we conclude from $q(\int d^2\theta) = 1$ that

$$e^{i\alpha J_0} W(X_i) e^{-i\alpha J_0} = W(e^{i\alpha q_i} X_i) = e^{i\alpha} W(X_i) \ , \tag{2.9}$$

i.e. the superpotential is homogeneous if the X_i are each counted with weight q_i. Because of the identification (2.9), in the fixed point theory chirality (2.2) is a null vector condition $G^*_{-\frac{1}{2}}|X\rangle = 0$. Using the commutation relations of the algebra, we find

$$0 = \langle X|\, G_{\frac{1}{2}} G^*_{-\frac{1}{2}}\, |X\rangle = \langle X|\,(2L_0 - J_0)\,|X\rangle \ , \tag{2.10}$$

so that the conformal weight is determined by the charge, $2h = q$. This is the simplest of the null vector conditions in N=2 representation theory[20]. Note that (2.10) is a relation in the global superconformal algebra, and holds in any dimension[3]. In particular in three dimensions one can have superpotential terms up to X^4, such that the chiral fields have anomalous dimensions determined by their R-charge. Unfortunately, the nonchiral integration measure is no longer scale invariant, however unless the scale dimension of the kinetic term can be smaller than that of free field theory the fixed point will be determined by the superpotential. A classification of fixed points analogous to the 2d one should hold.

There is yet more information that we can derive from the superpotential. Consider the unitarity plot in fig. 1.

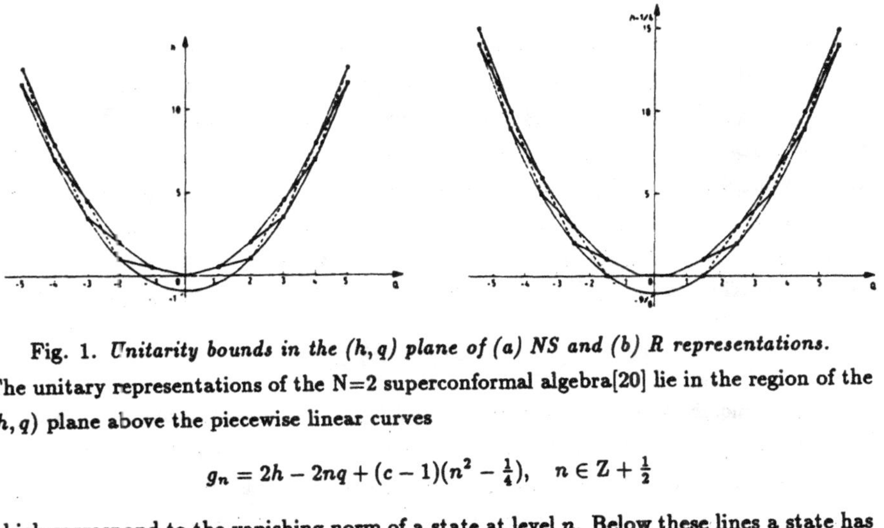

Fig. 1. *Unitarity bounds in the (h,q) plane of (a) NS and (b) R representations.* The unitary representations of the N=2 superconformal algebra[20] lie in the region of the (h,q) plane above the piecewise linear curves

$$g_n = 2h - 2nq + (c-1)(n^2 - \tfrac{1}{4}), \quad n \in \mathbb{Z} + \tfrac{1}{2}$$

which correspond to the vanishing norm of a state at level n. Below these lines a state has negative norm (with the possible exception of states on other g_m, $m < n$ curves), and the

[3] I would like to thank V. Pasquier and H. Saleur for asking how much of the analysis extends to higher dimensional theories.

theory is not unitary. These straight line segments terminate at their intersection with the curve

$$f_{1,2} = 2(c-1)h - q^2 + c$$

representing a null vector at level one in the space spanned by $L_{-1}|h,q\rangle$, $G^*_{-\frac{1}{2}}G_{-\frac{1}{2}}|h,q\rangle$, and $J_{-1}|h,q\rangle$. The null lines $g_{\pm1/2}$ passing through the origin are the null curves for (anti)chiral fields. They terminate at $2h = q = c$. Thus if there is a representation with this value of q in every model we may determine c by identifying the highest weight chiral field of maximal charge. The existence of this representation in every model is a consequence of spectral flow[21]. That is, by acting with the charge shifting operator $e^{ia\hat{q}}$ ($\hat{q}$ is conjugate to J_0) there is a continuous family of N=2 representations labelled by the relative U(1) charge a, with states $e^{ia\hat{q}}|\Phi\rangle$ and operators $e^{ia\hat{q}}\mathcal{O}e^{-ia\hat{q}}$. Clearly all expectation values in these theories are the same. The Laurent expansions of the supercurrents conjugated by $e^{ia\hat{q}}$ are given by

$$G(z) = \sum_{n\in\mathbf{Z}+\frac{1}{2}} G_{n+a}z^{-n-a-3/2}\ , \quad G^*(z) = \sum_{n\in\mathbf{Z}+\frac{1}{2}} G_{n-a}z^{-n+a-3/2}\ .$$

The NS sector is $a = 0$, the R sector $a = \pm\frac{1}{2}$ depending on whether one wants G_0 or G_0^* to annihilate the ground state. Spectral flow by integer amounts clearly leaves the theory unchanged, but may mix up the various representations; for instance, the vacuum representation is carried into the highest weight representation built on $|a\rangle = e^{ia\hat{q}}|0\rangle$. This state has dimension $h = a^2 \cdot \frac{c}{2}$; we see that the existence of the highest charge chiral field ϵ is guaranteed by spectral flow from the identity representation to $a = 1$, and so exists in every model. We are done then if we can identify the chiral field of highest charge. This is easily done. The equations of motion following from the effective Lagrangian (2.6) are

$$\frac{\partial W}{\partial X_i} = D^*\bar{D}^*(\partial K/\partial X_i)\ ,$$

so that any monomial in chiral fields that is a multiple of the equations of motion cannot be highest weight – it will be a $G^*_{-\frac{1}{2}}$ descendant of nonchiral and antichiral fields. Simple enumeration of the list of highest weight fields then completes the task. Mathematically, we denote the set of monomials by $\mathbf{C}[X_i]$; the set of chiral fields is this ring quotiented by the ideal $\{\vec{\nabla}W\}$ generated by the equations of motion

$$\mathcal{R} = \frac{\mathbf{C}[X_i]}{\{\vec{\nabla}W\}}\ .$$

From a result of [22] (p.101) we deduce that

$$c = \sum (1 - 2q_i) , \tag{2.11}$$

a formula given in [14][15]. The essence of the proof[22] computes the characteristic polynomial $\chi_{\nabla W}(t) = \sum_m a_m t^m$ (a_m is the number of fields of degree m) of the map $\nabla W : \mathbf{C}^r \to \mathbf{C}^r$ for the ring of chiral fields in two steps.[4] First, note that if W is equal to a sum of powers one easily computes $\chi = \prod \frac{t^{d-n_i}-1}{t^{n_i}-1}$, with $q_i = \frac{n_i}{d}$ and d the least common denominator of the q_i. Second, one can convert any weighted homogeneous function into a strictly homogeneous one via the map $X_i = Y_i^{n_i}$, which has characteristic function $\prod \frac{t^{n_i}-1}{t-1}$; then using $\chi_{f\circ g} = \chi_f \chi_g$ we have

$$\chi_{\nabla W} = \prod \frac{t^{d-n_i} - 1}{t^{n_i} - 1} \tag{2.12}$$

because we may always deform a strictly homogeneous map to a sum of powers. The highest term has degree $\sum (d - 2n_i)$ where W has degree d. The formula (2.12) suggests we scan the space of field theories by choosing sets of weights q_i; however not all weights correspond to quasihomogeneous functions. See [23] for a partial list of conditions for this to be the case.

Another important property of the chiral fields is evident from fig. 1; they are all related to Ramond ground states via spectral flow. Thus, computing the Witten index[24] $tr(-)^F$ is equivalent to counting highest weight chiral fields. This indicates the topological nature of the chiral fields, especially with respect to their spacetime interpretation, since the index is the Euler character of spacetime. The index is simply the order μ of the ring $\mathcal{R}$; this is calculated from χ in the limit $t \to 1$: $tr(-)^F = \mu = \prod (q_i^{-1} - 1)$. The characteristic polynomial χ is essentially the U(1) character-valued index.

A way to picture $\mathcal{R}$ is to construct its *Newton diagram* (that's Isaac Newton!). Let $\mathcal{K} = \{\mathbf{k} | X^{\mathbf{k}} \equiv X_1^{k_1} \cdots X_r^{k_r} \in \mathcal{R}\} \subset \mathbf{Z}_+^r$ be the set of monomials in $\mathcal{R}$. One easily computes $\mathcal{K}$ graphically from W[22]. The terms in ∇W demark a set of poyhedral faces in $\mathbf{Z}_+^r$. Translating these faces (the relations $\nabla W = 0$) by any vector in $\mathbf{Z}_+^r$ multiplies the relations by all possible monomials and fills out the excluded region complementary to $\mathcal{K}$. The region

[4] The map $\vec{X} \to \vec{\nabla} W$ is simply the Nicolai map in supersymmetric quantum mechanics, so obviously the degree of this map is the index. The refined information here is the spectrum of charges of the ground states (which could in principal be obtained from a character valued index in supersymmetric QM).

in $\mathbf{R}^r_+$ occupied by $\mathcal{K}$ is bounded by some polyhedron $\Gamma(W)$ called the Newton diagram of W. The index is merely the number of integral points enclosed by $\Gamma(W)$. There is in fact much more topological data embodied in the Newton polyhedron[25] important in the resolution of singularities[26]; most of this data can be computed from the polyhedron by combinatorial methods.

By way of illustration, let us enumerate the set of LG fixed points that have $c < 1$ (cf. Arnold [22]). By our relation between c and the maximal dimension chiral field, all chiral fields must be relevant operators. First, if there is only one chiral field, the superpotential is $W = X^n, n \geq 2$. The field X has charge $1/n$, and the ring of chiral highest weights consists of $1, X, X^2, ..., X^{n-2}$, stopping there because $X^{n-1} = D^*\bar{D}^*(X^* + ...)$ is a descendant of X^*. The central charge is then $\frac{n-2}{n}$. Note that $n = 2$ is the universality class of the massive free superfield, the trivial fixed point with $c = 0$. Thus adding any number of such massive fields doesn't affect the fixed point structure. Now consider two-field potentials $W(X,Y)$. If the superpotential is divisible by X^2 or Y^2, there are flat directions in the ordinary potential, and the ring $\mathcal{R}$ is not finite. We have assumed implicitly in the above that this did not occur, but there is nothing wrong with this possibility. Consider the analogous situation in the Ashkin-Teller model, the Z_2 orbifold line of the Gaussian model. This has the effective potential

$$V = x^4 + y^4 + \alpha x^2 y^2 \ .$$

At $\alpha = 0$ we have the tensor product of two Ising models, and the perturbation is the product of the mass terms $\bar{\psi}\psi$ in the two theories (of course we have no way of knowing in the effective theory that this operator is exactly marginal, but we take this as given). Generically the potential has only the D_4 symmetry of the Askin-Teller line; however, when $\alpha = 2$ the potential acquires a full $SO(2)$ symmetry, and it is natural to identify this point with the intersection of the Gaussian and Ashkin-Teller lines[12]. It is then natural to conjecture that a similar phenomenon occurs in the N=2 context; the degeneration of the superpotential corresponds to the appearance of extra naively marginal operators (the ones that generate translations in the flat directions), signalling incidence with another part of the moduli space. The degeneration of the superpotential causes the ring $\mathcal{R}$ to become of infinite order, corresponding to the ring of functions of the degenerate subspace. Clearly this forces the coordinates for the degenerate space to jump to $h = q = 0$ due to the relation $c = q_{max}$. It is conceivable that this could provide a mechanism for topology change in string theory[27][28]. Nevertheless we leave for future work the extension to potentials without isolated critical points (apart from some comments in section 6). The nondegenerate two field superpotentials are (a) $X^n + Y^m$, (b) $X^n + XY^m$, or (c) $X^nY + XY^m$. In each of these cases the charges of (X, Y) are (a) $(\frac{1}{n}, \frac{1}{m})$, (b) $(\frac{m-1}{nm}, \frac{1}{m})$, and (c) $(\frac{m-1}{nm-1}, \frac{n-1}{mn-1})$. We find $c < 1$ only in the cases

{n}	d	Superpotential	Type
(1,n)	2n	$X_1^{2n} + X_2^2$	$A_{2n-1},\ n \geq 1$
(2,2n+1)	4n+2	$X_1^{2n+1} + X_2^2$	$A_{2n},\ n \geq 1$
(2,n-2)	2n-2	$X_1^{n-1} + X_2^2 X_1$	$D_n,\ n \geq 4$
(3,4)	12	$X_1^4 + X_2^3$	E_6
(4,6)	18	$X_1^3 X_2 + X_2^3$	E_7
(6,10)	30	$X_1^5 + X_2^3$	E_8

Table 1. *Classification of $c < 1$ superpotentials.*

Here d is the least common denominator of the charges $q_i = \frac{n_i}{d}$. The ring of the singularity consists of monomials whose degrees are the exponents of the associated algebra (in accord with the known properties of the $c < 1$ modular invariants). All other $c < 1$ superpotentials can be put into one of the above forms through field redefinitions (cf. Arnold[22]). Writing superpotentials involving three or more variables as above[22], the c formula shows that $c \geq 1$ in these cases except for tensoring the ADE models with the trivial theory $W = X^2$.

3. Some moonshine: singularities at low c

The $c < 1$ models have an ADE classification[29]. How are the superpotentials in table 1 related to ADE? Each of these Lie groups is associated with a discrete subgroup Γ of SU(2): the cyclic groups $\mathcal{C}_n$ with A_n, dihedral groups $\mathcal{D}_{n-1}$ with D_{n+1}, the tetrahedral group $\mathcal{T}$ with E_6, octahedral group $\mathcal{O}$ with E_7, and the icosahedral group $\mathcal{I}$ with E_8. These groups leave fixed some regular solid, which we may project onto the surface of the sphere $\mathbf{P}_1$. A nice way to think about these groups is to try to tile the sphere with spherical triangles, where the covering group consists of reflections in the sides of the triangle, see fig. 2.

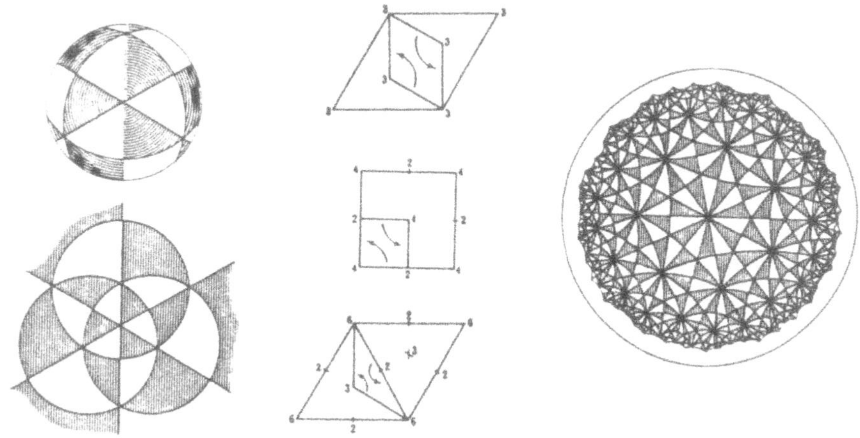

Fig. 2. *Tilings of the sphere, plane, and hyperbolic disc by triangles.*

It turns out that such a tiling is only possible for triangles with interior angles $\frac{\pi}{p}, \frac{\pi}{q}, \frac{\pi}{r}$, where p, q, r are the lengths of the legs of a D- or E-type dynkin diagram[30] (the A_n case is degenerate because the cyclic group is a subgroup of $U(1) \subset SU(2)$). The orientation preserving subgroup of index two in this reflection group, consisting of even numbers of reflections, is the regular solid group. The vertices of the triangles coincide with the edge centers, face centers, and vertices of the associated regular solid. The appearance of these 'triangle groups' is a feature which extends beyond the $c < 1$ models, as we will see. It turns out that the $W = 0$ is the relation among the canonical generators of the ring of invariant polynomials of these groups. Using projective coordinates (u, v) for $\mathbf{P}_1$, $SU(2)$ acts via

$$\begin{pmatrix} u \\ v \end{pmatrix} \to \begin{pmatrix} \alpha & \beta \\ -\bar{\beta} & \bar{\alpha} \end{pmatrix} \begin{pmatrix} u \\ v \end{pmatrix} , \qquad \alpha\bar{\alpha} + \beta\bar{\beta} = 1 .$$

Now pick a discrete subgroup, *e.g.* C_n, generated by $\begin{pmatrix} \omega & 0 \\ 0 & \omega^{-1} \end{pmatrix}$, $\omega^{2n} = 1$. The ring of invariant polynomials is generated by u^n, v^n, uv. Letting

$$X(u, v) = uv , \quad Y(u, v) = \frac{u^n + v^n}{2i} , \quad Z(u, v) = \frac{u^n - v^n}{2} ,$$

we have

$$X^n + Y^2 + Z^2 = 0$$

as the relation between the three invariant polynomials. As another example, consider the icosahedron. Icosahedra have 12 vertices, 20 faces, and 30 edges. Let $a_i = \frac{u_i}{v_i}$ be the projective coordinate of vertex i, $i = 1, ..., 12$ of a spherically projected icosahedron. Then

$$\prod_{i=1}^{12}(u - a_i v) \equiv X(u,v)$$

is an invariant polynomial, since the group action just permutes the vertices of the discrete solid. Construct a similar polynomial $Y(u,v)$ for the 20 face centers b_i and $Z(u,v)$ for the 30 edge centers c_i. These polynomials obey the relation $X^5 + Y^3 + Z^2 = 0$. The other cases are similar, but with one extra subtlety: in general the polynomials v, e, and f built from the vertex, edge, and face centers (the vertices of the fundamental triangle) are not invariant under the group, but only projectively invariant; they transform by phases under the action of the center $\Gamma/[\Gamma,\Gamma]$. They do however obey the relation

$$f^p + e^q + v^r = 0 , \tag{3.1}$$

where $\frac{\pi}{p}, \frac{\pi}{q}, \frac{\pi}{r}$ are the angles of the fundamental triangle of the tesselation of the sphere. One must then construct absolute invariants X, Y, and Z out of v, e, and f that transform without phases; the relation between these is induced from (3.1) and is precisely the equation $W = 0$, where W is given in table 1[31]! Note that the map $(u,v) \rightarrow (X,Y,Z)$ embeds $\mathbf{C}^2/\Gamma$ into $\mathbf{C}^3$ as an algebraic hypersurface, hence the name singularity theory – the orbifold quotient at the origin is the simplest type of algebraic surface singularity, the kind used in the blowing up constructions involving Eguchi-Hanson spaces. Finally, the curve Σ in the weighted projective space $\mathbf{P}(q_X, q_Y, q_Z)$ obtained by identifying $(X,Y,Z) \sim (\lambda^{q_X} X, \lambda^{q_Y} Y, \lambda^{q_Z} Z)$, with λ in the multiplicative complex numbers $\mathbf{C}^*$, is always $\mathbf{P}_1$ for each $c < 1$ model! It is probably not a coincidence that a convenient description of the N=2 discrete series uses the supersymmetric GKO coset SU(2)/U(1)=$\mathbf{P}_1$, and that the N=2 algebra is built out of SU(2) parafermions[32].

For $c = 1$, there are only three LG superpotentials:

$\{d; n\}$	Superpotential	Signature	Type
(3;1,1,1)	$X_1^3 + X_2^3 + X_3^3 + \alpha X_1 X_2 X_3$	$(1; \infty, \infty, \infty)$	$\hat{E}_6$ or P_8
(4;1,1,2)	$X_1^4 + X_2^4 + X_3^2 + \alpha X_1^2 X_2^2$	$(1; \infty, \infty)$	$\hat{E}_7$ or X_9
(6;1,2,3)	$X_1^6 + X_2^3 + X_3^2 + \alpha X_1^4 X_2$	$(1; \infty)$	$\hat{E}_8$ or J_{10}

Table 2. $c = 1$ *superpotentials.*

The signature $(g;\mathbf{e})$ lists the genus g and orders e_i of the branch points of the curve Σ in weighted projective space. These are orbifolds of tori, $\mathbf{C}/\tilde{\Gamma}$, where $\tilde{\Gamma}$ is a discrete subgroup of ISO(2) (also known as the Euclidean group E(2)) consisting of a lattice of translations together with SO(2) rotations by angles $\frac{2\pi}{3}$, $\frac{2\pi}{4}$, or $\frac{2\pi}{6}$. One can again associate a triangle group with each of these models; the first corresponds to tiling the plane by reflections in equilateral $(\frac{\pi}{3},\frac{\pi}{3},\frac{\pi}{3})$ triangles, the second using $(\frac{\pi}{2},\frac{\pi}{4},\frac{\pi}{4})$ triangles, the last using $(\frac{\pi}{2},\frac{\pi}{3},\frac{\pi}{6})$ triangles. Incidentally, these are the fundamental domains in the root lattices of the rank two groups SU(3), SO(5), and G_2. The group $\tilde{\Gamma}$ is the orientation preserving subgroup of index two in this triangle group. The equation $W = 0$ is the relation among the ring of regular functions on a line bundle of degree −3 ($\hat{E}_6$), −2 ($\hat{E}_7$), or −1 ($\hat{E}_8$) over a torus[33]. These functions map $\mathbf{C}^2$, quotiented by the action of certain Heisenberg groups, into $\mathbf{C}^3$. Another description of the singularity is given by taking the total space of the line bundle and contracting the zero section to a point. If the degree is less than −3 the ring of invariants has more than three generators. Finally, the algebraic curve defined by $W = 0$ in the weighted projective space $\mathbf{P}(q_i)$ is a torus in each case. Of course each of these models has a marginal operator (it must by the relation between c and the highest charge chiral field); the perturbation parametrized by α in the table. This parameter is the radius of the covering torus, the angle of this torus being fixed by the requirement of Z_3, Z_4, or Z_6 symmetry. The special point $\alpha = 0$ is a point of enlarged symmetry, where the superpotential admits an extra $S_3 \ltimes Z_3 \times Z_3$, $Z_2 \times Z_4$, or Z_3 symmetry, respectively. These are the self-dual radii where the torus model admits an enlarged holomorphic algebra. The points $4\alpha^3 = 27$ in the $\hat{E}_6$ and $\hat{E}_8$ cases, and $\alpha^2 = 4$ in the $\hat{E}_7$ example, are points where W degenerates; we expect these to be points of enhanced symmetry where another part of $c = 1$ moduli space meets these orbifold lines. Note that once again, we may regard the torus as E(2)/U(1), and because in the parafermionic decomposition of the $c = 1$ N=2 algebra the parafermions have $h = 1$, the N=2 algebra implies that these parafermions are in fact free fields.

Because we can construct the $c = 1$ models by other means, we have a tool to better understand the amount of information supplied by the effective lagrangian. In these examples, the chiral fields X_i are none other than the twist fields of the orbifold. To see this, note that the $c = 1$ Gaussian model consists of a free complex boson and a free complex (Dirac) fermion. The N=2 U(1) charge is just the fermion number, and is hence integral for any untwisted field. The ground states in the sector twisted by $2\pi k/p$ carry charge k/p (and also have $2h = q$; cf. [34]) and are thus created by chiral fields. The states in the untwisted sector correspond to nonchiral fields; for example, all the momentum and winding states, created by exponential fields in the usual parametrization, are D-terms since they have $h > \frac{q}{2}$. Of course, their spectrum depends on the moduli of the torus – the parameter α in the superpotential – but computing it purely from the LG theory would require solving for the exact renormalization of the fields in the neighborhood of the fixed point. It is only because we have the exact solution that we know something about the D-term spectrum. The information contained in the effective potential is the topological structure of the spacetime: the Witten index (*i.e.* cohomology), the Virasoro central charge (a combination of the dimension and the first Chern number), etc. Note that in *any* orbifold model, the twist fields are chiral. Thus we might be able to write down the effective theory on the basis of our knowledge of the orbifold group alone.

What about $c > 1$? Arnold[22] lists 14 'exceptional' singularity germs that correspond to $c = \frac{d+2}{d}$ for certain integers d:

$\{d;\mathrm{n}\}$	Superpotential	Signature	Int. Graph	Type
(24;9,8,6)	$X_1^2X_3 + X_2^3 + X_3^4$	(0;2,3,9)	(3,3,4)	Q_{10}
(18;7,6,4)	$X_1^2X_3 + X_2^3 + X_2X_3^3$	(0;2,4,7)	(3,3,5)	Q_{11}
(15;6,5,3)	$X_1^2X_3 + X_2^3 + X_3^5$	(0;3,3,6)	(3,3,6)	Q_{12}
(16;5,4,6)	$X_1^2X_3 + X_2X_3^2 + X_2^4$	(0;2,5,6)	(3,3,4)	S_{11}
(13;4,3,5)	$X_1^2X_3 + X_2X_3^2 + X_1X_2^3$	(0;3,4,5)	(3,4,5)	S_{12}
(12;4,4,3)	$X_1^3 + X_2^3 + X_3^4$	(0;4,4,4)	(4,4,4)	U_{12}
(30;8,6,15)	$X_1^3X_2 + X_2^5 + X_3^2$	(0;2,3,8)	(2,4,5)	Z_{11}
(22;6,4,11)	$X_1^3X_2 + X_1X_2^4 + X_3^2$	(0;2,4,6)	(2,4,6)	Z_{12}
(18;5,3,9)	$X_1^3X_2 + X_2^6 + X_3^2$	(0;3,3,5)	(2,4,7)	Z_{13}
(20;5,4,10)	$X_1^4 + X_2^5 + X_3^2$	(0;2,5,5)	(2,5,5)	W_{12}
(16;4,3,8)	$X_1^4 + X_1X_2^4 + X_3^2$	(0;3,4,4)	(2,5,6)	W_{13}
(42;14,6,21)	$X_1^3 + X_2^7 + X_3^2$	(0;2,3,7)	(2,3,7)	K_{12}
(30;10,4,15)	$X_1^3 + X_1X_2^5 + X_3^2$	(0;2,4,5)	(2,3,8)	K_{13}
(24;8,3,12)	$X_1^3 + X_2^8 + X_3^2$	(0;3,3,4)	(2,3,9)	K_{14}

Table 3. *Hyperbolic triangle singularities.*

The signature $(g;\mathrm{e})$ is as in table 2 the genus g and orders of branching e_i of the curve $W = 0$ in $\mathbf{P}(\mathrm{n})$. The column 'Int. Graph' will be explained in the next section. The subscript in the last column refers to the number μ of chiral fields. Six of these are tensor products of $c < 1$ models; the others are new, isolated fixed points (since there are no $h = \frac{1}{2}$ chiral fields in the spectrum). It is easy to verify that they are not supersymmetric coset models based on compact groups[35]. $W = 0$ is again the relation between the generators (X_1, X_2, X_3) of the ring of automorphic forms on special Riemann surfaces (special because the ring of automorphic functions generically has more than three generators). Here n_i is the degree of the form X_i. These Riemann surfaces are the quotient of the Poincaré disc $\mathbf{D} = SU(1,1)/U(1)$ (the unit disc with hyperbolic metric) by discrete subgroups Γ of SU(1,1) generated by reflections in the edges of hyperbolic triangles, just as the $c < 1 (c = 1)$ groups are generated by reflections in spherical (Euclidean) triangles. There is again an associated surface singularity described this time by the total space of the anticanonical $\mathbf{C}^*$ bundle over $\mathbf{D}$, minus the zero section, quotiented by the action of $\tilde{\Gamma}$. The smooth covering of the surface $W = 0$ in weighted projective space is a curve of genus greater than one. For example, one of these singularities is associated with $W = X^7 + Y^3 + Z^2$; the smooth covering of $W = 0$ in $\mathbf{P}(2,3,7)$ has covering group $PSL(2, \mathbf{F}_7)$ and is the curve of genus 3 with maximal symmetry.

The common thread in these three sets of examples is the appearance of triangle groups Γ, together with the associated ring of invariant functions or forms on the Riemann surface $\Sigma = \mathcal{M}/\tilde{\Gamma}$, where $\mathcal{M} = G/U(1)$ is the covering space $\mathbf{P}_1 = SU(2)/U(1)$, $\mathbf{C} = ISO(2)/U(1)$, or $\mathbf{D} = SU(1,1)/U(1)$, Γ is a discrete subgroup of G and $\tilde{\Gamma}$ is the orientation preserving subgroup of index two consisting of even numbers of reflections. Σ is simply the identification of a pair of fundamental triangles along corresponding edges. There are two algebraic varieties naturally associated to W – the surface singularity $W = 0$ in $\mathbf{C}^3$ which is an isolated quotient singularity, and the variety $W = 0$ in $\mathbf{P}(\mathbf{n})$, a compact curve. The singular two-manifold is the quotient of the canonical, trivial, or anticanonical $\mathbf{C}^*$ bundle over $\mathbf{P}_1$, $\mathbf{C}$, or $\mathbf{D}$ – with the zero section contracted to a point – by the action of $\tilde{\Gamma}$. The above singularities are not the only examples, just ones that have a description in terms of three LG fields and are thus given by a single equation in weighted projective space which describes a sphere branched over three points. The N=2 algebra in each of the above cases may also be decomposed using a parafermion representation for the appropriate space. In fact, every $N = 2$ model has a parafermionic description[32][36], where we factor out the (bosonised) N=2 U(1) current. Specifically, we write the N=2 currents as

$$\begin{aligned} J &= i\sqrt{c}\partial\varphi \qquad & G &= \sqrt{2c}\,\psi e^{i\varphi/\sqrt{c}} \\ T &= -\tfrac{1}{2}(\partial\varphi)^2 + T_\psi \qquad & G^* &= \sqrt{2c}\,\psi^* e^{-i\varphi/\sqrt{c}} \end{aligned} \tag{3.2}$$

where ψ, ψ^* are the parafermions of level k for the group

$$\begin{array}{ll} SU(2) & c<1 \quad (\epsilon\equiv+1) \\ ISO(2) & c=1 \quad (\epsilon\equiv 0) \\ SU(1,1) & c>1 \quad (\epsilon\equiv-1) \end{array}$$

such that $c=\frac{k}{k+2\epsilon}$. In the latter two cases k may be rational, in which case we write $k=\frac{R}{S}$. For $c=1$ ψ is a complex boson on a torus of radius $\sqrt{k}$. Fields in the theory may be decomposed as

$$\Phi=\Phi_{\rm pf}\, e^{i(q\varphi+\bar{q}\bar{\varphi})/\sqrt{c}}\ .$$

We will have occasion to use this parafermionic representation in the discussion of orbifolds in section 5; in particular it manifests the spectral flow properties of the theory.

For the case $\mathcal{M}=\mathbf{C}$, *i.e.* $c=1$, one knows from the independent construction that the cft is a sigma model on the orbifold Σ. The twist fields of the orbifold model are none other than our chiral LG fields X, Y, and Z. Similarly, I claim that the $c<1$ and 'exceptional' $c>1$ models are also nonlinear sigma models on Σ, but with nonstandard target metric, with the LG fields the twist fields for the three branch singularities. The supersymmetry index[24] is not the same as the corresponding NLSM (except for the A_1 theory), but then the metric is 'singular' because of the branch points. We will see below that the orbifolds of these models by the symmetries of W are identical to the NLSM on the corresponding projective curve. Thus the original 'unorbifold' is the twist of the NLSM by the dual symmetry; this will be explained in detail in section 5. Normally the sigma model on $\mathbf{P}_1$ has a mass gap, so for this idea to work there must be a nontrivial interplay between the twist procedure and renormalization, since the usual NLSM on $\mathbf{P}_1$ has nonzero beta function (and is in the universality class of the trivial fixed point) while the twisted theory has vanishing beta function (*i.e.* is in the universality class of some minimal model). Note also that without a detailed solution of the quantum theory, the superpotential contains only limited information, namely the spectrum of chiral fields and their algebra. All information about D-terms is nontopological and depends on solving for the renormalized theory at the fixed point. For instance, in the orbifold case the D-terms include all the twist invariant exponentials of the torus coordinate, which contain the information about the metric structure of the target. In particular the metric determined by the Kahler potential need not be the SO(3) invariant one for which the mass gap appears, and in the $c=1$ case it is the flat metric outside the fixed points. Of course, in weak coupling the kinetic term is the Kahler potential of the target, and so we should regard the nonchiral fields as containing the spectral (*i.e.* metrical) information. Thus in a sense our LG fields are not describing the spectrum of the Laplacian here, but rather some algebraic characteristics of the space, and perhaps the neighborhoods of some special singular points on it.

4. Singularity Theory

The classification of the simplest N=2 qft's parallels that of algebraic surface singularities. Let me digress for a moment to explain the rudiments of algebraic singularity theory[22][37]. The structure appears in the discussion of the next section in that the compactification of the corresponding singularities is the manifold of some of the NLSM's we are interested in. Our discussion will be general, with examples drawn from the theory of algebraic surfaces.

Consider an algebraic hypersurface $\mathcal{V}$ with isolated singularity defined by the set of points $X \in \mathbf{C}^r$ such that $W(X) = 0$. This means that at some point p (chosen to be the origin of coordinates) $W = 0$ coincides with $\nabla W = 0$, so the embedding degenerates. We can resolve the singularity by considering a generic *deformation* (or *unfolding*) of W

$$W(X,g) = W(X) + \sum_{\alpha=1}^{\mu=|\mathcal{R}|} g^{\alpha} p_{\alpha}(X) ,$$

where p_α form a basis for the ring of chiral fields $\mathcal{R}$ of section 2. Then generically the vanishing locus $\mathcal{V}_g$ of W_g is nonsingular. The choice of W and basis for $\mathcal{R}$ is actually a choice among represenatives in the space of functions modulo smooth coordinate redefinitions. In the qft, the response of the measure under shifts of the coordinates gives the equation of motion, so the determination of $\mathcal{R}$ in singularity theory and in qft stem from the same source. Generically the hypersurface $\mathcal{V}_g \subset \mathbf{C}^r$ has no singularities, but on a submanifold of g-space called the *discriminant* Δ, $\mathcal{V}_g$ is still singular. The discriminant harbors all the singularities into which $\mathcal{V}$ may be deformed analytically, and which will all be 'simpler' than $\mathcal{V}$ (or of similar singularity type if the perturbation is marginal or irrelevant). If we consider a neighborhood B of the origin of the space of g^α, then we can form a bundle[37] with fiber $\mathcal{V}_g$ and base $B - \Delta$, *i.e.* the set of smooth surfaces. For example, if $W = X_1^2 + \ldots + X_r^2$, then $\mathcal{R}$ consists of only the constant function, $W(X,g) = W(X) + g_0$, and $\mathcal{V}_g$ is diffeomorphic to the space of vectors of length ≤ 1 in the tangent bundle of S^{r-1}

(since $\sum X_i^2 = g_0$ is a compact r–1 dimensional sphere for X_i real and is noncompact in the other directions). The prototype is $r = 2$ when $\mathcal{V}_g$ is a branch cut in the complex plane $X_1^2 = X_2^2 + g_0$. The discriminant Δ is the point $g_0 = 0$ in this example, where the branch points coalesce. The structure of the singularity when $g_0 \to 0$ is generic: a middle dimension homology sphere S^{r-1} *vanishes* (*e.g.* for $r = 2$ a circle surrounding the branch cut shrinks away). Since the singular point $X = 0$ of W splits into $\mu = |\mathcal{R}| = \dim B$ critical points of quadratic type for a generic perturbation $g \in B - \Delta$, $\mathcal{V}_g$ has the structure of $(S^{r-1})^\mu$ and $H_{r-1}(\mathcal{V}_g)$ is a rank μ lattice. The condition which defines Δ is a single equation, so Δ is complex codimension one in $\mathbf{C}^\mu$. Therefore the space of constant perturbations g_0 generically intersects Δ in some number of points p_i (see fig. 3).

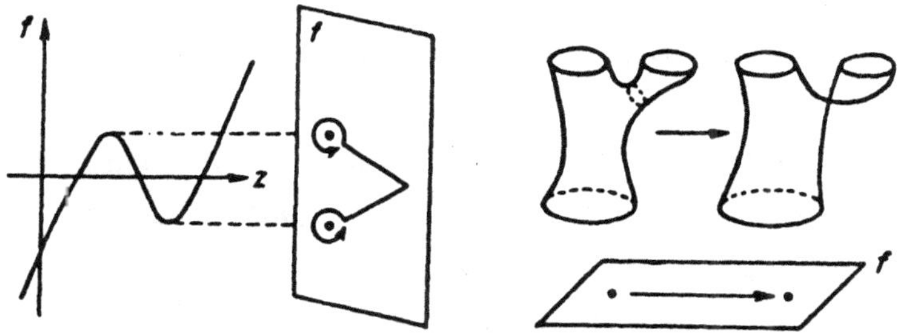

Fig. 3 *Deformation of a singularity and its monodromy.*

At each p_i one of the μ S^{r-1} cycles vanishes (contracts to a point). Geometrically we are scanning through the level sets of W to find one where $\nabla W = 0$. The fundamental group of $B - \Delta$ therefore acts on the middle dimension homology of the fiber $\mathcal{V}_g$ as automorphisms preserving the intersection form in H_{r-1}. This monodromy is generated by parallel transport of $H_{r-1}(\mathcal{V}_g)$ along paths around the p_i in the space of g_0 (see the figure). For the ADE$\hat{E}$ surface singularities the intersection matrix is none other than (minus) the Cartan matrix of the corresponding algebra. The generalization to other singularities of table 3 finds the Cartan matrix of a dynkin diagram with three legs whose lengths are specified in the column 'Int. Graph'.

Another structure assodiated to the singularity is its *resolution*. A resolution is a map $\pi : \mathcal{V}' \to \mathcal{V}$ from a nonsingular variety $\mathcal{V}'$ to the singular variety $\mathcal{V}$ which is an isomorphism over the singular part $\pi : \mathcal{V}' - \pi^{-1}(0) \xrightarrow{\sim} \mathcal{V} - 0$. The singular fiber $\pi^{-1}(0)$ is that which we 'glue in' to smooth the singularity. The resolution of the singularity has a structure similar to the monodromy matrix: for the ADE$\hat{E}$ singularities the graph of the resolution is the same as that of the monodromy matrix, *i.e.* we glue in a collection of $\mathbf{P}_1$'s intersecting in the pattern of the dynkin diagram. The resolution of the A_{n-1} singularity is the simplest example, and will occur often in the sequel. By taking the polynomials $X = uv$, $Y = u^n$, $Z = v^n$, we can write the singular surface $\mathcal{V}$ as $W = X^n - YZ = 0$. Now let $T_1, ..., T_{n-1}$ be coordinates in $(\mathbf{P}_1)^{n-1}$. Define a surface $\mathcal{V}'$ in $\mathbf{C}^3 \times (\mathbf{P}_1)^{n-1}$ by

$$XT_1 = Y, \quad XT_{i+1} = T_i \ (1 \le i \le n-2), \quad X = T_{n-1}Z \ . \tag{4.1}$$

Then letting $\pi : \mathbf{C}^3 \times (\mathbf{P}_1)^{n-1} \to \mathbf{C}^3$ be the natural projection, $\pi : \mathcal{V}' \to \mathcal{V}$ is a resolution of $\mathcal{V}$. This means that away from the origin $\mathcal{V}'$ is isomorphic to $\mathcal{V}$, but at the origin it is blown up into a sequence of $\mathbf{P}_1$'s intersecting in the pattern of the Dynkin diagram (see fig. 4a). To see the configuration of the singular fiber, suppose T_j is the first nonzero coordinate in $(0,0,0; T_1, ..., T_{n-1}) \in \mathbf{C}^3 \times (\mathbf{P}_1)^{n-1}$, the parametrization of the blowup. From the equations (4.1) for $\mathcal{V}'$ we see that $T_i = \infty$ for $i > j$, and T_j parametrizes the jth $\mathbf{P}_1$ of the resolution. Thus the singular fiber consists of $n-1$ $\mathbf{P}_1$'s intersecting (at $T = 0$ and $T = \infty$) as shown in the figure.

For more complicated singularities, *e.g.* those of table 3, it consists of excising the singular point and gluing in a collection of curves intersecting in a starlike pattern (see fig. 4b), with $\mathbf{P}_1$'s along the chains; the central curve Σ in the star is the curve described by $W = 0$ in $\mathbf{P}(\mathbf{n})$ (think of scaling down to the origin along $W(X) = W(\lambda^q X) = 0$), and the branch points of this curve are symptoms of further cyclic A_n quotient singularities of $\mathcal{V}$ that intersect Σ at the branch points and must be blown up – hence the chains and starlike pattern.

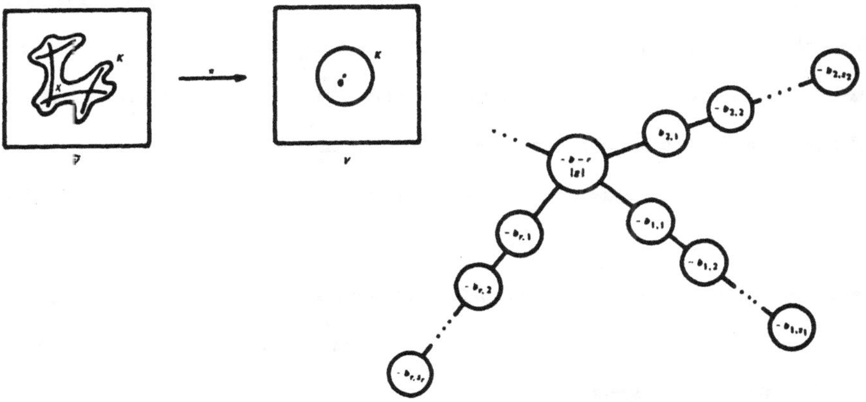

Fig. 4. ***Blowup of A_{n-1} singularity, and generic resolution graph of surface singularity.***

To illustrate, let us describe the resolution of the K_{14} singularity. Due to the C^* symmetry, the neighborhood of the origin can be described as a C^*-bundle over the projective curve Σ given by $W = 0$ in $P(n)$ (in this case $X_1^3 + X_2^8 + X_3^2 = 0$ in $P(8,3,12)$). Weighted projective spaces generically have orbifold singularities (for details see section 5), in the case at hand the singularities of $P(8,3,12)$ are a Z_3 orbifold singularity along $X_1 = 0$ and a Z_4 quotient along $X_2 = 0$. The curve Σ intersects these, and therefore has branch points, at $(0,1,e^{\pm i\pi/2})$ and $(-1,0,1)$; hence the signature $(0;3,3,4)$ (the genus of Σ can be established by formulae in [38][39]). Arnold[22] noted that the resolutions graphs of the singularities in table 3 coincide with the intersection graphs of the singularity with the same c value, and dubbed it 'strange duality'. This duality was explained mathematically[40]; later we will see this phenomenon appearing as a physical duality in certain LG constructions of K3. The resolution of a general singular hypersurface is of course more complicated[26]. A standard method uses the Newton polyhedron $\Gamma(W)$; briefly, there is a canonical procedure that associates to each $\Gamma(W)$ a dual polyhedron $\Gamma^*(W)$. Then to a certain simplicial decomposition of Γ^* one associates a canonical resolution $\pi : \mathcal{V}' \to \mathcal{V}$. Roughly speaking the subdivision adds the extra local functions needed to resolve $\mathcal{V}$; it breaks the singularity up into pieces (like the P_1's of the A_{n-1} resolution) glued together along the common boundaries of the simplices. The Euler character and other topological data of the resolution are determined combinatorially from the simplicial decomposition[26].

Of course not all singularity types are described by quasihomogeneous functions. Those that are correspond to 2d cft's. The nonhomogeneous singularities might describe particularly interesting RG trajectories of nonconformal theories. The connection between LG theories and algebraic singularities seems to stem mostly from the fact that a class of examples of both objects are characterized by quasihomogeneous polynomials. Quasihomogeneous singularities enjoy covariance under the $\mathbf{C}^*$ symmetry $X_i \to \lambda^{q_i} X_i$. Generally, affine varieties of dimension n with $\mathbf{C}^*$ action are equivalent to positively graded algebras of dimension n[39] (the dimension of a finitely generated graded algebra A is the order of the pole of its Poincare power series $P_A(t)$ at $t = 1$). A finitely generated means that there exists $\phi : \mathbf{C}[X_1 ... X_n] \to A$ for some n. Let $\mathcal{V} = \{X \in \mathbf{C}^n | f(X) = 0 \ \forall f \in \ker\phi\}$ be the algebraic variety associated to A; that ϕ can be made a graded homomorphism implies that $\ker\phi$ is generated by weighted homogeneous polynomials. The ring of automorphic forms relative to a Fuchsian group has dimension two; it turns out that any surface singularity with $\mathbf{C}^*$ action has the same simple geometrical interpretation as in the previous section[39]:

Theorem: *If $\mathcal{V}$ is a complex surface with isolated singularity at 0 and $\mathbf{C}^*$ action then there exists a simply connected Riemann surface U, a line bundle $\mathcal{L}$ over U, and a discrete group Γ of automorphisms of U such that (a) $\mathcal{V}_0 = \mathcal{V} - \{0\} =$ (total space of $\mathcal{L})/\Gamma$, and (b) $\Sigma = \mathcal{V}_0/\mathbf{C}^* = U/\Gamma$.*

The coordinate ring of $\mathcal{V}$ is isomorphic to the ring of automorphic forms on Σ. A class of higher dimensional examples may be generated in this way as well. When the ring of automorphic forms has three generators X_1, X_2, X_3 the associated surface singularity embeds in $\mathbf{C}^3$. Such singularities have been classified by Wagreich[39]; besides those listed in tables 1-3, the list includes[33][39]

{d; n}	Equation	Signature	Type
(18;2,6,9)	$X_1^9 + X_2^3 + X_3^2 = 0$	(0;2,2,2,3)	$J_{3,0}$
(14;2,4,7)	$X_1^7 + X_2^3 X_1 + X_3^2 = 0$	(0;2,2,2,4)	$Z_{1,0}$
(12;2,4,5)	$X_1^4 X_2 + X_2^3 + X_1 X_3^2 = 0$	(0;2,2,2,5)	$Q_{2,0}$
(12;2,3,6)	$X_1^6 + X_2^4 + X_3^2 = 0$	(0;2,2,3,3)	$W_{1,0}$
(10;2,3,4)	$X_1^5 + X_2^2 X_3 + X_1 X_3^2 = 0$	(0;2,2,3,4)	$S_{1,0}$
(9;2,3,3)	$X_2^3 + X_3^3 + X_1 X_2^3 = 0$	(0;2,3,3,3)	$U_{1,0}$
(10;2,2,5)	$X_1^5 + X_2^5 + X_3^2 = 0$	(0;2,2,2,2,2)	N_{16}
(8;2,2,3)	$X_1^4 + X_2^4 + X_2 X_3^2 = 0$	(0;2,2,2,2,3)	$V_{1,0}$
(12;1,4,6)	$X_1^{12} + X_2^3 + X_3^2 = 0$	(1;2)	$J_{4,0}$
(10;1,3,5)	$X_1^{10} + X_1 X_2^3 + X_3^2 = 0$	(1;3)	$Z_{2,0}$
(9;1,3,4)	$X_1^9 + X_2^3 + X_1 X_3^2 = 0$	(1;4)	$Q_{3,0}$
(8;1,2,4)	$X_1^8 + X_2^4 + X_3^2 = 0$	(1;2,2)	$X_{2,0}$
(7;1,2,3)	$X_1^7 + X_1 X_2^3 + X_1 X_3^2 + X_2^2 X_3 = 0$	(1;2,3)	$S_{2,0}^*$
(6;1,2,2)	$X_1^6 + X_2^3 + X_3^3 = 0$	(1;2,2,2)	
(5;1,1,2)	$X_1^5 + X_2^5 + X_1 X_3^2 = 0$	(2;2)	
(6;1,1,3)	$X_1^6 + X_2^6 + X_3^2 = 0$	(2)	
(4;1,1,1)	$X_1^4 + X_2^4 + X_3^4 = 0$	(3)	

Table 4. *Remaining singularities which embed in* $\mathbf{C}^3$ *(see also Table 3).* These models all have $c = \frac{d+2}{d}$. The signature $(g; \mathbf{e})$ is again the genus and ramification order of the branch points on the corresponding RS Σ. We have left all moduli out of the defining polynomials. The resolution of the singular point involves inserting a collection of curves intersecting in a starlike graph: Σ at the center with chains of e_i $\mathbf{P}_1$'s radiating outward, similar to that of table 3.

Fixing a point in the space $\mathbf{C}^{\mu-1}$ of all couplings to nonconstant perturbations, the p_i are simply the critical values of $W(X, g)$ and the corresponding critical points X are the minima of the bosonic potential V associated to W. We learn that the vanishing cycles are associated with the potential wells of the qft, but rather indirectly. Physically this is because all the topology of the hypersurface is contained in the singularity where the measure of the naive LG theory is peaked. Moreover, deformation of a quasihomogeneous singularity is related to the structure of RG flows in the qft. First of all, note that perturbation by elements of $\mathcal{R}$ with weighted degree larger than one don't change the singularity type as one readily sees by the scaling behavior of $W = 0$ near $X = 0$. Of course in the qft this is just the property that the fixed point is stable under perturbation by irrelevant operators. Perturbation of W by constants does nothing to the qft. The remaining perturbations are the relevant and marginal deformations of W. Marginal perturbations maintain the singularity type in algebraic geometry, and maintain exact conformal invariance in the qft by the R-symmetry (technically there will be some adjustment of the D-terms). Perturbations by linear and quadratic terms take us along RG flows to the trivial fixed point; excluding these, we find a wide variety of perturbations that generate flows to other RG fixed points. This is in marked contrast to the typical situation in qft, where we must fine tune to reach an infrared fixed point because renormalization usually induces mass terms at some order in the expansion, and we are driven to the trivial fixed point unless we carefully adjust the initial conditions of the RG flow. However in N=2 the space of perturbations leading to a massless theory is easy to find due to the nonrenormalization theorem – just leave out linear shifts and mass terms and the theory must flow to a nontrivial fixed point. Simplicity of singularities in the sense of algebraic geometry is thus the genericity of a fixed point in qft: the low c fixed points are ubiquitous because one always finds them under typical perturbations. The nonrenormalization theorem partially reduces the problem of mapping out the space of field theories to one of functional analysis. More difficult is to understand the relationship of different orbifolds with the same superpotential, flows between models with different field content, etc.

In conformal perturbation theory the flows may be calculated as in [4][41][6]. If only a single scaling field is added to the Lagrangian, the $O(g^2)$ term in β_g vanishes due to charge conservation (nontrivial quadratic terms appear when considering perturbations by two or more fields). Thus the beta function has the structure

$$\beta_g \sim (h - \tfrac{1}{2})g + Cg^3 + O(g^5, gg') \ .$$

One reflection of the nonrenormalization theorem is that $\beta_g \propto g$, *i.e.* if we leave a term out of the perturbation it is not generated by other couplings. In general the consequences of the nonrenormalization theorem for beta functions and RG flows is an interesting question that lies beyond the scope of this article. I merely wish to remark that the existence of only wavefunction renormalization implies interesting (and nontrivial) functional relations between the beta functions of F- and D-terms. Note that, in contrast to the lower N situation, in N=2 the existence of a nontrivial weak coupling fixed point depends on the sign of C. From the success of the LG theory we expect it to turn out favorably, but to my knowledge it has yet to be calculated. At any rate, we have a fixed point at $g_* \sim \sqrt{\frac{h}{2C}}$, with $\delta c \sim O(g_*^4)$.

5. Various varieties

One of the unifying threads above seems to be that the variety $W = 0$, when considered as an equation in a weighted projective space, is related to the spacetime being described by the qft. Why should this be so? After all, with the canonical kinetic term the minima of the ordinary bosonic potential $V = \sum_i |\nabla_i W|^2$ are at the zeros of ∇W, not W. The authors of [16][5] proposed that if one changes variables from X_i to a scale Λ and $r-1$ scale invariant coordinates Y_i,

$$W = X_1^{n_1} + \ldots + X_r^{n_r} = \Lambda(1 + Y_2^{n_2} + \ldots + Y_r^{n_r}) \,,$$

W takes the form of a Lagrange multiplier Λ times the potential written in inhomogeneous coordinates, provided one ignores the kinetic term of Λ. Integration over the Lagrange multiplier then enforces $W = 0$. Ignoring the kinetic term of Λ is equivalent to investigating the regime of extremely strong coupling for this field; typically the fixed point couplings are of order one (actually $1/\sqrt{d}$). This is all right because irrelevance of the kinetic terms should imply that one is still in the universality class of the nonlinear sigma model. While the foregoing presents an appealing picture, unfortunately integration over the remaining auxiliary fields of Y_i also forces $\partial W/\partial Y_i = 0$ as well, provided one allows some small amount of dynamics in order to render the functional integral meaningful.[6] For example, using $K(X_i, X_i^*) = \sum_i X_i^* X_i$, one finds after integrating the auxiliary fields a bosonic potential

$$V = f \cdot |W(y)|^2 + \sum_{i=2}^{r} g_i \cdot |\nabla_i W(y)|^2$$

where f, g_i are nonsingular functions of the lowest components λ, y_i of the superfields Λ, Y_i, with the kinetic term still $(\partial x)^2$ in homogeneous coordinates. In fact, for quasihomogeneous functions $W = 0$ whenever $\nabla W = 0$; the change of variables simply exchanges the former equation for one of the latter. This should be expected on physical grounds; if the potential is peaked at the origin, then changing variables should not alter this fact much. But $W = 0$ has solutions where the fields take on arbitrarily large values, where the measure should be exponentially damped; so we still must explain why the fields effectively take values on the

---

[5] Whom I wish to thank for correspondence concerning their proposal.

[6] Without such an addition there are some fields which disappear from the action upon integrating over auxiliary fields. In components $S = \int F_\Lambda(1 + \sum y_i^{n_i}) + \lambda(\sum n_i F_i y_i^{n_i - 1})$+fermionic terms, and obviously integration over all fields is ill-defined.

NLSM manifold. At strong coupling, where one should approach the picture advocated in [16], the fields are actually concentrated at the origin. The point is that the minimal kinetic term need not lie in the universality class of the NLSM, even after the change of variables above. So rather, let us show conversely that the NLSM on a projective algebraic variety can be written so that it belongs to one of the universality classes with superpotential W.

First we describe the NLSM on weighted projective space. Let $q_i = \frac{n_i}{d}$, where d is the least common denominator of the q_i. If there are r LG fields, we define $\mathbf{n} = \{n_1...n_r\}$ and the weighted projective space[42] $\mathbf{P}(\mathbf{n})$ is $\mathbf{C}^r$ modulo the identification of $(X_1, ..., X_r)$ with $\lambda(X_1, ..., X_r) = (\lambda^{n_1}X_1, ..., \lambda^{n_r}X_r)$. The standard complex projective space $\mathbf{P}_{r-1}$ is simply $\mathbf{P}(1,1,...,1)$. $\mathbf{P}(\mathbf{n})$ may be thought of as the quotient $\mathbf{P}_{r-1}/G$, where $G = \prod_i \mathbf{Z}_{n_i}$ acts by $g(X_1, ..., X_r) = (g_1X_1, ..., g_rX_r)$, $g_i^{n_i} = 1$ on the homogeneous coordinates of $\mathbf{P}_{r-1}$. One can write the nonlinear sigma model on $\mathbf{P}(\mathbf{n})$ in terms of dynamics on $\mathbf{C}^r$ with the $\mathbf{C}^*$ symmetry gauged[18]

$$S = \int d^2z d^4\theta \Big[\sum_{i=1}^{r}(X_i^*)^{n_i} e^V (X_i)^{n_i} - V\Big] \ ;$$

solving for the nondynamical gauge field V by its equation of motion we find the action

$$S = \int d^2z d^4\theta \log\Big[\sum_{i=1}^{r}(X_i^* X_i)^{n_i}\Big] \ . \tag{5.1}$$

This action still has the $\mathbf{C}^*$ gauge symmetry, which we could remove by going to inhomogeneous coordinates, but we will not do so. Instead, let us add the superpotential[43]

$$\int d^2z (d^2\theta \Lambda W(X_i) + h.c.) \tag{5.2}$$

which enforces $W = 0$ in the weighted projective space; choosing the gauge $\Lambda = 1$ (a kind of unitary gauge) then gives the desired superpotential, but with a rather peculiar kinetic term. The R symmetry of the fixed point theory has been turned into an internal symmetry. Note that one gets the theory twisted by Gepner's orbifold group[13] $G_0 = G \cap \mathbf{C}^*$. There is no way to add relevant operators to (5.2) without destroying the C^* gauge symmetry one started with. The theory will have only integer charge states, just as the naive LG theory projected by the symmetry $X \to g(X)$. But up to discrete ambiguity, the superpotential determines the fixed point structure of the kinetic terms by universality. If there is only one universality class, as is true for $c < 1$, then the model (2.6) is in the same universality class as the minimal kinetic term. We find a realization of the speculation in [14], that the fixed point theory lives on $W = 0$ in projective space whereas the naive classical Lagrangian does not; the scale symmetry is gained near criticality by universality. Generally, however, the LG theory and its integer charge projection reside in different universality classes. This is not so for $c < 1$ only because the orbifold theory is identical to the original dual theory. The above construction clearly extends to intersections of polynomials in products of weighted projective spaces[16], provided that the number of constraints is less than or equal to the number of projective spaces in the tensor product, and that the gauge $\Lambda_i = 1$ is compatible with the $\mathbf{C}^*$ symmetries. There is a close interplay between the potential and kinetic terms; the gauge $\Lambda = 1$ would be impossible without the introduction of the constraint. The singularity of $K(X, X^*)$ at the origin is responsible for pushing the measure away from the singularity and onto the full manifold in projective space. The probability measure for $\vec{X} = 0$ is small because of the singularity of K, coresponding to the fact that this is not a point of $\mathbf{P}(\mathbf{n})$. In this sense the authors of [16] are correct in that there is a choice of kinetic terms which realizes the nonlinear sigma model on the weighted projective variety, but it is certainly not restricted to manifolds of vanishing first Chern class, and is not clearly the same universality class as the one they describe (although I believe this is likely, the difference being that the kinetic term implicit in [16] is farther from the fixed point). The gauge $X_1 = 1$ (inhomogeneous coordinates) is closest to what is described in [16], but looks like an asymmetric limit where the dynamics of one of the LG fields has been removed. Our construction gives the NLSM on an algebraic variety in projective space, but with the metric induced from the embedding. This is not the Ricci-flat metric in the case of Calabi-Yau manifolds. However, we only care that K is in the universality class of this metric.

Having a description of the LG orbifold as the NLSM on $\mathbf{P}(\mathbf{n})$, we can now interpret the naive LG theory geometrically as the 'unorbifold' obtained by undoing the twisting with respect to the symmetries of W. That is, if the LG orbifold describes the NLSM, then an orbifold of the NLSM describes the naive LG theory. The orbifold group is a discrete subgroup of the $U(1) \times \overline{U(1)}$ group of complex structure rotations (so in Calabi-Yau spaces it destroys the holomorphic c-form). The moduli of the NLSM manifold coming from the twisted sectors are charged under this 'dual' orbifold group, so the theory will only have the required symmetry at special points in the moduli space where these fields are absent from the action. In particular the radius is often fixed at about the string scale. Note that the spacetime described by the naive LG theory is compact, as in the nonsupersymmetric examples explored in the introduction. The spacetime is not the singular hypersurface in $\mathbf{C}^r$, but rather its quotient by the symmetry $\mathbf{C}^* \ltimes \tilde{G}_0$.

Now let us describe the singularity structure of weighted projective spaces. The space $\mathbf{P}(\mathbf{n})$ has orbifold singularities on a subvariety $\mathbf{P}_{sing}$, but is not a global orbifold (*i.e.*, the quotient of a manifold by a discrete group) because G does not act isometrically on $\mathbf{P}_{r-1}$. For example, consider the space $\mathbf{P}(1,n,n)$. The projective equivalence is $(X_1,X_2,X_3) = (\lambda X_1, \lambda^n X_2, \lambda^n X_3)$; for $X_1 \neq 0$, X_2 and X_3 are good projective coordinates for $\mathbf{P}_1$, but for $X_1 = 0$ there are n values of λ which solve the equation $X_i' = \lambda^n X_i$, $i = 2,3$, *i.e.* as λ runs over $\mathbf{C}^*$, X_i covers $\mathbf{C}^*$ n times (these points were distinguished when $X_1 \neq 0$), and so $X_1 = 0$ is an n-fold orbifold point. In general, let $I(X) = \{j|X_j \neq 0\}$; then $X \in \mathbf{P}_{sing}$ IFF g.c.d.$\{n_i; i \in I(X)\} > 1$. In other words, just as in our example, if some nontrivial subset of the n_i have a common divisor, then when the coordinates not associated to that subset vanish there is an orbifold singularity. A hypersurface defined by the vanishing of a single polynomial equation is singular if it intersects a singularity in the ambient space. We encountered this already in the resolution of the singularities of tables 1-4. Consider for example the equation $X_1^2 + X_2^3 + X_3^7 + X_4^{42} = 0$ in $\mathbf{P}(21,14,6,1)$, which is the projectivisation of the singularity K_{12} of table 3. The singular varieties in the ambient affine space are $X_3 = X_4 = 0$ with sevenfold branching, $X_2 = X_4 = 0$ with threefold branching, and $X_1 = X_4 = 0$ with twofold branching; we see the resolution graph with three branches of lengths 2,3,7. Taking into account the projective equivalence, the variety $W = 0$ intersects each of these subspaces in precisely one point. The resolution modifies the Euler character of the space because the chains of curves change H_2. Now $h_{1,1}$ is just the number of deformations of the complex structure of the surface $W = 0$, plus one for the Kahler form itself. The number of algebraic deformations of the complex structure is just the number of deformations of degree one of the defining polynomial W. Blowing up a singularity of type A_n, D_n, or E_n adds n to the Euler number. In the example at hand, there are 7 algebraic deformations of the complex structure coming from the polynomial deformations, and 7+3+2 nonalgebraic deformations from repairing the singularities, for a total of 19. This together with the Kahler form and $h_{0,0} = h_{2,0} = h_{0,2} = h_{2,2} = 1$ gives 24, the Euler number of K3. Another example is $X_1^2 + X_2^3 + X_3^9 + X_4^{18} = 0$ in $\mathbf{P}(9,6,2,1)$, the projectivisation of $J_{3,0}$ of table 4; the singularities are at (1,–1,0,0) of type A_2, and (0,α,1,0), $\alpha^3 = -1$, of type A_1, hence the four-legged resolution graph with legs of length 2,2,2,3. There are 14 polynomial deformations, minus 4 for the excised singular points, plus $1 \cdot 3 + 3 \cdot 2$ from the Eguchi-Hanson resolution, for a total of 19. The case of $X_1^2 + X_2^3 + X_3^{10} + X_4^{15} = 0$ is somewhat special, since there the singular set in the ambient weighted projective space is a curve rather than a point; however this problem may be avoided by using the D modular invariant for X_3[44].

In fact, for each entry in tables 3-4 there is a uniform construction of a K3 surface obtained by compactifying the singularity[40][33]. From $W(X_i; g^\alpha)$ construct a new polynomial

$$\widehat{W}(X_i; g^\alpha) = W(X_i; 0) + \sum_{\alpha,\, q(p_\alpha)\leq 1} g^\alpha p_\alpha(X_i) Y^{n_\alpha} ,$$

where $q(p_\alpha) = \frac{n_\alpha}{d}$ and d is the weighted degree of W. For $g^\alpha = \delta_{\alpha 0}$ we have $W = W(X_i; 0) + Y^d$; one readily finds all of Gepner's examples of K3 constructions including D modular invariants in tables 3-4 (and of course orbifolds of tori using table 2). The D invariant for Y can be interpreted as another way of projectivizing the surface singularity $\mathcal{V}$. $\widehat{W}$ is weighted homogeneous, so it defines a surface $\hat{\mathcal{V}}_g$ in $\mathbf{P}(n_1, n_2, n_3, 1)$. This surface has two parts: for $Y \neq 0$ the surface is isomorphic to the deformed surface $\mathcal{V}_g$ in $\mathbf{C}^3$. For $Y = 0$, the 'curve at infinity' that one adds to compactify $\mathcal{V}_g$ is precisely the projective curve $\Sigma = \{X | W(X, \bar{g}) = 0 \text{ in } \mathbf{P}(\mathbf{n})\}$ of the previous section ($\bar{g}$ are the marginal perturbations of W), which is the central curve in the resolution graph of $\mathcal{V}$. The legs of the resolution graph come from the branch points of Σ, *i.e.* quotient singularities of type $A_{e_i - 1}$. The preceeding compactification is the same procedure by which one obtains $\mathbf{P}_n$ as $\mathbf{C}^n$ with $\mathbf{P}_{n-1}$ added at infinity. Thus the homology of the compactification $\hat{\mathcal{V}}_g$ combines that of the singularity $\mathcal{V}_g$ with that of the resolution. This is in fact general and not restricted to surfaces. The space $\mathcal{S} = \{(g_0, g_1, ..., g_{\mu-2}) \in \mathbf{C}^{\mu-1} | \hat{\mathcal{V}}_g$ has only ADE singularites$\}$ of all relevant and marginal perturbations of W may be compactified; $\mathcal{S} - \{0\}/\mathbf{C}^*$ is a coarse moduli space for K3 surfaces of the same type as $\hat{\mathcal{V}}_g$. (Coarse just means that the true moduli space is the quotient of this space by a discrete group ($\mathrm{Aut}(H_2)$ in particular); 'the same type' means the surface has a divisor of the same type as the curve at infinity Σ.)

One easily sees that the marginal deformations of $\hat{W}$ do not account for all the moduli of K3 surfaces; there are almost always complex structure moduli arising from the resolution of the singluarities of $\hat{\mathcal{V}}_g$, as well as the perturbations of the Kahler form. In the qft, these extra moduli appear in the twisted sectors when constructing the G_0-orbifold of the naive LG theory, which we have argued is the NLSM on $\hat{\mathcal{V}}$. Let us first examine the G_0-projection in discrete series tensor products, then generalize to the generic LG theory. The discrete series may be regarded as super-G/H models[35] for $\frac{SU(2)}{U(1)}$, which boil down to the bosonic coset $\frac{SU_{n-2}(2)\times U_2(1)}{U_n(1)}$, or in the parafermionic description we have

$\frac{SU_{n-2}(2)}{U_{n-2}(1)} \ltimes \frac{U_{n-2}(1)\times U_2(1)}{U_n(1)}$. Fields $\Phi_{lms}^{\overline{lms}}$ carry $SU_{n-2}(2)$ labels $l, \bar{l}$, $U_n(1)$ labels $m, \bar{m}$, and $U_2(1)$ labels $s, \bar{s}$, with $h = \frac{l(l+2)-m^2}{4n} + \frac{s^2}{8}$ and $q = \frac{m}{2n} + \frac{s}{2}$. Thus the (anti)chiral representations have $l = m(-m)$, $\bar{l} = \bar{m}(-\bar{m})$. The N=2 U(1) symmetry acts in the space of m, s quantum numbers. We are interested in the $\mathbf{Z}_d$ twisting of the tensor product theory. To find the spectrum of the tensor product theory it is simplest to look at the torus partition function. The twisted theory has

$$
\begin{aligned}
\mathcal{Z} &= \sum_{x,y\in \mathbf{Z}_d} \prod_{i=1}^{r} \mathcal{Z}^{(i)}(x,y) \\
\mathcal{Z}(x,y) &= \tfrac{1}{2} \sum_{m,l_s,\bar{l}_s} L_{l\bar{l}} N_{s\bar{s}} \chi^{l}_{m+2y,s} \bar{\chi}^{\bar{l}}_{m,\bar{s}}\, e^{2\pi i \frac{(m+y)x}{n}} \, .
\end{aligned}
\tag{5.3}
$$

We perform a similar twisting on the $s, \bar{s}$ indices to align the sum over 2d fermion boundary conditions so that states are either all NS or all R in each factor, and we project only onto $(-)^{\Sigma F_i + \bar{F}_i}$. A little effort shows that the above is equivalent to Gepner's procedure[7]. Basically y measures the string twisting – strings only close up to some element of G_0, and x puts an element of G_0 in the trace. Clearly, the sum over x in $\mathcal{Z}$ sets $\sum_{i=1}^{r} q_i \in \mathbf{Z}$. Thus only integer charge chiral fields will survive the projection; however there could be fractional charge twisted chiral fields coming from the twisted sectors. A trivial example is $r = 1$, where the Z_n orbifold is identical to the original ADE theory, just expressed in terms of twisted chiral fields. When c is integral, fractional charge does not appear because $\sum_i \frac{1}{n_i}$ is always integral by the c formula (2.11). There will then be at least one chiral or twisted chiral modulus of the conformal field theory corresponding to the Kahler form of the target. For example, in the Fermat model $W = X_1^r + ... + X_r^r$ one has always $\tilde{\phi}_{Kahler} = \prod_{i=1}^{r} \Phi_{1,1,0}^{1,-1,0}$; in a slightly more nontrivial example, in the $W = X_1^2 + X_2^3 + X_3^7 + X_4^{42}$ model every chiral modulus field $\phi = \prod_i \Phi_{lms}^{lms\ (i)}$ yields a twisted chiral field $\tilde{\phi} = \prod_i \Phi_{lms}^{l,-m,s\ (i)}$ (this is *not* a generic feature). For $c = 3$, all the moduli coming from deformations of the Kahler form, as well as some deformations of the complex structure, are twisted chiral fields (except in certain torus constructions where $h_{2,1} = 0$, so that we reverse our identification of complex structure and kahler form deformations in terms of chiral and twisted chiral fields). When the hypersurface $\mathcal{V}$ passes through the orbifold singularities of the ambient projective space, then extra cohomology appears in their resolution. The singularities are always cyclic quotient singularities due to the structure of $\mathbf{P}_{sing}$. For the general theory

[7] I would like to thank H. Ooguri for an illuminating discusion on this matter.

of resolution of such singularities, see [26]. The homology of the resolution tells us the structure of massless N=2 representation coming from the twisted sectors. The general N=2 theory will have structure similar to (5.3). Using the parafermionic decomposition (3.2) of the theory, the characters may be written as

$$\begin{aligned} \chi_{h,q} &= \sum_{j \bmod 2R} C_{h,Q(k+2\epsilon)+j}\, \vartheta_{(j+2\epsilon Q)(R+2\epsilon S),R(R+2\epsilon S)}(\tau, \tfrac{\theta}{k+2\epsilon}) \\ \vartheta_{a,b}(\tau,\theta) &= \sum_{n\in\mathbf{Z}} exp\big[2\pi i\tau \tfrac{b}{2}(n+\tfrac{a}{2b})^2 + 2\pi i b\theta(n+\tfrac{a}{2b})\big] \end{aligned} \tag{5.4}$$

and $C_{h,q}$ is the parafermion character. Now we wish to compute the $\mathbf{Z}_{d/\alpha}$ orbifold projection of the theory, where $\alpha|d$. This is easy, because the twisting is with respect to a discrete subgroup of the R-symmetry; hence it only involves the N=2 U(1) structure. Suppose we have fields $\Phi_{h,q} = \Phi^{\rm pf}_{h,q} e^{\frac{i}{\sqrt{c}}(q\varphi+\bar{q}\bar{\varphi})}$; we wish to retain all the fields with $\frac{q+\bar{q}}{2} \in \frac{\alpha}{d}\mathbf{Z} \equiv \frac{1}{\beta}\mathbf{Z}$ – these will comprise the untwisted sector of the orbifold theory. The twisted sectors, as in any orbifold, consist of all those fields which are local with respect to the G-invariant untwisted fields but nonlocal with respect to the G-noninvariant untwisted fields. Now since

$$e^{\frac{i}{\sqrt{c}}(a\varphi+\bar{a}\bar{\varphi})} e^{\frac{i}{\sqrt{c}}(b\varphi+\bar{b}\bar{\varphi})} \sim (z-w)^{-\frac{ab}{c}} (\overline{z-w})^{-\frac{\bar{a}\bar{b}}{c}} e^{\frac{i}{\sqrt{c}}[(a+b)\varphi+(\bar{a}+\bar{b})\bar{\varphi}]} ,$$

if $a = \bar{a} \in \frac{1}{\beta}\mathbf{Z}$ is an invariant untwisted field then locality means that $\frac{ab-\bar{a}\bar{b}}{c} \in \mathbf{Z}$, and so $(b-\bar{b}) \in \beta c\mathbf{Z}$. But this is precisely what we get when we spectral flow by β units! From the characters one easily computes that a unit of spectral flow $\theta \to \theta+\pi$ shifts $q \to q + \frac{2}{k+2\epsilon}$. Finally we arrive at the formula for the orbifold partition function

$$\mathcal{Z}_{\rm orb} = \sum_{\substack{h,\bar{h},q,\bar{q} \\ \frac{\beta(q+\bar{q})}{2}\in\mathbf{Z},\ \frac{q-\bar{q}}{\beta c}\in\mathbf{Z}}} N_{h\bar{h}} \chi^{h}_{q} \overline{\chi}^{\bar{h}}_{\bar{q}} \ .$$

The $(-)^{F+\bar{F}}$ projection is incorporated by the requirement $j - \bar{j} \in 4\mathbf{Z}$ in the sums (5.4) defining the characters. The tensor product case is readily seen to follow this structure: projection on untwisted charges, and the twisted sectors arising from holomorphic spectral flow. This is, in fact, the end result of Gepner's procedure[13]. The orbifold process just changes the radius of the boson φ by a factor $\frac{\alpha}{d}$, since one projects onto momenta which divide $\frac{d}{\alpha R}$, then add strings which only close up to a multiple of $\frac{R\alpha}{d}$. To understand the massless twisted sector states one needs to compute the resolution of $\mathcal{V}$. This may be done using rather general methods of algebraic geometry[26], an issue which I will pursue in a separate publication[45]. It is important to note that the quotient of the hypersurface singularity by the action of $\mathbf{C}^*$ is not always a manifold, so the resulting string theories need not always have a manifold interpretation, although the theory of resolutions goes through. Of course we are not necessarily interested in smoothing the singularity; but we do need to know the topology of the 'quasismooth' variety $\mathcal{V}$ at the singularity.

For $c = 1$ we obtain a torus from each of the models of table 2. In the theory of abelian orbifolds, one begins with a theory carrying an 'electric' (abelian) G symmetry. The twist fields carry a dual 'magnetic' G symmetry; twisting by this symmetry reproduces the original unorbifolded theory. But of course, which theory one calls the 'unorbifold' is a matter of convention. The naive LG theory describes a torus orbifold, where X_i are the twist fields of the torus as described above; it is the theory parametrized by the 'magnetically' charged fields. The twisting by G is the flat torus whose natural parametrization uses the torus coordinate (and its 'electrically' charged exponentials) as dynamical field. We assign a phase to each twist field X_i, then twist by it to obtain the flat torus theory. The twisted fields of the LG orbifold are the states projected out when passing from the flat torus to its orbifold, and vice versa.

For $c = 2$, we are guaranteed to obtain a K3 surface. This surface has $h_{1,1} = 20$, $h_{2,0} = h_{0,0} = 1$. The (2,0) form on the manifold is the special representation at $2h = q = c = 2$ built purely from the N=2 U(1) current. For general integral c this is the (n,0) form on the manifold[20]. For nonintegral c this field exists but describes a spacetime (n,0) form with singularities; in a sense it is parafermionlike as a 2d field. For K3 its holomorphic part $\epsilon(z)$ is single-valued; the triplet of currents ϵ, ϵ^*, and J form the level one SU(2) current algebra of K3 quaternionic structures. The superpartners of ϵ, ϵ^* are the additional supersymmetry currents that make the N=4 structure of the K3 NLSM. The action of $\epsilon(\epsilon^*)$ on a (twisted)chiral field produces a (un)twisted antichiral field, so actually the N=4 massless matter multiplets come in SU(2) doublets (since $q = 2J_3^{SU(2)} = 1$) of one chiral and one twisted antichiral field (and of course there is always the conjugate representation) when described in terms of N=2 representations[46].

The astute reader is probably wondering why the deformations of the target Kahler form appear as twisted chiral fields and not in the kinetic D terms of the NLSM. For example in the torus models of table 2, the field $\epsilon(z)$ (*e.g.* $X_1X_2X_3(z)$ in the $\hat{E}_6$ case) is the holomorphic part of the marginal perturbation that we would identify with $DZ(z)$, the super-U(1) current of the usual parametrisation using the torus coordinate Z; the combinations $DZ\bar{D}Z$, $DZ\bar{D}^*Z^*$, *etc.* are the perturbations of the modulus of the torus and associated antisymmetric tensor field, which we usually call kinetic terms. However, in a 'good' quantum field theory – where the probability measure limits the field to a bounded region of configuration space so that the spectrum of scaling dimensions is gapped – the identity operator is the only field of $h \sim 0$. The hypersurface of the NLSM in (5.1),(5.2) is indeed compact. Any D-term Φ of $h \sim 0$ splits into (twisted) chiral representations because $D\Phi$ is becoming null. One can think of the Kahler deformations as the residues of such a field, which undergo operator mixing with the kinetic D-term (especially its upper components) under renormalization or as we move along a marginal line. In particular the nonrenormalization theorem will not apply to such fields.

We can now explain the 'strange duality' of table 3. Arnold noticed that the sum of $\mu = |\mathcal{R}|$ for the dual pairs is always 24, and they always have the same d. Pinkham[40] explained this by compactifying the singularity as above to obtain a K3 surface and showing that the intersection matrix $(-E_8)^{\otimes 2} \oplus \left(\begin{smallmatrix} 0 & 1 \\ 1 & 0 \end{smallmatrix}\right)^{\otimes 3}$ of K3 splits into two orthogonal sublattices, the intersection graphs of the two dual singularities. The compactification of the singularity involves the inclusion of a set of $\mathbf{P}_1$'s according to the dual configuration. In the qft, there is a similar duality; namely, in each K3 construction built from a LG potential in table 3 we have both the ring of chiral fields $\mathcal{R}$ and that of twisted chiral fields $\tilde{\mathcal{R}}$, whose orders add up to 24 (including the identity and (2,0) form). The marginal fields have the form $p_\alpha(X_i)Y^{n_\alpha}$ for chiral fields, $\tilde{\phi}_\alpha \tilde{Y}^{n_\alpha}$ for twisted chiral fields (here $Y = \Phi^{110}_{110}$, $\tilde{Y} = \Phi^{1,-1,0}_{110}$). In the six cases where the LG theory is a tensor product of minimal models one may explicitly construct the dual fields $\tilde{\phi}_\alpha$, and one finds the charge spectrum of the dual singularity (of course four of these examples are self-dual). The interpretation of the remaining eight is the same: the ring $\tilde{\mathcal{R}}$ of twisted chiral fields is the chiral ring of the dual singularity! The twisted chiral fields are the elements of H_2 needed to resolve the singularities of $\mathcal{V}$. Thus there are two LG descriptions of the same K3 surface: one in terms of chiral fields, the other in terms of twisted chiral fields, related by orbifold. This is a rather nontrivial example of the duality transformation of [17]. As a corollary, we find that the 'dual' theories of table 3 are orbifolds of one another. A more elementary example exhibiting this duality phenomenon is of course the discrete series, where the intersection and resolution graphs are identical because the orbifold of the discrete series is the same theory in dual variables.

6. Discussion

The main point in the foregoing is that there is a transcription of fixed points described by N=2 effective Lagrangians into the language of algebraic geometry, for which I have built a partial translation dictionary. Thus we may borrow the wealth of tools in algebraic geometry to understand more about qft, and perhaps exploit our qft intuition to develop new tools in algebraic geometry. With regard to the program outlined in the introduction, I think the foregoing represents a major advance. It should be possible to develop intuitions about string geometry in N=2 backgrounds because the 'points' have quasialgebraic properties. Since they naturally yield spacetimes with size of order the string scale, the LG models provide a useful playground for the study of the sigma model at strong coupling. A priori it is not at all obvious that the techniques of algebraic geometry should continue to hold in this regime.

We have seen the algebraic description of complex manifolds emerge from simple effective potential models. What I have in mind is to treat the LG fields as algebraic variables, with the ring of chiral fields a prototype of string geometry where 'points' are ideals in the ring of operators of a qft. One of my main motivations for [14][6] was to develop an intuition about such notions. It seems that there is a close relationship between the ring of chiral fields and the function ring of an algebraic variety. The function ring is $\mathbf{C}[X]/W$, the cohomology ring $\mathbf{C}[X]/\nabla W$ in the LG examples. However, the RG scaling fields are perhaps not the right variables; they describe global features of spacetime, being eigenfunctions of the Laplacian (however in the spacetimes described by LG models, one naturally obtains spacetimes with only a finite number of points, so local and global notions tend to get mixed up). We would rather have a quasilocal description in terms of 'points', coordinate neighborhoods, tangent spaces, etc. using a set of 2d quantum field variables. We don't arrive at the equivalence principal by studying global structure; the issue is to find a language in which to describe quasilocal physics in string theory. Such a language should manifest the 'position uncertainty principle' sketched in the introduction, which supplements the usual phase space uncertainty principle of quantum mechanics. The lack of such a language is a severe shortcoming of all present approaches.

An open question is how much of the space of N=2 quantum field theories has a LG description. The discussion of section 5 indicates that complete intersection varieties can be recast as LG theories only if one can identify the Lagrange multipliers of the constraints with some of the coordinates of projective space. An examination of some simple coset models[35] reveals that some have a LG description, some not.[8] There is of course a ring of chiral fields $\mathcal{R}$ in any model; the question is whether $\mathcal{R}$ is a function ring. Orlik and Randell[23] give a partial set of conditions for the collection of weights q_i of a generating set of fields to come from a quasihomogeneous polynomial. Also, one may use the method of [5], further developed in [6][14] to investigate any given model. Begin with the chiral fields X_i which are not contained in the operator products with other chiral fields. These are in some sense a 'generating set' for $\mathcal{R}$. For the case of coset models, in the weak coupling ($k \to \infty$) limit we look for $\frac{1}{2}$dim(G/H) LG fields whose charge goes to zero, corresponding to functions on a patch of $\mathbf{C}^{\frac{1}{2}\dim(G/H)}$. In fact, from the form

$$G^* = \frac{2}{k+g} \sum_{\alpha \in G/H} \psi_{-\alpha} E^{\alpha}$$

(for notation, see [35]) of the coset supercurrent, the chirality condition $G^*(z)\Phi_{\lambda}^{\Lambda,\bar{\Lambda}} \sim$ finite is satisfied by any field with $\bar{\Lambda} = 0$ and highest weight under the generators in $\mathbf{g} - \mathbf{h}$. Then we try to identify the boundary of $\mathcal{R}$ as the Newton polyhedron of some function $W(X_i)$. That is, when applying X_i to $p_\alpha \in \mathcal{R}$ we have schematically

$$X_i(z)p_\alpha[X(w)] \sim (z-w)^{h_\Phi - h_i - h_\alpha}\Phi + \dots ,$$

and either $\Phi \in \mathcal{R}$ (in which case $h_\Phi - h_i - h_\alpha = 0$) or else $\Phi \sim D^*\bar{D}^*\Phi'$ for some Φ'. If $\Phi' = X_i^*$ we try to set $\frac{\partial W}{\partial X_i} = D^*\bar{D}^*X_i^* + \dots$ and integrate to find W. Furthermore the OPE $X_j X_j^* \sim 1$ suggests $W = G^*_{-\frac{1}{2}}\bar{G}^*_{-\frac{1}{2}}1 = 0$! Thus although $W = 0$ doesn't appear as a dominant location of the probability measure, we in some sense obtain it in that fields outside the ring $\mathcal{R}$ are redundant. The above procedure works for some coset models, *e.g.* $\frac{SU(3)}{SU(2)\times U(1)}$, but not others, *e.g.* $\frac{SU(3)}{U(1)^2}$. For $\frac{SU(3)}{SU(2)\times U(1)}$ the fundamental fields come from the decomposition $3 = 2^{(+1)} \oplus 1^{(-2)}$; there is a chiral field $\phi = \Phi^{3}_{2,1}$ (suppressing antiholomorphic quantum numbers) of charge $q = \frac{1}{k+3}$ and another $\chi = \Phi^{3}_{1,2}$ of charge $q = \frac{2}{k+3}$. From these it is easy to build candidate $q = 1$ polynomials for W. For $\frac{SU(3)}{U(1)^2}$ there are three basic chiral fields, two with charge $\frac{2}{k+3}$ and one with $q = 1$, so for even k there is no candidate $q = 1$ polynomial for the superpotential. Of course one can make $\frac{SU(3)}{U(1)^2}$ from $\frac{SU(3)}{SU(2)\times U(1)} \times \frac{SU(2)}{U(1)}$, which both have LG descriptions, by the orbifold that identifies the SU(2) quantum numbers in the two factors. Alternatively, many of the G/H models have

[8] A preliminary investigation of these models was carried out in collaboration with M. Douglas (see also [47]).

extra U(1) symmetries (essentially whenever G/H is not a Hermitian symmetric space) and might be described by the more general Lagrangian (2.3),(2.5). One may conjecture that all N=2 fixed points come in this way, from LG theories and their orbifolds; however I think such speculation is rather premature until we have further evidence. On the other hand, we might also suitably expand the use of singularity theory to take account of a larger class of N=2 fixed points. For instance[22][48], the remaining Cartan groups B,C,F,G arise in singularity theory as singularities on spaces with boundary (for BCF) or invariant under the permutation group S_3 (for G); in the qft these are the orbifolds

Singularity	Covering	Equation	Twist
B_n	A_{2n-1}	$\pm X_1^n + X_2^2 = 0$	$X_1 \leftrightarrow -X_1$
C_n	D_{n+1}	$\pm X_1^n + X_1 X_2 = 0$	$X_2 \leftrightarrow -X_2$
F_4	E_6	$\pm X_1^2 + X_2^3 = 0$	$X_1 \leftrightarrow -X_1$
G_2	D_4	$(W_{D_4} = X_1^3 + X_2^3 = 0)$	${}^{X_1 \to \omega X_1}_{X_2 \to \omega^{-1} X_2}, \omega^3 = 1$

Table 5. *Orbifolds of simple singularities.*

The twisting removes fields from the ring $\mathcal{R}$, so that naively it is not a function ring. But for instance in the BCF cases the chiral ring is instead

$$\mathcal{R} = \frac{\mathrm{C}[X]}{\{X_1 \partial_1 W, \partial_2 W\}},$$

i.e. we look at the ring of functions invariant under the twist group[49]. Of course this only gives the twist-invariant states of the untwisted sector of the orbifold. The full chiral ring generally involves contributions from the twisted sectors; for instance, in the first two lines of table 5 the twisted sectors extend $B_n \to D_{n+1}$, $C_n \to A_{2n-1}$ because the D_{n+1} models are the $X \to -X$ orbifolds of the A_{2n-1} models. If it turns out that all the G/H models have an interpretation as orbifolds of LG models, we might expect a geometrical interpretation analogous to that of the minimal models where the LG theory is related to the projective embedding of a NLSM on (an orbifold of) G/H or some associated flag variety. Some of the numerology of [50] suggests that there is a Feigin-Fuchs type free field representation of the super coset models. These authors develop the method of quantisation of orbits, realising the complex coordinates on G/H and their conjugates as (constrained) free fields. This may assist in the geometrical interpretation of these theories.

In sections 4-5 we discussed two methods of smoothing singularities: the deformation of the defining polynomial W, which smooths the singular middle-dimensional homology

H_{r-1} of a hypersurface $\mathcal{V} = \{W = 0\}$ in $\mathbf{C}^r$ into a collection of spheres; and resolution by blowing up, $\pi : \mathcal{V}' \to \mathcal{V}$, where a minimal resolution and its topology are determined via the Newton diagram of W, and the homology of the exceptional divisor $\pi^{-1}(0)$ is inserted. Both forms of resolution play an important role in recent work of Candelas, Green, and Hubsch[28] on the connectivity of the moduli space of threefolds. For surfaces both methods of resolution only affect H_2; however for threefolds, deformation changes H_3 whereas blowing up points changes H_2. A method was proposed whereby, passing through a singular variety, the topology changes; on one side the singularity is approached by deformation, on the other side by blowing down. It is important to determine whether the same phenomenon occurs in string theory. The disappearance of homology on the NLSM manifold is not by itself cause for topology change, as torus orbifolds tell us; the string tends to remember what to put at the singular points. In other words, although the manifold is singular, the loop space is not. The question we must answer is, why is the deformation boundary related to the blow-down of the resolution *in string theory*? The nodal variety in $\mathbf{P}_4$ examined in the first section of [28] deforms to a nonsingular quintic in $\mathbf{P}_4$. The associated naive LG model has a nonisolated singularity (*i.e.* flat direction) when the superpotential degenerates to the nodal variety. This agrees with the proposal of section 2. The issue at hand is whether the orbifolds of the LG models are connected. However the fact that the spectrum of the LG chiral fields jumps and naively becomes continuous is consistent with this possibility.

There are several ways topology can change in 2d field theory. In LG theories states can move continuously in and out from the boundary of field space by changing the asymptotics of the potential[24][6], and moreover such processes occur while moving a finite distance in coupling constant space. This kind of topology change happens when we travel along an RG trajectory, and we can see the states decouple. Perhaps the degeneration of the potential is the equivalent phenomenon when we travel along a marginal line. Instead of isolated potential minima running off to infinity, a flat direction opens up to infinity, allowing states to leak off. The shift in topology could happen by a finite number of D-terms degenerating toward $h = 0$. Then if half of the states of these multiplets decouple, we are left with extra (twisted) chiral fields; topology would disappear by the reverse process.

Topology can also change if the spectrum becomes continuous. Naive intuition dictates that if some piece of the spacetime manifold becomes small, then although the momentum eigenstates are becoming widely gapped the winding modes are becoming dense. This certainly applies to vanishing H_1. In examples there is often an alternate interpretation in terms of 'large' spacetimes[51]. It does seem that the spectrum becomes continuous when the LG potential degenerates, because the Newton polyhedron is unbounded (the ring $\mathcal{R}$ is not finite). However, this may be an artifact of parametrization; for instance, in torus models the polynomials in the torus coordinate are all of logarithmic scaling dimension just as in the degenerate LG potential, but the proper exponential eigenstates have discrete spectrum if the torus has finite radius. The equivalent analysis of the other side of the singularity, the degeneration by blowing down of H_2, is something I intend to return to elsewhere. The resolution of singularities along the lines of [26] carries over to the LG theories as well[45], and it seems that the qft structure is not too different from that of the algebraic geometry. If so, then we will have a complete set of tools for the analysis of topology change in qft.

Acknowledgements

I would like to thank T. Banks, L. Dixon, M. Douglas, W. Fulton, T. Klassen, H. Ooguri, H. Riggs, N. Seiberg, and S.H. Shenker for helpful discussions.

Note added: While the manuscript was in preparation, I received a paper from W. Lerche, C. Vafa, and N. Warner, *Chiral Rings in N=2 Superconformal Theories*, Harvard preprint HUTP-88/A065, which overlaps substantially with the present work.

References

[1] D. Friedan, *Phys. Rev. Lett.* **45** (1980) 1057; *Ann. Phys.* **163** (1985) 318; C. Lovelace, *Nucl. Phys.* **B273** (1986) 413; D. Friedan and S. Shenker, talk at the Aspen Summer Inst., August 1984, unpublished; E.S. Fradkin and A.A. Tseytlin, *Phys. Lett.* **158B** (1985) 316; *Nucl. Phys.* **B261** (1985) 1; D. Friedan, C. Callan, E. Martinec, and M. Perry, *Nucl. Phys.* **B262** (1985) 593; A. Sen, *Phys. Rev.* **D32** (1985) 2102; *Phys. Rev. Lett.* **55** (1985) 1846; P. Candelas, G. Horowitz, A. Strominger, and E. Witten, *Nucl. Phys.* **B258** (1985) 46.

[2] T. Banks and E. Martinec, *Nucl. Phys.* **B294** (1987) 733; D. Friedan, *Phys. Lett.* **162B** (1985) 102.

[3] D. Friedan in Ref. [1]; C. Callan and Z. Gan, *Nucl. Phys.* **B272** (1986) 647.

[4] A.B. Zamolodchikov, *Sov. J. Nucl. Phys.* **46** (1988) 1090; see also J. Cardy and A. Ludwig, *Nucl. Phys.* **B285[FS19]** (1987) 687; A.M. Polyakov, preprint SLAC-TRANS-0222 (1986).

[5] A.B. Zamolodchikov, *Sov. J. Nucl. Phys* **44** (1986) 529.

[6] D. Kastor, E. Martinec, and S.H. Shenker, *Renormalization Group Flows in N=1 super-Discrete Series*, Nucl. Phys. B[FS], in press; E. Martinec, talk at the US/USSR conference on strings and QCD, Yerevan, June 1988. In this unpublished talk the author emphasized the irrelevance of kinetic terms and the role of the superpotential in determining the fixed point in N=2 supersymmetric Landau-Ginsburg models, and discussed the possibility of a relation between LG models and algebraic varieties.

[7] V. Drinfeld, *Quantum Groups*, talk at the 1986 ICM, Berkeley; L. Faddeev, N. Reshitikhin, and L. Takhtadjan, *Quantization of Lie Groups and Lie Algebras*, Steklov Inst. (Leningrad) preprint LOMI-E14-87 (Sept. 1987).

[8] G. Moore and N. Seiberg, *Classical and Quantum Conformal Field Theory*, IAS preprint IASSNS-HEP-88/39, Comm. Math. Phys. in press.

[9] G.E. Andrews, R.J. Baxter, and J.P. Forrester, *J. Stat. Phys.* **35** (1984) 193.

[10] D. Huse, *Phys. Rev.* **B30** (1984) 3908.

[11] V. Pasquier, *Nucl. Phys.* **B285[FS19]** (1987) 162; *J. Phys.* **A20** (1987) L217, L221, L1229; *J. Phys.* **A20** (1987) 5707.

[12] P. Ginsparg, *Nucl. Phys.* **B295[FS21]** (1988) 153.

[13] D. Gepner, *Nucl. Phys.* **B296** (1988) 757 (see also the preprint version); *Phys. Lett.* **199B** (1987) 380; *String Theory on Calabi-Yau Manifolds: the Three Generations Case*, Princeton preprint (Dec. 1987); *Nucl. Phys.* **B311** (1988) 191.

[14] E. Martinec, *Algebraic Geometry and Effective Lagrangians*, Phys. Lett. **217B** (1989) 431.

[15] C. Vafa and N. Warner, *Catastrophes and the Classification of Conformal Field Theories*, Phys. Lett. **218B** (1989) 51.

[16] B. Greene, C. Vafa, and N. Warner, *Calabi-Yau Manifolds and Renormalization Group Flows*, Harvard preprint HUTP-88/A047 (Nov. 1988).

[17] S.J. Gates, C. Hull, and M. Rocek, *Nucl. Phys.* **B248** (1984) 157.

[18] See, *e.g.*, S.J. Gates, M. Grisaru, M. Rocek, and W. Siegel, *Superspace, or 1001 Lessons in Supersymmetry*, Benjamin-Cummings (1983).

[19] J. Polchinski and N. Seiberg, to appear.

[20] W. Boucher, D. Friedan, and A. Kent, *Phys. Lett.* **172B** (1986) 316.

[21] D. Friedan, A. Kent, S.H. Shenker, and E. Witten, unpublished (1986); A. Schwimmer and N. Seiberg, *Phys. Lett.* **184B** (1987) 191.

[22] V.I. Arnold, *Singularity Theory*, Lond. Math. Soc. Lecture Note Series #53.

[23] P. Orlik and R. Randell, *The Structure of Weighted Homogeneous Polynomials*, in Proc. Symp. Pure Math. **30** (1977) 57; A. Kushnirenko, *Usp. Mat. Nauk* **32#3** (1977) 169-70 (in Russian).

[24] E. Witten, *Nucl. Phys.* **B202** (1982) 253.

[25] A. Kushnirenko, *Polyedres de Newton et Nombres de Milnor*, *Inv. Math.* **32** (1976) 1; A.N. Varchenko, *Zeta Function of Monodromy and Newton's Diagram*, *Inv. Math.* **37** (1976) 253; V.I. Danilov, *Newton Polyhedra and the Vanishing Cohomology*, Funct. *Anal. and Appl.* **13** (1979) 103.

[26] M. Oka, *On the Resolution of Hypersurface Singularities*, in Adv. Stud. Pure Math. **8** (1986) 405; V.I. Danilov, *Toric Varieties*, *Russian Math. Surveys* **33** (1978) 97-154; D. Markushevitch, M. Olshanetsky, and A. Perelomov, *Reolution of Singularities in Superstring Compactification*, preprint ITEP 86-138.

[27] P. Green and T. Hubsch, *Phys. Rev. Lett.* **61** (1988) 1163; *Comm. Math. Phys.* **119** (1988) 431.

[28] P. Candelas, P. Green, and T. Hubsch, *Connected Calabi-Yau Compactifications*, Texas preprint (1988).

[29] Z. Qiu, *Phys. Lett.* **198B** (1987) 497; F. Ravanini and S.K. Yang, *Phys. Lett.* **195B** (1987) 202; D. Gepner in Ref. [13].

[30] L. Ford, *Automorphic Functions*, Chelsea 1953.

[31] P. DuVal, *Homographies, Quaternions, and Rotations*, Oxford Press (1964); K. Lamotke, *Regular Solids and Isolated Singularities*, F. Viewig and Sohn (1985).

[32] A.B. Zamolodchikov and V.A. Fateev, *Sov. Phys. JETP* **63** (1986) 913; Z. Qiu, *Phys. Lett.* **188B** (1987) 207.

[33] K. Saito, *Regular Systems of Weights and Associated Singularities*, Adv. Stud. in Pure Math. **8** (1986) 479.

[34] L. Dixon, J. Harvey, C. Vafa, and E. Witten, *Nucl. Phys.* **B261** (1985) 678; *Nucl. Phys.* **B274** (1986) 285; L. Dixon, D. Friedan, E. Martinec, and S.H. Shenker, *Nucl. Phys.* **B279** (1987) 13; S. Hamidi and C. Vafa, *Nucl. Phys.* **B279** (1987) 465.

[35] Y. Kazama and H. Suzuki, *New N=2 Superconformal Field Theories and Superstring Compactification*, Tokyo preprint UT-Komaba 88-8 (July 1988).

[36] L. Dixon, J. Lykken, and M. Peskin, *N=2 Superconformal Symmetry and SO(2,1) Current Algebra*, SLAC PUB-4884 (March 1989).

[37] J. Milnor, *Singular Points of Complex Hypersurfaces*, Ann. Math. Studies **61**; S.M. Husein-Zade, *Monodromy Groups of Isolated Singularities of Hypersurfaces*, Russian *Math. Surveys* **32** (1979) 23.

[38] P. Orlik and P. Wagreich, *Isolated Singularities of Algebraic Surfaces with* $\mathbf{C}^*$ *Action*, *Ann. Math.* **93** (1971) 205.

[39] P. Wagreich, *The Structure of Quasihomogeneous Singularities*, Proc. Symp. Pure Math. **40** (1983) 593; *Algebras of Automorphic Forms with Few Generators*, Trans. *AMS* **262** (1980) 367.

[40] H. Pinkham, *Singularités exceptionelles, la dualité étrange d'Arnold et les surfaces K3*, *C.R. Acad. Sci. Paris* **284A** (1977) 615; E. Looijenga, *The Smoothing Components of a Traingle Singularity. I*, Proc. Symp. Pure Math. **40** (1983) 173; E. Brieskorn, *The Unfolding of Exceptional Singularities*, *Nova Acta Leopoldina* **NF52 #240** (1981) 65.

[41] R.G. Pogossyan, *Study of the Vicinities of Superconformal Fixed Points in 2d Field Theory*, Yerevan preprint EPhN-1003(53)-87; Y. Kitazawa, N. Ishibashi, A. Kato, K. Kobayashi, Y. Matsuo, and S. Odake, *Nucl. Phys.* **B306** (1988) 425.

[42] I. Dolgachev, *Weighted Projective Varieties*, in SLN #956.

[43] J. Latorre and A. Lutken, *Constrained* $\mathbf{CP}^n$ *Models*, Nordita 88/42 (Dec. 1988).

[44] M. Oka, *Resolution of the 3d Brieskorn Singularities*, Adv. Stud. Pure Math **8** (86) 437.

[45] E. Martinec, to appear.

[46] T. Eguchi and A. Taormina, *Phys. Lett.* **210B** (1988) 125.

[47] D. Gepner, *Scalar Field Theory and String Compactifications* Princeton preprint PUPT-1115 (Dec. 1988).

[48] P. Slodowy, *Simple Singularities and Simple Algebraic Groups* SLN#815.

[49] D. Siersma, *Singularities of Functions on Boundaries, Corners, etc.*, Quart. J. Math. *Oxford* **33** (1981) 119.

[50] A. Gerasimov, A. Marshukov, A. Morozov, M. Olshanetsky, and S. Shatashvili, *Wess-Zumino-Witten Model as a Theory of Free Fields* ITEP preprint (March 1988).

[51] See, *e.g.*, M. Dine, P. Huet, and N. Seiberg, *Large and Small Radius in String Theory*, and references therein.

On the Explicit Construction of the Superstring Loop Measure in the Super-Schottky-Reggeon Formalism

J.L. Petersen[1],

Theory Division CERN, CH-1211 Genève 23, Switzerland

ABSTRACT

A technique for performing string perturbation theory in flat space time is reviewed with particular emphasis on the features arising as a consequence of the Ramond string complications. Details on results obtained so far are presented.

[1] On leave of absence from The Niels Bohr Institute, University of Copenhagen, Denmark

1 Introduction

Despite a massive effort on the part of a large number of groups and people in order to work out in detail (super-) string perturbation theory in flat space-time background, and despite the fact that the completion of such a programme would only represent a somewhat modest step towards a full understanding of string dynamics, many issues remain unclear (for a review, see [1]). Whereas the situation for the bosonic string is rather well understood, recent effort has concentrated on the Neveu- Schwarz-Ramond superstring. Techniques involved include direct considerations of the path integral [2,3,4] [5,6], an improved version of the dual resonance model oscillator formalism with ghosts included [7,8,9,10,11,12,13], considerations of algebraic geometry [14,15], exploitations of bosonization-techniques [16,17], relating determinants to θ-functions [18,19,20], a new operator formalism [21,22], a group theoretic approach [23,24,25,26] and techniques pertaining to special cases [27,28,29] (see also [1] for many more references).

Here we shall concentrate on an approach which attempts to unify the path integral and the operator (oscillator) approach. It is rather well developed for the bosonic string [30,29,31] and worked out in broad outline for the Neveu-Schwarz-Ramond superstring [32,33,34]. For the bosonic string it is closely related to the pure oscillator approach [8,9] as discussed in [35,36]. The Riemann surface structure of multiloop amplitudes emerges in a Schottky representation [37] from the process of "sewing" reggeons [38], a process which is seen as a way of gradually performing the path integral (see also [5]). However, emphasis is put on using a manifestly world-sheet supersymmetric formulation building on ideas of [39], the spirit in this respect being similar to [22,23]. Thus one looks for a super-Schottky description of super Riemann surfaces [40,41] which, however, turns out to differ from the one of [42].

The present approach consists in evaluating the Polyakov path integral in the BRST-formalism (in the Weyl-invariant situation of the critical dimension) with specified boundary conditions. Those are carefully chosen not to violate the complex structure (we think of closed strings) by choosing an external coherent state basis, thereby also making a translation to an oscillator language immediate. Thus, external currents are chosen to provide a classical (super-) string field, (super-) analytic on the world sheet with singularities encoding the coherent external state parameters [30,32,33,43,31]. Holomorphic [14,15] and superholomorphic factorization [44,4,45,22] are manifestly maintained.

The technique for building the measure on moduli space (counting tree diagram *Koba-Nielsen (KN) points on an equal footing with loop moduli*) consists in constructing (super-) Beltrami differentials which are distributions concentrated around certain closed contours and built as (antiholomorphic) derivatives of certain quasi-(super-) conformal transformations: ones that are (super-) conformal Killing vector fields, but discontinuous across the contour. Following [46] these find a natural place as ghost "zero-modes". They only couple to specific anti-ghost-modes which turn into the Schottky form of holomorphic quadratic differentials on the Riemann surface (super-holomorphic 3/2- differentials on the super-Riemann surface).

The result of working out loop amplitudes in such an approach consists in explicit expressions for the integrand on (super-) moduli space, in very particular choices of coordinates depending on the order in which 3-vertices are joined together to form a tree, and on how the branches of the tree are sewn to form loops (cf. also [38]). In this way moduli space gets covered with patches which are neighbourhoods of fully pinched surface-configurations. One may hope that such explicit expressions might be useful for better understanding the rather complicated structure of super moduli space.

In this approach there is a nearly obvious relation between sewing, factorization and the path integral [30]: On a punctured Riemann sphere corresponding to a particular string amplitude (with punctures corresponding to infinite external state "tubes", conformally mapped to points) draw a contour separating the surface in two. The full path integral may be thought of as (i) path integrating over fields living on only one or the other of the two parts with particular boundary conditions along the contour, and (ii) path integrating over the boundary conditions. These in turn refer to Laurent-expansions of the field on the contour with powers of opposite sign for the two halves of the surface. The first of the above steps builds amplitudes corresponding to the two halves, each with one external "reggeon" in a coherent state corresponding to the boundary condition: one reggeon is described by the positive power part of the Laurent-expansion on the contour, the other by the negative powers (in suitable coordinates). The second step is sewing - or in reverse - factorization. In sewing the two amplitudes together, a mapping must be provided to relate the coördinates on the two individual surfaces. This mapping plays the role of a propagator and introduces $(1|0)$ or $(1|1)$ (even|odd) additional moduli for a Neveu-Schwarz (NS) or Ramond (R) sewing. Building the Riemann surface by sewing is therefore a way of gradually doing the path integral (for a recent discussion, see [5]).

In sec. 2 we describe the comparatively simple situation of tree diagrams. In flat space-time the action is gaussian, and the result of the path integral is just given by the classical action evaluated at the classical field configuration with the specified boundary conditions (singularities). To write down the classical field in terms of singularities is trivial for the bosonic and the NS-string. For the R-string states, complications arise due to the square-root branch cuts [16,17,39]. Once the classical field is known the action is easily found by contour integrations.

In sect. 3 we briefly describe the treatment of BRST-invariant physical external states.

In sect. 4 the 3 NS-string vertex and the RR-NS vertex are treated with special emphasis on the differences: the 3 NS vertex depends on a $(0|1)$ modulus, the RR-NS vertex depends on no moduli.

In sect. 5 the NS- and R-sewings and their associated quasi-superconformal ghosts are described.

In sect. 6 we give some indications on how to perform loop calculations in the pure (non-modular invariant) NS-problem, and some general results are written down [32], obtained also in [9].

Sect. 7 describes how to treat the non-zero-mode part of the sewing for a completely general vacuum amplitude with g_{NS} NS-loops and g_R R-loops.

Finally in sect. 8 we venture to provide some details and techniques on how to perform the important ghost zero-mode-integrations providing the measure on moduli space. The present state of the art can handle a single R-loop together with an arbitrary number of NS-loops.

In sect. 9 we summarize the results.

2 Tree Diagrams and Notations

We consider a closed superstring with action [32,17,40]

$$\begin{aligned} S(X,B,C) &= \int dZd\overline{Z}\{\frac{1}{2}DX^{\mu}(Z,\overline{Z})\overline{D}X^{\mu}(Z,\overline{Z}) \\ &\quad +B(Z,\overline{Z})\overline{D}C(Z,\overline{Z})+\overline{B}(Z,\overline{Z})D\overline{C}(Z,\overline{Z})\} \end{aligned} \tag{2.1}$$

We shall be concentrating on the (super-) holomorphic part only, and hence we shall not even specify the meaning of $\overline{D}$ and $\overline{Z}$, that meaning being different for the heterotic and the type II superstring. $Z=(z,\theta)$, $dZ = dz d\theta$, $D=\partial/\partial\theta+\theta\partial/\partial z$ with $D^2=\partial/\partial z$ and θ is the grassmannian coordinate. The superfields for matter, ghosts and antighosts, which we shall often denote as the triple $\Phi=(X^{\mu},B,C)$, have components

$$\begin{aligned} X^{\mu}(Z) &= x^{\mu}(z)+\theta\psi^{\mu}(z) \\ B(Z) &= \beta(z)+\theta b(z) \\ C(Z) &= c(z)+\theta\gamma(z), \end{aligned} \tag{2.2}$$

and have superconformal dimensions 2, 3/2 and -1 respectively.

As explained above the result of performing the path integral with prescribed boundary conditions, i.e. singularities in $\Phi(Z)$, may be written as

$$W_N = \int d^{D_N}\mu \exp\{+S(\Phi_{Cl})\}(2\pi)^d\delta^d(\sum p_i)$$

$$\cdot \prod_{i=1}^{N} \exp\{-S(\Phi_i)\}\ , \ (d=10). \tag{2.3}$$

Here $d^{D_N}\mu$ is the measure on (tree-diagram-) moduli space which (see below) has $D_N = (N_B + 2N_F - 3|N_B + N_F - 2)$ (bosonic|fermionic) dimensions. N_B is the number of external bosons (Neveu-Schwarz or NS-string states) and $2N_F$ the number of external fermions (Ramond or R-string states). The factors $\exp\{-S(\Phi_i)\}$ are (divergent) external state normalizations removing some self contraction infinities in $S(\Phi_{Cl})$, (and some otherwise explicit Weyl-breaking dependence, cf. [31] and [30]).

Φ_{Cl} is expressed in terms of individual "reggeon" fields Φ_i, each specifying the singularities of a particular string state.

This is very simple for the bosonic string and the pure NS-string, which we consider first. In fact by sewing (sec. 5) we shall build

$$\Phi_{Cl}(Z) = \sum_{i=1}^{N} \Phi^{(i)}(Z), \ \Phi^{(i)}(Z) = V_i \Phi_i(Z). \tag{2.4}$$

Here a general "reggeon" field in the standard configuration (see below) is transported by a superconformal transformation V_i which in this NS-case is just an $OSp(2|1)$ transformation. For a superconformal field of superconformal dimension, h, and a superconformal map, V [40]

$$\begin{aligned} V : Z \rightarrow V(Z) = (V_B(Z), V_F(Z)) \quad , \quad DV_B(Z) &= V_F(Z) DV_F(Z) \\ V^{-1}\Phi_h(Z) &= (DV_F(Z))^{2h}\Phi_h(V(Z)). \end{aligned} \tag{2.5}$$

One parametrization of an $OSp(2|1)$ transformation is in terms of superspace points $U = (u, \mu)$, $V = (v, \nu)$ and a complex parameter, K

$$\begin{aligned} V(Z) &= (\tilde{z}, \tilde{\theta}) \\ \tilde{z} &= K^{-1}\frac{Z-U}{Z-V}(U-V), \quad U - V \equiv u - v - \mu\nu \ \ etc. \\ \tilde{\theta} &= K^{-1/2}(Z-V)^{-1}\{(\theta-\mu)(U-V) - (\mu-\nu)(Z-U)\} \end{aligned} \tag{2.6}$$

If the i'th external state is a coherent state with bosonic parameters $\{p_i, a_n^i, \beta_r^i, \gamma_r^i\}$ and fermionic parameters $\{\psi_r^i, b_n^i, c_n^i\}$, then the reggeon fields in the "standard" configuration take the form

$$\begin{aligned} X_i(Z) &= -ip_i \log z + i\sum_{n=1}^{\infty} \frac{a_n^i}{n} z^{-n} + i\theta \sum_{r=1/2}^{\infty} \psi_r^i z^{-r-1/2} \\ B_i(Z) &= \sum_{r=-1/2}^{\infty} \beta_r^i z^{-r-3/2} + \theta \sum_{n=-1}^{\infty} b_n^i z^{-n-2} \\ C_i(Z) &= \sum_{n=2}^{\infty} c_n^i z^{-n+1} + \theta \sum_{r=3/2}^{\infty} \gamma_r^i z^{-r+1/2}. \end{aligned} \tag{2.7}$$

Thus $V_i(0)$ is the KN-point Z_i. Since $\overline{D}$ on these vanish except at singularities, the classical action evaluates, by Stoke's theorem, to

$$S_{Cl}(\Phi_{Cl}) = \sum_{i,j}[\frac{1}{2}(X^{(i)}|X^{(j)}) + (B^{(i)}|C^{(j)})] \tag{2.8}$$

where the bilinear forms

$$\begin{aligned} (X^{(i)}|X^{(j)}) &= \oint_{Z_j} dZ\, DX^{(i)}(Z)X^{(j)}(Z) \\ (B^{(i)}|C^{(j)}) &= \oint_{Z_j} B^{(i)}(Z)C^{(j)}(Z) \end{aligned} \tag{2.9}$$

are symmetric and $OSp(2|1)$ invariant (technical complications arise for "zero-modes" labelled by p, b_n, $n = \pm 1, 0$, β_r, $r = \pm 1/2$, see refs.[30,32] and below). The super-contour integrals are evaluated by the super-Cauchy rule:

$$\oint_Z dZ' \frac{\theta' - \theta}{Z' - Z} F(Z') = F(Z) \tag{2.10}$$

(a factor $2\pi i$ is absorbed in dZ) and identities obtained by acting with D's on that.

The measure is obtained by the process of sewing and has a general form like ($N_B = N$ here)

$$\begin{aligned} d^{(N-3|N-2)}\mu &= \prod_{i=1}^{N-3} d\log K_i \prod_{i=1}^{N-2} d\vartheta_i \\ &\cdot \int \prod_{j=1}^{N-3} d\check{c}_{s_j} \prod_{i=1}^{N-2} d\check{\gamma}_{\vartheta_i} \prod_{n=-1}^{1} d\check{c}_n \prod_{r=-1/2}^{1/2} d\check{\gamma}_r \\ &\cdot \exp(B_{Cl}|\sum \check{C}_{s_i} + \check{C}_0). \end{aligned} \tag{2.11}$$

K_i and ϑ_i are the tree diagram bosonic and fermionic moduli, $\check{C}_{s_i}$ are the quasi (super-) conformal fields, expressed in terms of the modes

$$\{\check{c}_{s_i}, \check{\gamma}_{\vartheta_i}, \check{c}_n, \check{\gamma}_r\}, \tag{2.12}$$

discontinuous across the "sewing contours" C_{s_i} so that

$$(B_{Cl}|\check{C}_{s_i}) = \oint_{C_{s_i}} dZ\, B_{Cl}(Z)\check{C}_{s_i}(Z) \tag{2.13}$$

whereas the superconformal Killing vector field (SCKVF) – describing the ghost zero-modes – is given by

$$\check{C}_0(Z) \equiv [\check{c}_{-1}z^2 + \check{c}_0 z + \check{c}_1 + \theta(\check{\gamma}_{-1/2}z + \check{\gamma}_{1/2})]\vartheta(C_0) \tag{2.14}$$

with $\vartheta(C_0)$ vanishing outside the contour C_0 which encloses all reggeon singularities (or, equivalently a single spurious singularity, see sect. 8 below).

We use the *check* accent to denote the quasi superconformal and zero-mode degrees of freedom.

The general situation when Ramond string states are present requires more care due to the presence of square root branch points at arbitrary points in superspace. We introduce the following definitions:

1. We say $\Phi \in \mathcal{R}$ iff there exists a Laurent type expansions (*for the entire classical field*)
$$\begin{aligned} X(Z) &= q - ip\log z + i\sum_{n\neq 0}\frac{a_n}{n}z^{-n} + i\theta\sum_{n\in Z}\psi_n z^{-n-1/2} \\ B(Z) &= \sum_{n\in Z}\beta_n z^{-n-3/2} + \theta\sum_{n\in Z} b_n z^{-n-2} \\ C(Z) &= \sum_{n\in Z} c_n z^{-n+1} + \theta\sum_{n\in Z}\gamma_n z^{-n+1/2}. \end{aligned} \tag{2.15}$$

2. For any superconformal transformation, V (i.e. $DV_B = V_F DV_F$)
$$\mathcal{R}[V] \equiv \{\Phi | \exists\tilde{\Phi} \in \mathcal{R}:\ \Phi = V\tilde{\Phi}\}. \tag{2.16}$$

3. For any number of maps V_i, we define
$$\mathcal{R}[V_1, V_2, \ldots] = \bigcap_i \mathcal{R}[V_i]. \tag{2.17}$$

Thus the classical field belonging to a situation describing N_B external NS-strings and $2N_F$ external R-strings will have N_F branch cuts and belong to a space $\mathcal{R}[V_1, V_2, \ldots, V_{N_F}]$. We shall see that contrary to the NS-case, these V_i's in general do *not* belong to $OSp(2|1)$.

It is crucial to notice that the symmetry, or equivalently, the set of SCKVF's in this case gets increased. Indeed, whereas in the pure NS-problem we only had invariance under
$$\Phi_{Cl} \rightarrow V\Phi_{Cl},\ V \in OSp(2|1) \tag{2.18}$$
(all other superconformal transformations would introduce new singularities, thereby violating the boundary conditions) corresponding to (3|2) degrees of freedom ($osp(2|1)$ is generated by $(L_{-1}, L_0, L_1|G_{-1/2}, G_{1/2})$), we now have the following larger invariance:

Define the "super-dilation" (generated by L_0 and G_0)
$$d_{[K,\kappa]} : Z = (z,\theta) \rightarrow d_{[K,\kappa]}(Z) = (K(z + \theta\kappa\sqrt{z}), \sqrt{K}(\theta + \kappa\sqrt{z})). \tag{2.19}$$
The action (and the amplitude) is now invariant under
$$\begin{aligned} \Phi &\rightarrow (V_i \circ d_{[K,\kappa]} \circ V_i^{-1})\Phi \in \mathcal{R}[\tilde{V}_1, \ldots, \tilde{V}_{N_F}] \neq \mathcal{R}[V_1, \ldots, V_{N_F}] \\ \tilde{V}_j &\equiv V_i \circ d_{[K,\kappa]} \circ V_i^{-1} \circ V_j. \end{aligned} \tag{2.20}$$

Thus we have $(3|2+N_F)$ invariances resulting in the well-known $(N_B + 2N_F - 3|N_B + N_F - 2)$ moduli of the general tree-diagram.

Similarly we have more SCKVF modes. In fact for each Ramond-cut there is an additional contribution of

$$\Delta_i \check{C}_0(Z) = \theta \check{\gamma}_0^R \sqrt{z} \vartheta(C_0), \; i = 1, \ldots, N_F \tag{2.21}$$

each of these being transported by their respective V_i as a field with superconformal dimension -1, and $\check{\gamma}_0^R$ denoting the associated d.o.f.

3 External (Physical) States

Begin with the ghost part and use the notation of ref.[17]. We have the vacua $|q, q'>$ where q and q' are ghost numbers of (b, c) and (β, γ) respectively. In the general notation (a hat denotes a quantum oscillator operator as distinct from a c-number or Grassmann-number)

$$\hat{b}_n|q>=0, \; n > \epsilon q - \lambda; \quad \hat{c}_n|q>=0, \; n \geq -\epsilon q + \lambda \tag{3.1}$$

where $\epsilon = 1, \; \lambda = 2$ for the (b, c) and $\epsilon = -1, \; \lambda = 3/2$ for the (β, γ).

NS-coherent states are built from the vacuum $|0,0>$ whereas R-coherent states are built from $|0, 1/2>$.

A useful observation is [39,32,22]

$$\begin{aligned} |q+\epsilon> &= \delta(\hat{c}_{-\epsilon q+\lambda-1})|q> \\ |q-\epsilon> &= \delta(\hat{b}_{\epsilon q-\lambda})|q> \\ \delta(\hat{c}_n) &\equiv \int db_{-n} \exp(b_{-n}\hat{c}_n) \\ (&= \hat{c}_n \text{ for } \hat{c}_n \text{ fermionic}). \end{aligned} \tag{3.2}$$

For physical strings we may take ghosts to be in the $|1, -1>$ and $|1, -1/2>$ vacua respectively for NS- and R- strings [17].

Our formalism directly gives amplitudes in coherent state bases. For an arbitrary state, however, they are easily obtained by either (i) replacing coherent state parameters by oscillators and letting the ensuing operator act on the desired states (with suitable vacua to the left, this is the most common formalism, cf. [8,9,7,23,12] etc.); or (ii) equivalently we may fold by coherent state wave functions of the desired states, amounting to acting on our coherent state amplitude with certain differential operators, obtained by replacements like $\hat{a}^i_{-n} \rightarrow \partial/\partial a^i_n$ etc. in the constructions of the desired external states.

For Ramond cuts the fermionic zero-mode ψ_0^μ is identified by (however, cf. also [34])

$$<\Theta_1|\hat{\psi}_0^\mu|\Theta_2> = <\Theta_1|\Theta_2> \psi_0^\mu \tag{3.3}$$

where the R-vacua $< \Theta_1|$, $|\Theta_2>$ describe the corresponding pair of spinors. The operators $\hat{\psi}_0^\mu$ satisfy the Clifford algebra of the gamma matrices. In expressing the spinor states in a coherent state basis corresponding to $\hat{\psi}^{[i]} = \hat{\psi}^i + i\hat{\psi}^{i+5}$ (and the hermitean conjugate) one convinces oneself that the Grassmann parameters ψ_0 translate into antisymmetrized products of gamma matrices, and the spinor vacua to suitable spinors u_1, u_2 [39,32,34].

4 The 3 String Vertices

4.1 The 3-NS string vertex

See refs. [7,8,32,9,11,12]. By $OSp(2|1)$ invariance we may arrange that

$$Z_1 = (0,0), \quad Z_2 = (\infty, 0), \quad Z_3 = (1, \vartheta) \tag{4.1}$$

For simplicity we chose

$$V_1(Z) = Z, \quad V_2(z,\theta) = (1/z, i\theta/z), \quad V_3(Z) = (1-z, \vartheta - \theta). \tag{4.2}$$

But unlike for the 3-reggeon bosonic vertex, a fermionic modulus, ϑ remains. Our main point here is the fact that this gives rise to an associated superconformal ghost field describing the change in that modulus, while keeping the KN points at their form eq.(4.1). There are two possible forms of $\delta\theta$ that may be employed. We think in terms of the conformal dimension $-1/2$ field $\gamma(z)/\sqrt{dz}$:

$$1/\sqrt{dz} \quad \text{and} \quad z/\sqrt{dz} \tag{4.3}$$

related by $z \to 1/z$ and vanishing at ∞ and 0 respectively. Thus the first may be used with C_ϑ an anticlockwise circle of radius < 1 and vanishing inside (case 1), and the second with a clockwise circle of radius > 1 and vanishing outside (case 2).

At this point, consider in general the form of a (ghost-) displacement field: Put $\tilde{Z} = (z + \delta z, \theta + \delta\theta)$ and get [46] $C(Z) \sim \tilde{Z} - Z = \delta z + \theta\delta\theta$. However, both δz and $\delta\theta$ may depend in general on both z and θ. Requiring $\tilde{Z}$ to be a superconformal map gives

$$D(\delta z) = \delta\theta + \theta D(\delta\theta) \tag{4.4}$$

Putting $\delta z(z,\theta) = \delta z_B + \theta\delta z_F$ one finds $\theta\delta z_F = \theta\delta\theta$. Hence [1]

$$C(Z) = \delta z_B + 2\theta\delta\theta \tag{4.5}$$

So we take

$$\check{C}_\vartheta(Z) = 2\theta\check{\gamma}_\vartheta\vartheta(C_\vartheta)\begin{cases} 1 & \text{(case 1)} \\ z & \text{(case 2)} \end{cases} \tag{4.6}$$

When integration over the SCKVF and this ghost mode, $\check{\gamma}_\vartheta$, is performed there arise 3 bosonic and 3 fermionic antighost delta functions. This structure is generic.

[1] I am very grateful to K.O. Roland for discussions on this factor of 2.

4.2 The RR-NS 3-string vertex

See [23,12,32,22]. There are no moduli in this case because we may use the extra invariance eq.(2.20) to put $Z_3 = (1,0)$. Thus, compared to the 3-NS vertex, this extra SCKVF mode results in the "measure factor" being replaced by the "constant"

$$d^0\mu[RR - NS] = \int d\check{c}_{-1} d\check{c}_0 d\check{c}_1 d\check{\gamma}_{-1/2} d\check{\gamma}_{1/2} d\check{\gamma}_R \exp\{(B_{Cl}|\check{C}_{0,R})\} \qquad (4.7)$$

with

$$\check{C}_{0,R}(Z) = \check{C}_0(Z) + \theta\check{\gamma}_R\sqrt{z}\vartheta(C_0) \qquad (4.8)$$

$\check{\gamma}_R$ being the d.o.f. associated with the extra invariance. This form again gives rise to 3 fermionic and 3 bosonic delta functions in the anti-ghost modes, when integrations over the $\check{C}_{0,R}$ modes have been performed. But the origin of the 3'd bosonic delta function is quite different from the NS-case.

The square root cut causes some subtleties. Let us choose the Ramond reggeons to be at $Z_1 = (0,0)$ and $Z_2 = (\infty, 0)$. The contribution to ψ^μ_{Cl} from those is easily parametrized as

$$\psi^{\mu(1+2)}_R(z) = i\sum_{n=-\infty}^{\infty} \psi^\mu_n z^{-n-1/2} \qquad (4.9)$$

where the terms with $n < 0$ refer to reggeon 2 at $z = \infty$ (the modes ψ_{-n} differ from the coherent state parameters of that reggeon by a phase found by considering the superconformal map $z \to 1/z$, $\theta \to i\theta/z$).

For reggeon 3 at $z = 1$ ($\theta = 0$) we would use the following form in a 3-NS vertex:

$$\psi^{\mu(3)}_{NS}(z) = i\sum_{r=1/2}^{\infty} \psi^{\mu(3)}_r (1-z)^{-r-1/2}. \qquad (4.10)$$

However, although this field has the required singularity near $z = 1$ it clearly does not belong to the proper space $\mathcal{R}$ (sect. 2). Thus we introduce a projection operator, projecting on to that space, with the requirement that the square root cut introduced by this projection should not modify the modes of reggeons 1 and 2. In the present case of a single R-cut, it is easy to provide an explicit form of this projection:

$$\begin{aligned}\psi^{\mu(3)}_R(z) &\equiv \mathcal{P}_{\mathcal{R}}(\psi^{\mu(3)}_{NS}(z)) \\ &= \oint dz' \left[\frac{z+z'}{2\sqrt{zz'}(z-z')}\right] \psi^{\mu(3)}_{NS}(z'). \end{aligned} \qquad (4.11)$$

One easily checks that this is indeed a projection (cf. [34] for a slightly modified treatment). In eq.(4.11) a careful treatment has been provided for the ψ^μ_0 zero-mode. This form is dictated by the requirement of it yielding the correct Ramond propagator which therfore appears in the integration kernel [32].

For the ghost and anti-ghost fields, similar projections have to be considered, but we leave them for the more explicit discussion in sect. 8.

For illustration we give the result of performing the evaluation of the action and the amplitude for the case where the ghosts belong to the physical vacuum (sec. 3). The wave function factors in eq.(2.3) do not know about the above projection to the $\mathcal{R}$ spaces and hence the cancellation of infinite self-contractions in this case leaves a finite contribution [32]. This is a new feature. A simple point splitting technique gives the following oscillator result:

$$
\begin{aligned}
& W[RR-NS] = \\
= \quad & (2\pi)^d \delta^d(\sum p_i) \prod_{i=1}^{3} (V_i'(0))^{p_i^2/2} \exp\{\sum_{n>0} \frac{1}{2} \frac{p_i a_n^i}{n!} \partial_z^n \log V_i'(z) \,|_{z=0} \} \\
& \cdot \exp \left\{ \frac{1}{2} \sum_{i \neq j} \sum_{n,m \geq 0} \frac{a_n^i}{n!} \frac{a_m^j}{m!} \partial_z^n \partial_y^m \log(V_i(z) - V_j(y)) \,|_{z=y=0} \right. \\
& + \frac{1}{4} \sum_{r,s \geq 1/2} \frac{\psi_r^3}{(r-1/2)!} \frac{\psi_s^3}{(s-1/2)!} \partial_y^{(r-1/2)} \partial_x^{(s-1/2)} \\
& \left[\frac{\frac{1}{2}}{x-y} \left(\sqrt{\frac{x+1}{y+1}} + \sqrt{\frac{y+1}{x+1}} - 2 \right) \right] |_{y=x=0} \\
& - \sum_{r,n \geq 0} \frac{\psi_n}{n!} \frac{\psi_r^3}{(r-1/2)!} \partial_z^n \partial_y^{(r-1/2)} \frac{z+y+1}{2\sqrt{y+1}(z-y-1)} \,|_{z=y=0} \\
& + \sum_{r \geq 1/2, n<0} \frac{\psi_n}{(-n)!} \frac{\psi_r^3}{(r-1/2)!} \partial_z^{(-n)} \partial_y^{(r-1/2)} \frac{zy+z+1}{2\sqrt{y+1}(1-zy-z)} \,|_{z=y=0} \\
& \left. + \sum_{n>0} \psi_{-n} \psi_n \right\} \qquad (4.12)
\end{aligned}
$$

Here we have had in mind the choices $V_1(z) = z$, $V_2(z) = 1/z$, $V_3(z) = 1 - z$. This expression[32] satisfies the overlap conditions [23] and agrees with [12] when allowance is made for the different (and simpler) choice of V_i's employed here.

5 Sewing

Consider sewing two tree diagrams. One diagram is expressed in terms of classical fields $(\Phi_A + \Phi_\alpha)(Z^I)$, where Φ_A denotes the contribution from any number of external string states (singularities) *not* to be sewn, and Φ_α denotes the contribution from the reggeon to be sewn. Z^I is the world sheet super variable for that diagram. Similarly the other diagram is described by $(\Phi_B + \Phi_\beta)(Z^{II})$.

Sewing Φ_α and Φ_β to form a new diagram involves the following:

1. Variables Z^I and Z^{II} must be identified by a "propagator" mapping. This involves some ambiguity, part of which is associated with the overall invariances of the resulting diagram, and part of which should be viewed as giving rise to different coordinations of the same thing. To fix matters, choose KN-points Z^I_A and Z^{II}_B arbitrarily among the KN-variables of the $A-$ and $B-$ sets, and choose coordinates such that

$$Z^I_\alpha = (0,0),\ Z^{II}_\beta = (\infty,0),\ Z^I_A = (\infty,0),\ Z^{II}_B = (0,0). \qquad (5.1)$$

Then the possible "relative mappings" for the propagators are, respectively for NS and R-sewings:

$$\begin{aligned} NS:\ P = d_{[K]} &\ :\ z \to Kz,\ \theta \to \sqrt{K}\theta \\ R:\ P = d_{[K,\kappa]} &\ :\ z \to K(z + \theta\kappa\sqrt{z}),\ \theta \to \sqrt{K}(\theta + \kappa\sqrt{z}). \end{aligned} \qquad (5.2)$$

The phase choice of $\sqrt{K}$ will become a spin structure choice in the loop problem (sects. 6,7,8).

2. Now sewing is the rule of gaussian integration:

$$\begin{aligned} \int \mathcal{D}(\Phi_\alpha, \Phi_\beta) &\cdot \exp\{+S[\Phi_A + \Phi_\alpha]\}\exp\{+S[\Phi_B + \Phi_\beta]\} \\ &\cdot \exp\{-S[\Phi_\alpha + P\Phi_\beta]\} \\ &= \exp\{+S[\Phi_A + P\Phi_B]\}. \end{aligned} \qquad (5.3)$$

In the integrand the first two factors are the amplitudes to be sewn, the third is the "sewing"- i.e. path integral weight. The result is the full, sewn amplitude. In the process the V_i^B's of the B-field have been changed to PV_i^B. This is how the final V_i's of the full amplitude are gradually being built up by successive sewings from the "trivial" V_i's of the 3-vertices.

3. When Ramond-cuts are present in Φ_A and/or Φ_B we must first write

$$\Phi_A + \Phi_\alpha = \mathcal{P}_\mathcal{R}(\Phi_A) + \tilde{\Phi}_\alpha \qquad (5.4)$$

(and similarly for the other amplitude) where the projection operator projects from the sort of Ramond-space (sect. 2) relevant for Φ_A, Φ_α, $\Phi_A + \Phi_\alpha$, to the one relevant for $\Phi_A + P\Phi_B$. The crucial observation is that this implies a shift in Φ_α, the path integral variable, *compatible* with the singularities of that field; i.e. the extra cuts that have to appear after the projection, may be described by the type of Laurent expansion that defines Φ_α, i.e. the allowed location of the singularities of Φ_α. [32].

4. Corresponding to the NS- and R-propagators eq.(5.2), we must construct quasi-superconformal ghost fields generating the shifts $K \to K + dK$ and $\kappa \to \kappa + d\kappa$ in order to calculate the measure. The result is found (cf. eq.(4.5)) to be [32]

$$\begin{aligned} NS: \check{C}_K(Z) &= \check{c}_K \cdot z\vartheta(C_K) \\ R: \check{C}_{K,\kappa}(Z) &= [\check{c}_K \cdot z + \check{\gamma}_\kappa(-\kappa z + 2\theta\sqrt{z})]\vartheta(C_{K,\kappa}) \end{aligned} \quad (5.5)$$

where the contours C_K and $C_{K,\kappa}$ include all $P\Phi_B$-singularities and exclude all Φ_A-singularities, i.e. run around the tube connecting the two diagrams to be sewn. The associated measure factor, say for R-sewing, is then

$$(d\mu)_{K,\kappa} = d\log K d\kappa \int d\check{c}_K d\check{\gamma}_\kappa \exp\{\oint_{C_{K,\kappa}} dZ B_{Cl}(Z)\check{C}_{K,\kappa}(Z)\} \quad (5.6)$$

which is easy to evaluate to a product of a fermionic and a bosonic delta function in B_{Cl}-field modes.

5. The sewing is a path integral entirely analogous to the one in the tree diagram, sect. 2. For the same reason then we get δ-functions arising from zero-mode integrations: momentum conservation in the sewing, and δ-functions in the lowest antighost sewing modes. We shall mostly imagine that these δ-functions have been used to integrate out antighost modes b_n, β_n with $|n| \le 1$ in one of the reggeons. In the case of tree diagrams, the same modes of the other sewing reggeon may be integrated out and we are left with the requisite antighost δ-functions for the entire diagram. In the case of loops (sect. 8) these last integrations provide what we refer to as the "measure factor" in loop calculations.

We see that adding a 3 vertex always gives rise to 1 more bosonic and 1 more fermionic modulus. But their origin is quite different for NS- and R-sewing. This means that the requirement of duality works out in a rather non-trivial way. Thus, we may consider the well-known example of the 2R-2NS amplitude. It may be obtained either by NS-sewing or by R-sewing, the two procedures giving rise to very different coodinate descriptions, indeed two very different choices for the (1|1) moduli. It is gratifying to check [32] by explicit computation that the two calculations give rise to the same (well-known) physical amplitude.[2]

6 Sewing Loops from a General NS-Tree

We first consider this simple but non modular invariant case. Out of a total of 2^{2g} spin structures (a concept strictly speaking only making sense

[2]K.O. Roland has recently revised the calculation and found a missing factor of 2 in both cases in [32].

near pinching limits, or, when the super moduli vanish) this corresponds to the subset of 2^g spin structures (around b-cycles) arising from the different possible phase choices of $\sqrt{K_i}$ in the fermionic part of the sewing propagator (cf. eq.(6.2), eq.(5.2)). The remaining spin structures (around a-cycles) require sewings of any combination of R- and NS-reggeons (cf. sect. 7,8).

The calculational techniques represent a very straightforward generalization of the ones developed for the bosonic string [30]. Indeed, the results obtained for the most general N-string g-loop amplitude [32] were also obtained by a simple supersymmetrization of the corresponding bosonic result [9].

We decompose matter, ghost and antighost superfields for an NS-tree amplitude with $2g+N$ external lines as

$$\begin{aligned} \Phi_{tot}^{Cl} &= \sum_{i=1}^{g}(\Phi_{\alpha_i}+\Phi_{\beta_i})+\Phi_e \\ \Phi_e &= \sum_{k=1}^{N}\Phi^{(k)}=\sum_{k=1}^{N}V_k\Phi_k \end{aligned} \tag{6.1}$$

An advantage in the formalism is that the detailed form of Φ_e expressed in terms of external fields is of no concern. We may even regard the total Φ_e as describing one single reggeon.

Reggeons Φ_{α_i} and Φ_{β_i} are to be sewn together to form g loops. This is done using the g "propagator" $OSp(2|1)$ transformations P_i, with fixed points Z_{α_i} and Z_{β_i} and with multipliers K_i. These concepts are defined by

$$P_i=\Gamma_i^{-1}\circ d_{K_i}\circ\Gamma_i \tag{6.2}$$

where Γ_i is of the form eq.(2.6) mapping Z_{β_i} and Z_{α_i} to $(0,0)$ and $(\infty,0)$, and d_{K_i} is a dilation with multiplier K_i.

The loop amplitude is then constructed as

$$\begin{aligned} &\int d\check{C}\int\mathcal{D}[\Phi_{\alpha_i},P_i^{-1}\Phi_{\beta_i}]\exp[+S(\Phi_{tot}^{Cl})]\exp[(B_{tot}^{Cl}|\check{C})] \\ &\cdot\prod_{i=1}^{g}\exp[-S(\Phi_{\alpha_i}+P_i^{-1}\Phi_{\beta_i})] \end{aligned} \tag{6.3}$$

where $\check{C}$ is the sum of all the quasi-superconformal ghost modes introduced by the various sewings. In sect. 8 we shall treat in some detail the evaluation of $\int d\check{C}\exp[(B_{tot}^{Cl}|\check{C})]$.

The integral over loop modes is gaussian. It is convenient to distinguish between "zero-modes" (momentum modes and ψ_0^μ modes for the matter fields; ghost/antighost "zero-modes", see sect. 8) and non-zero-modes. The integral over momentum modes gives rise to a (super-) period matrix and is the only part of the result that breaks manifest (super-) holomorphic factorization [14,15,1,30,32]. The integral over non-zero-modes (performed first) involves the $OSp(2|1)$-mapping valued $2g \times 2g$ matrix A, easily obtained from eq.(6.3) acting between the bilinear forms eq.(2.8)

$$\mathsf{A} = \begin{pmatrix} \Pi - I & \Pi - P^{-1} \\ \Pi - P & \Pi - I \end{pmatrix} \tag{6.4}$$

where

$$\begin{aligned} \Pi_{ij} \equiv 1 \quad &, \quad I_{ij} \equiv \delta_{ij} \\ P_{ij} \equiv P_i \delta_{ij} \quad &, \quad P^{-1}_{ij} \equiv P_i^{-1}\delta_{ij} \\ i,j \;&=\; 1,\ldots,g \end{aligned} \tag{6.5}$$

Expressing eq.(6.3) in terms of A it is seen to act between a row and a column of the fields (for the zero-modes, a modification occurs [30]) $\{|\Phi_{\alpha_i}\rangle,\ |\Phi_{\beta_i}\rangle\}$. Standard techniques [37,30,32,8,9] allow to evaluate det(A) in terms of the spectrum of the elements, γ, of the the (super-) Schottky group, S, generated by the propagators, $\{P_i,\ i = 1,\ldots,g\}$. This spectrum is almost the same whether A acts between dim 0 matter super fields or between dimension 3/2 and $-1/2$ anti-ghost and ghost super fields. However, the lowest eigenvalue is absent in the ghost case (see the expression eq.(6.13)). The analogous feature for the bosonic string has been discussed [30,8] and found [29] to be crucial for consistency with general results from algebraic geometry [27], modular invariance in particular.

In addition to producing the determinants, the gaussian integration results in evaluating the action on some shifted, critical fields, expressed in terms of A^{-1}. This last operator has the effect of converting the original tree diagram external field Φ_e into an "automorphized" form, making it well defined on the ensuing (super-) Riemann surface (automorphic with respect to the Schottky group in the covering space).

If we restrict ourselves to external string states with ghosts in the physical vacua of sect. 3, the final result may be written as [32]

$$\begin{aligned} W_{N,g} \;=\;& (2\pi)^d \delta^d(\textstyle\sum p_i)(\det 2\hat{T})^{-d/2} \\ &\cdot \prod_{(\gamma)}{}' \prod_{n=1}^{\infty} \left[\frac{1-K_\gamma^{n-1/2}}{1-K_\gamma^n}\right]^d \cdot \prod_{(\gamma)}{}' \prod_{n=2}^{\infty} \left[\frac{1-K_\gamma^{n-1/2}}{1-K_\gamma^n}\right]^{-2} \\ &\cdot \exp\{\frac{1}{2}\int DX_e^{Aut}\overline{D}X_e^{Aut}\} \\ &\cdot \exp\left\{\frac{1}{4}\sum_{i,j}\left(\int[\omega_i\overline{D}X_e^{Aut} + DX_e^{Aut}\overline{\omega}_i]\right.\right. \\ &\left.\left.(2\pi\hat{T})^{-1}_{ij}\int[\omega_j\overline{D}X_e^{Aut} + DX_e^{Aut}\overline{\omega}_j]\right)\right\} \\ &\cdot d\mu_g d\mu_N \\ d \;\equiv\;& 10. \end{aligned} \tag{6.6}$$

Here

$$X_e^{Aut}(Z) = \sum_{\gamma \in S} \gamma X_e^{Aut} \tag{6.7}$$

is the automorphized external matter field. The ω_i's are the superholomorphic half-forms (cf. eq.(2.5) for notation)

$$\omega_i(Z) = \sum_{\gamma}{}^{(i)} \left[\frac{\theta - \gamma_F(Z_{\alpha_i})}{Z - \gamma(Z_{\alpha_i})} - \frac{\theta - \gamma_F(Z_{\beta_i})}{Z - \gamma(Z_{\beta_i})} \right] dZ \tag{6.8}$$

($\sum{}^{(i)}$ omits elements $\gamma \in S$ with a rightmost P_i^n, $n \neq 0$; similarly for ${}^{(i)}\sum$),

$$\hat{T}_{ij} \equiv \frac{1}{2\pi i}(T_{ij} - \overline{T}_{ij}) \tag{6.9}$$

$$\begin{aligned} T_{ij} = \int_Z^{P_i(Z)} \omega_j &= \frac{1}{2\pi i}(\delta_{ij} \log K_i - \\ &- {}^{(i)}\sum_{\gamma}{}'^{(j)} \log \left[\frac{Z_{\beta_i} - \gamma(Z_{\alpha_j})}{Z_{\beta_i} - \gamma(Z_{\beta j})} \cdot \frac{Z_{\alpha_i} - \gamma(Z_{\beta_j})}{Z_{\alpha_i} - \gamma(Z_{\alpha_j})} \right] \Bigg) \end{aligned} \tag{6.10}$$

(The prime means the identity is absent for $i = j$; $\overline{T}$ is a similar expression for anti-holomorphic modes; in general $\overline{T}_{ij} \neq T_{ij}^*$). In eq.(6.6) we have treated DX_e^{Aut} etc. as forms, including a factor of dZ. The primed products over (γ) denote products over cyclic classes of primary elements [37,30].

The measure factors may be expressed in various ways, for example as

$$\begin{aligned} d\mu_g &= \prod_{i=1}^{g} \frac{dK_i}{K_i^{3/2}} \left(\frac{1 - K_i^{1/2}}{1 - K_i} \right)^{-2} \prod_{i=1}^{g} \frac{1}{Z_{\alpha_i} - Z_{\beta_i}} \prod_{i=2}^{g-1} dZ_{\alpha_i} dZ_{\beta_i} \cdot dZ_{\alpha_g} d\theta_{\alpha_1} \\ d\mu_N &= \prod_{k=1}^{N} \frac{dZ_k}{DV_{k,F}(0)} \end{aligned} \tag{6.11}$$

where we used $OSp(2|1)$ invariance to fix $Z_{\beta_g} = (0,0)$, $Z_{\beta_1} = (\infty, 0)$, $Z_{\alpha_1} = (1, \theta_{\alpha_1})$. The calculation of $d\mu_g$ will be treated in some detail in sect. 8. A manifest $OSp(2|1)$ generalization is easily written down [9].

Using contour integration techniques

$$\begin{aligned} \int d^2Z \overline{D} X^{(k)}(Z) D X^{(l)}(Z) &= - \oint_{Z_l} dZ X^{(k)}(Z) D X^{(l)}(Z) \\ D X^{(l)}(Z) &= - \oint_{Z_l} dZ' \frac{\theta' - \theta}{Z' - Z} D X^{(l)}(Z') \\ X^{(k)}(Z) &= + \oint_{Z_k} dZ'' \log[Z'' - Z] D X^{(k)}(Z'') \end{aligned} \tag{6.12}$$

the two exponential factors in eq.(6.6) may be expressed in terms of oscillator modes of the external (coherent) states. In the calculation one must remember the peculiar properties of momentum modes under conformal transformations[30,32]. The result is [32]

$$\prod_{k=1}^{N} \exp\left\{\sum_{m=0}^{\infty} \frac{p_k A_m^{(k)}}{[m]!} D_{Z'}^{2m} \log DV_{k,F}(Z')\right\} |_{Z'=0}$$
$$\cdot \exp\left\{\frac{1}{2}\sum_{\gamma\in\mathcal{S}}{}' \sum_{n,m=0}^{\infty}\sum_{k,l=1}^{N} \frac{A_n^{(k)}}{[n]!} D_Z^{2n} \frac{A_n^{(l)}}{[m]!} D_{Z'}^{2m} \log[V_k(Z) - \gamma V_l(Z')]|_{Z=Z'=0}\right\}$$
$$\cdot \exp\left\{\frac{1}{4}\sum_{n,m=0}^{\infty}\sum_{k,l=1}^{N} \frac{A_n^{(k)}}{[n]!} D_Z^{2n} \frac{A_m^{(l)}}{[m]!} D_{Z'}^{2m}\right.$$
$$\left.\Omega_i(V_k(Z))(2\pi\hat{T})_{ij}^{-1}\Omega_j(V_l(Z'))|_{Z=Z'=0}\right\}, \tag{6.13}$$

where $\sum'_{\gamma\in\mathcal{S}}$ means that the identity element is absent from the sum whenever $k = l$; $[n]$ denotes the integer part of n, and the sum over n, m extends over all positive half odd integers and over all non-negative integers, and we use the notation [9,32]

$$A_n^{(k)} = \begin{cases} a_n^{(k)} & n \text{ integer} \\ \psi_n^{(k)} & n \text{ half odd integer} \end{cases} \tag{6.14}$$

and $a_0^{(k)} \equiv p_k$ for notational convenience. The first Abelian superintegrals, Ω_i, satisfy $D\Omega_i = \omega_i$ and may be given by

$$\Omega_i(Z) = \sum_{\gamma\in\mathcal{S}} {}^{(i)} \log \frac{Z - \gamma(Z_{\alpha_i})}{Z - \gamma(Z_{\beta_i})}. \tag{6.15}$$

The convergence of the exponent of eq.(6.13) rests on momentum conservation: $\sum a_0^{(k)} = 0$. The expression may then further be related [32] in a very simple way to the NS-super-prime form [9]. Notice that the term $\gamma = I$ represents a convenient oscillator expression for the tree diagram (cf. [7,23,47]).

7 The Non-Zero Mode Part of the Most General Vacuum Diagram

To calculate the most general vacuum diagram built by sewing g_R loops of R-reggeons and g_{NS} loops of NS-reggeons, it is convenient to distinguish between zero-modes and non-zero-modes.

The zero-modes may further be separated into (i) the momentum modes giving rise to some generalized form of period matrix (a contribution breaking (super-) holomorphic factorization), (ii) the fermionic matter zero-mode ψ_0^μ, and (iii) the quasi superconformal ghost modes and the (super-) Teichmüller anti-ghost modes to which they couple. This last part is of particular interest since it is here that the measure on (super-) moduli space resides. It will be treated in some detail in the next section.

In this section we treat the (infinitely many) non-zero-modes in the most general case. It is interesting that apparently a very simple structure emerges, even though the functional dependence on (moduli|supermoduli) is quite complicated.

For simplicity we restrict ourselves to coordinates corresponding to building the $g = g_R + g_{NS}$ loop vacuum amplitude from a g-point NS-tree amplitude by attaching g_{NS} 3-NS-vertices and g_R RR-NS vertices to the legs, thereby using NS-sewings only, cf. fig.1. This has the simplifying consequence that the corresponding tree diagram is built entirely using $OSp(2|1)$ transformations, V_i. The loop amplitude is then constructed using the propagator transformations (cf. eq.(5.2)) P_i^{NS}, P_j^R characterized as follows:

1. For the g_{NS} NS-loops, $P_i^{NS} \in OSp(2|1)$, $i = 1, \ldots, g_{NS}$ with fixed points $Z_{\alpha_i}^{NS}$, $Z_{\beta_i}^{NS}$ and multipliers K_i^{NS}.

2. The tree diagram classsical fields belong to the space (cf. sect. 2), $\mathcal{R}[V_1^R, \ldots, V_{g_R}^R]$ with a V_i^R for each R-cut, i.e. for each *pair* of R-reggeons to be sewn into a loop. Then

$$P_j^R = V_j^R \circ d_{[K_j^R, \kappa_j]} \circ (V_j^R)^{-1}, \; j = 1, \ldots, g_R. \tag{7.1}$$

   In the coordination employed here the V_i^R's are $OSp(2|1)$-maps constructed in the process of building the tree amplitude (cf. sect. 8 for examples), but the "super-dilations", $d_{[K,\kappa]}$ and hence the R-propagators, P_i^R are not.

The super-Schottky group as usual consists of all products of all powers of the generators, $\{P_i^{NS}, P_i^R\}$. It should be noticed that therefore we are led to consider a somewhat incomplete uniformization of the super-Riemann surface: Our uniformizing space is *not* the super Riemann sphere, but a double cover of that, glued together at the Ramond-branch points in a rather complicated way, as described by the space $\mathcal{R}[V_1^R, \ldots, V_{g_R}^R]$. Thus our treatment differs from that of [42] and others in which the super-Schottky group is taken to be a subgroup of $OSp(2|1)$. Our point is that such a subgroup does not seem to naturally arise from the sewing picture.

Formally, however, most of the analysis in the previous section goes through, and the integration over non-zero-modes gives rise to a determinant of the matrix A, and again that may be expressed in terms of the spectrum of elements in this new super-Schottky group. However, whereas this spectrum was trivial to write down for an $OSp(2|1)$ map γ in terms of the multipliers – it is given as $\{(K_\gamma^n | K_\gamma^{n-1/2}),\ n \in \mathsf{N}\}$, corresponding to (bosonic|fermionic) eigenfields – it is not immediately obvious how to write it down in the general case when $\gamma \notin OSp(2|1)$.

Nevertheless, we now argue that a very similar structure emerges also here:

When γ is conjugate to a Ramond propagaor, P_j^R the spectrum is trivially obtained as in the 1-loop problem:

$$\{((K_j^R)^n|(K_j^R)^n),\ n \text{ integer}\}. \tag{7.2}$$

When $\gamma \notin OSp(2|1)$ we may still introduce the concept of fixed points ($\gamma(Z_{F.P.}) = Z_{F.P.}$) and convince ourselves that generically there are two (since this is the case when all odd super-moduli vanish). Further, when γ is not conjugate to any Ramond propagator, one easily sees that the fixed points are regular points of the map. Now observe that when $U = (u, \mu)$ is a regular fixed point of any superconformal map γ, then

$$\gamma_B'(U) + \mu\gamma_F'(U) = (D\gamma_F(U))^2 \tag{7.3}$$

and that *is* the multiplier for $\gamma \in OSp(2|1)$. Sufficiently close to the fixed point, U, γ "looks like" an $OSp(2|1)$ transformation with fixed point U and multiplier given by

$$K_\gamma^{1/2} \equiv D\gamma_F(U). \tag{7.4}$$

The phase here corresponds to a spin structure choice, see below. Eigenfunctions must vanish or be singular at U. Take the one with a principal (cf. [41]) zero at U (we think of scalar fields for simplicity). Clearly then the eigenvalue must be K_γ^{-1} (if U is attractive). Denote the eigenfunction by $\Gamma_B^{-1}(Z)$ and find $\Gamma_F^{-1}(Z)$ such that the two together constitute a superconformal map:

$$D\Gamma_B^{-1}(Z) = \Gamma_F^{-1}(Z) D\Gamma_F^{-1}(Z). \tag{7.5}$$

One verifies that $\Gamma_F^{-1}(Z)$ is then an eigenfield with eigenvalue $K_\gamma^{-1/2}$ and indeed that

$$\gamma \equiv \Gamma \circ d_{K_\gamma} \circ \Gamma^{-1} \tag{7.6}$$

from which the spectrum is trivially written down as

$$\{(K_\gamma^n|K_\gamma^{n-1/2}),\ n \text{ integer}\}. \tag{7.7}$$

The different phase choices of $K_\gamma^{1/2}$ follow uniquely by eq.(7.4) from the spin structure choice of the phases of the $(K_i^{NS})^{1/2}$, $(K_i^R)^{1/2}$.

It is now obvious how to write down the non-zero-mode contribution to the most general vacuum amplitude. It is given by

$$\prod_{i=1}^{g_R}\prod_{n=2}^{\infty}\left[\frac{1 \pm (K_i^R)^n}{1-(K_i^R)^n}\right]^{d-2} \cdot \left[\frac{1 \pm K_i^R}{1-K_i^R}\right]^{d}$$
$$\cdot \ \tilde{\prod_{(\gamma)}}{}' \prod_{n=2}^{\infty}\left[\frac{1-K_\gamma^{n-1/2}}{1-K_\gamma^n}\right]^{d-2} \cdot \left[\frac{1-K_\gamma^{1/2}}{1-K_\gamma}\right]^{d}. \tag{7.8}$$

Here $\prod_{(\gamma)}^{\sim}{}'$ is only over those cyclic classes of primary elements for which that primary element is *not* a Ramond-generator. The contribution from those form the first line in eq.(7.8).

It should be pointed out that the functional dependence of K_γ on the (moduli|supermoduli) is completely specified by eq.(7.4) but will in general be fairly complicated. One would expect the integration over supermoduli to lead to a tractable result, but it might not be simple. Also notice that the analysis was performed without providing an explicit form of the eigenfunctions. This was fortunate, since these must be very complicated (involving infinite accumulations of cuts for example).

8 The Measure Factor: Ghost Zero Modes

We treat here in detail the situation depicted in figs. 1,2,3: The case $g_R = 1$, g_{NS} arbitrary, $g = g_{NS} + g_R = g_{NS} + 1$. This case too does not provide a modular invariant result, however, the techniques developed are hoped to pave the way towards the completely general situation. The principal lesson is that most of the serious complications due to the presence of the Ramond-cut, can be circumvented by manipulations of contour integrals.

8.1 Parametrization of the tree

We first build the $g+2$ string tree in fig. 1. The dashed line connecting Z_{α_g} and Z_{β_g} indicates that these two strings are R-ones. All the sewing contours, $C_{A_i} = -C_{\vartheta_{i-1}}$ (see below) refer to NS-sewings. This represents the the simplest choice in our formalism (for discussions of other possibilities in the bosonic case, cf.[35]). In fig.2 we have attached 3-NS-vertices to the $g-2$ strings $Z_{\alpha_2}, \ldots, Z_{\alpha_{g-1}}$, and the loop diagram is completed by sewing Z_{α_i}, Z_{β_i} together, $i = 1, \ldots, g = g_{NS} + 1$. In fig. 3 the KN-points and the various sewing contours are indicated in the (super-)Z-plane.

For the purpose of the calculation we shall find it convenient to use also the mapped variable, $W = (w, \phi)$:

$$z = \frac{z_{\alpha_g} w}{w - (1 - z_{\alpha_g})} \quad ; \quad \theta = i\frac{\sqrt{z_{\alpha_g}(1 - z_{\alpha_g})}}{w - (1 - z_{\alpha_g})}\phi$$

$$w = \frac{1 - z_{\alpha_g}}{z - z_{\alpha_g}} z \quad ; \quad \phi = -i\theta\frac{\sqrt{z_{\alpha_g}(1 - z_{\alpha_g})}}{z - z_{\alpha_g}}$$

$$D_\theta\phi = -i\frac{\sqrt{z_{\alpha_g}(1 - z_{\alpha_g})}}{z - z_{\alpha_g}} \quad ; \quad D_\phi\theta = i\frac{\sqrt{z_{\alpha_g}(1 - z_{\alpha_g})}}{w - (1 - z_{\alpha_g})}. \tag{8.1}$$

Koba-Nielsen points Z_{α_i}, Z_{β_i} are mapped to KN-points W_{α_i}, W_{β_i}.

In the Z-plane the tree diagram is built by using for the sewings always simple dilations with multiplier A_i. Thus at the level of fig. 1 we have used

$OSp(2|1)$ invariance as well as superdilation invariance for the R-cut (sect. 2) to put

$$\begin{aligned} \theta_{\alpha_g} = 0 \quad &; \quad z_{\beta_g} = 0,\ \theta_{\beta_g} = 0 \\ z_{\alpha_1} = 1,\ \theta_{\alpha_1} \equiv \vartheta_{\alpha_1} \quad &; \quad z_{\beta_1} = \infty,\ \theta_{\beta_1} = 0. \end{aligned} \tag{8.2}$$

Further

$$\begin{aligned} z_{\alpha_i} = A_i \cdot A_{i-1} \cdot \ldots \cdot A_2 \quad &, \quad i = 2, \ldots, g \\ \theta_{\alpha_i} = \sqrt{z_{\alpha_i}} \vartheta_{\alpha_i} \quad &, \quad i = 1, \ldots, g-1 \end{aligned} \tag{8.3}$$

where ϑ_{α_i}, $i = 1, \ldots, g-1$ are the odd moduli of the 3-NS-vertices. Our moduli at this point are the $\{A_i\}$ and the $\{\vartheta_{\alpha_i}\}$,*not* the KN-points.

Associated with these moduli are the quasi-superconformal ghost fields

$$\begin{aligned} \check{C}_{A_i}(Z) = z\check{c}_{A_i}\vartheta(C_{A_i}) \quad &, \quad i = 2, \ldots, g \\ \check{C}_{\vartheta_{\alpha_i}}(Z) = 2\sqrt{z_{\alpha_i}}\theta\check{\gamma}_{\vartheta_{\alpha_i}}\vartheta(C_{\vartheta_{\alpha_i}}) \quad &, \quad i = 1, \ldots, g-1 \end{aligned} \tag{8.4}$$

where we have taken $C_{\vartheta_{\alpha_i}} = -C_{A_{i+1}}$ as indicated ($\vartheta(C)$ vanishes *outside* C, $\vartheta(-C)$ vanishes *inside* C) corresponding to the choice, case 1, in eq.(4.6).The i'th 3-NS-vertex has been subjected to a total dilation by z_{α_i}, whence the factor $\sqrt{z_{\alpha_i}}$ in $\check{C}_{\vartheta_{\alpha_i}}$.

Next we attach the $g-2$ 3-NS-vertices of fig. 2. We choose to sew using propagators with fixed points Z_{α_i} and $(0,0)$ and multipliers H_i. Figuring out the various associated conformal transformations (using eq.(2.6)) gives

$$\begin{aligned} z_{\beta_i} = z_{\alpha_i}/[1 - H_i - \sqrt{H_i}\vartheta_{\beta_i}\vartheta_{\alpha_i}] \quad &, \quad i = 2, \ldots, g-1 \\ \theta_{\beta_i} = \sqrt{z_{\alpha_i}}(\sqrt{H_i}\vartheta_{\beta_i} + \vartheta_{\alpha_i})/(1 - H_i) \quad &, \quad i = 2, \ldots, g-1 \end{aligned} \tag{8.5}$$

where again ϑ_{β_i} are the basic odd moduli of the new 3-NS-vertices.

The associated superconformal ghost fields are found to be (using case 2 in eq.(4.6))

$$\begin{aligned} \check{C}_{H_i}(Z) &= \check{c}_{H_i} \cdot (Z - Z_{\alpha_i})z/z_{\alpha_i}\vartheta(C_{H_i}) \\ \check{C}_{\vartheta_{\beta_i}}(Z) &= \check{\gamma}_{\vartheta_{\beta_i}} \cdot 2(Z - Z_{\alpha_i})z_{\alpha_i}^{-3/2}H_i^{-1/2}(\theta z_{\alpha_i} - \theta_{\alpha_i}z)\vartheta(C_{H_i}) \end{aligned} \tag{8.6}$$

with $C_{\vartheta_{\beta_i}} = C_{H_i}$.

For each A_i, H_i, ϑ_{α_i} and ϑ_{β_i} we get the following differential factors associated with the sewings

$$\begin{aligned} d\log A_i \quad &, \quad i = 2, \ldots, g \\ d\log H_i \quad &, \quad i = 2, \ldots, g-1 \\ d\vartheta_{\alpha_i} \quad &, \quad i = 1, \ldots, g-1 \\ d\vartheta_{\beta_i} \quad &, \quad i = 2, \ldots, g-1. \end{aligned} \tag{8.7}$$

Thus for the tree we have $(2g-3|2g-3)$ moduli in accordance with the general rule $(N_B+2N_F-3|N_B+N_F-2)$ (sect. 2) for $N_B=2(g-1)$, $N_F=1$.

It is now straightforward to transform all these expressions to the W-variable, using eq.(8.1). In particular we find

$$W_{\alpha_1}=(1,-i\vartheta_{\alpha_1}\sqrt{\frac{z_{\alpha_g}}{1-z_{\alpha_g}}})\quad,\quad W_{\beta_1}=(1-z_{\alpha_g},0)$$
$$W_{\beta_g}=(0,0)\quad,\quad W_{\alpha_g}=(\infty,0). \tag{8.8}$$

Thus, in the W-variable, all (tree diagram) fields belong to the space $\mathcal{R}$ (sect. 2). Therefore it will be more convenient to use transformations to that variable in the calculations. We shall not give all the explicit expressions which are, however, easy to work out.

To finish the treatment of the tree diagram we write down the SCKV zero-mode field relevant to the $\mathcal{R}$-space

$$\check{C}_0(W)=[\check{c}_{-1}w^2+\check{c}_0w+\check{c}_1+\phi(\check{\gamma}_{-1/2}w+\check{\gamma}_{1/2}+\check{\gamma}_0\sqrt{w})]\vartheta(C_0). \tag{8.9}$$

Here we have introduced the contour C_0 running around all the physical singularities, or, equivalently, around the spurious singularity, Z_0 (or W_0 in the W-plane). See further below.

8.2 The antighost sewing fields

When we introduce the antighost sewing modes, labelled for each NS-reggeon by $\{b^i_n,\ \beta^i_r,\ n=\pm1,0,\ r=\pm1/2\}$ $i=1,\dots,g_{NS}$, and for the R-reggeon by $\{b^R_n,\ \beta^R_n,\ n=\pm1,0\}$, we meet with two different kinds of subtleties.

First, because of the (super-) Riemann-Roch theorem [48,41] there is no superconformal dim 3/2 super-field on the (super-) Riemann sphere (with or without Ramond cuts) that has only those modes, i.e. only such singularities at one single KN-point. Inevitably, if such singular behaviour is imposed, the corresponding B-field will be singular elsewhere. As in the case of the bosonic string [30] and in the pure NS-case [32], we deal with this problem by allowing all "individual" B-reggeon fields to have the necessary spurious singularity at some common point, Z_0 (W_0 in the W plane). The δ-functions in B arising from the $\check{C}$-zero-modes eq.(8.9) will then precisely make sure that the total singularity here is vanishing. Obviously, for consistency the final result should be independent of Z_0.

The second subtlety concerns the fact that all fields must be projected to the space $\mathcal{R}$ (in the W variable) corresponding to the present situation in which we have a single R-cut.

It turns out to be very convenient to parametrize the problem in terms of the "full" unprojected sewing fields

$$B_\alpha+P^{-1}B_\beta \text{ or } PB_\alpha+B_\beta \tag{8.10}$$

in which we imagine that the ghost zero-modes in the sewing of that loop have already been made use of to ensure that the fields eq.(8.10) have no spurious singularitiy, but instead are singular both at Z_α and Z_β but with one single set $\{b_n\ \beta_r\}$ of degrees of freedom. We first write down the expressions relevant before projection to $\mathcal{R}$. Afterwards we come back to that projection as well as to the question of the form of the "individual" fields B_α and B_β, then with some spurious singularity.

Thus for the Ramond B-field we simply have

$$\begin{aligned}
(B^R_{\alpha_g} + P_R^{-1}B^R_{\beta_g})(W) &= \sum_{n=-1}^{1}\{\beta^R_n w^{-n-3/2} + \phi b^R_n w^{-n-2}\} \\
(P_R B^R_{\alpha_g} + B_{\beta_g})(W) &= \sum_{n=-1}^{1}\{\beta^R_n(P_R) w^{-n-3/2} + \phi b^R_n(P_R) w^{-n-2}\} \\
\beta^R_n(P_R) &\equiv K^n_R(\beta^R_n - \kappa_R b^R_n) \\
b^R_n(P_R) &\equiv K^n_R(b^R_n - \kappa_R n\beta^R_n)
\end{aligned} \qquad (8.11)$$

and these fields already belong to $\mathcal{R}$.

For the NS-fields B_{α_i}, B_{β_i} , $i = 1, \ldots, g_{NS}$ it is similarly trivial to write down expressions if $W_{\alpha_i} = (0,0)$ and $W_{\beta_i} = (\infty, 0)$. By the superprojective transformation eq.(2.6) we then get in the general case

$$\begin{aligned}
&(B_{\alpha_i} + P_i^{-1}B_{\beta_i})(Z) = \\
&\sum_{r=-1/2}^{1/2} \beta^i_r (Z - Z_{\alpha_i})^{-r-3/2}[(Z - Z_{\beta_i})/(Z_{\alpha_i} - Z_{\beta_i})]^{r-3/2} \\
&+ \ \{(\theta - \theta_{\alpha_i}) - (Z - Z_{\alpha_i})\frac{(\theta_{\alpha_i} - \theta_{\beta_i})}{(Z_{\alpha_i} - Z_{\beta_i})}\} \\
&\cdot \ \sum_{n=-1}^{1} b^i_n (Z - Z_{\alpha_i})^{-n-2}[(Z - Z_{\beta_i})/(Z_{\alpha_i} - Z_{\beta_i})]^{n-2} \\
&i \ = \ 1, \ldots, g_{NS} = g - 1
\end{aligned} \qquad (8.12)$$

and the field $(P_iB_{\alpha_i} + B_{\beta_i})(Z)$ looks the same with $\beta^i_r \to K^r_i\beta^i_r$, $b^i_n \to K^n_i b^i_n$. Here we have the convenient situation, that in the W-plane these NS-fields look exactly the same with $Z \to W$ throughout (after a trivial change of variables with (super-) Jacobian 1).

From the generic fields $(B_\alpha + P^{-1}B_\beta)$ and $(PB_\alpha + B_\beta)$ given above in eq.(8.11) and eq.(8.12), we project onto $\mathcal{R}$ and construct "individual" reggeon fields $(B_\alpha)^R_{W_0}$, $(B_\beta)^R_{W_0}$ with a spurious singularity at W_0 as follows

$$\begin{aligned}
(B_\alpha)^R_{W_0}(W) &= \oint_{W_\alpha} dW' p(W, W'; W_0)(B_\alpha + P^{-1}B_\beta)(W') \\
(B_\beta)^R_{W_0}(W) &= \oint_{W_\beta} dW' p(W, W'; W_0)(PB_\alpha + B_\beta)(W') \\
p(W, W'; W_0) &= \left(\frac{w' - w_0}{w - w_0}\right)^3 \frac{1}{w - w'}\{-\phi + \sqrt{\frac{w}{w'}}\phi'\}.
\end{aligned} \qquad (8.13)$$

Here we have chosen a $W_0 = (w_0, \phi_0) = (w_0, 0)$. One easily verifies that (i) these fields lie in $\mathcal{R}$, (ii) $(B_\alpha)^R_{W_0}$ $((B_\beta)^R_{W_0})$ have the required singularities near W_α (W_β) but are regular at W_β (W_α) and have a spurious singularity at W_0.

As we shall see below, explicit evaluation of these projections can be completely circumvented.

8.3 The loop sewing overlaps associated with C_i

Associated with each loop sewing is a contour, C_i, $i = 1, \ldots, g$, and, in the covering space, the contours C_i', images of C_i under P_i. There are quasi-superconformal fields $\check{C}_i(Z)$ discontinuous across C_i, as well as the factors

$$d\kappa_R d\log K_R \prod_{i=1}^{g_{NS}} d\log K_i \tag{8.14}$$

$(K_R \equiv K_g)$. It is easy to work out the overlaps $(B_{Cl}|\check{C}_i)$ for two reasons:(i) since $\check{C}_i$ vanishes outside C_i we need only worry about B_{α_i} in the overlap, in fact, any field with the singularities of B_{α_i} (in the neighbourhood of Z_{α_i}) will do. Thus the projection onto $\mathcal{R}$ becomes irrelevant and we may in fact use directly the fields $(B_{\alpha_i} + P^{-1}B_{\beta_i})$; (ii) this overlap is independent of which variable is used for its evaluation, thus we may use variables for which $Z_{\alpha_i} = (0,0)$ and $Z_{\beta_i} = (\infty, 0)$, so that in fact the $\check{C}_i$ are given by eq.(5.5) directly. The integrals over $\check{c}_i$ $(\equiv \check{c}_{K_i})$, $i = 1, \ldots, g_{NS}$ will produce δ-functions in the modes b_0^i with a factor 1, and integrating over $\check{c}_R$ and $\check{\gamma}_{\kappa_R}$ will give δ-functions in b_0^R and β_0^R and a factor 1/2. Thus in the sequel we may forget all those modes.

8.4 The $\mathcal{R}$-projection

Consider the generic overlap

$$\begin{aligned}
((B_\alpha)^R_{W_0}|\check{C}) &= \oint_C dW (B_\alpha)^R_{W_0} \check{C}(W) \\
&= \oint_C dW \oint_{Z_\alpha} dW' p(W', W; W_0) B_\alpha(W') \check{C}(W) \\
&\equiv \oint_{Z_\alpha} dW' B_\alpha(W') \check{C}^R_{W_0}(W') \\
\check{C}^R_{W_0}(W') &\equiv \oint_C dW \check{C}(W) p(W, W'; W_0)
\end{aligned} \tag{8.15}$$

where $B_\alpha(W)$ denotes any field with the correct singularities near $Z = Z_\alpha$ (in particular we might use $(B_\alpha + P^{-1}B_\beta)$). Now $p(W, W'; W_0)$ is singular at $W = W'$ and $W = W_0$ and has a cut from $(0,0)$ to $(\infty, 0)$ eq.(8.13). However, often we shall meet cases where C does not enclose W_0 (see fig. 3) and where the cut complication drops out (see below). Then we simply have

$\check{C}^R_{W_0}(W) = \check{C}(W)$, and the unpleasant complications of the $\mathcal{R}$ projection drop out – this is another way of justifying the treatment of the previous subsection. In fact the only exception will be in the overlaps with $\check{C}_0$.

Indeed, considering fig. 3 we notice that the $\mathcal{R}$ projection will be irrelevant for the contours C_{H_i} and $C_{\vartheta_{\alpha_i}}$. Further, for the contour C_{A_i}, $i = 2, \ldots, g_{NS}$ we first deform that to a union of contours

$$C_{A_i} \sim C_{H_i} \bigcup C_{H_{i+1}} \bigcup \cdots \bigcup C_{H_{g_{NS}}} \bigcup C_{A_g}. \tag{8.16}$$

But if we imagine evaluating the overlap associated with C_{A_g} first, integrating over $\check{C}_{A_g}$ to provide a δ-function in B-fields, then similarly treating $C_{H_{g-1}}$, $C_{A_{g-1}}$ etc., we see that in the above decomposition of C_{A_i}, only C_{H_i} will provide a non-vanishing contribution. This is because the functional form of all the $\check{C}_{A_i}$'s are the same. A similar argument applies to the $C_{\vartheta_{\alpha_i}}$ (thinking of the order of evaluation as being reversed).

8.5 The remaining $(B|\check{C})$ overlaps

Using the above with $\check{C}_0$ given by eq.(8.9) we evaluate

$$\begin{aligned}
(\check{C}_0)^R_{W_0}(W) &= \check{c}_{-1}w^2 + \check{c}_0 w + \check{c}_1 + \phi\check{\gamma}^R_0(w) \\
\check{\gamma}^R_0(w) &\equiv \sum_{n=-1}^{1} \check{\gamma}^R_n w^{-n+1/2} \\
\check{\gamma}^R_{-1} &\equiv \frac{3}{8} w_0^{-1/2}\check{\gamma}_{-1/2} - \frac{1}{8} w_0^{-3/2}\check{\gamma}_{1/2} \\
\check{\gamma}^R_0 &\equiv \frac{3}{4} w_0^{1/2}\check{\gamma}_{-1/2} + \frac{3}{4} w_0^{-1/2}\check{\gamma}_{1/2} + \check{\gamma}_0 \\
\check{\gamma}^R_1 &\equiv -\frac{1}{8} w_0^{3/2}\check{\gamma}_{-1/2} + \frac{3}{8} w_0^{1/2}\check{\gamma}_{1/2}.
\end{aligned} \tag{8.17}$$

This shows that we may change variables

$$(\check{\gamma}_{-1/2},\ \check{\gamma}_0,\ \check{\gamma}_{+1/2}) \rightarrow (\check{\gamma}^R_{-1},\ \check{\gamma}^R_0,\ \check{\gamma}^R_{+1}) \tag{8.18}$$

with Jacobian $= 1/8$. This we shall do in the following, thereby explicitly achieving independence of the position, W_0, for the spurious singularity, and otherwise significantly simplifying the computation.

It is now a completely straightforward exercise in super-contour integration and super-determinant evaluation to complete the calculation. After all b_0 and β_0 modes have been gotten rid of, the calculation proceeds as follows:

The Ramond antighost d.o.f.'s $b^R_{\pm 1}$, $\beta^R_{\pm 1}$ couple to $\check{C}_{A_g}$ and $\check{C}_0$. From

$$(B^R_{\alpha_g}|\check{C}_{A_g}) = -b^R_1 \check{c}_{A_g}(1 - z_{\alpha_g})^{-1} \tag{8.19}$$

and

$$((B^R_{\alpha_g} + P_R^{-1} B^R_{\beta_g})|\check{C}_{A_g}) \equiv 0 \tag{8.20}$$

we immediately write down

$$(B^R_{\alpha_g} + B^R_{\beta_g}|\check{C}_{A_g}) = [(K_R - 1)b^R_1 + K_R\kappa_R\beta^R_1]\check{c}_{A_g}(1 - z_{\alpha_g})^{-1}. \tag{8.21}$$

This technique is used throughout. Integrating over $b^R_{\pm 1}$, $\beta^R_{\pm 1}$ (it is convenient to perform a linear change of variables) one finds a factor of 1 together with δ-functions in $\check{c}_1$, $\check{\gamma}^R_{\pm 1}$. These may subsequently be ignored. Thus the contribution from the Ramond-loop is simply

$$4\frac{(1-K_R)^2}{(1-K_R)^2}\frac{dK_R}{K_R}\frac{dA_g}{A_g}d\kappa_R\frac{1}{(1-z_{\alpha_g})}. \tag{8.22}$$

The factor $1 = (1-K_R)^2/(1-K_R)^2$ is to emphasize the difference to the NS-loop contributions below.

The remaining antighost d.o.f.'s from the i'th NS loop, $b^i_{\pm 1}$, $\beta^i_{\pm 1/2}$ are taken care of by considering the overlaps along $C_{\vartheta_{\alpha_i}}$, $C_{\vartheta_{\beta_i}}$, C_{H_i}, C_{A_i} – all deformed to coincide. The result of integrating over $\{b^i_{\pm 1}, \beta^i_{\pm 1/2}, \check{c}_{A_i}, \check{c}_{H_i}, \check{\gamma}_{\vartheta_{\alpha_i}}, \check{\gamma}_{\vartheta_{\beta_i}}\}$ is

$$4^{-1}\frac{(1-K_i)^2}{(1-K_i^{1/2})^2}\frac{dK_i}{K_i^{3/2}}\frac{dA_i}{A_i}\frac{dH_i}{H_i^{3/2}}d\vartheta_{\alpha_i}d\vartheta_{\beta_i}, \quad i = 2,\ldots,g_{NS} \tag{8.23}$$

(using $z_{\beta_i}/(Z_{\alpha_i} - Z_{\beta_i}) = H_i^{-1}$, cf. eq.(8.5)).

By our particular coordinate choice, NS loop no. 1 (the last in the calculation) is special. It is the only one overlapping with the remaining $\check{C}_0$ modes, thereby feeling the presence of the R-cut. The contribution turns out to be

$$8\frac{\sqrt{1-z_{\alpha_g}}}{2-z_{\alpha_g}}\frac{dK_1}{K_1^{3/2}}\frac{(1-K_1)^2}{(1-K_1^{1/2})^2}\frac{1}{\sqrt{z_{\alpha_g}}}d\vartheta_{\alpha_1}. \tag{8.24}$$

Collecting all factors we finally find for the ghost zero-mode contribution in these coordinates to the $g_R = 1$ part of the vacuum amplitude:

$$\begin{aligned} & 32\cdot 4^{-g_{NS}}\frac{\sqrt{1-z_{\alpha_g}}}{2-z_{\alpha_g}}\frac{dK_R}{K_R}d\kappa_R \\ & \cdot \prod_{i=1}^{g_{NS}}\frac{(1-K_i)^2}{(1-K_i^{1/2})^2}\frac{dK_i}{K_i^{3/2}}\prod_{i=2}^{g}\frac{dA_i}{A_i^{3/2}}\prod_{i=2}^{g_{NS}}\frac{dH_i}{H_i^{3/2}} \\ & \cdot\ d\vartheta_{\alpha_1}\prod_{i=2}^{g_{NS}}d\vartheta_{\alpha_i}d\vartheta_{\beta_i} \\ z_{\alpha_g} \equiv & \prod_{j=2}^{g}A_j \end{aligned} \tag{8.25}$$

This calculation works in a coordinate patch around the pinching limit A_i, H_i, $K_i \to 0$. One observes the expected singular behaviour in that limit. The factors on the first line are associated with the Ramond loop. Apart from those this calculation agrees with the pure NS-result sect. 6. The result there for $d\mu_g$ eq.(6.11) is essentially that, after the (super-) Jacobian for the transformation to (super-) KN variables has been worked out and inserted. In the present case too, one might similarly change to super-KN variables, however, it would be less trivial to express the final answer in a way that manifestly expresses the symmetries of the problem: $OSp(2|1)$ and super-dilation invariance (cf. sect. 2). In contrast, the (moduli|supermoduli) employed here in eq.(8.25) are all manifestly invariant under these symmetries.

9 Conclusion

A method [32] based on sewing reggeons, for building superstring loop densities on supermoduli-space (in a super-Schottky parametrization) has been explained and investigated. It generalizes the one [30] developed and checked [29] for the bosonic string. The most general superloop amplitudes have not yet been dealt with in all detail. However, the treatment is completed for (i) the most general N-string g loop problem involving NS-strings only (sect. 6 eq.(6.6), eq.(6.11), eq.(6.13)); (ii) the non-zero-mode part of the most general vacuum amplitude with an arbitrary number of Ramond-loops, g_R, and NS-loops, g_{NS}; (iii) the ghost zero-mode (i.e. measure) part for the case $g_R = 1$, g_{NS} arbitrary. Ramond strings introduce essential complications, however, experience gained so far indicates that they remain tractable or can be circumvented.

Acknowledgements

I have benefited much from discussions with J.R. Sidenius, K.O. Roland and A.K. Tollstén.

References

[1] E. D'Hoker and P.H. Phong, Rev. Mod. Phys. **60** (1988) 917.

[2] S. Mandelstam in *Unified String Theories*, M.B. Green and D.J. Gross (eds.), World Scientific (1986);
in *Perspectives in String Theory*, Copenhagen 1987, P. Di Vecchia and J.L. Petersen (eds.), World Scientific (1988).

[3] E. D.'Hoker and P.H. Phong, Phys. Rev. Lett. **56** (1986) 912;
Nucl. Phys. **B269** (1986) 205.

[4] E. D.'Hoker and P.H. Phong, Nucl. Phys. **B278** (1986) 225;
Nucl. Phys. **B292** (1987) 317.

[5] J. Polchinski, Nucl. Phys. **B307** (1988) 61.

[6] H. Sonoda, Nucl. Phys. **B284** (1987) 157.

[7] A. Neveu and P. West, Phys. Lett. **168B** (1986) 192;
Nucl. Phys. **B278** (1986) 601;
Phys. Lett. **179B** (1986) 235;
Phys. Lett. **180B** (1986) 34;
Y. Kazama, A. Neveu, H. Nicolai and P. West, Nucl. Phys. **B278** (1986) 833.

[8] P. Di Vecchia, R. Nakayama, J.L. Petersen and S. Sciuto, Nucl. Phys. **B282** (1987) 179;
P. Di Vecchia, R. Nakayama, J.L. Petersen, J. Sidenius and S. Sciuto, Phys. Lett. **182B** (1986) 164;
P. Di Vecchia, M. Frau, A. Lerda and S. Sciuto, Nucl. Phys. **B298** (1988) 526;
Phys. Lett. **199B** (1987) 49;
P. Di Vecchia. K. Hornfeck, M. Frau, A. Lerda and S. Sciuto, Phys. Lett. **211B** (1988) 301.

[9] P. Di Vecchia, K. Hornfeck, M. Frau, A. Lerda and S. Sciuto, Phys. Lett. **206B** (1988) 643;
Phys. Lett. **205B** (1988) 250;
A. Lerda in *Perspectives in String Theory*, Copenhagen 1987, P. Di Vecchia and J.L. Petersen (eds.), World Scientific (1988).

[10] C. Cristofano, F. Nicodemi and R. Pettorino, Phys. Lett. **200B** (1988) 292;
Naples preprint, to be published in Int. Jour. of Mod. Phys. A.

[11] U. Carow-Watamura and S.Watamura, Nucl. Phys. **B288** (1987) 500;
Nucl. Phys. **B301** (1988) 132;Nucl. Phys. **B302** (1988) 149;
U. Carow-Watamura, Z.F. Ezawa and S. Watamura Nucl. Phys. **B319** (1989) 187.

[12] A. LeClair, Nucl. Phys. **B297** (1988) 603;
Nucl. Phys. **B303** (1988) 189;
A. LeClair, M.E. Peskin and C.R. Preitschopf, SLAC preprints SLAC-PUB-4306, 4307 and 4464 (1988).

[13] T. Kobayashi, H. Konno and T. Suzuki, UTHEP-175 and UTHEP-181 preprints (1988);
H. Konno, UTHEP-183 preprint (1988).

[14] A.A. Belavin and V.G. Knizhnik, Phys. Lett. **168B** (1986) 201.

[15] E. Gava, R. Iengo, T. Jayaraman and R. Ramachandran, Phys. Lett. **168B** (1986) 207;
J.-B. Bost and Th. Jolicoeur, Phys. Lett. **174B** (1986) 273;
R. Catenacci, M. Cornalba, M. Martellini and C. Reina, Phys. Lett. **172B** (1986) 328.

[16] V.G. Knizhnik, Phys. Lett. **160B** (1985) 403.

[17] D. Friedan, E. Martinec and S. Shenker, Nucl. Phys. **B271** (1986) 93.

[18] V.G. Knizhnik, Phys. Lett. **180B** (1986) 247.

[19] E. Verlinde and H. Verlinde, Nucl. Phys. **B288** (1987) 357;
Phys. Lett. **192B** (1987) 95;
THU-87/26 preprint (1987).

[20] L. Alvarez-Gaumé, J. Bost, G. Moore, P. Nelson and C. Vafa, Comm. Math. Phys. **112** (1987) 503;
L. Alvarez-Gaumé, G. Moore and C. Vafa, Comm. Math. Phys. **106** (1986) 40;
L. Alvarez-Gaumé, G. Moore, P. Ginsparg and C. Vafa Phys. Lett. **171B** (1986) 155;
L. Alvarez-Gaumé, G. Moore, P. Nelson, C. Vafa and J. Bost, Phys. Lett. **178B** (1986) 41.

[21] N. Ishibashi, Y. Matsuo and H. Ooguri, Mod. Phys. Lett. **A2** (1987) 119;
L. Alvarez-Gaumé, C. Gomez and C. Reina, Phys. Lett. **190B** (1987) 55;
C. Vafa, Phys. Lett. **190B** (1987) 47;
S. Mukhi and S. Panda, Phys. Lett. **203B** (1988) 387.

[22] L. Alvarez-Gaumé, C. Gomez, G. Moore and C. Vafa, Nucl. Phys. **B303** (1988);
L. Alvarez-Gaumé, C. Gomez, P. Nelson, G. Sierra and C. Vafa, CERN TH.5018 (1988).

[23] A. Neveu and P. West Phys. Lett. **193B** (1987) 187;
Phys. Lett. **194B** (1987) 200;
Comm. Math. Phys. **114** (1988) 613;
Phys. Lett. **200** (1988) 275;
Comm. Math. Phys. **119** (1988) 585.

[24] P. Di Vecchia, K. Hornfeck and M. Yu, Phys. Lett. **195B** (1987) 557.

[25] P. West, Phys. Lett. **205B** (1988) 38.

[26] M.D. Freeman and P. West, Phys. Lett. 205B (1988) 30.

[27] A.A. Belavin, V.G. Knizhnik, A. Morozov and A. Perelomov, Phys. Lett. **177B** (1986) 324;
A. Morozov, Phys. Lett. **184B** (1987) 171;
A. Kato, Y. Matsuo and S. Okabe, Phys. Lett. **179B** (1986) 241.

[28] V.G. Knizhnik, Comm. Math. Phys. **112** (1987) 567;
Phys. Lett. **196B** (1987) 473.

[29] J.L. Petersen, K.O. Roland and J.R. Sidenius,Phys. Lett. **205B** (1988) 262;
K. Roland, Nucl. Phys. **B313** (1989) 432.

[30] P. Di Vecchia, R. Nakayama, J.L. Petersen, S. Sciuto and J.R. Sidenius, Nucl. Phys. **B287** (1987) 621;
J.L. Petersen and J.R. Sidenius, Nucl. Phys. **B301** (1988) 247;
J.R. Sidenius in *Perspectives in String Theory*, Copenhagen 1987, P. Di Vecchia and J.L. Petersen (eds.), World Scientific (1988).

[31] H. Dorn and H.-J. Otto, Humboldt-Universität preprint (1989).

[32] J.L. Petersen, J.R. Sidenius and A.K. Tollstén, Phys. Lett. **213B** (1988) 30;
NBI-HE-88-29 preprint (1988), to be published in Nucl. Phys. **B**;
NBI-HE-88-40 preprint (1988), to be published in Phys. Lett. **B**.

[33] J.L. Petersen, NBI-HE-88-65 preprint (1988), to appear in *Proceedings of the Landau Birthday Symposium in Copenhagen 13-17 June, 1988.*

[34] B.E.W. Nilsson, A.K. Tollstén, A. Wätterstam, Göteborg-preprint 88-52 (1988).

[35] K.O. Roland NBI-HE-88-55 preprint (1988).

[36] M. Bochicchio and A. Lerda, to be published in Phys. Lett. **B**.

[37] C. Lovelace, Phys. Lett. **32B** (1970) 703;
V. Alessandrini, Nuovo Cimento **2A** (1971) 321;
C. Montonen, Nuovo Cimento **19A** (1974) 69.

[38] H. Sonoda, Nucl. Phys. **B311** (1988) 401;
Nucl. Phys. **B311** (1988) 417.

[39] V.G. Knizhnik, Phys. Lett. **178B** (1986) 21.

[40] D. Friedan in *Unified String Theories*, M.B. Green and D.J. Gross (eds.), World Scientific (1986).

[41] M.A. Baranov, Ya.I. Manin, I.V.Frolov and A.S.Schwarz, Comm. Math. Phys. **111** (1987) 373;
M.A. Baranov, I. Frolov and A.S. Schwarz, Teor. Math. Phys. **70** (1987) 64;
A.A. Rosly, A.S. Schwarz and A.A. Voronov, ITEP preprints 107,115 (1987).

[42] L. Crane and J.M. Rabin, Comm. Math. Phys. **113** (1988) 601.

[43] P. Mansfield, Nucl. Phys. **B283** (1987) 351.

[44] M.A. Baranov and A.S. Schwarz, Int. Journ. Mod. Phys. **A2** (1987) 1773.

[45] H. Sonoda, Nucl. Phys. **B302** (1988) 104;
LBL-24409 preprint (1988).

[46] S.B. Giddings and E. Martinec, Phys. Lett. **B278** (1986) 91;
E. Martinec Nucl. Phys. **B281** (1987) 157.

[47] A. Clarizia and F. Pezzella, Nucl. Phys. **B298** (1988) 636.

[48] H.M. Farkas and I. Kra, *Riemann Surfaces*, Springer Verlag, N.Y. (1980).

Figure Captions

Figure 1: A particular construction of the tree with 2 R-strings (connected by a dashed line) at KN-points Z_{α_g} and Z_{β_g}, and $g_{NS}+1=g$ NS-strings. Sewing-contours C_{A_i} and $C_{\vartheta_{\alpha_i}}$ are indicated.

Figure 2: The tree fig. 1, after $g-2$ additional 3-NS-vertices have been attached to the branches of fig. 1 by NS-sewings along contours C_{H_i} ($= C_{\vartheta_{\beta_i}}$). The loop amplitude – a tree with tadpoles – is obtained by sewing strings α_i and β_i together, along sewing contours C_i (to be identified with C_i').

Figure 3: The Z-plane picture corresponding to fig. 2. KN-points and sewing contours are indicated. The Ramond-cut connecting Z_{α_g}, Z_{β_g} is shown. So is the spurious singularity at Z_0.

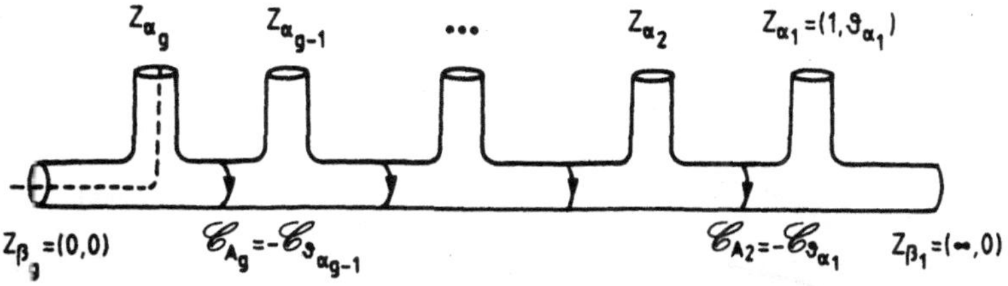

Z_{α_g}
$Z_{\alpha_{g-1}}$
•••
Z_{α_2}
$Z_{\alpha_1}=(1,\vartheta_{\alpha_1})$
$Z_{\beta_g}=(0,0)$
$\mathscr{C}_{A_g}=-\mathscr{C}_{\vartheta_{\alpha_{g-1}}}$
$\mathscr{C}_{A_2}=-\mathscr{C}_{\vartheta_{\alpha_1}}$
$Z_{\beta_1}=(\infty,0)$

Fig. 1

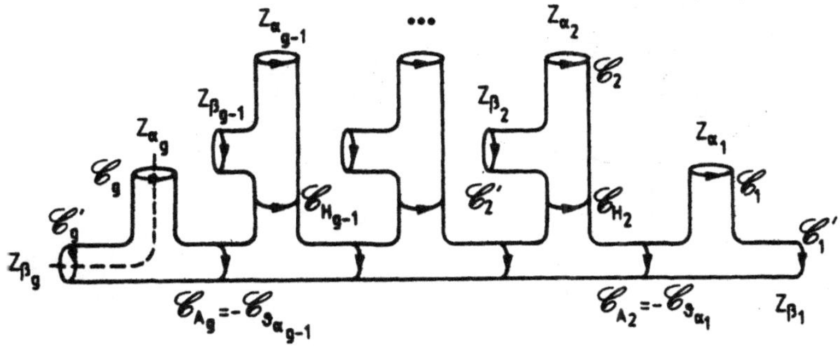

$Z_{\alpha_{g-1}}$
•••
Z_{α_2}
$\mathscr{C}_2$
$Z_{\beta_{g-1}}$
Z_{β_2}
Z_{α_g}
Z_{α_1}
$\mathscr{C}_g$
$\mathscr{C}_1$
$\mathscr{C}_{H_{g-1}}$
$\mathscr{C}_2'$
$\mathscr{C}_{H_2}$
$\mathscr{C}_g'$
$\mathscr{C}_1'$
Z_{β_g}
$\mathscr{C}_{A_g}=-\mathscr{C}_{\vartheta_{\alpha_{g-1}}}$
$\mathscr{C}_{A_2}=-\mathscr{C}_{\vartheta_{\alpha_1}}$
Z_{β_1}

Fig. 2

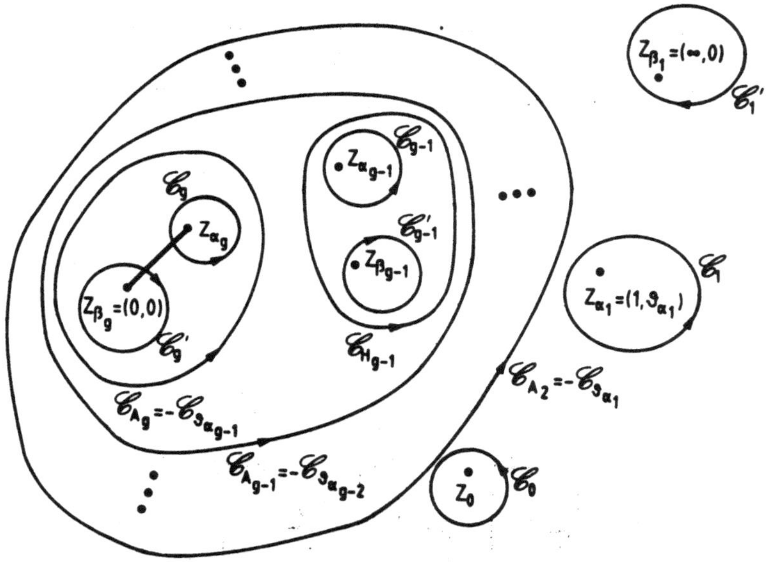
$Z_{\beta_1}=(\infty,0)$
$\mathscr{C}_1'$
$\mathscr{C}_{g-1}$
$Z_{\alpha_{g-1}}$
$\mathscr{C}_g$
Z_{α_g}
$\mathscr{C}_{g-1}'$
$Z_{\beta_{g-1}}$
$\mathscr{C}_1$
$Z_{\beta_g}=(0,0)$
$Z_{\alpha_1}=(1,\vartheta_{\alpha_1})$
$\mathscr{C}_g'$
$\mathscr{C}_{H_{g-1}}$
$\mathscr{C}_{A_2}=-\mathscr{C}_{\vartheta_{\alpha_1}}$
$\mathscr{C}_{A_g}=-\mathscr{C}_{\vartheta_{\alpha_{g-1}}}$
$\mathscr{C}_{A_{g-1}}=-\mathscr{C}_{\vartheta_{\alpha_{g-2}}}$
Z_0
$\mathscr{C}_0$

Fig. 3

SUPERSYMMETRIC SUPERSTRING COMPACTIFICATION

John H. Schwarz

California Institute of Technology, Pasadena, CA 91125

Abstract

The Goddard–Kent–Olive construction is used to derive various conformal and superconformal theories, including a large class of $N = 2$ models recently discovered by Kazama and Suzuki. Various approaches to compactification based on exactly soluble conformal field theories, including Gepner's proposal for using the $N = 2$ minimal models, are sketched. Recent progress in understanding $N = 2$ models and Calabi–Yau spaces using mathematical techniques of singularity theory is described.

1. Introduction

In the four years that have passed since string theory emerged as a popular approach to unification, we have learned a great deal about many facets of the program.[1] Two general categories of issues have emerged as critical and received much scrutiny in recent times. One is the quest for a genuine nonperturbative formulation of quantum string theory. A very broad range of ideas and approaches have been put forward, but as yet none has emerged as the consensus favorite. The second category of questions, which is the subject of this paper, concerns the search for classical solutions of the heterotic string theory, with particular emphasis on those that are most promising for achieving phenomenological success.

Four years ago string theory was touted for its uniqueness. In a certain sense that is still a viable point of view today. There are only three theories—type I, type II, and heterotic—that appear to be internally consistent when analyzed in perturbation theory. Each is completely free of parameters or other arbitrariness. What has become clear in the intervening years is that this uniqueness at the level of the fundamental equations (whatever they are) is not reflected at the level of classical solutions. A bewildering proliferation of classical solutions have been discovered by a variety of techniques. However, there is a unifying principle. Four dimensions can be taken to be flat Minkowski space with the remaining degrees of freedom described by an arbitrary conformal field theory having suitable central charges, supersymmetry, and modular invariance. All classical solutions that have been proposed, whether or not expressed in these terms, can be interpreted as particular constructions of the internal conformal field theory.

Unfortunately, at the level of perturbative analysis, this program has some disturbing arbitrariness. There is no compelling theoretical reason to separate off four-dimensional space-time or to require that it be a Minkowski space. Moreover, the possibilities for the internal conformal field theories are very numerous. The number can be reduced greatly by imposing some plausible physical requirements such as $N = 1$ supersymmetry (to solve the hierarchy problem) or a particular number of families.

One of the first proposals, Calabi–Yau compactification, still looks the most promising. From the abstract conformal field theory point of view it corresponds to a class of (2,2) superconformal models with central charge $c = 9$. The abstract approach treats related orbifold compactifications as part of the same category. The detailed procedure for turning an arbitrary (2,2) superconformal model with $c = 9$ into an $N = 1$, $D = 4$ heterotic string solution with E_6 families of quarks and leptons has been worked out by Gepner. Many examples can be constructed from Kac–Moody algebra cosets, using the method of Goddard, Kent, and Olive. A supersymmetric extension of this method has been formulated by Kazama and Suzuki and applied to the construction of a new class of $N = 2$ superconformal models, which can also be used in the construction of the internal $c = 9$ model.

Gepner gave convincing evidence that certain of his models correspond to known Calabi–Yau compactifications. This seemed miraculous at the time, since his construction is entirely algebraic and uses exactly soluble minimal models. Calabi–Yau spaces on the other hand are very complicated geometric structures, none of which has a known metric. The origins of this 'miracle' are better understood now in view of recent developments applying the mathematical techniques of singularity theory to the description of $N = 2$ models.

2. Superconformal Symmetry

Conformal Field Theory

This section briefly reviews some of the central ideas in conformal field theory that are required in the sequel. For more throrough and systematic discussions the reader is referred to the review articles in ref. [2].

It is convenient to Euclideanize the world sheet so that its metric $h_{\alpha\beta}$ becomes positive definite. Having done this, we can now introduce complex coordinates z (in local patches) and regard the world sheet as a Riemann surface. The residual gauge invariances are conformal mappings $z \rightarrow \tilde{z}(z)$ and similarly for $\bar{z}$. Thus we are led to consider conformally invariant two-dimensional field theory.[3] These transformations are generated by the energy–momentum tensor components $T(z)$ and $\bar{T}(\bar{z})$. More precisely, for an infinitesimal transformation $z \rightarrow z + f(z)$, an operator A transforms by

$$\delta A = [T_f, A] + [\bar{T}_{\bar{f}}, A] \ , \tag{2.1}$$

where

$$T_f = \oint \frac{dz}{2\pi i} f(z) T(z), \tag{2.2}$$

and similarly for $\bar{T}_{\bar{f}}$.

It is convenient to introduce mode expansions for a conformal dimension h field of the form

$$\phi(z) = \sum_{-\infty}^{\infty} \frac{\phi_n}{z^{n+h}} \quad . \tag{2.3}$$

In particular,

$$T(z) = \sum_{-\infty}^{\infty} \frac{L_n}{z^{n+2}} \quad . \tag{2.4}$$

More generally, conformal fields are functions of z and $\bar{z}$ and thus have a pair of dimensions $(h, \bar{h})$. However, to simplify writing I will suppress $\bar{z}$ dependences. If the ghosts are included (so that the conformal anomaly cancels), $T(z)$ has dimension (2, 0) and $\bar{T}(\bar{z})$ has dimension (0, 2). Similarly, $G(z)$ has dimension (3/2, 0) and $\bar{G}(\bar{z})$ has dimension (0, 3/2).

Under a finite transformation $z \rightarrow \tilde{z}(z)$, a conformal dimension h field transforms as follows

$$\phi(z) \rightarrow \left(\frac{\partial \tilde{z}}{\partial z}\right)^h \tilde{\phi}(\tilde{z}) \quad . \tag{2.5}$$

One sometimes says that ϕ is an 'h-form,' since $\phi(z)(dz)^h$ is invariant. The infinitesimal transformation is determined by the commutation relations $[L_m, \phi_n]$. Equivalent information can be expressed in terms of operator product expansions. The basic idea is that in defining a commutator, such as $[T_f, \phi(w)]$, the operator products are only well-defined for radially ordered operators. In other words, $T_f\phi(w)$ is well-defined if the integration contour satisfies $|z| > |w|$, while $\phi(w)T_f$ is well-defined for $|w| > |z|$. The commutator is therefore properly defined as the difference of these terms, which can be rewritten as a single contour integral

$$[T_f, \phi(w)] = \oint_w \frac{dz}{2\pi i} f(z) T(z) \phi(w) \tag{2.6}$$

in which the contour encircles the point w in the usual counterclockwise sense. In view of Cauchy's theorem, the result is determined entirely by the poles inside the contour. The previous result for $[L_m, \phi_n]$ therefore is easily seen to be equivalent to the operator product expansion (OPE)

$$T(z)\phi(w) \sim \frac{h\phi(w)}{(z-w)^2} + \frac{\partial\phi(w)}{z-w} + \ldots \quad . \tag{2.7}$$

The symbol ∂ now means differentiation with respect to w, of course. The dots represent nonsingular terms. Similarly, the superconformal algebra corresponds to the OPE's

$$\begin{aligned} T(z)T(w) &\sim \frac{c}{2(z-w)^4} + \frac{2T(w)}{(z-w)^2} + \frac{\partial T(w)}{z-w} + \dots \\ T(z)G(w) &\sim \frac{3}{2}\frac{G(w)}{(z-w)^2} + \frac{\partial G(w)}{z-w} + \dots \\ G(z)G(w) &\sim \frac{2c}{3(z-w)^3} + \frac{2T(w)}{z-w} + \dots \quad . \end{aligned} \tag{2.8}$$

The algebra of bosonic coordinates $X^\mu(z)$ and fermionic partners $\psi^\mu(z)$ can also be expressed as OPE's. For example,

$$\psi^\mu(z)\psi^\nu(w) \sim \frac{\eta^{\mu\nu}}{z-w} + \dots \tag{2.9}$$

$$X^\mu(z)X^\nu(w) \sim -\eta^{\mu\nu}\ln(z-w) + \dots \quad . \tag{2.10}$$

Since the latter formula is not single-valued (and requires an infrared cutoff), strictly speaking $X^\mu(z)$ is not a conformal field. However, its derivative $\partial X^\mu(z)$ has single-valued OPEs that are cutoff independent. Therefore it is a conformal field (with $h = 1$). Also, the normal-ordered exponential $: \exp ik \cdot X :$, which appears as a factor in vertex operator formulas for emission of string states of momentum k^μ, is a conformal field of dimension $\frac{1}{2}k^2$.

Particularly interesting examples of conformal fields are the two-dimensional currents associated with a Lie group symmetry in a conformal field theory.[4)] Using current conservation one can show that there is a holomorphic component $J^A(z)$ and an antiholomorphic component $\bar{J}^A(\bar{z})$, just as for T and G. As above, we consider $J^A(z)$ only. The zero modes J_0^A are the generators of a Lie algebra G with

$$[J_0^A, J_0^B] = i f_{ABC} J_0^C . \tag{2.11}$$

The algebra of the currents $J^A(z)$ is an infinite-dimensional extension of this, known as an affine Lie algebra or as a Kac–Moody algebra $\hat{G}$. These currents have conformal dimension $h = 1$, and therefore their mode expansion is

$$J^A(z) = \sum_{-\infty}^{\infty} \frac{J_n^A}{z^{n+1}}. \tag{2.12}$$

The Kac–Moody algebra is given by the OPE

$$J^A(z)J^B(w) \sim \frac{k\delta^{AB}}{2(z-w)^2} + \frac{if_{ABC}J^C(w)}{z-w} + \ldots \tag{2.13}$$

or the equivalent commutation relations

$$[J_m^A, J_n^B] = \frac{1}{2}km\delta^{AB}\delta_{m+n,0} + if_{ABC}J_{m+n}^C. \tag{2.14}$$

The parameter k in the Kac–Moody algebra, called the 'level,' is analogous to the parameter c in the conformal algebra. For a $U(1)$ Kac–Moody algebra, $\widehat{U}(1)$, it can be absorbed in the normalization of the current. However, for a non-Abelian group G, it has an absolute meaning once the normalizations are carefully specified. Rather than giving general formulas, let me simply choose $f_{ABC} = \epsilon_{ABC}$ in the case of $SU(2)$. With this normalization convention, the algebra admits unitary representations if and only if k is a positive integer. Unitarity is essential for us, since we will use such algebras to define the positive-definite Hilbert space of physical states. An important formula, due to Sugawara and Sommerfield,[5] gives the energy–momentum tensor associated with an arbitrary Kac–Moody algebra*:

$$T(z) = \frac{1}{k+\tilde{h}_G} \sum_{A=1}^{\dim G} :J^A(z)J^A(z): \quad . \tag{2.15}$$

In the case of simply-laced algebras ($G = A, D, E$) the dual Coxeter number $\tilde{h}_G$ equals c_A, where c_A is the quadratic Casimir number of the adjoint representation

---

* $\tilde{h}_G$ is the dual Coxeter number of the Lie group G. Its value is $n+1$ for $A_n = SU(n+1)$, $2n-1$ for $B_n = SO(2n+1)$ – except that it is 2 for $SO(3)$, $n+1$ for $C_n = USp(n)$, $2n-2$ for $D_n = SO(2n)$, 4 for G_2, 9 for F_4, 12 for E_6, 18 for E_7, and 30 for E_8.

defined (with our normalization conventions) by

$$f_{ABC}f_{A'BC} = c_A\delta_{AA'}. \tag{2.16}$$

The associated central charge is

$$c = \frac{k \dim G}{k + \tilde{h}_G}. \tag{2.17}$$

For example, in the case of $\widehat{SU}(2)_k$, $\tilde{h} = 2$ and $c = 3k/(k+2)$. These theories can be formulated as Wess–Zumino–Witten models,[6] which are σ models for the group manifold with a topological term having a coefficient proportional to k.

Spin Fields and Supersymmetry

So far our discussion of superstrings has treated the bosonic (NS) sector and fermionic (R) sector separately. All the operators we have introduced, whether bosonic or fermionic on the world sheet, are bosonic in space-time. They only map NS states to NS states and R states to R states. Clearly, if we wish to be able to describe fermion emission or supersymmetry we also need operators that map NS states to R states and vice versa, *i.e.*, they should be fermionic operators in the space-time sense. Such operators are necessarily a little tricky, since they switch the $\psi(z)$ boundary conditions from periodic to antiperiodic (or vice versa). In other words, $\psi(z)$ acquires a square-root branch point on the world sheet at the point where a fermionic operator acts.

To keep the discussion as elementary as possible, let me describe the construction in the light-cone gauge for flat ten-dimensional space-time. The manifest symmetry is a transverse spin(8), with the fields $\psi^i(z)$ belonging to the $\mathbf{8_v}$ representation. Since this group has rank 4, representations can be described by points in a four-dimensional weight lattice. For example, the $\mathbf{8_v}$ is given by $(\pm1,0,0,0)$, $(0,\pm1,0,0)$, $(0,0,\pm1,0)$, $(0,0,0,\pm1)$. These weights can be regarded as corresponding to the fields $\frac{1}{\sqrt{2}}(\psi^1 \pm i\psi^2)$, $\frac{1}{\sqrt{2}}(\psi^3 \pm i\psi^4)$, etc. A convenient trick is to use the method of bosonization to re-express these fields in terms of four free bosonic fields $H^\alpha(z)$, $\alpha = 1,2,3,4$. Thus

$$\frac{1}{\sqrt{2}}(\psi^1 \pm i\psi^2) =: e^{\pm iH^1(z)} : c_\pm \text{ , etc.} \tag{2.18}$$

where $c_\pm$ is a cocycle operator, required to obtain the correct statistics. In general, to any four-dimensional weight vector w^α, we can associate an operator $: e^{iw\cdot H(z)} : c_w$, which has conformal dimension $h = \frac{1}{2}w\cdot w$.

The group spin(8) has two 8-dimensional spinor representations, denoted $\mathbf{8_s}$ and $\mathbf{8_c}$. Their weight vectors are given by $(\pm 1/2, \pm 1/2, \pm 1/2, \pm 1/2)$, with an even number of +'s for $\mathbf{8_s}$ and an odd number for $\mathbf{8_c}$. Thus we can introduce an $\mathbf{8_s}$ spinor S^a, having conformal dimension $h = 1/2$, by using the corresponding weight vectors for each of the eight components $a = 1,2,\ldots,8$. A second spinor $S^{\dot a}$ belonging to the $\mathbf{8_c}$ representation can be defined in an analogous fashion. Now $S^a(z)$ is a free fermion field on the world sheet, just like $\psi^i(z)$.[11] Thus

$$S^a(z)S^b(w) \sim \frac{\delta^{ab}}{z-w} + \ldots \tag{2.19}$$

Moreover, these operators are fermionic in the space-time sense, since they map tensor spin(8) representations into spinor ones, and vice versa.*

Let us consider now the OPE of S^a and ψ^i. As explained above, a square-root branch cut is to be expected. Indeed, using the bosonization formulas one can show that

$$\begin{aligned} S^a(z)\psi^i(w) &\sim \frac{\gamma^i_{a\dot a}S^{\dot a}(w)}{(z-w)^{1/2}} + \ldots \\ S^{\dot a}(z)\psi^i(w) &\sim \frac{\gamma^i_{\dot a a}S^{a}(w)}{(z-w)^{1/2}} + \ldots , \end{aligned} \tag{2.20}$$

where $\gamma^i_{a\dot a}$ are Dirac matrices that describe the invariant coupling of the three inequivalent eight-dimensional representations of spin(8). By the triality symmetry of

* The 8-component spinor $S^a(z)$ can be identified with the surviving components of a covariant spinor $\theta(z)$ introduced in ref. [12] after choosing a light-cone gauge.

spin(8), one also must have

$$S^a(z)S^{\dot a}(w) \sim \frac{\gamma^i_{a\dot a}\psi^i(w)}{(z-w)^{1/2}} + \dots . \tag{2.21}$$

In ten dimensions the supersymmetry charge is a 16-component Majorana–Weyl spinor. In terms of the spin(8) subgroup of the Lorentz group it decomposes into $\mathbf{8_s}$ and $\mathbf{8_c}$ pieces, which we denote Q^a and $Q^{\dot a}$. In this notation the supersymmetry algebra takes the form

$$\begin{aligned} \{Q^a, Q^b\} &= 2p^+\delta^{ab} \\ \{Q^{\dot a}, Q^{\dot b}\} &= 2p^-\delta^{\dot a\dot b} \\ \{Q^a, Q^{\dot b}\} &= \sqrt{2}\gamma^i_{a\dot b}p^i . \end{aligned} \tag{2.22}$$

It is now natural to identify this algebra as describing the zero-mode sector of conformal fields. Let us try the following identifications

$$\begin{aligned} Q^a(z) &= \sqrt{2p^+}S^a(z) \\ Q^{\dot a}(z) &= \frac{1}{\sqrt{p^+}}\gamma^i_{\dot a a}S^a(z)\partial X^i(z). \end{aligned} \tag{2.23}$$

Using formulas given previously, it is clear that this gives the OPE's

$$\begin{aligned} Q^a(z)Q^b(w) &\sim \frac{2p^+\delta^{ab}}{z-w} + \dots \\ Q^a(z)Q^{\dot b}(w) &\sim \sqrt{2}\gamma^i_{a\dot b}\frac{\partial X^i(w)}{z-w} + \dots . \end{aligned} \tag{2.24}$$

The formula for $Q^{\dot a}(z)Q^{\dot b}(w)$ can also be worked out, but obviously is more complicated. We will settle here for noting that it contains a term $\frac{2}{p^+}T(w)\delta^{\dot a\dot b}/(z-w)$. This is the only term that contributes to the zero-mode anticommutator. It gives the desired result when one makes the identification $L_0 \to p^+p^-$, which is correct on mass shell. Thus the zero modes of $Q^a(z)$ and $Q^{\dot a}(z)$ do satisfy (2.22).

Since each of the eight $Q^{\dot{a}}$ has conformal dimension 3/2, the algebraic structure defined by Q^{a} and $Q^{\dot{a}}$ is an $N = 8$ superconformal algebra. This is true (more or less by definition), but perhaps not terribly useful. The additional terms in $Q^{\dot{a}}(z)Q^{\dot{b}}(w)$ give rise to new operators, and by the time the algebra closes an infinite sequence of distinct operators are generated. This is inevitable for $N > 4$.

For four-dimensional superstring solutions, it is possible to have fewer supersymmetry charges, since some of the supersymmetry can be broken by the compactification of six dimensions. An analysis similar to that given above can be repeated for such cases. This will be explored in the next section.

A covariant version of the constructions of this section has also been found.[13)] It involves a number of additional technicalities, such as bosonization of the superconformal ghosts and the introduction of 'picture-changing operators.'

The GKO Construction

The physical string spectrum must give a unitary representation of the conformal or superconformal algebra. More specifically, in the case of the heterotic string the left-moving sector must give a $c_L = 24$ unitary representation of the $N = 0$ conformal symmetry algebra and the right-moving sector must give a $c_R = 12$ unitary representation of the $N = 1$ superconformal symmetry algebra. In the case of four dimensions, two units of c_L are provided by the two transverse X coordinates, so that 'internal' degrees of freedom should provide $c_L = 22$. Similarly, three units of c_R are provided by the transverse X and ψ coordinates, so that the internal superconformal symmetry algebra should have $c_R = 9$. The conformal symmetry algebra can be 'accidentally' enlarged for specific vacuum configurations. Indeed, this happens for solutions with space-time supersymmetry. We will be especially interested in the case of $N = 2$ superconformal symmetry, which corresponds to the case of $N = 1$ space-time supersymmetry in four dimensions.

In view of these considerations, the crucial mathematical problem is the classification of unitary representations of conformal symmetry algebras. These will be essential ingredients in the construction of explicit string theory solutions. In particular, the classification of $N = 2$ representations with $c_R = 9$ will be of special interest. Unfortunately, this is an unsolved mathematical problem. However, quite a bit is known and the subject is developing rapidly. All unitary representations are known for $N = 0$ with $c < 1$, for $N = 1$ with $c < 3/2$, and for $N = 2$ with $c < 3$. These particular examples, known as minimal models, are quite interesting. They are exactly soluble in the sense that it is possible (at least in principle) to calculate correlation functions for arbitrary products of fields of these models on an arbitrary Riemann surface.

A method of constructing unitary representations of the Virasoro algebra was invented by Goddard, Kent, and Olive.[14)] This method is sufficiently powerful to construct all of the minimal models as well as large classes of additional models. It certainly is not able to give all irreducible unitary representations, however. It is known that the Virasoro algebra admits such representations for all $c > 1$, but the GKO construction always gives rational values of c.

The idea is really quite simple to explain. It generalizes the Sugawara-Sommerfield (SS) construction to cosets. This construction already gives many unitary representations of the Virasoro algebra, since there is the freedom to choose a group G, level k, and highest weight states. For example, in the case of $\widehat{SU}(2)_k$, the highest weight states can have any $SU(2)$ spin j satisfying $j \leq k/2$. Even so, it is easy to see that it gives a quite limited class of representations. In particular, $c_G \geq \text{rank}(G)$. Thus all choices give $c \geq 1$, and except for the first few values of k in $\widehat{SU}(2)_k$ they have $c \geq 2$. There are interesting models not realized in this way, but many of them can be realized by a coset generalization of the SS construction.

Suppose that the Kac–Moody algebra $\widehat{G}_k$ has a subalgebra $\widehat{H}_l$. The level l is determined by the embedding of H in G. For example, if the simple roots of H are a subset of the simple roots of G, then $l = k$. If $\widehat{G}$ is a direct product of the form $\widehat{G}_{k_1} \times \widehat{G}_{k_2}$ and $\widehat{H}_l$ is the diagonal subgroup, then $l = k_1 + k_2$. The SS construction associates to this subalgebra an energy–momentum tensor $T_H(z)$ with central charge

$$c_H = \frac{l \dim H}{l + \tilde{h}_H}. \tag{2.25}$$

Now consider the difference of two energy–momentum tensors

$$T(z) = T_G(z) - T_H(z). \tag{2.26}$$

This operator also defines a representation of conformal symmetry, which is trivially unitary, since it is realized on a subspace of the positive-definite representation space of $\widehat{G}_k$. The central charge of $T(z)$ is

$$c = c_G - c_H. \tag{2.27}$$

As a specific example, consider the coset model given by[15)]

$$\frac{\widehat{SU}(2)_k \times \widehat{SU}(2)_l}{\widehat{SU}(2)_{k+l}} \tag{2.28}$$

This defines a chiral algebra with central charge

$$\begin{aligned} c &= \frac{3k}{k+2} + \frac{3l}{l+2} - \frac{3(k+l)}{k+l+2} \\ &= 1 - \frac{6l}{(k+2)(k+l+2)} + \frac{2(l-1)}{l+2} \quad . \end{aligned} \tag{2.29}$$

The particular cases $l = 1$ and $l = 2$ give the minimal models of the $N = 0$ and $N = 1$ conformal algebras.

N=2 Superconformal Symmetry

By definition, an N-extended superconformal symmetry algebra is one containing N dimension-3/2 fermionic generators $G^\alpha(z)$, $\alpha = 1, 2, \ldots, N$. It is also required that the OPE $G^\alpha(z)G^\beta(w)$ contain the term $2T(w)\delta^{\alpha\beta}/(z-w)$, where $T(w)$ is energy–momentum tensor.

Recall that the $N=1$ superconformal algebra is given by the OPE's

$$\begin{aligned} T(z)T(w) &\sim \frac{c}{2(z-w)^4} + \frac{2T(w)}{(z-w)^2} + \frac{\partial T(w)}{z-w} + \cdots \\ T(z)G(w) &\sim \frac{3}{2}\frac{G(w)}{(z-w)^2} + \frac{\partial G(w)}{z-w} + \cdots \\ G(z)G(w) &\sim \frac{2c}{3(z-w)^3} + \frac{2T(w)}{z-w} + \cdots . \end{aligned} \tag{2.30}$$

A generalization of this algebra that involves two supercurrents $G^1(z)$ and $G^2(z)$ was discovered in 1976 by Ademollo *et al.*[16] In addition to T, G^1, G^2, it also contains a dimension-one current $J(z)$. J can be regarded as defining an Abelian Kac–Moody algebra $\widehat{U}(1)$.

The $N=2$ superconformal algebra is given by

$$\begin{aligned} T(z)T(w) &\sim \frac{c}{2(z-w)^4} + \frac{2T(w)}{(z-w)^2} + \frac{\partial T(w)}{z-w} + \cdots \\ T(z)G^\alpha(w) &\sim \frac{3}{2}\frac{G^\alpha(w)}{(z-w)^2} + \frac{\partial G^\alpha(w)}{z-w} + \cdots \\ T(z)J(w) &\sim \frac{J(w)}{(z-w)^2} + \frac{\partial J(w)}{z-w} + \cdots \\ G^\alpha(z)G^\beta(w) &\sim \left(\frac{2c}{3(z-w)^3} + \frac{2T(w)}{z-w}\right)\delta^{\alpha\beta} + i\left(\frac{2J(w)}{(z-w)^2} + \frac{\partial J(w)}{z-w}\right)\epsilon^{\alpha\beta} + \cdots \\ J(z)G^\alpha(w) &\sim i\epsilon^{\alpha\beta}\frac{G^\beta(w)}{z-w} + \cdots \\ J(z)J(w) &\sim \frac{c/3}{(z-w)^2} + \cdots . \end{aligned} \tag{2.31}$$

It is sometimes convenient to consider the charge ± 1 combinations

$$G^\pm(z) = \frac{1}{\sqrt{2}}(G^1(z) \pm iG^2(z)), \tag{2.32}$$

which satisfy

$$J(z)G^\pm(w) \sim \pm\frac{G^\pm(w)}{z-w} + \cdots . \tag{2.33}$$

It also follows that $G^+(z)G^+(w)$ and $G^-(z)G^-(w)$ are nonsingular, whereas

$$G^+(z)G^-(w) \sim \frac{2c}{3(z-w)^3} + \frac{2J(w)}{(z-w)^2} + \frac{2T(w) + \partial J(w)}{z-w} + \cdots \ . \tag{2.34}$$

This algebra has been studied extensively in the recent literature.[17)]

A one parameter twisting of the $N = 2$ superconformal algebra is given by the boundary conditions

$$G^{\pm}(z) = e^{\pm 2\pi i\eta} G^{\pm}(e^{2\pi i} z). \tag{2.35}$$

This implies that $T(z)$ and $J(z)$ are periodic and therefore have integer modes. The case $\eta = 0$ (or any $\eta \in Z$) corresponds to the NS sector, and $\eta = 1/2$ (or any $\eta \in Z + 1/2$) corresponds to the R sector. Now it is easy to show[18)] that for all values of η the algebras are isomorphic. This is demonstrated by observing that the operators

$$\begin{aligned} G_\eta^{\pm}(z) &= z^{\pm\eta} G^{\pm}(z) \\ T_\eta(z) &= T(z) + \eta J(z) + \frac{c\eta^2}{6z^2} \\ J_\eta(z) &= J(z) + \frac{c\eta}{3z} \end{aligned} \tag{2.36}$$

satisfy the $N = 2$ OPE's for all η. In particular the zero mode eigenvalues h and q are related by

$$\begin{aligned} h_\eta &= h + \eta q + \eta^2 c/6 \\ q_\eta &= q + c\eta/3. \end{aligned} \tag{2.37}$$

By considering the 'spectral flow' as η is continued from zero to one-half, for example, one can establish a one-to-one correspondence between NS and R states. This is a strong hint of a connection between $N = 2$ superconformal symmetry and space-time supersymmetry. Such a connection was proposed in Refs. 19 and 20 and will be explored further.

Any unitary representation of the algebra necessarily has $h \geq 0$ for all of its states. Requiring that $h_\eta \geq 0$ for all η we deduce the interesting inequality

$$h \geq \frac{3q^2}{2c}, \tag{2.38}$$

which is valid for all states in a unitary representation of the $N = 2$ algebra. Additional inequalities can be deduced from the positive definiteness of the anticommutator

$$\{G^+_\lambda, G^-_{-\lambda}\} = 2L_0 + 2\lambda J_0 + \frac{c}{3}(\lambda^2 - \frac{1}{4}). \tag{2.39}$$

This implies that

$$h \geq |\lambda q| + \frac{c}{6}(\frac{1}{4} - \lambda^2). \tag{2.40}$$

Therefore in the NS sector

$$h \geq \frac{1}{2}|q|,\ \frac{3}{2}|q| - \frac{c}{3},\ \ldots \tag{2.41}$$

and in the R sector

$$h \geq \frac{c}{24},\ |q| - \frac{c}{8},\ \ldots\ . \tag{2.42}$$

By definition, a chiral $N = 2$ superfield Φ is one for which $G^\pm_{-1/2}\Phi$ is null. Using (2.39) with $\lambda = \pm 1/2$, this implies that $h = |q|/2$, saturating the first inequality in (2.41). Combining this with (2.38) or the second inequality of (2.41) gives $h \leq c/6$ for a chiral superfield.

Space-Time Supersymmetry

Let us examine more closely how $N = 1$ space-time supersymmetry in four dimensions is related to $N = 2$ superconformal symmetry. The discussion that follows is similar to one given by Banks *et al.* for a covariant formulation.[21)] The supercharge is a four-component Majorana spinor, which can be decomposed into Q^a and

$Q^{\dot{a}}$, just as we did for the ten-dimensional case. As in that case, we can write the supersymmetry algebra in the form

$$\begin{aligned}\{Q^a, Q^b\} &= 2p^+\delta^{ab} \\ \{Q^{\dot{a}}, Q^{\dot{b}}\} &= 2p^-\delta^{ab} \\ \{Q^a, Q^{\dot{b}}\} &= \sqrt{2}\gamma^i_{a\dot{b}}p^i\,.\end{aligned} \tag{2.43}$$

Alternatively, it is convenient to replace the labels a, $\dot{a}$ by helicities $\pm 1/2$, so that

$$\begin{aligned}\{Q^{1/2}, Q^{-1/2}\} &= 2p^+ \\ \{Q^{\dot{1}/2}, Q^{-\dot{1}/2}\} &= 2p^- \\ \{Q^{1/2}, Q^{\dot{1}/2}\} &= p^1 + ip^2, \text{ etc.}\end{aligned} \tag{2.44}$$

Let us bosonize the fermions ψ^1 and ψ^2 as before by setting

$$\frac{1}{\sqrt{2}}(\psi^1 \pm i\psi^2) =: e^{\pm iH} :\,, \tag{2.45}$$

where H is a canonically normalized scalar field and cocycle factors are omitted. Now, we want to identify

$$Q^{\pm 1/2}(z) = \sqrt{2p^+}S^{\pm}(z)\,, \tag{2.46}$$

where $S^{\pm}(z)$ are free fermi fields satisfying

$$\begin{aligned}S^+(z)S^-(w) &\sim \frac{1}{z-w} + \cdots \\ S^{\pm}(z)S^{\pm}(w) &\sim O(1)\,.\end{aligned} \tag{2.47}$$

Previously this was achieved in terms of $\exp i\vec{w}\cdot\vec{H}$ based on spinor representations of the transverse $SO(8)$. The corresponding construction, $:\exp \pm\frac{i}{2}H:$, now gives operators of dimension $1/8$. Therefore let us set

$$S^{\pm}(z) =: e^{\pm(i/2)H(z)} : \Sigma^{\pm}(z)\,, \tag{2.48}$$

where $\Sigma^{\pm}(z)$ are dimension $3/8$ fields constructed from the internal $(c=9)$ fields.

The question now arises what the OPE's of the $\Sigma^{\pm}$ fields must be in order that the $S^{\pm}$ have the desired algebra. One finds that (2.47) requires that

$$\Sigma^{\pm}(z)\Sigma^{\pm}(w) = (z-w)^{-1/4}O^{\pm}(w) + \cdots \tag{2.49}$$

where $O^{\pm}(w)$ has dimension 1/2. We also require that

$$\Sigma^{+}(z)\Sigma^{-}(w) \sim (z-w)^{-3/4}I + \frac{1}{2}(z-w)^{1/4}J(w) + \cdots \tag{2.50}$$

in order that $S^{+}(z)S^{-}(w)$ in (2.47) come out right. The $J(w)$ term does not contribute to the singular part of this product, but secondary terms that trail the leading one by integer powers are inevitable, in general. Also the necessity for $J(w)$ will become apparent in a moment.

Let us construct $Q^{\pm 1/2}$ by analogy with the ten-dimensional formulas:

$$Q^{\pm 1/2} = \frac{1}{\sqrt{p^{+}}}(\partial X^{1} \pm i\partial X^{2})S^{\mp}\,. \tag{2.51}$$

Now we must examine the OPE's of these operators. The claim is that the $Q^{\pm 1/2}$ are proportional to the $G^{\pm}(z)$ operators of the $N=2$ superconformal algebra and that $J(w)$ corresponds to the $U(1)$ current in that algebra. The coefficient of $(z-w)^{-1}$ in $J(z)J(w)$ controls the central charge as in the last line of (2.31).

In constructing models with $N=1$ space-time supersymmetry there is an additional requirement, analogous to a GSO projection condition, that must be imposed. Specifically, the value of the total $U(1)$ charge, defined by the zero mode of $J(z)$, must be an odd integer for physical states. For example, massless states have $h=1/2$ and $U(1)$ charges $Q=\pm 1$. The supersymmetry operators $Q^{\pm 1/2}$ map the two charge states into one another, i.e., they form a supermultiplet.

N = 2 Minimal Models

The $N=2$ algebra also has an infinite sequence of minimal models.[23,19] As one would expect, their conformal anomalies accumulate at $c=3$, the value corresponding to the representation based on X^1, X^2, ψ^1, ψ^2 described in the preceding section.

The construction of $N = 2$ minimal models can be described by combining an $SU(2)$ parafermionic theory having Z_k discrete symmetry with a free scalar field.[24] The parafermion theory[25] is a generalization of the critical Ising model, which corresponds to the case $k = 2$. An alternative construction of the $N = 2$ minimal models, using an $N = 1$ extension of the GKO method, will be described in the next section. It should also be noted that yet another derivation, using bosonic fields only, also exists.[26] The trick that makes it possible to describe an interacting conformal field theory in terms of free bosons is to include vertex-operator-like terms in the formula for the energy–momentum tensor.[27]

The parafermionic theory can be obtained from two distinct GKO constructions. The first one is based on the coset

$$\frac{\widehat{SU}(2)_k}{\widehat{U}(1)} . \tag{2.52}$$

This construction utilizes previous work on $SU(2)$ Wess–Zumino–Witten models.[6] The corresponding central charge is

$$c = \frac{3k}{k+2} - 1 = \frac{2(k-1)}{k+2} , \tag{2.53}$$

since a $\widehat{U}(1)$ algebra always contributes $c = 1$ via the SS construction.

The parafermionic theory can also be obtained by a GKO construction based on the coset

$$\frac{\widehat{SU}(k)_1 \otimes \widehat{SU}(k)_1}{\widehat{SU}(k)_2} . \tag{2.54}$$

This again gives a conformal field theory with central charge

$$c = 2(k-1) - \frac{2(k^2-1)}{k+2} = \frac{2(k-1)}{(k+2)} . \tag{2.55}$$

Adding one for the free scalar gives

$$c_2 = \frac{3k}{k+2} , \tag{2.56}$$

for the $N = 2$ minimal models.

N=1 Extension of the GKO Construction

Super Kac-Moody algebras contain dimension 1/2 fermionic currents $j^A(z)$ as superpartners of the usual dimension 1 bosonic currents $J^A(z)$. Since the $j^A(z)$ belong to the adjoint representation

$$J^A(z)j^B(w) \sim \frac{if_{ABC}j^C(w)}{z-w} + \ldots . \tag{2.57}$$

They have free fermion commutation relations, so choosing a convenient normalization,

$$j^A(z)j^B(w) \sim \frac{k\delta^{AB}}{2(z-w)} + \ldots . \tag{2.58}$$

The unitary representation of a super Kac-Moody algebra with the lowest possible level is given by setting the bosonic currents $J^A(z)$ equal to the fermion bilinears

$$J_f^A(z) = -\frac{i}{k} f^{ABC} j^B(z) j^C(z) \quad . \tag{2.59}$$

In this case it is easy to see that the conformal anomaly is just that of $\dim G$ free fermions

$$c_f = \frac{1}{2}\dim G \quad , \tag{2.60}$$

and the level of the Kac-Moody algebra is

$$k_f = c_A(G) \quad . \tag{2.61}$$

Using this special representation, the general representation of a super Kac-Moody algebra can be obtained.[28] Letting

$$J^A(z) = J_f^A(z) + \hat{J}^A(z) \quad , \tag{2.62}$$

one sees that $\hat{J}^A(z)$ defines a Kac-Moody algebra that is independent of the fermion fields. In other words, $\hat{J}^A(z)j^B(w)$ and $\hat{J}^A(z)J_f^B(w)$ are nonsingular. Therefore, if we consider a level $\hat{k}$ representation of the algebra given by the $\hat{J}^A(z)$, we obtain a representation of the $J^A(z)$ algebra with level

$$k = \hat{k} + c_A(G) = \hat{k} + \tilde{h}_G \tag{2.63}$$

and central charge

$$c_G = \frac{\hat{k}\ \dim G}{\hat{k} + \tilde{h}_G} + \frac{1}{2}\dim G \quad . \tag{2.64}$$

The allowed values of $\hat{k}$ are $0, 1, 2, \ldots$, where the choice $\hat{k} = 0$ implies setting $\hat{J}^A = 0$. For simplicity it is assumed that G is simply-laced so that $c_A(G) = \tilde{h}_G$, where $c_A(G)$ is the Casimir number defined in (2.16). The corresponding energy–momentum tensor associated with the group G is

$$T_G(z) = \frac{1}{k}(:\hat{J}^A(z)\hat{J}^A(z): - :j^A(z)\partial j^A(z):) \quad . \tag{2.65}$$

Note that this is a generalization of the construction of the $\widehat{SU}(2)_{k+2}$ algebra. Just as in that case, we can also give an explicit formula for the supercurrent:

$$G_G(z) = \frac{2}{k}\left(j^A(z)\hat{J}^A(z) - \frac{i}{3k}f_{ABC}j^A(z)j^B(z)j^C(z)\right) \quad . \tag{2.66}$$

It is now easy to generalize the GKO construction to super KM algebras.[29] Let $J^a(z)$, $a = 1, \ldots, \dim H$, denote the generators of $\hat{H}$ as before. Then, just as for $\hat{G}$, we can write

$$J^a(z) = \hat{J}^a(z) - \frac{i}{k}f_{abc}j^b(z)j^c(z) \quad , \tag{2.67}$$

where f_{abc} are the H structure constants and $j^a(z)$ are $\dim H$ of the free fermions. The remaining $\dim G - \dim H$ free fermions are denoted $j^{\bar{a}}(z)$. The H supercurrent is

$$G_H(z) = \frac{2}{k}\left(j^a(z)\tilde{j}^a(z) - \frac{i}{3k} f_{abc} j^a(z) j^b(z) j^c(z)\right) \quad , \tag{2.68}$$

and the coset model is defined by the difference

$$\begin{aligned} G(z) &= G_G(z) - G_H(z) \\ &= \frac{2}{k}\left(j^{\bar a}(z)\hat{j}^{\bar a}(z) - \frac{i}{3k} f_{\bar a\bar b\bar c} j^{\bar a}(z) j^{\bar b}(z) j^{\bar c}(z)\right) \quad . \end{aligned} \tag{2.69}$$

Just as in the ordinary GKO construction, one can easily show that this has nonsingular OPE's with the H currents $\tilde{J}^a(w)$ and $j^a(w)$. Therefore, it is guaranteed to define an $N = 1$ superconformal theory with central charge

$$c = c_G - c_H \quad , \tag{2.70}$$

where

$$c_G = \frac{3}{2}\dim G - \frac{\tilde{h}_G}{k}\dim G \tag{2.71}$$

and similarly for c_H. Thus for the coset model

$$c = \frac{3}{2}\dim G/H + \frac{1}{k}(\tilde{h}_H \dim H - \tilde{h}_G \dim G) \quad . \tag{2.72}$$

The corresponding energy–momentum tensor is

$$\begin{aligned} T = &\frac{1}{k}\Big(\hat{j}^{\bar a}\hat{j}^{\bar a} - \frac{\hat{k}}{k} j^{\bar a}\partial j^{\bar a} + \frac{2i}{k} f_{\bar a\bar b\bar c}\hat{j}^{\bar a} j^{\bar b} j^{\bar c} \\ &- \frac{1}{k} f_{\bar a pq} f_{\bar b pq} j^{\bar a}\partial j^{\bar b} - \frac{1}{k^2} f_{\bar a\bar b\bar c} f_{\bar a\bar d\bar e} j^{\bar b} j^{\bar c} j^{\bar d} j^{\bar e}\Big). \end{aligned} \tag{2.73}$$

The formulas simplify considerably for $f_{\bar a\bar b\bar c} = 0$ (a symmetric space). In this case one simply has

$$G(z) = \frac{2}{k} j^{\bar{a}}(z)\hat{J}^{\bar{a}}(z) \tag{2.74}$$

and

$$T(z) = \frac{1}{k}\left(\hat{J}^{\bar{a}}\hat{J}^{\bar{a}} - \frac{\hat{k}}{k} j^{\bar{a}}\partial j^{\bar{a}} + \frac{2i}{k} f_{a\bar{b}\bar{c}}\hat{J}^{a} j^{\bar{b}} j^{\bar{c}}\right) \quad . \tag{2.75}$$

In the case of $SU(2)$, $\dim(G/H) = 2$ and $k = \hat{k} + 2$, so that

$$c = \frac{3\hat{k}}{\hat{k}+2} \quad , \tag{2.76}$$

providing a new derivation of the $N = 2$ minimal models. Not surprisingly, the modular-invariant partition functions of the minimal models have the same ADE classification that we presented for $\widehat{SU}(2)$ models.

New N=2 Models

In the preceding section, we have seen that it is possible to associate unitary representations of the $N = 1$ superconformal algebra with cosets by a generalized GKO construction. Here I wish to address the question of when the theory admits a second supercurrent and thus furnishes an $N = 2$ model. We saw that this happens in the case of $SU(2)/U(1)$ corresponding to the $N = 2$ minimal models. The question is what is the most general choice of G/H for which the $N = 1$ construction in fact defines an $N = 2$ algebra.

This question has been examined by Kazama and Suzuki.[29] In their second paper, they clearly spell out the conditions under which an $N = 2$ algebra is realized.[30] They find that there is a second supercurrent of the form

$$G^2 = \frac{2}{k}\left(h_{\bar{a}\bar{b}} j^{\bar{a}}\hat{J}^{\bar{b}} - \frac{i}{3k} S_{\bar{a}\bar{b}\bar{c}} j^{\bar{a}} j^{\bar{b}} j^{\bar{c}}\right) \tag{2.77}$$

provided that the following conditions are satisfied:

(*i*) $h_{\bar{a}\bar{b}} = -h_{\bar{b}\bar{a}}$ and $h_{\bar{a}\bar{p}}h_{\bar{p}\bar{b}} = -\delta_{\bar{a}\bar{b}}$

(ii) $h_{a\bar{p}} f_{\bar{p}\bar{b}e} = h_{\bar{b}\bar{p}} f_{\bar{p}\bar{a}e}$

(iii) $f_{\bar{a}\bar{b}\bar{c}} = h_{\bar{a}\bar{p}} h_{\bar{b}\bar{q}} f_{\bar{p}\bar{q}\bar{c}} + 2$ perms

(iv) $S_{\bar{a}\bar{b}\bar{c}} = h_{\bar{a}\bar{p}} h_{\bar{b}\bar{q}} h_{\bar{c}\bar{r}} f_{\bar{p}\bar{q}\bar{r}}$.

Note that (i) implies that $h_{\bar{a}\bar{b}}$ is a complex structure for the coset manifold G/H.

To analyze the implications of these equations, let us define

$$\phi_e = h_{\bar{a}\bar{b}} f_{\bar{a}\bar{b}e} \tag{2.78}$$

and consider

$$X_{cd} = f_{cde} \phi_e \quad . \tag{2.79}$$

Since $f_{cd\bar{e}} = 0$, by the H group property,

$$X_{cd} = h_{\bar{a}\bar{b}} f_{\bar{a}\bar{b}E} f_{cdE} \quad . \tag{2.80}$$

The Jacobi identity allows us to cycle the indices $\bar{a}, \bar{b}, c$ on the f's. Also, using the antisymmetry of $h_{\bar{a}\bar{b}}$, one obtains

$$X_{cd} = -2h_{\bar{a}\bar{b}} f_{c\bar{a}E} f_{\bar{b}dE} = -2h_{\bar{a}\bar{b}} f_{c\bar{a}\bar{e}} f_{\bar{b}d\bar{e}} = -2h_{\bar{a}\bar{e}} f_{c\bar{a}\bar{b}} f_{\bar{b}d\bar{e}} \quad , \tag{2.81}$$

where the last step uses property (ii). Now interchanging the labels $\bar{b}$ and $\bar{e}$ we see that X_{cd} is equal to its negative and hence vanishes. It therefore follows that ϕ_e can only be nonzero if e corresponds to a $U(1)$ factor in H. Thus we conclude that a necessary condition for $N = 2$ superconformal symmetry is that H contain a $U(1)$ factor.

Let us now consider the special case of a symmetric space ($f_{\bar{a}\bar{b}\bar{c}} = 0$). Manipulations similar to those of the previous paragraph allow one to show that

$$X_{\bar{c}\bar{d}} = f_{\bar{c}\bar{d}e} \phi_e = g h_{\bar{c}\bar{d}} \quad , \tag{2.82}$$

where g is the Casimir number for the group G (assumed simple) defined in (2.16), where it was called c_A. For hermitian symmetric spaces there is just one $U(1)$ factor.

Let $e = 0$ correspond to this $U(1)$ factor, so that the only nonzero component of ϕ is ϕ_0. Then we see that $h_{a\bar{b}} \propto f_{a\bar{b}0}$, with a normalization determined by condition (i). Thus we are able to associate an $N = 2$ superconformal model to every hermitian symmetric space. These were classified by Cartan and are listed in the book of Helgason.[31] The table lists the irreducible hermitian symmetric spaces (for compact G) and the associated $N = 2$ central charges. There are several cases that give the special value $c = 9$ irreducibly. There are also a number of examples based on cosets other than $SU(2)/U(1)$ that give $c < 3$. These always have a central charge corresponding to a particular minimal model and provide alternative constructions of that minimal model.

G/H	$c_{G/H}$
$SU(m+n)/SU(m) \times SU(n) \times U(1)$	$3\hat{k}mn/(\hat{k}+m+n)$
$SO(n+2)/SO(n) \times SO(2)$ for $n = 1$, $SO(3)/SO(2)$	$3\hat{k}n/(\hat{k}+n) \quad n \geq 2$ $3\hat{k}/(\hat{k}+2)$
$SO(2n)/SU(n) \times U(1)$	$\frac{3}{2}\hat{k}n(n-1)/(\hat{k}+2n-2)$
$Sp(n)/SU(n) \times U(1)$	$\frac{3}{2}\hat{k}(n+1)/(\hat{k}+n+1)$
$E_6/SO(10) \times U(1)$	$48\hat{k}/(\hat{k}+12)$
$E_7/E_6 \times U(1)$	$81\hat{k}/(\hat{k}+18)$

Table.[29] Hermitian symmetric spaces and the associated $N = 2$ central charges.

3. Four-Dimensional Models

Calabi–Yau Spaces

A special class of six-dimensional manifolds was shown by Candelas *et al.* to be preferred for compactification of the $E_8 \otimes E_8$ heterotic string from ten to four dimensions.[32,33] These manifolds, called Calabi–Yau spaces, arose as a consequence of requiring that the four-dimensional model have $N = 1$ space-time supersymmetry. This is a desirable outcome, since supersymmetry is the most promising approach to understanding why the energy scale characterizing weak symmetry breaking is much lower than the unification scale (the so-called 'gauge hierarchy' problem). Of course, we would feel a lot more comfortable about this assumption if it had experimental support. In my opinion, that is the most important question that could be answered by the new accelerators that are beginning operation.

Let us suppose that ten-dimensional space-time decomposes into the direct product of four-dimensional Minkowski space and a six-dimensional compact space K : $M^{10} \to M^4 \times K$.* Let us furthermore denote the coordinates of M^4 by x and those of K by y. A supersymmetry transformation parameter in ten dimensions is a 16-component Majorana–Weyl spinor $\epsilon(x, y)$ that decomposes as follows

$$\epsilon(x,y) = \sum_{\alpha=1}^{4} \epsilon_\alpha(x)\eta_\alpha(y) + \dots , \tag{3.1}$$

where ϵ_α and η_α are four-component spinors and the dots represent higher modes that are not important at low energy. A given ϵ_α describes a four-dimensional supersymmetry if and only if the corresponding $\eta_\alpha(y)$ is covariantly constant on the internal manifold K. Therefore, to obtain $N = 1$ supersymmetry in four dimensions we must have one (and only one) covariantly constant spinor $\eta(y)$ on K. The existence of such

---

* An exact description of our universe should involve a cosmologically evolving space instead. With few exceptions,[34] it is assumed that this is good enough for purposes of particle physics.

a spinor implies that

$$0 = [D_i, D_j]\eta = R_{ijkl}\Gamma^{kl}\eta = 0, \tag{3.2}$$

where $\Gamma_{kl} = \Gamma_{[k}\Gamma_{l]}$, and Γ_k are Dirac matrices on K. This in turn implies that the Ricci tensor R_{ij} vanishes. In other words, K is Ricci flat. Being based on a low-energy point-particle limit, this analysis is a little too specific, however. All that is really required is that K admit a Ricci-flat metric.

An important concept is the 'holonomy group' of the manifold K. This is the group that describes the possible transformations of various spinors or tensors under parallel transport around closed curves γ on K. The holonomy group is generated by matrices of the form

$$U_\gamma = P \exp \int_\gamma \omega \cdot dx, \tag{3.3}$$

where ω is the spin connection of K and P denotes path ordering of the exponential. In general ω belongs to the algebra of $SO(6)$, and hence the holonomy group is $SO(6)$ or its covering group $SU(4)$. However, the matrices U need not span the whole group, so for special manifolds the holonomy group can be a subgroup of $SU(4)$. The coordinates can be chosen in such a way that the covariantly constant spinor η introduced above takes the form $\eta = (0, 0, 0, \eta_0)$. The requirement that η be invariant then implies that the holonomy group can act only on the first three components. Thus it must be $SU(3)$, or a subgroup thereof. Requiring that there be no additional conserved spinors implies that it must be exactly $SU(3)$.

A Calabi–Yau space is a special case of a Kähler manifold, which in turn is a special case of a complex manifold. A complex manifold admits an 'almost complex structure' $J^i{}_j$ whose square (as a matrix) is -1. This allows the choice (in local coordinate patches) of a holomorphic coordinate system $(z^a, \bar{z}^{\bar{a}})$ $a, \bar{a} = 1, 2, 3$. In this basis the nonvanishing components of J are $J^a{}_b = i\delta^a_b$ and $J^{\bar{a}}{}_{\bar{b}} = -i\delta^{\bar{a}}_{\bar{b}}$. If, moreover, J is covariantly constant, then the manifold is called 'Kähler.' A Kähler metric satisfies $g_{ab} = g_{\bar{a}\bar{b}} = 0$, $g_{a\bar{b}} = g_{\bar{b}a}$.

On a Kähler manifold, the holonomy group transforms holomorphic and antiholomorphic coordinates separately. Therefore, the most general possibility (for $2N$ real dimensions) is $U(N)$. Since $U(N) = SU(N) \otimes U(1)$, we see that supersymmetry requires a Kähler manifold whose holonomy group is $SU(3)$. In other words, the $U(1)$ factor is absent. The curvature two-form associated with the $U(1)$ factor is an element of $H^2(K; R)$, the second de Rham cohomology group. The requirement that it should vanish (up to an exact form) means that it must be a trivial cohomology class. Thus this class, called the first Chern class $c_1(K)$, must vanish. The famous theorem conjectured by Calabi and proved by Yau is that a Kähler manifold with vanishing first Chern class admits a Ricci-flat metric.[35)] In string theory applications this particular metric is selected only in a certain approximation (the size of K large compared to the Planck scale), but not by the exact equations.[36)]

A Kähler manifold admits a covariantly constant two form k_{ij}, called the Kähler form. It relates the complex structure and the metric by the relation $J^i{}_j = g^{ik} k_{kj}$. In particular, for holomorphic coordinates $g_{a\bar{b}} = ik_{a\bar{b}}$ and $k_{ab} = k_{\bar{a}\bar{b}} = 0$. In the particular case of a Calabi–Yau space K, the Kähler form is given in terms of the spinor η by $k_{ij} = \bar{\eta}\Gamma_{ij}\eta$.

In discussing differential geometry of complex manifolds, it is convenient to decompose the exterior derivative into a holomorphic and antiholomorphic part: $d = \partial + \bar{\partial}$. Correspondingly, differential forms can be represented as (p,q)-forms $\psi_{a_1 \ldots a_p \bar{a}_1 \ldots \bar{a}_q}$ having p antisymmetric holomorphic indices and q antisymmetric antiholomorphic indices. The de Rham cohomology groups $H^n(K; R)$ which described closed n-forms modulo exact n-forms can be decomposed into cohomology groups for (p,q)-forms:

$$H^n = \bigoplus_{p+q=n} H^{p,q}. \tag{3.4}$$

The dimensionality of the vector spaces H^n are given by Betti numbers b_n. Those for the $H^{p,q}$ by Hodge numbers $h^{p,q}$. These are related by

$$b_n = \sum_{p+q=n} h^{p,q}. \tag{3.5}$$

In particular, the Euler characteristic $\chi(K)$ is given by

$$\chi = \sum_n (-1)^n b_n = \sum_{p,q} (-1)^{p+q} h^{p,q}. \tag{3.6}$$

The Hodge numbers have two symmetries: the holomorphic-antiholomorphic interchange symmetry $h^{p,q} = h^{q,p}$ and the Poincaré duality symmetry $h^{p,q} = h^{N-p,N-q}$.

In the case of Calabi–Yau spaces,

$$\begin{aligned} h^{0,0} &= h^{3,0} = 1 \\ h^{1,0} &= h^{2,0} = 0 \end{aligned} \tag{3.7}$$

in general. The two remaining independent Hodge numbers, $h^{1,1}$ and $h^{2,1}$, depend on the particular example. The Euler characteristic becomes

$$\chi = 2(h^{1,1} - h^{2,1}). \tag{3.8}$$

A fundamental requirement in heterotic string compactification is that the four-form $tr(F_\wedge F) - tr(R_\wedge R)$ formed from the Yang–Mills two-form F and the curvature two-form R should be exact (trivial in cohomology). The $D = 10$ heterotic string (with supersymmetry) can have as its gauge group $SO(32)$ or $E_8 \times E_8$, corresponding to the two self-dual sixteen-dimensional lattices. In the $SO(32)$ case $tr(F_\wedge F)$ is evaluated in the fundamental 32-dimensional representation, which is 1/30 of the value in the adjoint representation. In the $E_8 \times E_8$ case we define it as 1/30 of the value in the adjoint representation. To achieve the triviality of the four-form above we must relate the geometry of K (characterized by R) to the choice of specific background gauge fields (characterized by F). A particular choice that works out very nicely is achieved as follows. First decompose one of the two E_8's with respect to a maximal $SU(3) \times E_6$ subgroup so that the adjoint representation of E_8 breaks up as follows

$$\mathbf{248} = (\mathbf{3}, \mathbf{27}) \oplus (\overline{\mathbf{3}}, \overline{\mathbf{27}}) \oplus (\mathbf{8}, \mathbf{1}) \oplus (\mathbf{1}, \mathbf{78}). \tag{3.9}$$

Now identify the holonomy group with the $SU(3)$ factor. This means that background gauge fields are taken nonvanishing for the $SU(3)$ subgroup only and these are equated to the spin connection of K. This is often described as 'embedding the spin connection in the gauge group.'

This construction has the following important consequences, which I will not attempt to prove here. The unbroken gauge symmetry is $E_6 \times E_8$, but the E_8 describes a 'hidden sector,' which can be ignored for many purposes. The massless states form $N = 1$ supermultiplets. In addition to the Yang–Mills and supergravity multiplets, these include chiral Wess–Zumino multiplets. Specifically, there are $h^{2,1}$ such multiplets that transform as a **27** of E_6 and $h^{1,1}$ such multiplets that transform as a $\overline{\mathbf{27}}$ of E_6. Since **27** and $\overline{\mathbf{27}}$ multiplets can be expected in general to pair up and acquire a high mass, it is natural to define the net number of E_6 generations of quarks and leptons as $|N_{\mathbf{27}} - N_{\overline{\mathbf{27}}}|$. This coincides with one-half of the Euler characteristic of K.

Let me conclude this section with two examples of Calabi–Yau spaces. Both of them are given by complex projective spaces CP^n supplemented by polynomial constraints.[37] The space CP^n can be described by $n+1$ complex coordinates up to an overall scaling $(z_1, \ldots, z_{n+1}) \equiv \lambda(z_1, \ldots, z_{n+1})$. Various techniques from algebraic geometry can be used to calculate the Chern class $c_1(K)$ and the Hodge numbers in each case. I will not attempt to explain them.

The first example (called Q) is given by the intersection of CP^4 and the quintic hypersurface[32]

$$z_1^5 + z_2^5 + z_3^5 + z_4^5 + z_5^5 = 0 \,. \tag{3.10}$$

This manifold has $c_1(Q) = 0$, $h^{2,1} = 101$, $h^{1,1} = 1$. Thus it is a Calabi–Yau space that gives a model with 100 net generations. Q has a very large discrete symmetry group generated by $z_i \rightarrow e^{2\pi i/5} z_i$ for each z_i separately, as well as arbitrary permutations of the z_i. However, a common scaling of all the z_i is trivial. Therefore the group is

$$G = (S_5 \times Z_5^5)/Z_5, \tag{3.11}$$

which has 75,000 elements. There are $h^{2,1} = 101$ polynomial deformations of the complex structure obtained by modifying the quintic constraint equation by perturbations of the form $z_1^3z_2^2(20)$, $z_1^2z_2^2z_3(30)$, $z_1^4z_2(30)$, $z_1^2z_2z_3z_4(20)$, $z_1z_2z_3z_4z_5(1)$. The one anti-generation corresponds to the Kähler class. (This means that $H^{1,1}$ is generated by the Kähler form.)

A second example is the three-generation ($|\chi| = 6$) manifold of Tian and Yau.[38] For later purposes it is convenient to use a description of this manifold due to Schimmrigk[39], which is provided by a hypersurface S in $CP^2 \times CP^3$. Describing CP^2 by coordinates (x_1, x_2, x_3) and CP^3 by coordinates (z_0, z_1, z_2, z_3), one introduces two polynomial constraints

$$\begin{aligned} z_0^3 + z_1^3 + z_2^3 + z_3^3 &= 0 \\ z_1x_1^3 + z_2x_2^3 + z_3x_3^3 &= 0. \end{aligned} \tag{3.12}$$

The manifold S turns out to have $c_1(S) = 0$, $h^{2,1} = 35$, $h^{1,1} = 8$. Thus $\chi(S) = -54$, and it describes 27 generations. The discrete symmetry in this case is

$$G = S_3 \times (Z_3 \times Z_9^3)/Z_9. \tag{3.13}$$

The S_3 factor is a permutation symmetry of pairs (x_i, z_i), and an element (r_0, r_1, r_2, r_3) of $Z_3 \times Z_9^3$ is given by

$$\begin{aligned} z_i &\to e^{2\pi i r_i/3} z_i \qquad i = 0,1,2,3 \\ x_i &\to e^{-2\pi i r_i/9} x_i \qquad i = 1,2,3. \end{aligned} \tag{3.14}$$

The Z_9 factor in the denominator of (3.13) is generated by the trivial group element (1,1,1,1). Altogether G has 1458 elements.

The discrete symmetry G of the Calabi–Yau space S has a $Z_3 \times Z_3$ subgroup H. The first Z_3 is generated by the even permutations in the S_3 factor of G. The second Z_3 is generated by $(r_0, r_1, r_2, r_3) = (0, 3, 6, 0)$. One can now define a new Calabi–Yau space by forming the quotient S/H and resolving the orbifold singularities that arise because the second Z_3 does not act freely. The resulting CY manifold has $\chi = -6$ and gives a model with three generations. It corresponds to one particular point in the multidimensional moduli space of topologically equivalent Calabi–Yau manifolds. Different constructions are known that give the same topological Calabi–Yau space at other points of its moduli space.[40] It is not yet known for sure, but this collection of manifolds may be the only Calabi–Yau's with $|\chi| = 6$, making them potentially important for phenomenology. If so, the next question is which (if any) points in moduli space are dynamically selected. We will return to this question later.

Compactification Based on N=2 Conformal Models

Gepner[22,41] has investigated superstring compactifications that give models with $(10-2n)$-dimensional Poincaré symmetry by describing internal degrees of freedom as a sum of $N = 2$ minimal models with

$$c = \sum_i \frac{3k_i}{k_i + 2} = 3n. \tag{3.15}$$

These constructions can be carried out both for Type II superstrings and for heterotic strings. In the latter case modular invariance of partition functions (and hence of loop amplitudes) can be implemented by formulating an analog of embedding the spin connection in the gauge group. This basically involves using the same sum of minimal models for the left-movers. In the case of four dimensions, the remaining $22 - 9 = 13$ units of c_L are contributed by the level-one Kac–Moody algebra $\widehat{SO}(10) \otimes \widehat{E}_8$. (At level one the central charge is equal to the rank of the algebra.) The $SO(10)$ symmetry actually becomes enlarged to E_6, so that one has $E_6 \times E_8$ gauge symmetry, as in the case of Calabi–Yau compactification. In general, there is some additional 'accidental' gauge symmetry as well.

Gepner has explored two examples in considerable detail. The first example is based on five copies of the $k = 3$ model. (Note that each one has $c = 9/5$.) The second uses one copy of the $k = 1$ model and three copies of the $k = 16_E$ model $(1 + 3 \cdot 8/3 = 9)$. These are referred to succinctly as 3^5 and $1 \cdot 16_E^3$. The subscript E means that the E_7 exceptional affine invariant is used in the construction. He then gives overwhelming circumstantial evidence that these two models correspond to the

Q and S Calabi–Yau spaces described in the preceding section. The evidence includes a count of the numbers of generations and antigenerations as well as an analysis of the discrete symmetries.

Let us examine the discrete symmetries, recalling that the kth minimal model (with A_{k+1} affine invariant) has Z_{k+2} symmetry. Also including the permutation symmetry, the 3^5 model has $S_5 \times Z_5^5$ symmetry. However, it turns out that a Z_5 subgroup that acts simultaneously on all five factors must be modded out, so one ends up with the same discrete group as for the Calabi–Yau space Q. In similar fashion one might expect the $1 \cdot 16_E^3$ model to have $S_3 \times (Z_3 \times Z_{18}^3)$ symmetry. However, for the E_7 invariant, the Z_{18}'s get reduced to Z_9 and, as before, one Z_9 subgroup acts trivially and must be modded out. Thus one ends up with the same discrete symmetry found for the Calabi–Yau space S. There is also a counterpart of the $Z_3 \times Z_3$ action described for S in the preceding section that can be used to construct a three-generation model.

One remarkable feature of Gepner's results is that the minimal models are exactly soluble, whereas Calabi–Yau spaces are very unwieldy in general. The construction is completely algebraic, and so it is not evident how geometric structure arises. If one could write a formula for the metric tensor or curvature tensor of the manifold in terms of the conformal field theory, one would have solved a mathematical problem that is usually assumed to be hopeless. Of course, this may not be possible. As in the circle example discussed in the previous section, one conformal field theory could correspond to Calabi–Yau spaces at several different values of their moduli. However, this does not seem to be the case for the examples discussed. Rather, they seem to be analogous to the $R = 1$ value of the circle compactification. The evidence for this is two-fold. First, the large discrete symmetry groups that we have found are not likely to be realized at other values of the respective moduli spaces. Secondly, the models have extra $U(1)$ gauge symmetries beyond the expected $E_6 \otimes E_8$. There is one associated with each contributing minimal model, but one gets used up in extending $SO(10)$ to E_6. Thus the number of $U(1)$ factors is one less than the number of contributing minimal models (four for 3^5 and three for $1 \cdot 16_E^3$). In other examples the gauge symmetry can be extended even further. This is reminiscent of the extra gauge symmetry of the circle at $R = 1$.

Thus the minimal model constructions correspond to Calabi–Yau compactifications at special points of their moduli space where there is a large discrete symmetry group and enhanced gauge symmetry. One wonders whether this makes them too special to be of much interest, or whether these special features could make them physically preferred. Although it is not yet known how to calculate such things, it seems conceivable that nonperturbative effects could induce a potential that depends on the moduli in such a way that the theory would 'roll' to such special points. This would make them particularly good candidates for phenomenology. But even if this is not so, their study still seems to be a useful exercise, since they have so many realistic features. Also, many of these features only depend on the topology of the space and not on the particular choice of moduli. Of course, one is not restricted to minimal models only. If one were to use one of the Kazama–Suzuki models that gives $c = 9$ irreducibly, there would not be any extra 'accidental' gauge symmetry.

It is an interesting challenge to try to make a complete classification of (2,2) superconformal models with $c = 9$. This is a formidable task, although it does not look as hopeless now, as it did a few years ago.* The first step is to examine what can be done with minimal models only. The equation

$$c = \sum_i \frac{3k_i}{k_i + 2} = 9 \tag{3.16}$$

has 168 solutions, which have been enumerated in ref. [43]. However, there is additional freedom in choosing modular invariants of the contributing minimal models. Allowing arbitrary combinations of A, D, and E invariants gives 1176 possibilities. The 228 that only use A and E invariants are tabulated by Lütken and Ross,[44] who give the number of generations and anti-generations of E_6 multiplets in each case.

* A simpler problem is the classification of (4,4) superconformal models with $c = 6$, corresponding to K3 surfaces or their orbifold limits.[42]

There is some redundancy, however, since the E_6 and E_8 minimal models are reducible. Many more models could also be formed by using the new $N = 2$ models and by modding out by discrete symmetries. Most cases correspond to Calabi–Yau spaces, but some are orbifolds.[45] (There may even be examples with both interpretations!)

Recently there has been progress in evaluating the Yukawa couplings in these models.[46],[47] The $(\mathbf{27})^3$ coupling turns out to be given exactly by the lowest-order (large radius) field theory approximation,[32] whereas the $(\overline{\mathbf{27}})^3$ is not. The knowledge of these couplings should make it possible to evaluate mass ratios and other quantities of physical interest. Then we can examine how close models of this type can come to agreeing with experiment.

Singularity Theory Classification of N=2 Superconformal Models

New insights into the classification of (2,2) superconformal models have been obtained recently[48,49] using mathematical methods of singularity theory[50] (also known as catastrophe theory). In particular, this work explains how to construct the Calabi–Yau spaces corresponding to all Gepner-type compactifications.[51] This subject is new and developing fast. We will settle here for a brief description of some of the basic concepts.

Many interesting $N = 2$ $d = 2$ theories can be described by an action of the form

$$S = \int d^2x d^4\theta K(\Phi_i, \bar{\Phi}_i) + \left\{ \int d^2x d^2\theta W(\Phi_i) + c.c. \right\} \quad . \tag{3.17}$$

In this expression the fields Φ_i are chiral $N = 2$ superfields, meaning that they are annihilated by certain supercovariant derivatives. In terms of physical states $|\Phi_i\rangle = \Phi_i\,|0\rangle$, this means that the NS sector descendants $G^a_{-1/2}\,|\Phi_i\rangle$ are null. The superpotential W is a holomorphic function of the Φ_i. Because of the presence of the "F term" (the one containing W), these systems can be regarded as $N = 2$ Landau–Ginsburg models.

The main idea is to study the renormalization group flow under scale transformations of the $2D$ metric $g \to \lambda^2 g$ as $\lambda \to \infty$. The kinetic term (known as the "D

term") contains only irrelevant operators, and thus W determines the fixed point of the RG flow. Specifically, at a fixed point the chiral fields scale according to

$$\begin{aligned} \Phi_i &\to \lambda^{\omega_i}\Phi_i \\ W(\lambda^{\omega_i}\Phi_i) &= \lambda W(\Phi_i) \quad . \end{aligned} \tag{3.18}$$

In this case, one says that W is quasihomogeneous with weights ω_i. The fixed point describes a conformally invariant model in which the conformal dimensions of Φ_i are $(h_i, \bar{h}_i) = (\frac{1}{2}\omega_i, \frac{1}{2}\omega_i)$. The main result is that a quasihomogeneous function W uniquely characterizes an $N = 2$ superconformal model up to field redefinitions (with a finite nonzero Jacobian) and the addition of trivial quadratic terms in new fields. The analysis requires that Φ^n have a dimension n times that of Φ, which is valid as a consequence of the $N = 2$ algebra. By studying the scaling of the partition function, and comparing with the Weyl anomaly formula of Polyakov,[52)] one can show that the central charge is $c = 6\beta$, where

$$\beta = \sum_i (\frac{1}{2} - \omega_i) \tag{3.19}$$

is called the "singularity index" of W.

An important notion in singularity theory is "modality." Roughly speaking, this is the number of parameters that characterize the model. It is not quite the same thing as the number of physical moduli. Remarkably, the classification of modality $m = 0$ singularities precisely corresponds to the $N = 2$ minimal models! The same ADE classification discussed earlier was known previously in singularity theory, since there is a prescription for associating Dynkin-like diagrams to singularity types. The $m = 0$ classification is

$$\begin{array}{lcc} A_n: & x^{n+1} & k = n-1 \\ D_n: & x^{n-1} + xy^2 & k = 2n-4 \\ E_6: & x^3 + y^4 & k = 10 \\ E_7: & x^3 + xy^3 & k = 16 \\ E_8: & x^3 + y^5 & k = 28 \ . \end{array} \tag{3.20}$$

These give the usual central charges, namely $c = 3k/(k+2)$. Singularities can be classified as elliptic, parabolic, or hyperbolic. The ones given here are all the elliptic ones, which thus correspond precisely to minimal $N = 2$ models. Furthermore, $\beta = 1/2\,(c = 3)$ for parabolic singularities, and $\beta > 1/2\,(c > 3)$ for hyperbolic ones. An obvious problem is to find the superpotentials that correspond to the various Kazama–Suzuki models.

The power of this approach is illustrated by the fact that there are three reducible minimal models, which are readily identified. Namely, since $A_m \otimes A_n$ corresponds to $x^{m+1} + y^{n+1}$,

$$\begin{aligned} D_4 &\approx A_2 \otimes A_2 \\ E_6 &\approx A_2 \otimes A_3 \\ E_8 &\approx A_2 \otimes A_4 \ , \end{aligned} \tag{3.21}$$

The first of the correspondences is proved by noting that a linear change of variables allows $x^3 + xy^2$ to be expressed as the sum of two cubes. The discrete symmetries of the minimal models are also readily understood. For example, the Z_{k+2} symmetry of the A_{k+1} model is generated by $x \to \exp[2\pi i/(k+2)]x$. It is also evident that the E_7 model has $Z_3 \times Z_3 = Z_9$ symmetry.

Each of Gepner's models is characterized by a superpotential given by the appropriate sum of polynomials. For example, the 3^5 model corresponds to five A_4 models: $W = \sum_{i=i}^{5} \Phi_i^5$. We have already seen that the corresponding Calabi–Yau space is given by the hypersurface $W = 0$ in CP^4. This fact is derived in [51] by making an appropriate change of variables in the path integral

$$\int d\Phi_1 \ldots d\Phi_5 \; e^{i\int d^2x d^2\theta W(\Phi_i)} \quad . \tag{3.22}$$

Only some of Gepner's models correspond to complete-intersection Calabi–Yau spaces (CICY) given by polynomial constraints in products of complex projective spaces and classified in [37]. The appropriate generalization that accommodates all of Gepner's models, suggested by the structure of quasihomogeneous functions, involves "weighted projective spaces." The space $WCP^N_{k_1 \ldots k_{N+1}}$ is defined by the identification

$$[z_1, \ldots, z_{N+1}] \sim [\lambda^{k_1} z_1, \ldots, \lambda^{k_{N+1}} z_{N+1}] \quad . \tag{3.23}$$

One subtlety is that these have nontrivial fixed-point sets.

To define a space with three complex dimensions, we can consider $N-3$ homogeneous polynomial constraints of degrees $\ell_1, \ldots \ell_{N-3}$ in WCP^N. Remarkably, the condition for the vanishing of the first Chern class is simply that $c = 9$.

4. Concluding Remarks

I have discussed a variety of recent developments in conformal field theory and string theory. The emphasis has been on techniques that can be used to obtain classical solutions of the heterotic string theory with four-dimensional Poincaré symmetry and $N = 1$ space-time supersymmetry. I now feel compelled to address the question: what is all this good for? If we accept that we have the right theory (which is not completely obvious), the key question becomes whether there is a classical solution that is a sufficiently good approximation to the exact quantum ground state to make a convincing case that it is 'correct.' If so, by pursuing some of the techniques sketched here, we should be able to find it. If not, it will take a long time before string theorists can do meaningful detailed phenomenology.

It was argued several years ago that string theory is a 'strongly coupled' theory,[53] meaning that (nonperturbative) quantum effects would be important. This could cause one to worry that the study of classical ground states is a waste of time, but I think that would be an overreaction. The three-generation models clearly come quite close to giving the desired phenomenology; it would be foolish not to explore how far they can be pushed. Many realistic features emerge quite naturally: gravity, popular gauge groups and representations, chiral families of fermions, axions, symmetry breaking mechanisms, supersymmetry, etc. (I do not understand how those who say 'There is not a shred of experimental evidence for string theory' can ignore all these successes.) It would be a very strange coincidence if classical solutions were completely off the mark.

Certainly, nonperturbative phenomena will be crucial for a complete understanding. It is quite clear that the $N = 1$, $D = 4$ supersymmetry of the Calabi–Yau or orbifold solutions is not broken at any order in perturbation theory. Also, the dilaton and other massless states do not acquire a mass. Even breaking of the electroweak symmetry group is not expected. I think it is reasonable to expect these things to happen in the complete nonperturbative quantum theory, however. An encouraging result is the recent demonstration that the string perturbation expansion diverges.[54)] This suggests that not every classical solution need correspond to a quantum ground state, but that whenever one does the needed symmetry breaking and mass generation could occur. Also, there probably are instantons that give quantum tunneling between different classical vacua. These types of effects might lift the enormous degeneracy with which we are currently faced.

One question that has received much attention in recent years is why the cosmological constant is so small (less than 10^{-120} in Planck units). I don't know the answer, of course, but let me offer the following comments. Perhaps, as we have discussed, the correct string theory ground state is reasonably approximated by a classical solution with $N = 1$ supersymmetry in four dimensions. Since supersymmetry is unbroken at every order in the loop expansion, the cosmological constant undoubtedly vanishes at every order. The mystery that then needs to be understood is why the nonperturbative effects that break supersymmetry do not generate a cosmological constant at the same time. It seems to me that the problem must be addressed in the context of the complete theory and is very unlikely to be resolved by considerations that are not sensitive to Planck-scale physics. I have assumed that low-energy supersymmetry is required to solve the 'hierarchy problem.' This was the principle motivation for looking for supersymmetric solutions, although they do seem to fit in rather naturally. Still, it would be very reassuring and helpful to have supporting experimental evidence. It is not entirely clear to me that a 'string miracle' other than space-time supersymmetry might do the job. After all, we need one to eliminate the cosmological constant.

In conclusion, it would probably be foolhardy to predict dramatic phenomenological successes for string theory in the near term. Still, there are some encouraging possibilities that deserve to be pursued. We might get lucky!

I am grateful to L. Dixon, D. Gepner, W. Lerche, E. Raiten, and G. Rivlis for helpful discussions. This work was supported in part by the U.S. Department of Energy under Contract No. DE-AC0381-ER40050.

REFERENCES

1. M. B. Green, J. H. Schwarz, and E. Witten, "Superstring Theory," in 2 vols. (Cambridge Univ. Press, 1987).
2. M. Peskin, Proc. of the 1986 TASI Lectures, ed. H. Haber, (World Scientific, 1987); T. Banks, Proc. of the 1987 TASI Lectures, (World Scientific, 1988); P. Ginsparg, 1988 Les Houches Lectures, HUTP-88/A054.
3. A. A. Belavin, A. M. Polyakov, and A. B. Zamolodchikov, Nucl. Phys. **B241** (1984) 333.
4. For a review see: P. Goddard and D. Olive, Int. J. Mod. Phys. **A1** (1986) 303.
5. H. Sugawara, Phys. Rev. **170** (1968) 1659; C. Sommerfield, Phys. Rev. **176** (1968) 2019.
6. E. Witten, Commun. Math. Phys. **92** (1984) 455; V. Knizhnik and A. B. Zamolodchikov, Nucl. Phys. **B247** (1984) 83; D. Gepner and E. Witten, Nucl. Phys. **B278** (1986) 493.
7. J. L. Cardy, Nucl. Phys. **B270 [FS 16]** (1986) 186.
8. V. G. Kac and D. Peterson, Adv. Math. **53** (1984) 125; V. G. Kac, 'Infinite Dimensional Lie Algebras,' (Cambridge Univ. Press, 1985).

9. A. Capelli, C. Itzykson, and J.-B. Zuber, Nucl. Phys. **B280** [**FS 18**] (1987) 445; D. Gepner and Z. Qiu, Nucl. Phys. **B285** [**FS 19**] (1987) 423.
10. D. Gepner, Nucl. Phys. **B290** [**FS 20**] (1987) 10.
11. M. B. Green and J. H. Schwarz, Nucl. Phys. **B181** (1981) 502.
12. M. B. Green and J. H. Schwarz, Phys. Lett. **136B** (1984) 367; Nucl. Phys. **B243** (1984) 285.
13. D. Friedan, E. Martinec, and S. Shenker, Nucl. Phys. **B271** (1986) 93.
14. P. Goddard, A. Kent, and D. Olive, Phys. Lett. **152B** (1985) 88; Commun. Math. Phys. **103** (1986) 105; P. Goddard, W. Nahm, and D. Olive, Phys. Lett. **160B** (1985) 111.
15. J. Bagger, D. Nemeschansky, and S. Yankielowicz, Phys. Rev. Lett. **60** (1988) 389; D. Kastor, E. Martinec, and Z. Qiu, Phys. Lett. **200B** (1988) 434.
16. M. Ademollo *et al.*, Nucl. Phys. **B111** (1976) 77.
17. M. Kato and S. Matsuda, Phys. Lett. **184B** (1987) 184; G. Waterson, Phys. Lett. **171B** (1986) 77; S. Nam, Phys. Lett. **172B** (1986) 323.
18. A. Schwimmer and N. Seiberg, Phys. Lett. **184B** (1987) 191.
19. W. Boucher, D. Friedan, and A. Kent, Phys. Lett. **172B** (1986) 316.
20. A. Sen, Nucl. Phys. **B278** (1986) 289; Nucl. Phys. **B284** (1987) 423.
21. T. Banks, L. Dixon, D. Friedan, and E. Martinec, Nucl. Phys. **B299** (1988) 613.
22. D. Gepner, Nucl. Phys. **B296** (1988) 757.
23. P. Di Vecchia, J. L. Petersen, and M. Yu, Phys. Lett. **172B** (1986) 211; P. Di Vecchia, J. L. Petersen, M. Yu, and H. B. Zheng, Phys. Lett. **174B** (1986) 280.
24. A. B Zamolodchikov and V. A. Fateev, Zh. Eksp. Theor. Fiz. **90** (1986) 1553; Z. Qiu, Phys. Lett. **188B** (1987) 207.

25. A. B Zamolodchikov and V. A. Fateev, Sov. Phys. JETP **62** (1985) 215; **63** (1985) 913; D. Gepner and Z. Qiu, Nucl. Phys. **B285 [FS19]** (1987) 423.

26. K. Li and N. Warner, Phys. Lett. **211B** (1988) 101.

27. E. Kiritsis, Caltech preprint CALT-68-1508, 1988.

28. V. G. Kac and I. T. Todorov, Commun. Math. Phys. **102** (1985) 337.

29. Y. Kazama and H. Suzuki, Univ. of Tokyo preprint UT-Komaba 88-8(revised), Sept. 1988.

30. Y. Kazama and H. Suzuki, Phys. Lett. **216B** (1989) 112.

31. S. Helgason, "Differential Geometry, Lie Groups, and Symmetric Spaces," Academic Press (1978).

32. P. Candelas, G. Horowitz, A. Strominger, and E. Witten, Nucl. Phys. **B258** (1985) 46.

33. E. Witten, Nucl. Phys. **B258** (1985) 75; Nucl. Phys. **B268** (1986) 79; A. Strominger and E. Witten, Commun. Math. Phys. **101** (1985) 341.

34. I. Antoniadis, C. Bachas, J. Ellis, and D. V. Nanopoulos, Phys. Lett. **211B** (1988) 393.

35. F. Calabi, p. 78 in "Algebraic Geometry and Topology: A Symposium in Honor of S. Lefschetz" (Princeton Univ. Press, 1955); S. T. Yau, Proc. Natl. Acad. Sci **74** (1977) 1798.

36. D. Gross and E. Witten, Nucl. Phys. **B277** (1986) 1; M. T. Grisaru, A. Van de Ven, and D. Zanon, Phys. Lett. **173B** (1986) 423.

37. P. Candelas, A. M. Dale, C. A. Lütken, and R. Schimmrigk, Nucl. Phys. **B298** (1988) 493; P. Candelas, C. A. Lütken, and R. Schimmrigk, Nucl. Phys. **B306** (1988) 113.

38. S. T. Yau, p. 395 in "Symp. on Anomalies, Geometry, Topology," eds. W. A. Bardeen and A. R. White (World Scientific, 1985).

39. R. Schimmrigk, Phys. Lett. **193B** (1987) 175.
40. S. Kalara, P. K. Mohapatra, and R. N. Mohapatra, Phys. Rev. **D37** (1988) 3284.
41. D. Gepner, Phys. Lett. **199B** (1987) 380; Princeton Univ. preprint, Dec. 1987.
42. T. Eguchi and A. Taormina, Phys. Lett. **196B** (1987) 75; **200B** (1988) 315; **210B** (1988) 125.
43. M. Lynker and R. Schimmrigk, Phys. Lett. **208B** (1988) 216.
44. C. A. Lütken and G. G. Ross, Phys. Lett. **213B** (1988) 152.
45. T. Eguchi, H. Ooguri, A. Taormina, and S. Yang, Nucl. Phys. **B315** (1989) 193.
46. J. Distler and B. Greene, Nucl. Phys. **B309** (1988) 295; D. Gepner, Princeton Univ. preprint, April 1988.
47. A. Kato and Y. Kitazawa, Univ. of Tokyo preprint UT-535, Aug. 1988; G. Sotkov and M. Stanishkov, Phys. Lett. **215B** (1988) 674; M. Douglas and S. Trivedi, Caltech preprint CALT-68-1526.
48. C. Vafa and N. Warner, Phys. Lett. **218B** (1989) 51; W. Lerche, C. Vafa, and N. Warner, preprint HUTP-88/A065, Feb. 1989.
49. E. Martinec, Phys. Lett. **217B** (1989) 431.
50. V. I. Arnold, 'Singularity Theory,' London Math. Lect. Notes Series **53**, (Cambridge Univ. Press, 1981).
51. B. Greene, C. Vafa, and N. Warner, preprint HUTP-88/A047a, Nov. 1988.
52. A. M. Polyakov, Phys. Lett. **103B** (1981) 207, 211.
53. M. Dine and N. Seiberg, Phys. Rev. Lett. **55** (1985) 366; Phys. Lett. **162B** (1985) 299; V. Kaplunovsky, Phys. Rev. Lett. **55** (1985) 1036.
54. D. Gross and V. Periwal, Phys. Rev. Lett. **60** (1988) 2105.

PHYSICS WITH A FUNDAMENTAL LENGTH

G. Veneziano

CERN , Geneva

1. INTRODUCTION

Since the time physics was part of philosophy, physicists (philosophers) have been wondering about the infinite divisibility of space and time [1]. Is there a limit to the process of looking into smaller and smaller scales? Quantum field theory (QFT), as developed in the second part of this century, appears to answer this question in the negative. However, so far, QFT has failed to yield a consistent synthesis of the principles of general relativity and quantum mechanics: it is affected by infinities which, normally, but with important exceptions, can be swept under the rug in the so-called process of renormalization.

String theories supposedly avoid the above mentioned difficulties. How? A hint is that string theory, unlike point particle theory, contains a fundamental scale or length parameter. Although this fact had been stressed since the early days of dual string theory [2], the precise origin and meaning of this fundamental scale have long been clouded with mystery.

With the QCD reinterpretation of the old hadronic string the mystery is neatly resolved: a fundamental length, which is classically absent, emerges from quantization, through the so-called phenomenon of dimensional transmutation [3]. QCD loops diverge and the process of renormalization introduces a scale, the subtraction point. Renormalization group arguments guarantee that all physical quantities depend only on a particular combination of the subtraction point and of the coupling constant, which defines the fundamental scale Λ of QCD (and thus of its effective string theory).

The Scherk-Schwarz revolution of 1974 [4] takes again (super)string theory as a fundamental theory, rather than as the effective description of some non-perturbative dynamics of a local field theory. Thus, in the Theory of Everything (TOE) scenario for strings, the question of what the fundamental length stands for reappears in its full glory.

In this article I shall try to convey our present understanding of:

i) the origin of a fundamental scale in string theory

ii) the ways this scale may change our attitude towards a variety of problems in quantum field theories without affecting significantly their low energy predictions (when they exist!).

iii) the way the new scale entails large departures from field theory at distances of its order or less.

I shall add immediately that many of our considerations will not depend on whether or not String Theory is what provides a fundamental length. Similar results will hold for any quantum-relativistic theory endowed with a (nicely behaved) physical short-distance cut-off.

2. ORIGIN OF THE STRING SCALE PARAMETER

By far the most powerful, conceptually and technically simple, formulation of string theory available today is the one based on the first quantized approach of Polyakov [5]. We shall thus explain how the fundamental length emerges naturally in that approach. The material of the next two sections is largely borrowed from some recent lectures of mine [6].

The starting point of [5] is the classical string action, generalized to the case of a string is moving in a set of arbitrary background fields (for simplicity we describe the closed bosonic string here):

$$S = -1/2 \int d^2 \xi \; g^{1/2} \{ g^{\alpha\beta} \partial_\alpha X^\mu \partial_\beta X^\nu G_{\mu\nu}(X) + $$

$$+ {}^{(2)}R(\xi) \; \phi(X)/4\pi \} + \cdots\cdots \qquad (1)$$

Here, as usual, X^μ is the string coordinate, $g_{\alpha\beta}$ is the metric on the world sheet, $G_{\mu\nu}$ and ϕ are the target-space metric and dilaton background, and the dots indicate other possible backgrounds.

As everyone knows, the absolute normalization of the action is irrelevant at the classical level. There exists one conventional normalization but, since (1) is supposed to describe a TOE, we do not have to normalize our action with respect to anything else and we shall not.

At the quantum level, the action has to be transformed into a pure number whose absolute magnitude and phase are very relevant. Usually this is done by dividing the conventionally normalized action by h, but here, again because of the TOE nature of string theory, we do not want to introduce any constant until we are really forced to do so. Since the background fields are arbitrary, we may just define their (a priori arbitrary) normalization so that the quantum theory is defined a la Feynman by the path integral:

$$Z\,(\,G_{\mu\nu}\,,\,\phi,\,....\,) = \int Dg^{\alpha\beta}\;DX^{\mu}\quad \exp(\,-S_E(G_{\mu\nu}(X),\,\phi(X),\,..)\,) \qquad (2)$$

We have introduced no fundamental constant (and, in particular, no length) yet. The only new feature of (2), as opposed to (1) is that the background fields must be assigned dimensions such that S_E (E denotes that we went over to Euclidean metric) is dimensionless. If we call the dimensions of X length, G has dimension $(\text{length})^{-2}$, ϕ has no dimension, etc. Obviously, Z will depend on the background fields but not on any arbitrary fundamental constant (except for gauge-fixing parameters which will not affect physical predictions).

A crucial, yet poorly understood, feature of string theory comes in at this point. On one hand, quantum consistency imposes constraints on the possible backgrounds in which the string can consistently move: these constraints, related to conformal and Weyl invariance of the quantum theory, enforce the vanishing of certain β-functions [7]. On the other hand, one can construct, out of Z, an effective action (through a sort of Legendre transform [8]) and find that the same conditions also follow from requesting, as is customary in field theory, that the action be stationary:

$$\delta\Gamma/\delta G_{\mu\nu} = \delta\Gamma/\delta\phi = ... = 0 \quad (3)$$

Although these equations look very much like those of a classical field theory, I should stress that they came from quantizing a string theory: there is nothing like a classical field theory representing the low energy limit of a classical string theory [9].

Γ is, of course, very complicated, but one knows some of its very general properties at least in the limit in which only the massless backgrounds $G_{\mu\nu}$, ϕ (and the "torsion" $B_{\mu\nu}$) are kept and are slowly varying.

In general, Γ has a double infinite expansion:

$$\Gamma = \Sigma_g \Gamma_g \quad (4)$$

where g=0,1,2,... is the genus (number of handles) of the two-dimensinal Reimann surface described by the metric $g_{\alpha\beta}$. Because of the well-known relation,

$$4\pi\ (2\text{-}2g) = \int d^2\xi\ g^{1/2}\ {}^{(2)}R(\xi) \quad (5)$$

and of the way the dilaton field ϕ enters the action one finds,

$$\Gamma_g = \int dx\ G^{1/2} \exp(\ 2\phi(1\text{-}g))\ \ L\ (\partial\phi....) \quad (6)$$

This first expansion is the so-called string loop expansion: it is the equivalent, in string theory, of the usual field-theoretic loop expansion. The second expansion, which can be made genus by genus, is an expansion in derivatives of the background fields. It is also called the σ-model loop expansion and has no analog in field theory.

In particular, at the tree level (g=0), it reads:

$$\Gamma_0 = \int dx\ G^{1/2} \exp(2\phi)\ [(D\text{-}26)/12\ \text{-}1/4\ R\ +1/2\ G^{\mu\nu}\partial_\mu\phi\,\partial_\nu\phi\ +\cdot\cdot] \quad (7)$$

The possible ground states, or vacua, of string theory are the possible stationary points of Γ. One such ground state, at tree level, is flat D=26 space (D=10 for superstrings at any finite order) with a constant dilaton field. Since $G_{\mu\nu}$ has dimensions (length)$^{-2}$, a constant $G_{\mu\nu}$ introduces a fundamental length:

$$G_{\mu\nu} = \lambda_s^{-2}\,\eta_{\mu\nu} \quad , \quad \eta_{\mu\nu} = \text{diag}\,(-1, +1, +1, \ldots\ldots +1) \qquad (8)$$

where the component X^0 is to be identified eventually with ct.

We have thus discovered that only at the quantum level do we need a fundamental scale λ_s and this only because we have chosen a flat space time metric as our ground state! Incidentally, the observation that the appearance of a scale in Quantum Gravity is related to a vacuum expectation value of the metric had been made already, outside of the string context, by Fubini and coworkers [10].

One of the many, important roles of λ_s can be immediately appreciated by inserting (8) back into (2). We find that λ_s^2 is nothing but the Planck constant of string theory, the quantity one needs to convert the Nambu-Goto area action (which is formulated directly in flat space-time) into a pure number, a phase. If (8) is inserted into the expression for Γ, the derivative expansion becomes an expansion in $\lambda_s d/dx$, presumably a reliable expansion if the background fields change little (percentage-wise) over distances $O(\lambda_s)$.

We also notice a very important dimensionless number which, according to eq. (7), appears to control the relative importance of higher genus contributions:

$$\Gamma_g \simeq (\alpha_{SL})^g \quad ; \quad \alpha_{SL} = \exp(-2\phi) \qquad (9)$$

It is the string loop expansion parameter, related to the dilaton's expectation value [11], on which we shall come back later.

Incidentally, in the second quantized approach to string theory, one would regard Γ_0 as a classical action to be quantized by standard field theoretic methods. Doing that looks somewhat artificial in string theory: trees and loops come naturally together because of the necessity (related to unitarity) to sum over Reimann surfaces of arbitrary genus (this is one of the issues on which Dima has so much contributed, see other authors in this book).

Furthermore, the quantum length λ_s appears already in the would-be classical action, underlining the fact that Γ_0 itself is the result of string quantization.

For all these reasons the only sensible attitude in string theory seems to be to stick to the first quantized approach, at least as far as perturbation theory is concerned: in principle, this approach leads to a quantum relativistic theory of all interactions in D dimensions, which can be solved by quantizing a two-dimensional field theory: the practical and conceptual consequences of this statement can hardy be overestimated.

Unfortunately the program of computing higher genus contributions to Γ is far from completed. This is hardy surprising: one has to do the analog of developing the Feynman rules for QFT, something that took some 20 years of hard work.

3. THE PLANCK CONSTANT AS ULTRAVIOLET CUT-OFF

We want to come now to the "raison d'etre" of modern (super) string theory: the fact that it provides a consistent theory of quantum gravity. As we shall argue later on, this is due to the extremely nice ultraviolet behaviour of string loops which turn out to be cut-off, effectively, at momenta $O(\hbar/\lambda_s)$. In other words, the fundamental length, which we have already identified with Planck's constant, also plays the role of a short distance, ultraviolet cut-off.

This miracle is hardly unexpected: a string is a collection of harmonic oscillators and we know that, in the harmonic oscillator, Δx and Δp are both of order $(\hbar)^{1/2}$. Since, in string theory, the role of $\hbar$ is played by λ_s^2, Δx has to come out $O(\lambda_s)$. The Planck constant of the theory thus determines the spacial extent of the string and thus generates a form factor (as for any extended object) which damps large momenta.

If we wish to count at this point the number of fundamental, dimensionful constants in string theory, we find that there are only two: c, the speed of light and λ_s , the Planck constant. At first sight this looks like no gain with respect to having c, $\hbar$ and the string tension $1/\alpha'$, since, in order to come down from three to two, we had to choose a different system of units of mass ($\alpha' m$ replaces m and makes it a length). The gain, however, with respect to QFT, is that we have inherited, at no extra cost, a finite ultraviolet cut-off [12]!

As long as we are in a non-compact, Minkowski space, the actual value of λ_s is, of course, irrelevant. What the theory should determine are the ratios of λ_s (c) with some other physical length (speed) that we *call* cm. (or cm./sec.). In other words, the theory allows, in principle, to determine λ_s (c) in cm. (cm./sec.).

The situation changes drastically if some of the space-time dimensions are compact (as it has to be the case physically), say circles. In this case the ratio ρ of the compactification radius and λ_s is relevant i.e. it does matter how many Planck constants is the internal space large. Very amusing symmetries have been discovered [13] (for closed strings) relating large ρ to small ρ. They could lead [12] to a stabilization of the radius of the internal manifold to be (a precise multiple of) λ_s.

In the rest of this paper I shall use some string results and intuition, combined with the known field theoretic limits of string theory, to argue how the fundamental length λ_s can make itself felt, changing much of our present, field-theory-based (pre)conceptions on questions like infinities, regularizations, subtractions, quadratic and logarithmic divergences, running and renormalization group.

At the end, the usual low energy field theory result will emerge, even if somewhat reinterpreted and, in my opinion, finally free from possible criticism on the way of handling infinities. At the same time (as it is the case in the standard model vis a vis Fermi's theory) the new length parameter (the W-Z mass in the SM) is bound to affect physical predictions at scales less or equal to λ_s. These more speculative aspects of string theory will be discussed at the end.

4. PHYSICS MUCH ABOVE THE FUNDAMENTAL LENGTH

In this section we shall discuss the low energy, large distance limit of string theory arguing that the known results of Quantum Field Theory are recovered in a (perhaps) novel way. Quantum field theory thus emerges as an effective theory approximately valid at distances much larger than the string scale λ_s. The situation is thus reversed as compared to looking at string theory as being an effective theory for QCD.

We have already said that the tree level effective action of string theory closely resembles a classical field theory action, e.g. Einstein's theory of gravity coupled to a gauge theory. In order to show that there is a correspondence with QFT we have to argue that string loops reproduce, in some sense, at large distances, those of ordinary QFT.

I said "in some sense" since we know that, even at large distance, this cannot be, strictly speaking, the case. Indeed quantum gravity has no field theory limit in four dimensions and gauge theories have no field theory limit in more than four dimensions. All we can hope to recover is a regulated version of QFT loops which goes over to renormalized perturbation theory in the low energy limit for renormalizable interactions and to something new for non-renormalizable ones.

Let us consider, to be more specific, a string theory which has a perturbatively stable vacuum consisting of a number d of flat, infinite dimensions and a number d_c of compact dimensions. For the sake of simplicity, let us assume that the radii of the compact dimensions are all comparable and given by R_c. We shall also take $R_c > \lambda_s$.

There are two important features of string loops that have to be used in order to establish the connection with field theory.

i) In string loops virtual momenta are effectively cut-off at the string scale $\hbar/\lambda_S$. Thus loops are UV finite.

ii) The contribution of light particles circulating in the loop must reproduce, for internal momenta much smaller than $\hbar/R_C$, the field theoretic expression. Such expressions contain the effective coupling constant in d-dimensions given by:

$$\alpha_d = \alpha_{SL}/ V_C \tag{10}$$

where $V_C \simeq (R_C)^{d_C}$ is the volume of the compactified space.

As discussed in more detail in ref.[6], these two features imply the following consquences:

a) A typical string loop is "down" by a power of α_{SL} relative to a tree. Such a contribution comes from large (i.e.$O(\hbar/\lambda_S)$) internal momenta.

b) If the external momenta are of order $\hbar/\lambda_S$, the above is the actual order of the loop correction. Assuming α_{SL} to be small, we conclude that the bare (i.e. tree level) parameters have physical meaning: they are to be identified with the short distance effective couplings, masses, etc.

c) If the external momenta are much larger than $\hbar/\lambda_S$, $\hbar/R_C$, the problem becomes a string-theoretic one and cannot be tackled by field theory techniques (for some examples, see sect. 5).

d) If the external momenta are much smaller than $\hbar/\lambda_S$, $\hbar/R_C$, which is the case we are interested in for the moment, the loop correction can be enhanced relative to its typical value because of infrared singularities. It is precisely through these IR singularities that one recovers well-known QFT effects. Let us list some of them:

i) As a result of logarithmic enhancements one recovers:

- the running of gauge couplings in d=4 [6];

- the running of gravitational couplings in d=4 in the presence of a cosmological constant [6]. The latter phenomenon could be relevant in connection with the Cosmological Constant problem [14].

ii) As a result of power-like infrared singularities one recovers large effects forbidden at tree level e.g.

- light-light scattering, which, in string theory can occur at tree level, but is depressed by four powers of λ_s d/dx. At one loop, however, the graph contributing to $F_{\mu\nu}{}^4$ in the effective action is badly infrared singular with the electron mass playing the role of IR (mass) regulator. One thus recovers the field theoretic result $O(\alpha^2(p/m_e)^4)$.

- the famous Adler-Bell-Jackiw triangle anomaly in d=4 (and similar ones for other values of d). It is well known [15], indeed, that the ABJ anomaly can be seen either as a consequence of UV divergences or as due to a singular IR behaviour in the massless limit.

iii) Non-renormalizable interactions are no problem whatsoever. To the contrary, their bad UV behaviour has a counterpart in their better IR limit. Consequently, there are no large quantum gravity corrections in d=4, if the cosmological constant is zero, or gauge coupling renormalization if d>4. In particular, the running of gauge couplings stops at the (de)compactification scale $\hbar/R_c$, which thus emerges as the true Grand Unification scale. This unique property of d=4 may have something to do [16] with why compactification to d=4 should be preferred over compactification, say, to d=6 or to no compactification at all.

iv) Non-asymptotically free theories (such as QED or ϕ^4) have all the right to exist in string theory, while they are likely to be trivial in QFT [17]. As explained, the bare coupling constant basically coincides with the coupling at the string or compactification scale. Hence, by definition, if there is a Landau pole in the field theory limit, such a pole is above the string scale, i.e. in the region where string or higher dimensional corrections cannot be neglected. Vice versa, the low energy coupling is bounded from above by the fact that it is $< O(1)$ at the string scale and that it decreases towards larger distances. This will allow to recover bounds [18] on elementary Higgs masses and couplings.

In conclusion, all the good results of QFT are recovered by string theory. Furthermore, string theory makes sense of some "bad" field theories (including one we know for sure to belong to Nature, gravity) and of the Kaluza-Klein idea because of its beautiful UV cut-off, hence, in the end, because of its fundamental length parameter.

We shall now see how this parameter is likely to affect physics around and below the Planck scale in a much more drastic way.

5. PHYSICS BELOW THE FUNDAMENTAL LENGTH

In the last few years a few attempts have been made to understand string theory at short distances i.e. in the region where one expects it to differ drastically from field theory.

It is still not clear how to define off-shell correlation functions in string theory. For the time being one is obliged to proceed differently from QFT in order to explore short distances. So far two methods have been tried: i) High energies and ii) High temperatures.

i) High Energy

The situation in this approach has been recently reviewed by the author [19]. I shall recall here the main conclusions. Hereafter d will stand for the number of non-compact dimensions.

a) The high energy collision of gravitationally interacting particles (e.g. two gravitons) is characterized by a classical interaction range, the so-called Schwarzschild radius;

$$R_S = (G E)^{1/(d-3)} \tag{11}$$

where G is Newton's constant. String theory enters in two ways: i) It introduces a new scale, the already much-mentioned string length parameter λ_S. ii) It relates G and λ_S through the (closed) string unification formula [20]:

$$\lambda_P^{(d-2)} = \alpha_d \ \lambda_S^{(d-2)} \quad , \quad \lambda_P^{(d-2)} = G\hbar \tag{12}$$

It is easy to see that, even at energies much larger than $\hbar/\lambda_S$, R_S can be smaller than λ_S for a sufficiently small α_d, i.e. for a Planck length which is much smaller than the string length.

The two cases have to be distinguished carefully:

i) $R_S > \lambda_S$. This case includes the infinite energy limit at fixed Newton constant. One finds that expected Classical General Relativity effects can be recovered from flat space-time calculations. These effects dominate in this regime while string-scale phenomena remain screened inside R_S , which plays the role of a horizon. In order to arrive at this result a non-trivial resummation of leading terms to all orders in the string loop expansion is necessary. If one works at fixed genus, as E goes to infinity, one obtains the result that the S-matrix is trivial [21], in striking disagreement with CGR expectations. The point is, of course, that the genus that dominates at a given energy depends on the energy and goes to infinity with it.

To be sure, in this case calculations can only be made order by order in the parameter R_S /b, assumed to be small. As b approaches R_S , strong classical relativity effects start to emerge. Although technical complications prevent so far obtaining definite results, it seems very likely [22] that one goes over to a situation in which the collision is accompanied by the formation of a (Kerr) black hole [23]. If this is the case Hawking radiation with a thermal spectrum should be seen in the final state.

ii) $R_S < \lambda_S$

This case corresponds to large, but not extreme, energies or, if we prefer, to strings which are large in Planck-length units. We expect (and find!) string-size effects to be very relevant in this regime. In particular:

a) If we work at fixed c.m. angle, we find [24] that the Fourier transform from impact parameter to momentum transfer is always dominated by scales larger than λ_S.

b) If we work at fixed impact parameter we find that, unlike in the situation for $R_S > \lambda_S$, nothing dramatic happens as we reach values of $b < R_S$. The string length parameter appears to limit [12] the possible size of black holes to be larger than λ_S. Equivalently, because of Hawking's relation:

$$k_B T_{Haw.} = \hbar / R_S \tag{13}$$

one finds that the black hole's temperature cannot exceed the limiting string (Hagedorn) temperature:

$$k_B T_{Hag.} = \hbar / \lambda_S \tag{14}$$

A background and λ_S-independent characterization of an irrelevant metric deformation ΔG can be written as:

$$G^{-2} \partial^2 \Delta G >> 1 \tag{15}$$

ii) High temperature.

Recent investigations [25] of string thermodynamics at temperatures above $T_{Hag.}$ suggest that there is a drastic reduction in the number of degrees of freedom as compared to the field-theoretic case. This is evidentiated by the behaviour of the free energy F (say at genus one) in a given volume V as a function of the temperature:

$$F_{string}\,(k_B T V)^{-1} \quad \simeq \quad (\alpha' k_B T\,)/\lambda_s{}^d \tag{16}$$

as opposed to:

$$F_{field\ th.}\,(k_B T V)^{-1} \quad \simeq \quad (k_B T/\hbar)^{d-1} \tag{17}$$

It is amusing that the factor involved in going from the point to the string case is very similar to the one encountered [22] in the small impact parameter behaviour of the tree-level phase shift in the Planckian energy collision case, as if there was a correspondence between high T and small b.

This reduction of the relevant short distance degrees of freedom in string theory fits very well, of course, with the high energy collision results. It is also in line with the results of ref. [26] and with duality [13].

6. ENLARGED UNCERTAINTY AND EQUIVALENCE PRINCIPLES?

One can try to make abstraction from these result and ask whether or not distances shorter than λ_s can ever be probed in string theory. A few results give strong hints that they cannot and, at the same time, suggest (at least) two more general formulations of such impossiblity: a) An Enlarged Uncertainty Principle ; b) An Enlarged Equivalence Principle. Let us briefly discuss them in succession:

a) A simple way to summarize the findings of ref.[24] on the scale Δx probed in a string-string collision at energy E and scattering angle θ ,i.e. at momentum transfer $\Delta p = \theta E$, is given by the formula:

$$\Delta x \;=\; \hbar/\Delta p_s \;+\; \alpha' \Delta p_s \tag{18}$$

where:

$$\Delta p_s \;=\; \Delta p/\langle 1+g\rangle \tag{19}$$

and <g> is the average genus (loop order) contributing to the process in that kinematical regime. The first term in (18) looks like the usual uncertainty principle apart from the fact that the momentum transfer per loop Δp_S replaces the overall momentum transfer. The second term is a typical string effect. Obviously, the first term dominates over the second for $\Delta p_S < (\hbar/\alpha')^{1/2}$, while the opposite is true for $\Delta p_S > (\hbar/\alpha')^{1/2}$. In any event, the x+1/x structure of (18) yields a lower bound on Δx:

$$\Delta x > \lambda_S$$

This is the enlargement of the uncertainty principle caused by the string length parameter!

b) In a recent paper [27] one has tried to give an explicit argument in favour of the conjecture [9] that an enlarged equivalence principle lies behind string theory. The basic idea behind is simple: we know that string theory contains general relativity and, hence, that it obeys, at least at large scales, Einstein's equivalence principle. Mathematically this statement can be expressed as the invariance of the string partition function Z of eq. (2) under a change of the space-time metric $G_{\mu\nu}$ which corresponds to a General Coordinate Transformation (GCT).

The question is whether or not Z is invariant under a larger class of transformations which affect the metric non-trivally but only on very short scales, i.e. on scales much shorter than λ_S. In order to see if this is the case we have considered [27] the scattering of point-like and of string-like particles by massless, gravitational shock waves (which are precisely those relevant for High Energy superstring collisions) before and after adding ripples in the metric that live on scales much shorter than λ_S. We have found that, for point particles, the ripples influence the scattering process, while strings appear to be "blind" to such modifications. We thus have a confirmation that short distance degrees of freedom are irrelevant in string theory.

7. A SCALE-FREE (TOPOLOGICAL) PHASE?

I should finally mention an idea on a possible phase transition in string theory in which even the fundamental length λ_s disappears. We have seen that λ_s has come from having chosen, for phenomenological reasons, a string ground state consisting of Minkowski space time.

At very high temperatures and energy densities of the type possibly realized in the early universe, it is certain that Minkowski space-time was far from describing the average background in which strings were propagating. What could have been a more appropriate background?

An amusing possibility would be a background that is more symmetric than Minkowski and which does not force us to introduce a fundamental scale as a vacuum parameter (see again ref[10] for similar ideas outside of the string context). Two possibilities come to mind:

i) A non-trivial metric which is homogeneous of degree -2 in the coordinates:

$$G_{\mu\nu} \simeq X^{-2} \tag{21}$$

Since -2 is the physical dimension of $G_{\mu\nu}$ we do not need to introduce any dimensionful parameter in this case. The metric is naturally of the de Sitter type (if we also insist that it is homogeneous and isotropic) and is characterized by a pure number γ:

$$G_{\mu\nu} = \gamma\, X_0^{-2}\, \eta_{\mu\nu} \tag{22}$$

Study of string propagation in these metrics [28] shows that nowhere a fundamental length comes into play. However, what goes on depends rather crucially on the parameter γ. As is clear from the criterion (15), small γ corresponds to the dangerous case in which string effects may show up. Indeed one finds that, in this case, strings become highly unstable and blow up in size, a phenomenon reminiscent of Jean's instabilities and galaxy formation. Maybe this will lead to some sort of self-sustained inflation in string theory [29]. Another possibility is that, at γ <1, the instability we have found signals the transition to an even more dramatic background:

$$\text{ii)} \quad G_{\mu\nu} = 0 \tag{23}$$

In this case no metric-in-the-mean can be defined and the metric is entirely due to quantum fluctuations. This could correspond to a purely topological phase [30] for string theory. It also corresponds to a zero string-tension situation, which has been advocated to occur [31] above the limiting Hagedorn temperature. Maybe the early universe was in this topological phase and that the very concept of distance -and of a minimal scale- has only come out "later".

REFERENCES

1. Democritus, ca 400 B.C. ; Zenon, ca. 300 B.C.

2. see e.g. G. Veneziano, Phys. Reports, 9C (1974) 199.

3. S. Coleman and E. Weinberg, Phys. Rev. D7 (1973) 1888; D. J. Gross and F. Wilczek, Phys. Rev. D8 (1973) 3633.

4. J. Scherk and J. H. Schwarz, Nucl. Phys. B81 (1974) 118.

5. A. M. Polyakov, Phys. Lett. 103B (1981) 207, 211; see also M. Ademollo et al. , Nuovo Cim. 21A (1974) 77.

6. G. Veneziano, DST workshop on Superstring Theory (H.S. Mani and R. Ramachandran eds.) World Scientific Publ. Co. (1988) page 1.

7. see e.g. C. Lovelace, Phys. Lett. 135B (1984) 75; E.S. Fradkin and A.A. Tseytlin, Phys. Lett. 158B (1985) 316; ibid 160B (1985) 69; Nucl. Phys. B261 (1985)1; C. G. Callan D. Friedan, E.J. Martinec and M. J. Perry, Nucl. Phys. B262 (1985) 593;

8. T. Kubota and G. Veneziano, Phys. Lett. 207B (1988) 419 and references therein.

9. G. Veneziano, in "Strings and Gravitation", talk presented at the 5th Marcel Grossmann meeting (Perth, Aug.1988), Boston University preprint, BU-HEP-88-47 (1988).

10. V. de Alfaro, S. Fubini and G. Furlan, Nuovo Cim. A50 (1979) 523; ibid. B57 (1980) 227.

11. E. S. Fradkin and A. A. Tseytlin, ref. [7]; E. Witten, Phys. Lett. 149B (1984) 351.

12. G. Veneziano, Europhysics Lett. 2 (1986) 133.

13. K. Kikkawa and M. Yamasaki, Phys. Lett. 149B (1984) 357; N. Sakai

and I. Senda, Prog. Theor.. Phys. 75 (1986) 692; A. Giveon, E. Rabinovici and G. Veneziano, Nucl. Phys. B322 (1989) 167.

14. T.R. Taylor and G. Veneziano, Phys. Lett. 228B (1989) 311.

15. see e.g. T. Banks, Y. Frishman, A. Schwimmer and S. Yankielovicz, Nucl. Phys. B177 (1981) 157.

16. T.R. Taylor and G. Veneziano, Phys. Lett. 212B (1988) 147.

17. see e.g. L. Landau, "Niels Bohr and the Development of Physics", ed. W. Pauli (Pergamon Press London, 1955); K.G. Wilson, Phys. Rev. D6 (1972) 419; M. Aizenmann, Comm. Math. Phys. 86 (1982) 1; J. Frohlich, Nucl. Phys. B200 (1982) 281.

18. R. Dashen and H. Neuberger, Phys. Rev. Lett. 50 (1983) 1987. For a recent review, see e.g. P. Hasenfratz, Lattice '88 Conference, (Batavia, Ill. Sept. 1988), Nucl. Phys. (Proc. Suppl.) 9 (1989) 3.

19. G. Veneziano, "An Enlarged Uncertainty Principle from Gedanken String Collisions", talk presented at the Superstring '89 workshop, Texas A&M University (March 1989), CERN-TH. 5366/89, and references therein.

20. see e.g. D. J. Gross, J. A. Harvey, E. Martinec and R. Rohm, Nucl. Phys. B256 (1985) 253.

21. D. J. Gross and P. F. Mende, Phys. Lett. 197B (1987) 129; Nucl. Phys. B303 (1988) 407.

22. D. Amati, M. Ciafaloni and G. Veneziano, Int. Journ. Mod. Phys. 3A (1988) 1615; and in preparation.

23. see e.g. S. W. Hawking, Phys. Rev. D13 (1976) 191.

24. D. Amati, M. Ciafaloni and G. Veneziano, Phys. Lett. B216 (1989) 41.

25. J.J. Atick and E. Witten, Nucl. Phys.B310 (1988) 291; see also: Ya I. Kogan, JETP Lett. 45 (1987) 709; B. Sathiaplan, Phys. Rev. D35

(1987) 3277.

26. M. Karliner, I. Klebanov and L. Susskind, Int. Journ. Mod. Phys. A3 (1988)1981; T. Yoneya, " On the Interpretation of Minimal Length in String Theory", Univ. of Tokio preprint (1989); K. Konishi, G. Paffuti and P. Provero," Minimum Physical Length and the Generalized Uncertainty Principle in String Theory", Univ. of Pisa preprint, IFUP-TH 46/89.

27. M. Fabbrichesi and G. Veneziano, "Thinning out of Relevant Degrees of Freedom in Scattering of Strings", preprint CERN-TH.5509/89.

28. N. Sanchez and G. Veneziano, " Jeans-like Instabilities for Strings in Cosmological Backgrounds", preprint CERN-TH. 5491/89.

29. Y. Aharonov, F. Englert and J. Orloff, Phys. Lett. B199 (1987) 366; N. Turok, Phys, Rev. Lett. 60 (1988) 549.

30. See, e.g., E. Witten, "The Search for Higher Symmetries in String Theory", talk presented at the Royal Society Meeting on String Theory, London (Dec.1988).

31. See, e.g., P. Salomonson and B. S. Skagerstam, Physica A 158 (1989) 499.

THE CENTRAL CHARGE IN THREE DIMENSIONS

EDWARD WITTEN*

School of Natural Sciences,
Institute for Advanced Study,
Olden Lane,
Princeton, N.J. 08540

ABSTRACT

Many of the most elusive problems in string theory involve the search for the proper interpretation of ideas such as conformal invariance and duality that date back to the early days of the subject. In this paper, some modest new insights are offered about some of these old ideas.

* Research supported in part by NSF Grant 86-20266 and NSF Waterman Grant 88-17521.

1. Introduction

Many very current questions in string theory have their origins in concepts that were clearly formulated in the first epoch of string theory. Indeed, "dual models" were originally an attempt to implement certain phenomenological ideas about strong interactions. But the subject underwent a conceptual revolution in a relatively short time in the late 1960's and early 1970's, with many important developments that included the emergence of the string picture, the recognition of the important role of the Virasoro algebra and two dimensional conformal invariance, and the introduction of internal degrees of freedom and world-sheet supersymmetry. Many ideas from that period are still very fresh and current, and we have much to do to come to grips with them properly. This paper will be concerned with some modest new insights about some of those old themes.

The first old theme that we will consider is the anomaly in the Virasoro algebra. In its most oldest and most elementary manifestation, the Virasoro anomaly is a c-number anomaly term that arises in the commutation relations of diff $\mathbf{S}^1$, the group of diffeomorphisms of a circle. This is a one dimensional point of view about the Virasoro anomaly. The Virasoro anomaly also has a two dimensional interpretation, made manifest in world-sheet path integrals, in terms of the change in the effective action of a quantum field theory defined on a two dimensional manifold (a Riemann surface Σ) under local conformal transformations of the metric of Σ. This "local" two dimensional interpretation of the Virasoro anomaly also has a "global" counterpart [1,2] in which the Virasoro anomaly is related to a central extension of the mapping class group of a Riemann surface. The change from a one dimensional to a two dimensional viewpoint about the Virasoro anomaly is widely regarded as a significant advance in string theory.

Our goal in section (2) will be to describe a *three* dimensional interpretation of the Virasoro anomaly, which involves giving a three dimensional interpretation to the central extension just mentioned. This three dimensional interpretation of the Virasoro anomaly arises naturally in the three dimensional approach [3] to rational conformal field theory.

The second old theme that we will reconsider is duality. Duality originally was a phenomenological concept in hadron physics; string theory was born when it was realized, in various stages, that duality could be implemented via the use of two dimensional quantum field theory. The original concept of duality has for a close cousin the notion of duality of conformal blocks in two dimensional rational conformal field theory [4,5,2]. An important context for application of the latter concept is two dimensional current algebra of a compact group G, as was shown by Knizhnik and Zamolodchikov [6]. Two dimensional current algebra of a compact group G can be derived from three dimensional Chern-Simons gauge theory with the same symmetry group [3], so two dimensional duality is naturally also a key ingredient in understanding the Chern-Simons theories.

It is natural to wonder whether the concept of duality also applies for three dimensional gauge theories with a *non-compact* gauge group. In section three, we will consider this question for the particular non-compact gauge group $G = SL(2,\mathbb{R})$, and we will show that at least in this case, at least under a certain topological restriction, the concept of duality can be applied. There remains the challenge of finding the correct two dimensional interpretation of three dimensional gauge theory with gauge group $SL(2,\mathbb{R})$, and, possibly, with other non-compact gauge groups.

Section four is concerned with a slight generalization of presently known constructions of soluble conformal field theories. Constructing soluble conformal field theories is at least at present practically the same problem as constructing rational ones, that is, theories in which the quantum Hilbert space $\mathcal{H}$ is a finite sum of tensor products of Hilbert spaces of left-moving and right-moving degrees of freedom. Except for the possibility of including flat spaces and orbifolds thereof, all known soluble two dimensional conformal field theories are rational theories.

With the exception of toroidal compactifications and orbifolds, it appears that the so-called coset construction (starting with current algebra of compact symmetry groups) is the only presently known source of rational conformal field theories. The coset models are actually an illustration of the theme that many of our present concerns have roots in the early period of string theory. Particular cases of the coset models appeared in prescient early work on the use of current algebra to incorporate internal symmetry in string theory [7,8]. Their revival and systematic formulation and study has among other things led to a much better understanding of the conformal and superconformal discrete series [9].

To formulate coset models, one starts with a group G and a subgroup H, and one constructs a new model by "removing" the H degrees of freedom from G current algebra. This process is vaguely reminiscent of the formation of the homogeneous space or "coset space" G/H, and this is the origin of the name "coset models." However, in section four we will note that the coset models have a generalization to an arbitrary descending chain of compact groups $G = H_n \supset H_{n-1} \supset H_{n-2} \supset \ldots \supset H_1$ (the original coset construction corresponds to the case $n = 2$, $H_1 = H$). For $n = 3$ one would have three groups, say $G \supset H \supset K$. As there is no natural notion of a three step "coset space" $G/H/K$, the existence of this generalization of the coset construction shows that the name "coset model" cannot be taken very literally.

This paper is dedicated to the memory of two talented young physicists. Bruce McClain was a postdoctoral fellow at the University of Texas who in his tragically brief career had become known among other things for his work on string theory at finite temperature [10]. He was killed by a drunk driver in the state of Utah. V. G. Knizhnik in his short career had become known as one of the leading young Soviet physicists. I have already cited one of his important contributions above [6]. I would have been pleased to have their comments on the matters considered in this paper.

2. Three Dimensional Interpretation Of The Central Charge

To begin with, let us describe how it is that the Virasoro anomaly leads to the existence of a central extension of the mapping class group Γ_g of a Riemann surface Σ of genus g. Let m^i be a local system of coordinates on the moduli space of complex structures on Σ. If we are given a purely holomorphic conformal field theory, the partition function is (if we momentarily ignore the conformal anomaly) a holomorphic function $Z(m^i)$. In a more general rational conformal field theory, one does not have a single partition function Z, but a finite dimensional vector space $\mathcal{H}_\Sigma$ of possible partition functions; let Z_s, $s = 1 \dots N$ be a basis for this vector space.

Because of the conformal anomaly, the functions $Z_s(m^i)$ are only well-defined up to a multiplicative factor. Under a conformal transformation of the metric of Σ, one has

$$Z_s(m^i) \to e^F \cdot Z_s(m^i), \tag{2.1}$$

where F is proportional to the Virasoro central charge c and can be computed from the standard formulas for the conformal anomaly.

Modular invariance entails among other things the statement that the mapping class group Γ_g acts by linear transformations on the Z_s. Thus, modular invariance is the assertion that for every $f \in \Gamma_g$, there exists a matrix $M(f)$ such that

$$Z_s(f(m^i)) = \sum_t M(f)_{st} Z_t(m^i). \tag{2.2}$$

Because of the fact that (according to (2.1)) the Z_t are only defined up to a multiplicative factor, the same is true for the $M(f)$. As a result of this, the modular transformations only close up to a multiplicative factor. Thus, with a particular convention to fix the normalization of the $M(f)$'s, one will find that for

$f, g \in \Gamma_g$, one gets

$$M(f)M(g) = M(f,g) \cdot A(f,g), \tag{2.3}$$

where $A(f,g)$ is a non-zero complex number.

Equation (2.3) describes a central extension $\widehat{\Gamma}_g$ of the mapping class group Γ_g. Though the matrices $M(f)$ do not close to form a group, the matrices $\alpha M(f)$, for arbitrary $\alpha \in \mathbb{C}^*$ ($\mathbb{C}^*$ denotes the multiplicative group of non-zero complex numbers) are closed under multiplication and form a group $\widehat{\Gamma}_g$. This group is part of an exact sequence

$$1 \to \mathbb{C}^* \to \widehat{\Gamma}_g \to \Gamma_g \to 1, \tag{2.4}$$

where the homomorphism from the extended group $\widehat{\Gamma}_g$ to the original mapping class group Γ_g consists of identifying two matrices that are proportional to each other. The kernel of this homomorphism is $\mathbb{C}^*$ because an element of $\widehat{\Gamma}_g$ that is proportional to the identity matrix is an arbitrary, unique non-zero complex number.

An extension of a group Γ_g is by definition a group $\widehat{\Gamma}_g$ that has a surjective homomorphism to Γ_g, as in (2.4). An extension is said to be central if the kernel of this surjective homomorphism – $\mathbb{C}^*$ in (2.4) – is contained in the center of $\widehat{\Gamma}_g$. This condition is clearly obeyed in the case at hand, so $\widehat{\Gamma}_g$ is a central extension of the mapping class group Γ_g.

By evaluating the product $(M(f)M(g))\,M(h) = M(f)\,(M(g)M(h))$ one learns that if matrices $M(f)$ obeying (2.3) exist, then necessarily the A's obey

$$A(f,g)A(fg,h) = A(f,gh)A(g,h). \tag{2.5}$$

The extension $\widehat{\Gamma}_g$ should be considered trivial if by correct normalization of the $M(f)$'s one can eliminate the $A(f,g)$. Indeed, under $M(f)_{st} \to M(f)_{st} \cdot \Lambda(f)$ (with the Λ's being non-zero complex numbers) one has

$$A(f,g) \to A(f,g) \cdot (\Lambda(fg)\Lambda(f)^{-1}\Lambda(g)^{-1}) \tag{2.6}$$

The extension determined by the A's is trivial if and only if the Λ's can be chosen so that the $A(f,g) = 1$ for all f, g. The condition (2.5), together with the equivalence relation (2.6), precisely says that the $A(f,g)$ define an element of $H^2(\Gamma_g, \mathbb{C}^*)$. In general, the central extensions of any group G by an abelian group K correspond to elements of $H^2(G, K)$.

The extensions of the mapping class group that arise in conformal field theory are non-trivial. The extension defined by (2.3) depends only on the Virasoro central charge c, and not on other details of the rational conformal field theory under consideration, since F in (2.1) can be computed just from a knowledge of c.

In the case of a unitary conformal field theory, the M's are unitary matrices (relative to some unitary structure on the vector spaces $\mathcal{H}_\Sigma$), and the A's are then complex numbers of modulus one. Actually, it can be shown that for a conformal field theory of central charge c, the A's can be taken to be powers of $\exp(2\pi ic/24)$. We will sketch the abstract reason for this in the next subsection. The statements in the first two sentences of this paragraph are compatible, since in a unitary theory, c is real and hence $\exp(2\pi ic/24)$ is of modulus one.

2.1. Conventional Description Of The Central Extension

Before describing the three dimensional interpretation of the central extensions of Γ_g that arise in conformal field theory, I would first like for orientation to describe how this would be described conventionally without going to three dimensions.

Let $\mathcal{T}_g$ denote Teichmuller space in genus g. $\mathcal{T}_g$ is contractible, and moduli space is the quotient $\mathcal{M}_g = \mathcal{T}_g/\Gamma_g$. If it were the case that Γ_g acted freely on $\mathcal{M}_g$, then $\mathcal{M}_g$ would be a manifold and we could identify Γ_g as the fundamental group $\pi_1(\mathcal{M}_g)$. (In this case, $\mathcal{M}_g$, as the quotient of a contractible space by Γ_g, would be a model for the so called "classifying space" of Γ_g.) As it is, the action of Γ_g on $\mathcal{T}_g$ is not quite free, but has isotropy groups of finite order corresponding to Riemann surfaces with symmetries; as a result, moduli space is not a manifold but an orbifold (and $\mathcal{M}_g$ is a model of the classifying space only modulo torsion).

Despite the orbifold singularities of $\mathcal{M}_g$, one can interpret Γ_g as the fundamental group $\pi_1(\mathcal{M}_g)$ if one defines this fundamental group not in the naive way but in a way that takes into account the orbifold singularities correctly. In general, there is a notion (used in standard applications of orbifolds in string theory) in which one defines the fundamental group of an orbifold taking account of the singularities; roughly speaking, a loop that wraps r times around an orbifold singularity of order n is considered contractible only if r is divisible by n. In addition to this, we also must pick some basepoint P in defining $\pi_1(\mathcal{M}_g)$.

Now we would like to define geometrically an extension of Γ_g that will arise as the fundamental group of a certain orbifold that will be fibered over $\mathcal{M}_g$. Let $\mathcal{L}$ be (the total space of) a complex line bundle over $\mathcal{M}_g$ (for the time being any complex line bundle will do). $\mathcal{L}_g$ has the following properties: it is fibered over $\mathcal{M}_g$, with fiber a copy of the complex plane $\mathbb{C}$, and $\mathcal{M}_g$ is naturally embedded in $\mathcal{L}$ as "the zero section."

Let $\widehat{\mathcal{M}}_g$ be $\mathcal{L}$ with the zero section removed. $\widehat{\mathcal{M}}_g$ is fibered over $\mathcal{M}_g$ with fiber a copy of $\mathbb{C}^*$. In what follows it is fundamental that

$$\pi_1(\mathbb{C}^*) \cong \mathbb{Z}. \tag{2.7}$$

Let w be the projection $\widehat{\mathcal{M}}_g \to \mathcal{M}_g$.

Let $\widehat{\Gamma}_g$ be the fundamental group (in the sense of orbifolds) of $\widehat{\mathcal{M}}_g$. (In defining $\pi_1(\widehat{\mathcal{M}}_g)$ we pick some base point $\widehat{P}$ such that $w(\widehat{P}) = P$.) We claim that the group $\widehat{\Gamma}_g$ is a central extension of Γ_g, with kernel $\mathbb{Z}$ (which arises because of (2.7)). The required surjective homomorphism $\rho : \widehat{\Gamma}_g \to \Gamma_g$ is simply the map on the fundamental group induced from the projection map w. Thus, an element $\overline{x}$ of

$\pi_1(\widehat{\mathcal{M}}_g)$ is represented by a map $x : \mathbf{S}^1 \to \widehat{\mathcal{M}}_G$ with $x(0) = \widehat{P}$ (where 0 is a base point of $\mathbf{S}^1$). Then $y = w \circ x$ is a closed loop on $\mathcal{M}_g$, with $y(0) = P$, and represents an element $\overline{y}$ of $\pi_1(\mathcal{M}_g)$. We define $\rho(\overline{x}) = \overline{y}$.

That ρ is surjective can be seen as follows. Given a map $y : \mathbf{S}^1 \to \mathcal{M}_g$ representing an element of Γ_g, we must find a map $x : \mathbf{S}^1 \to \widehat{\mathcal{M}}_g$ such that $w(x) = y$. The existence of x follows from the fact that a complex line bundle over $\mathbf{S}^1$ is topologically trivial, so that there is no obstruction to lifting y to x.

To compute the kernel of the homomorphism $\rho : \widehat{\Gamma}_g \to \Gamma_g$, we must compute $\rho^{-1}(1)$ (where 1, the identity element of Γ_g, corresponds to the trivial map of $\mathbf{S}^1$ to $\mathcal{M}_g$ that sends the whole circle to the point P). The question is to determine the possible liftings of $1 : \mathbf{S}^1 \to \mathcal{M}_g$ to a map y of $\mathbf{S}^1$ to $\widehat{\mathcal{M}}_g$. Given that $w \circ y = 1$, y is simply a map $y : \mathbf{S}^1 \to \mathbb{C}^*$. Up to homotopy, the possible choices of y are determined by $\pi_1(\mathbb{C}^*)$, so the kernel of ρ is $\mathbb{Z}$.

Thus, for any complex line bundle $\mathcal{L}$ over $\mathcal{M}_g$, we have described a central extension by $\mathbb{Z}$ of the mapping class group,

$$1 \to \mathbb{Z} \to \widehat{\Gamma}_g \to \Gamma_g \to 1. \tag{2.8}$$

There is a bit of awkwardness in the notation here, since groups in this paper are written multiplicatively, but the group $\mathbb{Z}$ is almost always written additively.

Up to now, $\mathcal{L}$ is an arbitrary complex line bundle over $\mathcal{M}_g$. However, in view of the classification of such line bundles, the fundamental case is that in which $\mathcal{L}$ is the determinant line bundle of the $\bar{\partial}$ operator. Every other line bundle over $\mathcal{M}_g$ is a power of this one.

Now, we have constructed the group extension $\widehat{\Gamma}_g$ in an intrinsic (two dimensional) way, with no arbitrary choice of cocycle. To represent this extension by explicit formulas, one proceeds as follows. For every $f \in \Gamma_g$, pick $\widehat{f}$ such that $\rho(\widehat{f}) = f$. Then it will not necessarily be true that $\widehat{f} \cdot \widehat{g} = \widehat{fg}$. However,

$\rho(\widehat{f}\cdot\widehat{g})=\rho(\widehat{fg})$, so

$$\widehat{f}\cdot\widehat{g}=\widehat{fg}\cdot\alpha(f,g), \tag{2.9}$$

with $\alpha(f,g)\in\mathbb{Z}$. (Again, $\mathbb{Z}$ is being written multiplicatively.) The $\alpha(f,g)$ are a cocycle defining an element of $H^2(\Gamma_g,\mathbb{Z})$; they automatically obey the cocycle condition (2.6), and transform in the standard fashion of (2.5) under a change in the "lifting" $f\to\widehat{f}$.

Given the explicit description (2.9) of an extension of Γ_g by $\mathbb{Z}$, extensions by other groups can be described as follows. If F is any group, $u:\mathbb{Z}\to F$ is a homomorphism and

$$A(f,g)=u(\alpha(f,g)), \tag{2.10}$$

then the formula

$$\widehat{f}\widehat{g}=\widehat{fg}\cdot A(f,g) \tag{2.11}$$

defines an extension of Γ_g by F.

In particular, the extensions of Γ_g that arise in conformal field theory can be built in this way. One takes $F=\mathbb{C}^*$, and one defines a homomorphism $u:\mathbb{Z}\to\mathbb{C}^*$ by the formula $u(n)=\exp(2\pi inc/24)$ (where c is an arbitrary complex number, which in conformal field theory is the central charge in the Virasoro algebra). With these choices, (2.10) and (2.11) describe the extension of Γ_g that arises in a conformal field theory of central charge c. This explains why the values of the $A(f,g)$ arising in conformal field theory can be taken to all be powers of $\exp(2\pi ic/24)$.

Actually, it can be shown that the extension (2.8) of Γ_g is the "universal central extension" of Γ_g for $g\geq 4$. This means that any central extension of Γ_g by any abelian group F is obtained as in (2.10) and (2.11) for some unique choice of homomorphism $u:\mathbb{Z}\to F$. For small g, the extension (2.8) is not the universal central extension of Γ_g (for instance, for $g=1$, such a universal central extension does not exist, since $H_1(\Gamma_1) \neq 0$), but it is still the extension that arises in rational conformal field theory.

2.2. Three Dimensional Preliminaries

Having given a conventional account of the universal central extension of the mapping class group, we now proceed to our main interest, which is to describe this central extension in three dimensional terms.

It may be helpful to first explain the intuition behind the construction. Let M be an oriented three manifold, G a gauge group, A a connection on some G bundle E, and L the Chern-Simons functional

$$L = \frac{1}{4\pi} \int_M \mathrm{Tr} \left(A \wedge dA + \frac{2}{3} A \wedge A \wedge A \right). \tag{2.12}$$

One wishes to define three manifold invariants $Z_k(M)$ (for arbitrary positive integer k) by calculating the Feynman path integral over connections,

$$Z_k(M) = \int \mathcal{D}A \; \exp(ikL). \tag{2.13}$$

What makes these invariants computable is that one can get very strong information by "factorization," that is by cutting M on an embedded Riemann surface Σ of some genus g and summing over physical intermediate states.

More generally, one considers surgery. This means that one cuts on Σ to separate M into two three dimensional manifolds M_L and M_R, whose boundaries Σ_L and Σ_R are copies of Σ. Then one performs a diffeomorphism of Σ_L, by an element f of the mapping class group Γ_g, before gluing back M_L and M_R to get a new three manifold $M'(f)$. In this way, if one knows something about the action of the mapping class group on the physical intermediate states, one can compare the partition functions of various three manifolds $M'(f)$.

Framings Of Three Manifolds

The comments in the last few paragraphs were formulated in a rather naive way. Actually, as described in [3], to properly define the path integral (2.13), one needs certain additional data, namely a "framing" of M.

The tangent bundle T of an orientable three manifold is topologically trivial. I will not sketch here a proof of this fact, but merely note the following. The structure group of the tangent bundle of an oriented three dimensional manifold is $SL(3,\mathbb{R})$. (By choosing a Riemannian metric, one can reduce the structure group to $SO(3)$, but for our present purposes nothing is gained this way.) As explained in an elementary way in [11, pp. 402-9], the obstructions to triviality of a G bundle over an n dimensional manifold lie in the homotopy groups $\pi_k(G)$ for $k \leq n-1$. Since $\pi_0(SL(3,\mathbb{R})) = \pi_2(SL(3,\mathbb{R})) = 0$, the only obstruction to triviality of an $SL(3,\mathbb{R})$ bundle T over a three manifold comes from $\pi_1(SL(3,\mathbb{R})) \cong \mathbb{Z}_2$. Because of this non-zero obstruction, it is possible to have a non-trivial $SL(3,\mathbb{R})$ bundle over a three manifold. However, with a little elementary geometry it is possible to prove that the $\mathbb{Z}_2$ obstruction always vanishes in the case of the tangent bundle. We will not pause to explain this, since our approach will be in the first approximation to ignore $\mathbb{Z}_2$ torsion and then in a rigorous account to systematically kill all $\mathbb{Z}_2$ torsion information by a doubling procedure.

The fact that the tangent bundle of an orientable three manifold is trivial means that it is possible to parallelize an arbitrary orientable three manifold, picking three vector fields w_i, $i = 1\ldots 3$ that are everywhere linearly independent. (If a Riemannian metric has been picked, one can suppose that the w_i are everywhere orthonormal.) A trivialization $w = \{w_i\}$ of the tangent bundle of a manifold is called a framing.

The choice of framing of a three manifold is not unique, even up to homotopy. Let $v = \{v_i\}$ and $w = \{w_i\}$ be two orthonormal trivializations of the tangent bundle T of a three manifold. They differ by

$$v_i = \sum_j R_{ij} w_j, \tag{2.14}$$

where for each point $P \in M$, $R(P)$ is an element of $SL(3,\mathbb{R})$. In other words, R is a map of M to $SL(3,\mathbb{R})$. Such a map has a winding number

$$N(v,w) = \frac{1}{24\pi} \int_M \mathrm{Tr}(R^{-1}dR \wedge R^{-1}dR \wedge R^{-1}dR). \tag{2.15}$$

This winding number is associated with the homotopy group $\pi_3(SL(3,\mathbb{R})) \cong \mathbb{Z}$. It is possible to choose the map $R: M \to SL(3,\mathbb{R})$ to have arbitrary integer winding number, and therefore, given a framing w, one can find a second framing v (defined by (2.14)) with arbitrary integer value of the relative winding number $N(v,w)$.

For two framings v and w of an oriented three manifold M to be homotopic, it is certainly necessary that the relative winding $N(v,w)$ should vanish. This is not the only restriction, however. As explained in [11, pp. 402-9], the classification of the possible trivializations of a trivial bundle, with structure group G, over an n dimensional manifold, depends on all of the homotopy groups $\pi_k(G)$ for $k \leq n$. For the present case of $G = SL(3,\mathbb{R})$, in addition to the winding number invariant that is the object of interest, there is also torsion information in the comparison between two framings. This torsion information is associated with our familiar acquaintance $\pi_1(SL(3,\mathbb{R})) \cong \mathbb{Z}_2$. As noted earlier, for our first approach to the problem we will ignore this $\mathbb{Z}_2$ torsion problem; later, at the cost of some complication, we will give a precise way to eliminate it. Modulo torsion, the only topological information in the comparison of two framings is the relative winding number.

Embedded Riemann Surfaces

Now, let us return to the situation in which we wish to consider an embedded Riemann surface Σ in an oriented three manifold M, in the context of Chern-Simons gauge theory. We know that M is not a "bare" three manifold; it is presented with a framing of the tangent bundle T.

Along an embedded surface Σ (which we assume orientable), the tangent bundle to M splits as $S \oplus \epsilon$, where S will denote the tangent bundle to Σ, and ϵ is the normal bundle. The normal bundle to Σ is a trivial real line bundle, but unless Σ has genus one its tangent bundle S is topologically non-trivial (the obstruction being the Euler class of S). By a framing of Σ we will mean a trivialization of $S \oplus \epsilon$ (this would perhaps usually be called a stable framing of S). Obviously, a framing of M induces a framing of any embedded Riemann surface Σ.

Now, we want to carry out "surgery." In other words, we want to cut M along Σ, dividing M into two three manifolds M_L and M_R, with boundaries Σ_L and Σ_R, and make a diffeomorphism f of Σ_L before gluing Σ_L and Σ_R back together again to make a three manifold $M'(f)$. There is no problem in doing this for "bare" three manifolds. The problem arises in the context of Chern-Simons gauge theory in which one wishes to consider only *framed* three manifolds. In this case, since the framing of Σ_L would not in general be invariant under u, the framings of Σ_L and Σ_R do not agree after Σ_L is transformed by u, and therefore we do not know how to glue them together to get a *framed* three manifold $M'(f)$.

Therefore, in Chern-Simons gauge theory, we cannot use the mapping class group in carrying out surgery. We need a modification of this group that keeps track of the framings. The required modification turns out to be the universal central extension $\widehat{\Gamma}_g$ of the mapping class group Γ_g.

2.3. Three Dimensional Description Of The Central Extension

At this point, it is not too difficult to see what modification of the mapping class group is required. The problem was that the framing w of Σ was not invariant under the diffeomorphism $f : \Sigma \to \Sigma$. To cure the problem, it is necessary to consider not just a diffeomorphism $f : \Sigma \to \Sigma$, but a pair consisting of a diffeomorphism f and an instruction of how to compare the framing w to its transform under f. In this way, we will be led to a three dimensional description of the universal central extension of the mapping class group.

Modulo $\mathbb{Z}_2$ torsion, the precise definition of the desired group $\widehat{\Gamma}_g$ can be given as follows. We consider the basic object to be not just an oriented Riemann surface Σ but an oriented surface together with a framing w of $T = S \oplus \epsilon$. Here S is the tangent bundle of Σ, ϵ is a trivial real line bundle, and by a framing of Σ we mean a basis $w = \{w_i\}$ of the trivial rank three bundle T over Σ.

An automorphism of a pair (Σ, w) will be a pair (f, x) where f is a diffeomorphism $f : \Sigma \rightarrow \Sigma$, and x is an interpolation from w to $f^*(w)$. By $f^*(w)$ we mean the new framing that w is transformed into by f. By an interpolation x from w to $f^*(w)$ we mean the following. Very concretely, $f^*(w)$ is related to w by a formula

$$f^*(w_i) = \sum_j (R(f))_{ij} w_j, \tag{2.16}$$

where $R(f)$ is a map from Σ to $SL(3, \mathbb{R})$. Modulo $\mathbb{Z}_2$ torsion (which we will correct for later), such a map is homotopic to the identity. An interpolation x from w to $f^*(w)$ is to be, by definition, a homotopy from the identity map $1 : \Sigma \rightarrow SL(3, \mathbb{R})$ to the map $R : \Sigma \rightarrow SL(3, \mathbb{R})$.

More geometrically, this notion can be stated as follows. Consider the three manifold $W = \Sigma \times I$ where I is the unit interval. By an interpolation from w to $f^*(w)$ we mean a trivialization of the tangent bundle of W that agrees with w on $\Sigma \times \{0\}$ and with $f^*(w)$ on $\Sigma \times \{1\}$. Here we are tacitly identifying the trivial bundle ϵ that appears in the notion of framing of Σ with the tangent bundle to I.

In this geometrical language, it is easy to see that the pairs (f, x) are the symmetries of Riemann surfaces that make sense in Chern-Simons gauge theory. In the situation considered at the end of the last subsection, in which one cuts a three manifold M, with framing w, along a surface Σ and then makes a diffeomorphism f of Σ_L, the problem was to describe a framing of the new three manifold $M'(f)$ obtained by gluing Σ_L and Σ_R back together. This can be done with the use of the three manifold W of the last paragraph. One simply glues Σ_L onto $\Sigma \times \{1\}$ and glues Σ_R onto $\Sigma \times \{0\}$. The framings match on both Σ_L and Σ_R, so we obtain a framing of $M'(f)$.

The pairs (f, x) form a group $\widehat{G}$ in a fairly obvious way. Given two pairs (f, x) and (g, y), one has (in the sense of (2.16)) $R(fg) = R(f)R(g)$, so if $x(t)$ and $y(t)$, with $0 \leq t \leq 1$ are paths from the identity $1 : \Sigma \to SL(3, \mathbb{R})$ to $R(f)$ and $R(g)$, respectively, then $xy(t) = x(t)y(t)$ is the required path from 1 to $R(fg)$. So we define a group law via $(f, x) \cdot (g, y) = (g \circ f, xy)$.

The desired central extension $\widehat{\Gamma}_g$ of the mapping class group is now obtained by taking the group of components of $\widehat{G}$, in other words by identifying (f, x) with (g, y) if g is homotopic to f and y is homotopic to x.

Basic Properties

The first basic property of the group $\widehat{\Gamma}_g$ is that it is an extension of the mapping class group Γ_g. An extension K of a group G by definition is a group that has a surjective homomorphism to G

$$0 \to L \to K \to G \to 0 \quad (2.17)$$

with some kernel L that is called the kernel of the extension. In the case at hand the required homomorphism $\widehat{\Gamma}_g \to \Gamma_g$ is simply the forgetful map $(f, x) \to f$. We claim that the kernel of this homomorphism is $\mathbb{Z}$, so that we have an extension

$$0 \to \mathbb{Z} \to \widehat{\Gamma}_g \to \Gamma_g \to 0. \quad (2.18)$$

The claim that the kernel is $\mathbb{Z}$ amounts to the statement that for $f = 1$ the possible choices of x are labeled by the integers. Indeed, for $f = 1$, x is to be a path from the map $1 : \Sigma \to SL(3, \mathbb{R})$ to itself. Equivalently, x is a map $x : \Sigma \times \mathbf{S}^1 \to SL(3, \mathbb{R})$ that maps $\Sigma \times \{0\}$ to 1. Such a map x has a winding number in $\pi_3(SL(3, \mathbb{R}))$ (defined in (2.15)), and modulo $\mathbb{Z}_2$ torsion, which we have not yet come to grips with correctly, this winding number classifies the possible maps x. The winding number takes arbitrary integer values. So this explains, modulo the still mysterious torsion, why the kernel of the extension of Γ_g that we have constructed is $\mathbb{Z}$.

Though we will not prove it here, the group $\widehat{\Gamma}_g$ that we have constructed is the same extension that we constructed earlier, from a two dimensional point of view, in terms of the fundamental group of $\widehat{\mathcal{M}}_g$.

Surfaces of Genus One

The situation in genus one is very special and deserves special comment. It is known that the central extensions of mapping class groups that arise in rational conformal field theory are trivial in genus one. Let us understand why this is so from the present point of view.

A genus one surface Σ can be represented as the quotient of $\mathbb{R}^2$, which we will realize as the $p-q$ plane (p and q being two real variables), by the equivalence relations $p \rightarrow p+1,\ \ q \rightarrow q$ and $p \rightarrow p,\ \ q \rightarrow q+1$. It is possible to pick a framing w of the tangent bundle of $T = S \oplus \epsilon$ which is constant, that is, invariant under translations of p and q.

Any element f of the mapping class group of such a surface can be represented by a transformation of the type

$$p \rightarrow ap + bq, \qquad q \rightarrow cp + dq, \tag{2.19}$$

where a, b, c, and d are *constants* (and in fact integers, though this is not relevant at the moment). For a diffeomorphism of this form, and a translation-invariant framing, the map $R(f) : \Sigma \rightarrow SL(3, \mathbb{R})$ defined in equation (2.16) is constant (invariant under translations of p and q), and the interpolation $x(t)$ from the identity to $R(f)$ can similarly be chosen to be independent of p and q. This gives a *canonical* choice of the homotopy class of $x(t)$. This choice is preserved under the group law (since if $x(t)$ and $y(t)$ are independent of p and q then so is the product $x(t)y(t)$), so it gives in the genus one case a canonical trivialization of the central extension under investigation.

It is essential that the genus one central extension is not just trivial but has this canonical trivialization. In fact, since $H_1(\Gamma_1, \mathbb{Z}) \cong \mathbb{Z}_{12}$, this central extension has twelve trivializations, but the trivialization just described is the canonical choice and is the one that is usually used when explicit matrix representations of the genus one mapping class group are written down.

2.4. Elimination of $\mathbb{Z}_2$ Torsion

Our discussion so far is really only heuristic because we have systematically ignored questions of $\mathbb{Z}_2$ torsion that arise because $\pi_1(SL(3,\mathbb{R})) \cong \mathbb{Z}_2$. The above discussion was formulated as if the first non-trivial homotopy group of $SL(3,\mathbb{R})$ were π_3. We will now show how the discussion can be modified to take account of the non-vanishing $\pi_1(SL(3,\mathbb{R}))$ and reach the same conclusions.

The first place that $\pi_1(SL(3,\mathbb{R}))$ appeared in the above discussion was in connection with the question of the criterion for two framings v and w of an oriented three manifold M to be homotopic. We write $v \cong w$ to express the assertion that v and w are homotopic. Although the relative winding number (2.15) is an obstruction to the existence of such a homotopy, it is not the only obstruction; there is a $\mathbb{Z}_2$ obstruction coming from $\pi_1(SL(3,\mathbb{R}))$. It is convenient to propose a looser form of equivalence relation among framings with the property that the relative winding number is the only obstruction to equivalence.

If $v = \{v_i\}$ is a framing of an oriented three manifold M, that is a basis of everywhere linearly independent sections of the tangent bundle T, then we can regard $v \oplus v$ as a framing of the direct sum $T \oplus T$; here by a framing of an arbitrary vector bundle V we mean a trivialization of the bundle, or equivalently a set of sections that at every point $p \in M$ give a basis of the vector space V_p. We now propose for framings the weaker equivalence relation that $v \sim w$ if $v \oplus v \cong w \oplus w$. By doubling things in this way one kills the $\mathbb{Z}_2$ torsion that arises in determining whether two framings are equivalent; standard obstruction theory (as in [11, pp. 402–9]) shows that if v and w are framings of the tangent bundle of an oriented three manifold, then $v \sim w$ if and only if the relative winding number $N(v,w)$ vanishes.

So far the introduction of this weakened equivalence relation may sound like a nicety. But look back to equation (2.16) and the discussion following it. Because of the non-vanishing $\pi_1(SL(3,\mathbb{R}))$, the map $R(f)$ defined in (2.16) may not be homotopic to the identity, and therefore the interpolation x may not exist. To get around this problem while preserving the rest of the discussion, we modify the previous definition so that x not is not an interpolation from 1 to $R(f)$ but an interpolation from $1 \oplus 1$ to $R(f) \oplus R(f)$, both of these being maps from Σ to $SL(6,\mathbb{R})$. The doubling kills the $\mathbb{Z}_2$ obstruction to the existence of the interpolation x; it does not affect the definition of the relative winding number or create any new problems since the homotopy groups $\pi_k(SL(N,\mathbb{R}))$ are independent of N for $k \leq 3$ and $N \geq 3$.

At this point we must restate the criterion for equivalence of two automorphisms (f,x) and (g,y) of a framed surface (Σ, w). We say that (f,x) and (g,y) are equivalent if f is homotopic to g and $x \oplus x$ is homotopic to $y \oplus y$.

The doubling that we have carried out so far is adequate for the definition of the central extension of the mapping class group but not adequate for the application to three manifolds. Let us go back to the question of surgery. Cutting a three manifold M, with framing x, along a Riemann surface Σ and carrying out a diffeomorphism f on Σ_L, we used the interpolation x to compare the framings of Σ_L and Σ_R and thus to get a new framing on the three manifold $M'(f)$ that is made by gluing Σ_L and Σ_R back together. Here x was tentatively described above as a framing of the three manifold $\Sigma \times I$. But now we understand that in general because of the $\mathbb{Z}_2$ problem, we must take x to be a framing not of the tangent bundle T_0 of $\Sigma \times I$ but of $T_0 \oplus T_0$. Therefore, after the gluing, we get on $M'(f)$ not a framing of the tangent bundle T' but a framing of $T' \oplus T'$.

Thus, to proceed in developing the theory, we must relax our notion of what it means to "frame" a three manifold. Given a three manifold M with tangent bundle T, we must be willing to regard a framing of $T \oplus T$ as a "weak framing" of M. Actually, one wishes to be able to carry out repeated surgeries, and arguments along the lines of those above show that one must permit a doubling to occur at each step. So we arrive at the notion of a "weak framing" of a three manifold that seems most suitable for developing the theory. A weak framing of an oriented three manifold M is a framing of

$$T^{(r)} = \oplus_{i=1}^{2^r} T, \tag{2.20}$$

for any positive integer r. (Here $\oplus_{i=1}^{2^r} T$ denotes the direct sum of 2^r copies of T.) Moreover, two weak framings v and w, presented as framings of $T^{(r)}$ and $T^{(s)}$, are to be considered equivalent if there are positive integers m and n with $\oplus_{i=1}^{2^m} v$ homotopic to $\oplus_{j=1}^{2^n} w$ (here, of course, $r+m=s+n$). The three manifolds considered in Chern-Simons gauge theory should be oriented three manifolds with a weak framing. At this point, one should repeat all of the arguments of this section, permitting an arbitrary doubling whenever an object is said to exist or two objects are said to be homotopic. In this way one gets a precise way of formulating things in which one can proceed as if $\pi_1(SL(3,\mathbb{R}))$ were trivial. At the cost of some cumbersomeness, this puts our treatment of the central extension of the mapping class group on a precise, rigorous basis.

Finally, it is appropriate to modify the definition of the relative winding number so as to be invariant under doubling. Given two weak framings v and w of a three manifold M, presented as framings of $\oplus_{i=1}^{2^r} T$, we declare the relative winding number $N(v,w)$ to be 2^{-r} times the integral in (2.16). With this definition, $N(v,w)$ is unchanged if v and w are replaced by $v \oplus v$ and $w \oplus w$.

Special Cases

It may be useful to illustrate the notion of a weak framing with some comments on the exact solution of Chern-Simons gauge theory by surgery.

In the exact solution, it is convenient to begin with the fact that the partition function of $\mathbf{S}^2 \times \mathbf{S}^1$ is 1. This follows from the fact that the physical Hilbert space $\mathcal{H}_{\mathbf{S}^2}$ of $\mathbf{S}^2$ is one dimensional and that the partition function of $\mathbf{S}^2 \times \mathbf{S}^1$ can be computed by taking the trace of the identity operator in this one dimensional space. In this way of looking at things, an initial state on $\mathbf{S}^2$ is propagating in the $\mathbf{S}^1$ direction, and invariance under rotations of $\mathbf{S}^1$ is built in (it is because of this invariance that the partition function of $\mathbf{S}^2$ is the trace of the *identity* operator in the one dimensional space $\mathcal{H}_{\mathbf{S}^2}$).

In fact, the three manifold $\mathbf{S}^2 \times \mathbf{S}^1$ has up to homotopy a unique framing invariant under rotations of $\mathbf{S}^1$. This, then, is the framing for which the partition function of $\mathbf{S}^2 \times \mathbf{S}^1$ is 1.

In the exact solution by surgery, another key step is the assertion that the partition function of the three sphere is $Z(\mathbf{S}^3) = S_{0,0}$, where as explained in [3], $S_{0,0}$ is a certain matrix element of the modular group acting on the physical Hilbert space in genus one. This result is obtained by surgery relating $\mathbf{S}^3$ to $\mathbf{S}^2 \times \mathbf{S}^1$.

At this point, one may ask for what framing of $\mathbf{S}^3$ is it true that $Z(\mathbf{S}^3) = S_{0,0}$. Unlike $\mathbf{S}^2 \times \mathbf{S}^1$, $\mathbf{S}^3$ has no natural framing. If we identity $\mathbf{S}^3$ with the SU(2) manifold, it has a left-invariant framing and a right-invariant framing, but there is no natural choice between them. For what choice of framing of $\mathbf{S}^3$ is it true that $Z(\mathbf{S}^3) = S_{0,0}$?

The question is to be answered by figuring out what framing of $\mathbf{S}^3$ one gets by starting with the natural framing of $\mathbf{S}^2 \times \mathbf{S}^1$ and performing the standard surgery on an unknot to construct $\mathbf{S}^3$. One finds that this is precisely a case in which the map $R(f)$ of equation (2.16) does not exist. Therefore, the surgery building $\mathbf{S}^3$ from $\mathbf{S}^2 \times \mathbf{S}^1$ gives (starting with the natural framing of $\mathbf{S}^2 \times \mathbf{S}^1$) not a framing of the tangent bundle T of $\mathbf{S}^3$ but a framing of $T \oplus T$. Up to homotopy, there is a natural framing of $T \oplus T$ – take the left invariant framing of one copy of T, and the right invariant framing of the other. This is the weak framing of $\mathbf{S}^3$ for which its partition function is $S_{0,0}$.

<u>Spin Theories</u>

For clarity, I should perhaps point out that there is another way to deal with the $\mathbf{Z}_2$ questions. This alternative is simpler but is inequivalent physically.

The essential reason that in the above it was *necessary*, not just convenient, to start doubling things was that the map $R(f)$ introduced in (2.16) might not be homotopic to the identity. However, we could simply restrict ourselves to the subgroup of Γ_g consisting of f's such that $R(f)$ is homotopic to the identity. It can be shown that this subgroup is none other than the spin mapping class group $\Gamma_g^{\rm spin}$ consisting of elements of Γ_g that preserve a certain spin structure. If one restricts to $\Gamma_g^{\rm spin}$, the doubling procedure is unnecessary and the discussion of the central extension is valid in precisely the form originally stated.

This temptingly simple approach is appropriate for "spin theories" (as discussed in [12]) in which all Riemann surfaces and three manifolds are endowed with spin structures. It is not appropriate for purely bosonic theories.

3. Duality For Non-Compact Groups

In keeping with our mandate in this paper of trying to offer some modest new insights about old ideas in string theory, our next task is to discuss a slight extension of the notion of duality in rational conformal field theory [4]. One of V. G. Knizhnik's many celebrated papers was concerned with the application of duality in this sense to two dimensional current algebra [6], with which we will be concerned below.

Let us recall the relevant notions. A rational conformal field theory associates to an oriented Riemann surface Σ a Hilbert space $\mathcal{H}_\Sigma$, the space of conformal blocks. The $\mathcal{H}_\Sigma$ are subtle objects that depend on Σ only as an oriented and framed (in the sense of the last section) smooth surface.

Duality is the assertion that for any realization of Σ as a thickening of a trivalent graph, the $\mathcal{H}_\Sigma$ have a simple description of a particular type. For instance, in figure (1), a genus two surface Σ is depicted as a thickened version of a particular trivalent graph Γ. Duality is the assertion that $\mathcal{H}_\Sigma$ can be constructed from such a picture in the following way. For any given rational conformal field theory, there is a finite set Q of representations of the chiral algebra that can flow through any given line. There is also for every triple $a, b, c \in Q$ a finite dimensional Hilbert space $\mathcal{H}_{a,b,c}$ of possible "couplings" of the three representations a, b, c. By a labeling L of a graph Γ, we mean an assignment of a representation a, b, c, etc., to every line in the graph. Given a choice of labeling, we associate to every vertex V in the graph a Hilbert space $\mathcal{H}_V$ of couplings of the three representations attached to the three lines emanating from V. The "dual" description of $\mathcal{H}_\Sigma$ is

$$\mathcal{H}_\Sigma = \oplus_{L_a} \quad (\otimes_{V_i} \ \mathcal{H}_{V_i}) . \tag{3.1}$$

Here the L_a are the possible labelings of Γ, and the V_i run over all vertices in Γ. The real content of (3.1) – and the reason that it deserves the name "duality" – is the assertion that although the construction on the right hand side of (3.1) depends on the choice of a graph Γ of which the Riemann surface Σ is a thickening, the vector space $\mathcal{H}_\Sigma$ constructed this way in fact depends on Σ only and not on the choice of Γ.

In the event that the vector spaces $\mathcal{H}_{a,b,c}$ are all zero or one dimensional, the $\mathcal{H}_{V_i}$ appearing in (3.1) can be suppressed if we add the proviso that we only allow labelings in which whenever three representations a, b, c emanate from a vertex, they are such that the dimension of $\mathcal{H}_{a,b,c}$ is non-zero. In this case, duality reduces to the statement that $\mathcal{H}_\Sigma$ has a natural basis with one basis vector for every allowed labeling of Γ.

3.1. Duality From A Three Dimensional Point Of View

Our interest is now to think about duality from a three dimensional point of view. In three dimensions one begins with a gauge group G and the familiar Chern-Simons action

$$I = \frac{k}{4\pi} \int_M \mathrm{Tr} \left(A \wedge dA + \frac{2}{3} A \wedge A \wedge A \right) . \tag{3.2}$$

(We now include the "level" k explicitly in writing the action.) Canonical quantization of this action associates with every oriented surface Σ a Hilbert space $\mathcal{H}_\Sigma$. The association $\Sigma \rightarrow \mathcal{H}_\Sigma$ is formally similar to the association of Hilbert spaces to Riemann surfaces in rational conformal field theory. Indeed, the main observation in [3] was that in the case of a *compact* gauge group G, the Hilbert spaces $\mathcal{H}_\Sigma$ constructed by quantizing (3.2) coincide with those that are associated with a particular rational conformal field theory in two dimensions. The conformal field theory in question is G current algebra at level k. This is an important aspect of the fact that in general, two dimensional current algebra of a *compact* group G can be derived from the three dimensional Chern-Simons theory.

On the other hand, in three dimensions one can perfectly well consider the Chern-Simons theory with a non-compact gauge group G. In fact, there are many indications that it is worth while to do so, one of these indications being the fact that three dimensional quantum gravity is the Chern-Simons gauge theory of certain non-compact gauge groups [13,14]. Canonical quantization of Chern-Simons gauge theory with a non-compact gauge group will associate to a Riemann surface Σ a Hilbert space $\mathcal{H}_\Sigma$, but the $\mathcal{H}_\Sigma$ will be infinite dimensional.

It is natural to ask whether the three dimensional Chern-Simons theories with non-compact groups are related to some quantum field theory constructions in two dimensions. It seems likely that if so, the relevant two dimensional constructions are not known, or at least not well understood at present. One reason for this belief the three dimensional theory is a unitary theory even if the gauge group is non-compact, but two dimensional current algebra with for instance a semi-simple non-compact symmetry group is non-unitary. Therefore, the relationship of the three dimensional theory with non-compact symmetry to two dimensions must involve some new ideas – if there is such a relation.

Whether or not three dimensional Chern-Simons theory with non-compact gauge group is related to an interesting construction in two dimensions, it is very interesting to ask by what tools it can be effectively studied. In particular, does duality in the sense of (3.1) hold in the canonical quantization of (3.2) with a non-compact gauge group?

We will now show that the answer is "yes" at least in a very special case, in which the gauge group is $SL(2,\mathbb{R})$ and we consider only a certain special component of the classical phase space. The argument for this special case is very simple and is probably much more elementary than the algebraic geometry that enters to establish a similar result for the compact gauge groups.

3.2. Length-Twist Coordinates And Duality

To quantize (3.2), one first associates to a Riemann surface Σ the classical phase space $\mathcal{U}_\Sigma$ of critical points of (3.2) on the three manifold $\Sigma \times \mathbb{R}^1$. One then quantizes this phase space.

As was explained in [14], in the case of Chern-Simons gauge theory, the phase space $\mathcal{U}_\Sigma$ coincides with the moduli space of flat G connections on Σ. In the case that the gauge group is $G = SL(2,\mathbb{R})$, this moduli space has several components, labeled by the Euler class of the flat $SL(2,\mathbb{R})$ bundle. We will here consider only the case that the Euler class of the flat bundle has its maximum possible value. For genus greater than one, to which we now restrict ourselves, this component of $\mathcal{U}_\Sigma$ can be identified with $\mathcal{T}_\Sigma$, the Teichmuller space of Σ. While there are many equivalent definitions of Teichmuller space, for us it is convenient to think of Teichmuller space as the moduli space of metrics on Σ of constant curvature -1 metrics, up to diffeomorphisms that are continuously connected to the identity.

Our task is to quantize Teichmuller space, with the natural symplectic structure deduced from (3.2). This can be done in a remarkably simple way using length-twist coordinates. We recall how these are defined. To begin with, given a surface Σ, one picks a set of nonintersecting simple closed curves w_i such that cutting on those curves will divide the surface Σ in a collection of "pairs of pants." A possible choice is illustrated in figure (1). Now, a point P in Teichmuller space $\mathcal{T}_\Sigma$ represents a constant curvature metric g_P. In this constant curvature metric, it is not too difficult to show that any of the w_i can be deformed as simple closed curves to unique closed geodesics W_i. The length coordinates τ_i are the lengths of these geodesics.

Given a constant curvature -1 metric on Σ, a new one can be formed by cutting on the geodesics W_i, "twisting" by twist angles θ_i, and gluing the picture back together. One can pick a standard hyperbolic metric of given values of the length coordinates τ_i by requiring the hyperbolic metric to be invariant under a suitable mirror reflection (which is chosen to leave fixed the graph in the figure). Any hyperbolic metric of these fixed τ_i is obtained from the standard one just described by twisting by definite twist angles θ_j. The τ_i and θ_j are the length-twist coordinates of Teichmuller space. These variables have the range $0 \leq \tau_i < \infty$, $-\infty < \theta_j < \infty$.

Quantization of Teichmuller space is very easy in terms of length-twist coordinates because [15] the symplectic form is simply

$$\omega = k \cdot \sum_j d\theta_j \wedge d\tau_j. \tag{3.3}$$

The canonical commutation relations are therefore

$$[\theta_i, \theta_j] = [\tau_i, \tau_j] = 0, \quad [\theta_i, \tau_j] = -\frac{i}{k}\delta_{ij}. \tag{3.4}$$

These commutation relations can of course be implemented by regarding the τ_i as multiplication operators and representing θ_j as $-(i/k)d/d\tau_j$. If X is the product of the real half-lines $0 \leq \tau_i \ < \infty$, then the quantum Hilbert space $\mathcal{H}_\Sigma$ can be identified with $L^2(X)$.

Thus, for every choice of graph, we have obtained a simple description of the physical Hilbert space $\mathcal{H}_\Sigma$. A maximal set of commuting observeables are the τ_i, one for each line in the graph Γ of figure (1). The role of the τ_i is precisely analogous to the role of the labels that appear in the dual diagrams of ordinary current algebra.

It is, in fact, easy to see that, in further analogy with the labels in standard dual diagrams, the length coordinates τ_i correspond to representations of $SL(2,\mathbb{R})$. The length coordinates represent hyperbolic conjugacy classes in $SL(2,\mathbb{R})$, and such conjugacy classes correspond (by analogy with Verlinde's treatment [5] of compact groups) to continuous series unitary representations of $SL(2,\mathbb{R})$. It is remarkable that the "duality" can be seen so easily for $SL(2,\mathbb{R})$ gauge theory, at least for one component of the classical phase space, while almost nothing else about $SL(2,\mathbb{R})$ gauge theory is well understood.

4. Generalization Of The Coset Construction

Finally, in this section we will discuss a slight generalization of the "coset construction" of rational conformal field theories. The generalization is simple and may be known to some, but does not seem to have been formulated in print. This generalization of the coset construction fits into our overall program in this paper of offering some new reflections on old themes in string theory, because although the most spectacular applications of the coset construction are relatively recent [9], the idea dates at least in embryo to the early period [7,8]. A framework that might lead to other generalizations of the coset construction has recently been proposed [16]

Let us first recall the general idea behind the coset construction. A basic ingredient in a rational conformal field theory is a "chiral algebra" $\mathcal{A}$ consisting of mutually local holomorphic fields ϕ_λ closed under operator products.

A chiral subalgebra $\mathcal{B}$ is a subset of the ϕ_λ that is closed under operator products. Given a chiral algebra $\mathcal{A}$ and a chiral subalgebra $\mathcal{B}$, the coset construction amounts to a recipe for forming a new chiral algebra, which we denote as $\mathcal{A}/\mathcal{B}$. This algebra $\mathcal{A}/\mathcal{B}$, which one might loosely call the quotient of $\mathcal{A}$ by $\mathcal{B}$, is defined to consist of all local fields $\phi_\lambda \in \mathcal{A}$ that commute with $\mathcal{B}$ (or equivalently, whose operator products with fields in $\mathcal{B}$ have no short distance singularities). $\mathcal{A}/\mathcal{B}$ is a subalgebra of $\mathcal{A}$, and this makes it fairly obvious that it obeys all of the axioms of a chiral algebra if $\mathcal{A}$ does.

For any compact Lie group G, let $\mathcal{W}_{G;k}$ be the chiral algebra of G current algebra at level k. If H is a subgroup of G, then $\mathcal{W}_{H;k}$ is a subalgebra of $\mathcal{W}_{G;k}$. The coset construction is defined by saying that the G/H chiral algebra $\mathcal{W}_{G,H;k}$ is the quotient $\mathcal{W}_{G;k}/\mathcal{W}_{H;k}$. It is important to stress that $\mathcal{W}_{G,H;k}$ is a subalgebra of $\mathcal{W}_{G;k}$ (consisting of fields in $\mathcal{W}_{G;k}$ that commute with $\mathcal{W}_{H;k}$).

Now, suppose that we are given a chain of three groups $K \subset H \subset G$. In the fashion described above, we form the "quotient" of $\mathcal{W}_{H;k}$ by $\mathcal{W}_{K;k}$ to get a chiral algebra $\mathcal{W}_{H,K;k}$. This latter algebra is a subalgebra of $\mathcal{W}_{H;k}$ and therefore it is certainly a subalgebra of the larger algebra $\mathcal{W}_{G;k}$. Now, we form the "quotient" of the chiral algebra $\mathcal{W}_{G;k}$ by its subalgebra $\mathcal{W}_{H,K;k}$ to get a new algebra that we may call $\mathcal{W}_{G,H,K;k}$. In this way, one can generalize the coset construction to associate chiral algebras with a three step chain $K \subset H \subset G$. Concretely, $\mathcal{W}_{G,H,K;k}$ consists of all fields in $\mathcal{W}_{G;k}$ that commute with all those fields in $\mathcal{W}_{H;k}$ that commute with $\mathcal{W}_{K;k}$.

It is now easy to extend this process to an n step chain $H_1 \subset H_2 \subset H_3 \subset \ldots \subset H_n$. Suppose that, inductively, we have succeeded in defining for every $n-1$ step chain $H_1 \subset H_2 \subset \ldots \subset H_{n-1}$ a chiral subalgebra $\mathcal{W}_{H_{n-1},H_{n-2},\ldots,H_1;k}$ of $\mathcal{W}_{H_{n-1};k}$. As $\mathcal{W}_{H_{n-1};k}$ is a subalgebra of $\mathcal{W}_{H_n;k}$, it follows that $\mathcal{W}_{H_{n-1},\ldots,H_1;k}$ is necessarily a subalgebra of $\mathcal{W}_{H_n;k}$. It follows, therefore, that we can define a new chiral algebra which we may call $\mathcal{W}_{H_n,H_{n-1},\ldots,H_1;k}$ as the quotient of $\mathcal{W}_{H_n;k}$ by $\mathcal{W}_{H_{n-1},H_{n-2},\ldots,H_1;k}$. This is the promised generalization of the coset construction to an arbitrary ascending chain of groups $H_1 \subset H_2 \subset \ldots \subset H_n$.

In [2], it was shown that the G/H coset models are related to Chern-Simons gauge theory with gauge group $G \times H/T$, with T a finite subgroup of the center of $G \times H$. Presumably, a similar statement holds for the more general models associated with an arbitrary ascending chain of groups.

REFERENCES

1. D. Friedan and S. Shenker, Nucl. Phys. **B281** (1987) 509.
2. G. Moore and N. Seiberg, "Classical And Quantum Conformal Field Theory," IAS preprint HEP-88/35, to appear in Nucl. Phys. B.
3. E. Witten, "Quantum Field Theory And The Jones Polynomial," Comm. Math. Phys. **121** (1989) 351.
4. A. Belavin, A.M. Polyakov, and A. Zamolodchikov, Nucl. Phys. **B** (1984).
5. E. Verlinde, "Fusion Rules And Modular Transformations In 2d Conformal Field Theory," Nucl. Phys. **B300** (1988) 360.
6. V. G. Knizhnik and A. B. Zamolodchikov, Nucl. Phys. **B247** (1984) 83.
7. K. Bardakci and M. B. Halpern, Phys. Rev. **D3** (F1971) 2493.
8. M. B. Halpern, Phys. Rev. **D4** (1971) 2398.
9. P. Goddard, A. Kent, and D. Olive, Phys. Lett. **152B** (1985) 88.
10. B. McClain and B. Roth, Comm. Math. Phys. **111** (1987) 539.
11. E. Witten, "Topological Tools In Ten Dimensional Physics," in *Unified String Theories*, ed. M. B. Green and D. J. Gross (World Scientific, 1986) p. 400.
12. R. Dijkgraaf and E. Witten, "Topological Gauge Theories And Group Cohomology," preprint, to appear.
13. A. Achucarro and P. K. Townsend, "A Chern-Simons Action For Three Dimensional Anti-De Sitter Supergravity Theories," Phys. Lett. **180B** (1986) 89.
14. E. Witten, "2 + 1 Dimensional Gravity As An Exactly Soluble System," Nucl. Phys. **B311** (1988-9) 46.
15. S. Wolpert, "On The Symplectic Geometry Of Deformations Of A Hyperbolic Surface," Ann. Math. **117** (1983) 207; "On The Weil-Peterson Geometry Of The Moduli Space Of Curves."

16. M. B. Halpern and E. Kiritsis, "General Virasoro Construction On Affine $\mathcal{G}$," Berkeley preprint UCB-PTH-89/1, to appear in Mod. Phys. Lett. A; M. B. Halpern and J. P. Yamron, "Geometry Of The Virasoro Master Equation," IAS preprint (1989).

FIGURE CAPTIONS

1) A genus two surface realized as the thickening of a trivalent graph. The dotted lines point to a certain set of non-intersecting simple closed curves.

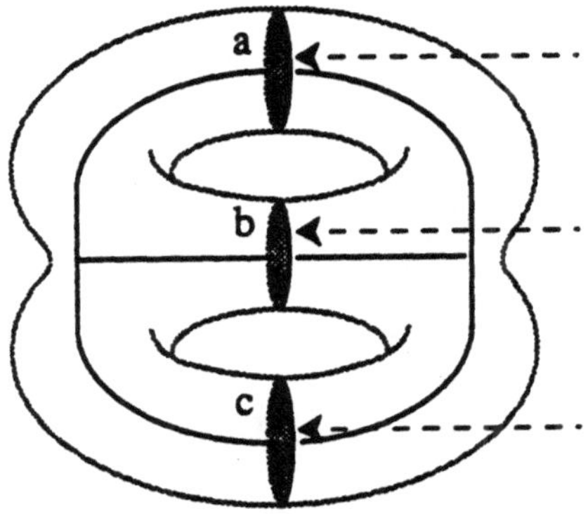

Covariant Formulations of the Superparticle
and
the Superstring

Lars Brink

Institute of Theoretical Physics
S-412 96 GÖTEBORG, Sweden

1 Introduction

Supersymmetry is the first real extension of space-time symmetry. It has given us great hope that we should be able to generalize ordinary geometry into a supergeometry and in this process obtain more unique and consistent models of physics. In some cases this has been achieved, but in most cases we still lack a natural and unique extention into a superspace.

The concept of superspace, i.e., a space with fermionic coordinates as well as bosonic coordinates,was introduced first in dual models by Montonen [1] in an attempt to construct multiloops in the Ramond-Neveu-Schwarz model [2]. This led eventually to the superconformal algebras and super-Riemannian spaces [3]. When supersymmetric field theories were discovered [4], it was soon realized that a superspace is the natural space in which to describe these models [5]. However, these descriptions, although in the end quite successful in establishing renormalization properties [6], always lacked a certain sense of naturalness. For each supermultiplet different ideas had to be used.

In supergravity theories [7], being extensions of truly geometric theories, the hopes were even higher and the results more discouraging. So far one has only managed to write superspace actions for the N=1 theory [8], and none of them is a natural extension of the Hilbert action. Superspace techniques were though eventually useful in describing the classical theories [9] and led to the really important result that any supergravity theory has infinitely many possible counterterms [10]. This is perhaps the most important result in supergravity theory, since it shows that we must expect these theories to diverge uncontrollably in the quantum case.

The most natural superspace arose when we realized that supersymmetric theories are streamlined to be described in the light-cone gauge [11]. Here one can eliminate all unphysical degrees of freedom and the physical ones fit nicely into one single superfield, making up all its components. This formalism was quite successful and led among other things to the proof of finiteness for the N=4 Yang-Mills theory [12]. However, the light-cone gauge is a gauge in which geometry is obscured and its virtues are more on the technical side than on the conceptual one.

Superspace techniques have become a natural framework to describe the Ramond-Neveu-Schwarz model [13] and its projection to the superstring [14]. It was also instrumental in the construction of new string models such as the SO(2) model [15] and the SU(2) one [16]. However, again when one tried to introduce superworld-sheet actions with superreparametrization invariance, only somewhat unnatural formulations were found, reflecting the same problem as in supergravity theories [17].

The superstring theory carries space-time supersymmetry and it is natural to ask if the space-time coordinates of this theory can be fit into a superspace. When attempting such a scheme, unexpected difficulties arise. In the original version one finds for example that it cannot be quantized covariantly, at least not in a straightforward way [18]. Several solutions to this dilemma have been put forward. They all involve adding new coordinates or momenta for the string or use an infinity of ghost coordinates, and none of them seems to be a very natural solution. In this report I will discuss these various attempts and compare their virtues.

It should be said that we might be studying an academic problem. In a string theory the world-sheet is really the important space, where the physics is defined, and the RNS model is built by demanding supersymmetry on this space. Space-time is somehow coming out as a solution and we must accept what we get. Furthermore we know that space-time will not really make sense in the very early universe at Planck energies, and if we look for a fundamental formulation that can be taken back to the earliest times we should certainly concentrate on the world-sheet physics. However, my hope is that a simultaneous study of the space-time and the world-sheet can lead us to the best set of variables quicker.

The perhaps most fascinating and profound discovery in string theory is the possible phase transition in the very early universe [19]. It gives us a logically consistent picture of how our universe, i.e., the string phase we are in now, was created with a finite energy density and temperature thus avoiding singularities. A truly fundamental description of physics should be able to describe both phases. However, it is most plausible that in the earlier phase, concepts such as space and time are not meaningful quantities. The really great question is to find the quantities that can describe both phases. Perhaps the problem above can shed light on this issue.

Let me finally raise one more question that at least bothers me. String theory is most naturally defined in a Euclidean space. All techniques based on Riemann surfaces are really Wick-rotated formulations.This fact was not too well understood in the days of dual models, but became clear in Polyakov's functional formulation [20]. Here it is also straightforward to compute the correct loop graphs, having factored out the modular transformations in the case of closed strings. If one tries to construct these loop graphs from covariant operator rules using unitarity (and hence a Minkowskian formulation) one obtains these loops with the modular group not divided out. (In a light-cone gauge treatment, which is inherently a Minkowskian treatment one gets the correct result.) I would like to interpret this fact to cast some shadow on the time variable.

In this report I will be somewhat brief in the detailed descriptions of the various attempts to quantize the superstring covariantly. It would take me too far to go into all details and I refer to the original papers for the details. Instead I will try to show the virtues and problems with each attempt.

2 The Minimal Covariant Action

To describe a string theory with space-time supersymmetry the natural coordinates are $x^\mu(\sigma,\tau)$ and $\theta^\alpha(\sigma,\tau)$, where x^μ and θ^α are vectors and spinors resp. under SO(1,d-1) with d being the dimension of space-time, which we take to be 10. The momentum density

$$\pi^\mu_\alpha = \partial_\alpha x^\mu - i\bar{\theta}\gamma^\mu\partial_\alpha\theta \quad (\alpha = (\tau,\sigma)) \tag{2.1}$$

is the natural supersymmetrically invariant extension of the momentum density used for bosonic strings. To construct an action the natural thing is to insert π^μ_α instead of $\partial_\alpha x^\mu$ in the action for the bosonic string. This is, however, not enough. To obtain a local fermionic symmetry which can eliminate unphysical fermionic degrees of freedom Green and Schwarz added an extra (invariant) term, a Wess-Zumino-Novikov-Witten term in the language of σ-models and suggested the action [21]

$$S = -\frac{1}{2}\int d\tau d\sigma\left[\sqrt{-g}g^{\alpha\beta}\pi_\alpha\cdot\pi_\beta - 2i\epsilon^{\alpha\beta}\partial_\alpha x^\mu\bar{\theta}\gamma_\mu\partial_\beta\theta\right]\ . \tag{2.2}$$

To understand the problems of this action it is easier and equally informative to study the point particle limit

$$S_p = -\frac{1}{2}\int d\tau\, e^{-1}\pi^\mu\pi_\mu \tag{2.3}$$

$$\pi^\mu = \dot{x}^\mu - i\,\bar{\theta}\gamma^\mu\,\dot{\theta} \tag{2.4}$$

and e is the einbein $\sim\ \sqrt{g_{\tau\tau}}$.
In a Hamiltonian formalism one gets the following primary constraints

$$p_e = 0 \tag{2.5}$$

$$\bar{\chi} = \bar{p}_\theta + i\bar{\theta}\not{p} = 0\ , \tag{2.6}$$

where

$$p^\mu = -\frac{1}{e}\pi^\mu\ .$$

The secondary constraint is

$$p^2 = 0\ . \tag{2.7}$$

In order to check if the constraints correspond to gauge symmetries of the action we must check if the constraint algebra, obtained by using the canonical Poisson brackets, closes. The critical bracket in the algebra turns out to be

$$\{\bar{\chi}_\alpha, \bar{\chi}_\beta\} = 2i\left(\gamma_0\not{p}\right)_{\alpha\beta}\ , \tag{2.8}$$

where the RHS clearly is not a constraint. The rank of this 16×16 matrix is 8 because of the constraint (2.7) and this fact shows that 8 of the 16 constraints $\overline{\chi}$ are not gauge constraints, i.e., they are second class constraints in Dirac's terminology and should be eliminated [18]. However, there is no covariant way of dividing up χ into two 8-component spinors. It is true that $\not{p}\chi$ is effectively an 8-component spinor because of (2.7), but there is no other vector satisfying (2.7) in the theory that can be used to project out the other 8-component spinor. This is the root of the problem. If we allow ourselves to break covariance there is no problem, and we can easily quantize the system in the light-cone gauge where only SO(8) covariance is maintained, since $16=8_s+8_c$ under the decomposition SO(1.9)$\rightarrow$SO(8). (The representations 8_s and 8_c are the two 8-dimensional spinor representations of SO(8).)

Various methods have been developed to treat systems with second-class constraints in the BRST-treatment. In the case above one can start by constructing a BRST-charge Q_B including all 16 constraints, hence introducing 16 ghost coordinates of bosonic type. This is an overcounting and must be compensated by a new set of 16 ghosts for ghosts, which in turn must be compensated by 16 ghosts for ghosts for ghosts and this procedure goes on ad infinitum. There is, in fact, a problem even with the set of constraints $p^2 = 0$ and $\not{p}\chi = 0$, since they are not independent of each other. A proper BRST formulation needs an infinity of ghosts. This fact shows up in a Lagrangian formulation by the effect that the gauge symmetry related to the constraint (2.6) (the κ-symmetry) does not close off-shell. Also here one can show that it leads to an infinity of ghosts [22]. Gauge-fixed actions which are quadratic in the infinity of coordinates (including the ghosts) have been constructed but it is still unclear if they properly take the second-class constraints into account.

We could, in fact, have been suspicious from the beginning. From supersymmetry we conclude that P^o is a positive operator in the quantum case, i.e., there are no negative energy states. However, a superstring clearly contains spinning states that for covariant descriptions need negative energy states. A rigorous deduction of the statement that $P^o > 0$ requires a proper time gauge quantization in a positive definite Hilbert space, which has not been done. However, in retrospect of the result above we can certainly trust it and use the reasoning above as a reference for attempts to quantize covariantly [23]. It should be mentioned that the obstructions above does not apply in the light-cone gauge, since all states have positive energy in light-cone variables.

Perhaps the most important aspect here is the theorem of Jordan and Mukunda [24], which states that no covariant commuting position operators for spinning particles can be defined on a Hilbert space spanned by positive energy states only. In quantizations of systems with only first class constraints the position operators are certainly commuting and the theorem above signals second-class constraints. In fact, by quantizing (2.3) in the light-cone gauge using the Dirac procedure we arrive at

$$[x^\mu, x^\nu] = \frac{-p_\rho}{2p^2}\bar{\theta}\gamma^{\mu\nu\rho}\theta \,. \tag{2.9}$$

Since covariance is already broken we can define a new position operator [25]

$$q^\mu = x^\mu + \frac{ip_\nu}{2p^+}\bar{\theta}\gamma^{\mu\nu+}\theta \,, \tag{2.10}$$

which does commute with itself and is canonically conjugate to p^μ.

If we are to use fields which are functions of positions we certainly need commuting position operators and the reasoning above must be kept in mind.

3 Covariantized Light-Cone Approach

There is a rather straightforward way to covariantly eliminate the second-class constraints by making the light-cone decomposition not with respect to a given frame but rather with respect to two null vectors which are not fixed a priori and are treated as dynamical variables [26]. More precisely, we introduce two null vectors n^μ and r^μ obeying

$$\begin{aligned} n^2 &= 0 \\ r^2 &= 0 \\ r \cdot n &= -1 \,. \end{aligned} \tag{3.1}$$

The pair (n^μ, r^μ) can now replace the (+,-) directions of the light-cone analysis.

In order to leave the dynamical content of the original system unchanged the new variables should be pure gauge. In the Hamiltonian formalism we then also impose the constraints that the momenta π_μ and σ_μ conjugate to n^μ and r^μ vanish.

$$\pi_\mu = \sigma_\mu = 0 \tag{3.2}$$

The variables (n^μ, r^μ), subject to (3.1) parametrize the coset space SO(1,9)/SO(8) since the stability group is isomorphic to SO(8) rotations.

The whole system is now described by constraints (2.5)-(2.7) and (3.1)-(3.2) together with the naive canonical Poisson brackets among the variables. There are new second-class constraints among the constraints (3.1)-(3.2) which further complicate the final Dirac brackets and hence the commutators. However, it is now straightforward to decompose the constraints $\bar{\chi}$ (2.6) into the second-class piece and the first-class one in a covariant manner. Indeed one can identify the second-class constraints as

$$\bar{\psi} = \bar{\chi}\not{n}\not{r} \tag{3.3}$$

while the first class constraints are

$$\overline{\varphi} = \overline{\chi}\not{p} \qquad (3.4)$$

According to the theorem we noted in the last section we still must expect to get a non-commuting position operator. Indeed we get

$$[x^{\mu}, x^{\mu}] = \frac{1}{4n \cdot p}\overline{\theta}\gamma^{\mu}\not{n}\gamma^{\nu}\theta \qquad (3.5)$$

As in the light-cone gauge one can though find a shifted position variable canonically conjugate to p_{μ} which is commuting.

$$q^{\mu} = x^{\mu} + \frac{i}{2}\frac{1}{n \cdot p}n_{\rho}p_{\sigma}\overline{\theta}\gamma^{\mu\rho\sigma}\theta - \frac{n^{\mu}}{2\,(n \cdot p)^{2}}\overline{p}_{\theta}\not{p}\theta \qquad (3.6)$$

Note that we have now evaded the theorem. The price we have had to pay is an increased number of variables. This is a heavy price, since it means that in a wave function representation or a field theory the functions will be functions of an increased number of (bosonic) variables. It also turns out that this extension of variables although minimal in a certain sense leads to fairly complicated commutators among the extra variables making the wave function representation somewhat obscure and the usefulness of the method doubtful.

4 Harmonic Superspace

We have seen in the preceding sections that light-cone gauge covariance (S0(8)) is very easily achieved in the quantization of the superstring, while the full one (SO(1,9)) is harder. This is very much reminiscent of the situation in supersymmetry, where simple N=1 supersymmetry is easy to implement, while the higher N supersymmetries are much more difficult. A way out of this dilemma was devised by Ogievetsky and his collaborators [27]. They reduce a global symmetry group G to a subgroup H by introducing harmonic variables u_H^G transforming covariantly under both G and H. By building in supersymmetry covariantly in H one can use the harmonic variables to implement the G symmetry covariantly. The price one has paid is the introduction of new fields u. The formalism is really borrowed from the vierbein formalism of gravity where the vierbein fields V_{μ}^{m} transform between G=GL(4,R) and H=SO(1,3) (for d=4).

A similar programme for the superstring has been carried out by Nissimov, Pacheva and Solomon [28]. They introduce new bosonic coordinates

$v_\alpha^{\pm 1/2}$ 2 Majorana-Weyl, spinors
u_μ^α 8 Lorentz vectors.
which satisfy the kinematical constraints.

$$\left(\overline{v}^{+\frac{1}{2}}\gamma^\mu v^{+\frac{1}{2}}\right)\left(\overline{v}^{-\frac{1}{2}}\gamma_\mu v^{-\frac{1}{2}}\right) = -1\ , \tag{4.1}$$

$$\left(\overline{v}^{\pm\frac{1}{2}}\gamma^\mu v^{\pm\frac{1}{2}}\right) u_\mu^\alpha = 0\ , \tag{4.2}$$

$$u_\mu^a u^{b\mu} = \delta^{ab} \tag{4.3}$$

The group SO(8) x SO(1,1) act on $u_\mu^a, v_\alpha^{\pm 1/2}$ as an internal group of local rotations with u_μ^a transforming as a vector under SO(8), (it could also be chosen as a spinor), and $v^{\pm 1/2}$ carry the charge $\pm 1/2$ under SO(1,1). Through a famous Fierz identity one can easily prove that the composite Lorentz vectors

$$u_\mu^\pm = v^{\pm\frac{1}{2}}\gamma_\mu v^{\pm\frac{1}{2}} \tag{4.4}$$

are light-like. With the help of $u_\mu^\pm$ and u_μ^a it is straightforward to disentangle Lorentz vectors and tensors covariantly to "light-cone components"

$$A^a \equiv u_\mu^a A^\mu;\ A^\pm = u_\mu^\pm A^\mu \tag{4.5}$$

As in the case in the preceding section we must now add an action for the harmonic variables making them pure gauge. Here we should note that both in the superparticle and in the superstring case we can use harmonic variables that depend on only one parameter, τ. We can fix the same harmonic frame at each value of σ in the string case.

As in the case in the preceding section it is now straightforward to decompose a 16-component spinor into two 8-component ones. For the two chiralities of SO(1,9) one has

$$\psi_\alpha = \left(\gamma^a\gamma^+ v^{-1/2}\right)\psi_a^{-\frac{1}{2}} + \left(\gamma^a\gamma^- v^{+1/2}\right)\psi_a^{+\frac{1}{2}} \tag{4.6}$$

$$\phi^\alpha = \left(\gamma^a v^{+1/2}\right)\phi_a^{-\frac{1}{2}} + \left(\gamma^a v^{-1/2}\right)\phi_a^{+\frac{1}{2}} \tag{4.7}$$

In this way we can again decompose the constraint (2.6) covariantly into its first- and second-class parts. One advantage here over the one in sect. 3 is that no new second-class constraints are introduced from the new (harmonic) variables. In fact, the formalism even allows us to eliminate the second-class ones altogether. Note that in the Dirac quantization procedure one adds gauge fixing conditions to the first-class constraints. The total set of these conditions must then constitute second-class constraints. One can turn this procedure around. On a set of second-class constraints one can attempt to divide them up into one set which is first-class and one which is to be regarded as gauge fixing conditions. In the case of the superstring augmented with harmonic variables one can in fact perform such a divison and be left with only first-class constraints.

In the BRST-description of this system one finds that a finite number of ghosts is sufficient.

Concluding this section we have found that the introduction of harmonic variables does allow us to quantize covariantly. Again the price we have had to pay is an increase in variables that the wave functionals depend on, although in the string case the new variables are constant in σ. This method has been quite successful and Nissimov, Pacheva and Salomon have reached a number of results. The possible objections to this method and the one in sect. 3 is that they are covariantized light-cone formalisms and would contain the same amount of information as light-cone gauge formulations.

5 Actions with only first-class constraints

There is, in fact, a way to write actions in terms of the original variables only which has been pursued by Siegel [29]. Let us go back to the constraints (2.5) and (2.6). Let me use Siegel's notation d for χ. The set of constraints

$$\mathcal{A} = p^2 \tag{5.1}$$

$$\mathcal{B} = \not{p} d\,. \tag{5.2}$$

are clearly first-class as was noted in (3.4). This system can, however, not be complete since the second-class constraints are simply dropped. Siegel then found another set of constraints that can be added, namely

$$C^{\mu\nu\rho} = \bar{d}\gamma^{\mu\nu\rho}d \tag{5.3}$$

The algebra (5.1)-(5.3) is closed. The new feature here is the appearance of constraints bilinear in fermionic operators, which hence cannot be solved. To compare this model with the original one in sect. 2 one can go to the light-cone gauge. The constraints (5.2) are then solved for, but the $C^{\mu\nu\rho}$'s must be treated in a BRST formulation. One finds two sections, one physical and one unphysical, and the physical one can be seen to agree with the one following from the action (2.3).

This formulation of the superparticle seems to violate the theorem of Jordan and Mukunda alluded to before. However, a detailed study shows that this is not the case [30]. Since the constraints (5.3) cannot be solved for they must be imposed on the physical states. The set of states must then be augmented with unphysical ones and one can show that among these ones there are states with negative energy. Hence the prerequisites for the theorem is avoided.

In the string case a convenient covariant gauge choice is to introduce

$$g_{\alpha\beta} = e^{\phi}\eta_{\alpha\beta} \tag{5.4}$$

and similar conditions for the possible other members of the 2-dimensional supergravity multiplet. This conformal gauge is a partial gauge fixing at the hamiltonian level and the remaining constraints are either treated in a BRST-formulation or used to project out the physical states. For the action (2.2) this procedure is hampered by the occurrence of the second-class constraints. In Siegel's approach he simply conjectures a set of constraints in the conformal gauge and then checks that it closes and that this formalism agrees with the older one in the light-cone gauge.

To start Siegel constructs operators that mimic the p^{μ}, d^{α} of the particle case.

$$\begin{aligned} P^{\mu}(\sigma) &= p_x^{\mu} + x'^{\mu} + i\bar{\theta}\gamma^{\mu}\theta' && (5.5)\\ D^{a}(\sigma) &= p_{\theta}^{a} + \gamma_{\mu}\theta\left(p_x^{\mu} + x'^{\mu}\right) + \frac{i}{2}\gamma^{\mu}\,\theta\,\bar{\theta}\,\gamma_{\mu}\theta' && (5.6)\\ \Omega^{a}(\sigma) &= i\theta'^{a}\,, && (5.7) \end{aligned}$$

which satisfy the Kac-Moody algebra

$$\begin{aligned} \left[P^{\mu}(\sigma)\,,P^{\nu}(\sigma')\right] &= i\delta\left(\sigma-\sigma'\right)\eta^{\mu\nu} && (5.8)\\ \left\{D^{a}(\sigma)\,,D^{b}(\sigma)\right\} &= 2\delta\left(\sigma-\sigma\right)\left(\gamma\cdot P(\sigma)\right)^{ab} && (5.9)\\ \left[D^{a}(\sigma)\,,P^{\mu}(\sigma')\right] &= 2\delta\left(\sigma-\sigma'\right)\left(\gamma^{\mu}\Omega(\sigma)\right)^{a} && (5.10)\\ \left\{D^{a}(\sigma)\,,\Omega^{b}(\sigma')\right\} &= i\delta'\left(\sigma-\sigma'\right)\delta^{ab}\,, && (5.11) \end{aligned}$$

the rest being zero.

From these operators one can construct a superconformal algebra with the generators

$$\begin{aligned} \mathcal{A} &= \frac{1}{2}P^{2} + \bar{\Omega}D && (5.12)\\ \mathcal{B}^{a} &= \left(\gamma\cdot PD\right)^{a} && (5.13)\\ \mathcal{C}^{ab} &= \frac{1}{2}\bar{D}^{[a}D^{b]} && (5.14)\\ \mathcal{D}^{\mu} &= i\bar{D}\gamma^{\mu}D\,. && (5.15) \end{aligned}$$

Through a lenghty calculation one can check that these generators close into an algebra, where the structure coefficients are field dependent. By going to the light-cone frame again one finds agreement with the standard superstring.

The algebra among the generators (5.11)-(5.14) is an extension of the Virasoro algebra. It is outside the standard classification of super-Virasoro algebras, in fact making up an N=16 algebra. The reason why it can work is the appearance of field dependent structure coefficients. Such occurrencies usually signal a partial gauge fixing. It is an intriguing question whether the algebra can be further extended to avoid the field dependence. A further complication is the fact that the generators (5.11)-(5.14) are not independent of each other. To find an independent set one must divide up the generators in light-cone components thus ruining the covariance.

The non-independence also complicates the BRST procedure. Again one is forced to introduce ghost for ghosts ad infinitum and such a formalism ought to coincide with the one from the original action (2.2).

The formalism described in this section is somewhat more general than the one in sect. 2. It allows in principle for a covariant quantization. In the case of the point-particle it does give an explanation why covariant superfield methods are working although it does not give a unified and natural framework for all supermultiplets as mentioned in the introduction. In the superstring case the advantages with this formalism is probably negligable. One could attempt an old-fashioned operator formalism without ghosts but it would be fairly awkward because of the complicated super-Virasoro algebra. In fact in such an attempt I see no clear way to implement unitarity. Introducing the ghosts, which I think is necessary would lead us back to the formalism in sect. 2.

6 Twistors

The formalism described so far have been rather conventional in the sense that they aim at describing superstring theory in terms of space-time coordinates augmented with other coordinates. However, if we are really trying to find variables that will be fundamental and can describe both phases of superstring theory we should search for variables from which space-time could be derived. We have very little guidance here. If the Poincaré group is the underlying symmetry group there are though a rather limited number of alternatives. (We may have to be more imaginative!) One such proposal which have been put forward by Bengtsson, Bengtsson, Cederwall and Linden [31] and originally by Shirafuji [32] is to use twistor variables.

There are two properties among twistors that are appealing in this connection.
(i) Twistors do substitute for x^μ and p^μ.
(ii) There is a close connection between twistors and division algebras and the dimensions in which superstrings can be described.

The basic relations are the local isomorphisms between the Lorentz groups SO(1,ν+1) and the groups SL$(2, \vec{K}_\nu)$, where

$$\vec{K}_\nu = \vec{R}, \vec{C}, \vec{H}, \vec{O} \; for \; \nu = 1, 2, 4 \; and \; 8 ,$$

the four division algebras. This means that for d=3,4,6 and 10 twistor formulations are possible and these are the dimensions in which the classical superstring can exist. It should be said here that d=10 is more complicated than the others since the octonionic algebra is non-associative. In the sequel I will for simplicity only discuss d=3, 4 and 6.

The starting-point now is to use the isomorphism above to write every Lorentz vector as a bispinor under SL$(2, \vec{K}_\nu)$

$$V^\mu \longrightarrow \vec{V}^{\alpha\dot{\alpha}} = \begin{pmatrix} \sqrt{2}V^+ & \vec{V} \\ \vec{V} & \sqrt{2}V^- \end{pmatrix}$$
$$\vec{V} \in \vec{K}_\nu \tag{6.1}$$

$$\vec{V} = \sum_{L=1}^{\nu} \vec{V}_i e_i \tag{6.2}$$

with e_i an orthonormal basis for $\vec{K}$.

In the point-particle case the free bosonic particle is described by the action

$$S = \int d\tau \left[\frac{1}{2} Sc \left(\dot{\vec{x}}^{\alpha\dot{\alpha}} \; \vec{p}_{\alpha\dot{\alpha}} \right) + V \; det \; \vec{p} \right] \tag{6.3}$$

where Sc means scalar part under the division algebra and V is an auxiliary field (determinant of the einbein, if we so wish). We immediately get the constraint

$$det \; \vec{p} = 0 \; , \tag{6.4}$$

which means

$$p_\mu p^\mu = 0 \; . \tag{6.5}$$

This can be solved by the constraint

$$\vec{\pi}_{\alpha\dot{\alpha}} = \vec{p}_{\alpha\dot{\alpha}} - \bar{\vec{\psi}}_\alpha \vec{\psi}_{\dot{\alpha}} = 0 \; , \tag{6.6}$$

where we introduce the bosonic spinor (twistor) $\vec{\psi}_{\dot{\alpha}}$ and its complex conjugate. The constraint (6.5) is automatic because of the famous Fierz identity, which we used in (4.4). We furthermore introduce the bosonic spinor $\vec{\omega}^\alpha$ according to

$$\vec{g}^\alpha = \vec{\omega}^\alpha - \vec{\psi}_{\dot{\alpha}} \vec{x}^{\alpha\dot{\alpha}} = 0 \tag{6.7}$$

We can now use ω and ψ as canonical variables instead of x and p with the Poisson bracket

$$\left\{\omega_i^{\alpha}, \psi_{j,\dot{\beta}}\right\} = \delta_{ij}\delta^{\alpha}_{\dot{\beta}} \tag{6.8}$$

Furthermore we can write an action which contains both sets

$$S = \int d\tau\, Sc\left\{\frac{1}{2}\vec{x}^{\dot{\alpha}\alpha}\vec{p}_{\alpha\dot{\alpha}} + \vec{\omega}^{\alpha}\vec{\bar{\psi}}_{\alpha} + \frac{1}{2}\vec{r}^{\dot{\alpha}\alpha}\vec{\pi}_{\alpha\dot{\alpha}}\left(\vec{p}, \vec{\psi}\right) + \vec{\bar{\gamma}}_{\alpha}\vec{g}^{\alpha}\left(\vec{\Omega}, \vec{\psi}, \vec{x}\right)\right\} \tag{6.9}$$

It contains enough gauge invariance to allow for a gauge $\vec{\omega} = \vec{\omega}_f, \vec{\psi} = \vec{\psi}_f$ and obtain the usual description in terms of x and p, or $\vec{x} = \vec{x}_f, \vec{p} = \vec{p}_f$ and obtain the twistor formulation.

In the twistor formulation the mass-shell condition $P^2 = 0$ is built in by construction. Instead the on-shell condition reads

$$u = \frac{1}{2}\left(\vec{\omega}^{\alpha}\vec{\bar{\psi}}_{\alpha} - \vec{\psi}_{\dot{\alpha}}\vec{\bar{\omega}}^{\dot{\alpha}}\right) = 0 \tag{6.10}$$

For d=4 this is the spin-shell condition. The twistor formulation hence reverses the order in which the Casmir invariants are treated. Off-shell means that the helicity is continuously varied. This fact means that interactions would be very much different in such a formulation and it is so far unclear if one can describe interactions in this language.

The twistor formalism is quite suitable also in the superparticle case and it lends itself naturally to a covariant quantization. As in the bosonic case one can start with a master action containing both the ordinary space-time coordinates and the twistors

$$S = \int d\tau\, Sc\left[\frac{1}{2}\vec{\pi}^{\dot{\alpha}\alpha}\vec{p}_{\dot{\alpha}\alpha} + \vec{\omega}^{\alpha}\vec{\bar{\psi}}_{\alpha} + \frac{i}{2}\vec{\xi}\vec{\bar{\xi}} + \frac{1}{2}\vec{r}^{\alpha\dot{\alpha}}\vec{\pi}_{\alpha\dot{\alpha}} + \vec{\bar{\gamma}}_{\alpha}\vec{g}^{\alpha} + \vec{\bar{\rho}}\vec{r}\right], \tag{6.11}$$

where

$$\vec{\pi}^{\dot{\alpha}\alpha} = \dot{\vec{x}}^{\dot{\alpha}\alpha} - i\left(\dot{\vec{\theta}}^{\dot{\alpha}}\vec{\bar{\theta}}^{\alpha} - \vec{\theta}^{\dot{\alpha}}\dot{\vec{\bar{\theta}}}^{\alpha}\right), \tag{6.12}$$

where

$$\vec{\pi}_{\alpha\dot{\alpha}} \equiv \vec{p}_{\alpha\dot{\alpha}} - \vec{\bar{\psi}}_{\alpha}\vec{\psi}_{\dot{\alpha}}, \tag{6.13}$$

$$\vec{g}^{\alpha} \equiv \vec{\omega}^{\alpha} - \vec{\psi}_{\dot{\alpha}}\left(\vec{x}^{\dot{\alpha}\alpha} - i\vec{\theta}^{\dot{\alpha}}\vec{\bar{\theta}}^{\alpha}\right), \tag{6.14}$$

$$\vec{r} = \vec{\xi} - \sqrt{2}\vec{\psi}_{\dot{\alpha}}\vec{\theta}^{\dot{\alpha}}. \tag{6.15}$$

Gauge invariance implies that we can either choose a gauge described by x and θ or one in terms of the twistor variables ω, ψ and ξ.

$$\{\omega_i^\alpha, \psi_{j\beta}\} = \delta_{ij}\,\delta_\beta^\alpha \tag{6.16}$$
$$\{\xi_i, \xi_j\} = -i\delta_{ij} \tag{6.17}$$

and one finds straightforwardly that there are only first class constraints. It is now fairly direct to quantize and to introduce wave functions (and fields) with twistors as coordinates. It is, as said above, a challenging task to try to write an interacting field theory this way.

The programme has so far not been fully implemented to the octonionic case (d=10), but there is good hope that it can be done. For strings there are further problems to overcome. Here one has two light-like vectors $\partial_+ x^\mu$ and $\partial_- x^\mu$ (according to the Virasoro conditions). The first guess would be to introduce two different twistors, one for each vector. However, it sounds like an overcounting to introduce two twistors for a vector. Some progress on this problem has recently been done by Cederwall [33].

The twistor approach is a challenging idea, which should be investigated to see if it can be used to describe interacting superstrings. It has the virtue of using a minimal set of variables out of which space-time can be constructed. However, it is a formalism in which we use representations of the Lorentz group and it is not clear to me that the original phase of superstrings need be described by that symmetry. In fact, there are lots of reasons to believe that the symmetry in the original phase is much bigger.

7 Conclusions

Quite a number of methods have been devised in order to implement a covariant quantization of the superstring. Apart from the twistor approach, all the methods use an enlarged phase space, either an infinite series of ghosts for ghosts or new bosonic coordinates. These formalisms are certainly going to give new insight into superstring theory. In fact they might be quite useful in one of the great issues in superstring theory, namely in the search to get an understanding of non-perturbative effects. We know that they must play a role, for example in seeking a true minimum among all the classical solutions, all conformal field theories with no anomalies. Here it is important to have as efficient a formalism as possible. I think this question justifies all the efforts made in order to find a covariant formalism.

The other, perhaps even more important and certainly deeper issue is to understand the possible phase transition in the early universe. For this issue I do not think we have the best formalism yet. It is not clear to me that we can use the standard coordinate to describe such a phase transition. In fact, we know that it occurs at Planck energies, where space-time really breaks down because of quantum fluctuations. Can we find more fundamental variables from which for lower energies space-time can be constructed? Is there a deeper level at which the phase transition can be understood from some physical principles? These are fascinating questions which we know very little about now. I think the search for answers to them is the real challenge in front of us in superstring physics.

References:

1. C. Montonen, Nuovo Cim. **19A** (1974), 69

2. P.M. Ramond, Phys. Rev. **D3** (1971), 2415
   A. Neveu and J.H. Schwarz, Nucl. Phys. **B31** (1971), 86; Nucl. Phys. **B31** (1971), 86; Phys. Rev. **D4** (1971) 1109

3. L. Brink and J.O. Winnberg, Nucl. Phys. **B103** (1976), 445
   D. Friedan, E. Martinec and S. Shenker, Nucl. Phys. **B271** (1986) 93

4. Yu. A. Golfand and E.P. Likhtman, JETP Lett. **13** (1971), 13
   J. Wess and B. Zumino, Nucl. Phys. **B70** (1974), 39

5. A. Salam and J. Strathdee, Nucl. Phys. **B76** (1974), 477

6. M.T. Grisaru, W. Siegel and M. Rocek, Nucl. Phys. **B159** (1979), 429

7. D.Z. Freedman, P. van Nieuwenhuizen and S. Ferrara, Phys. Rev. **D13** (1976), 3214
   S. Deser and B. Zumino, Phys. Lett. **62B** (1976), 335

8. J. Wess and B. Zumino, Phys. Lett. **74B** (1978), 51
   L. Brink and P. Howe, Phys. Lett. **88B** (1979), 81

9. L. Brink and P. Howe, Phys. Lett. **88B** (1979), 268

10. P.S. Howe and U. Lindström, Nucl. Phys. **B181** (1981), 487
R.E. Kallosh, Phys. Lett. **99B** (1981), 122

11. L. Brink, O. Lindgren and B.E.W. Nilsson, Nucl. Phys. **B212** (1983), 401

12. S. Mandelstam, Nucl. Phys. **B213** (1983), 149
L. Brink, O. Lindgren and B.E.W. Nilsson, Phys. Lett. **123B** (1983), 323

13. See the first reference in [3]

14. M.B. Green and J.H. Schwarz, Nucl. Phys. **B181** (1981), 502

15. M. Ademollo, L. Brink, A. D'Adda, R. D'Auria, E. Napolitano, S. Sciuto, E. Del Giudice, P. Di Vecchia, S. Ferrara, F. Gliozzi, R. Musto, R. Pettorini and J. Schwarz, Nucl. Phys. **B111** (1976), 77

16. M. Ademollo, L. Brink, A.D' Adda, R. D'Auria, E. Napolitano, S. Sciuto, E. Del Giudice, P. Di Vecchia, S. Ferrara, F. Gliozzi, R. Musto and R. Pettorini, Nucl. Phys. **B114** (1976), 297

17. P.S. Howe, J. Phys. A. Math. Gen. Vol. **12**, No. 3 (1979), 393

18. I. Bengtsson and M. Cederwall, ITP-Göteborg **1984-21**

19. B. Sundborg, Nucl. Phys.**B254** (1985), 583
M.J. Bowick and L.C.R. Wijewardhana, Phys. Rev. Lett. **54** (1985), 2485

20. A.M. Polyakov, Phys. Lett. **103B** (1981), 207, 211

21. M.B. Green and J.H. Schwarz, Phys. Lett. **136B** (1984), 367

22. M.B. Green and C.M. Hull, QMC-89-9 (1989)
R.E. Kallosh, Cern-TH-5355 (1989)
U. Lindström, M. Rocek, W. Siegel, P. van Nieuwenhuizen, E.A. van de Ven and J. Gates, ITP-Stony Brook (1989)

23. I. Bengtsson, M. Cederwall and N. Linden, Phys. Lett. **203B** (1988), 96

24. T.F. Jordan and N. Mukunda, Phys. Rev. **132** (1963), 1842

25. L. Brink and J.H. Schwarz, Phys. Lett. **100B** (1981), 310

26. L. Brink, M. Henneaux and C. Teitelboim, Nucl. Phys. **B293** (1987), 505

27. A. Galperin, E. Ivanov, S. Kalitzin, V. Ogievetsky and E. Sokatchev, Class. Quant. Grav. **1** (1984), 469; **2** (1985), 155

28. E. Nissimov, S. Pacheva and S. Solomon, Nucl. Phys. **B297** (1988), 369

29. W. Siegel, Nucl. Phys. **B263** (1985), 93

30. I. Bengtsson, Phys. Rev. **D39** (1989), 1158

31. A.K.H. Bengtsson, I. Bengtsson, M. Cederwall and N. Linden, Phys. Rev. **D36** (1987) , 1766
I. Bengtsson and M. Cederwall, Nucl. Phys. **B302** (1988), 81

32. T. Shirafuji, Progr. Theor. Phys. **70** (1983), 18

33. M. Cederwall, ITP-Göteborg **89-15** (1989)

The Space of Conformal Field Theories and the Space of Classical String Ground States

D. Friedan

Department of Physics and Astronomy, Rutgers University
P.O. Box 849, Piscataway, N.J. 08855-0849

ABSTRACT

A formal (and speculative) construction is given of the space of all conformal field theories of given conformal central charge – and also of the space of all classical ground states of string – using only intrinsic structure in the space of all closed Riemann surfaces.

1. INTRODUCTION

A succinct abstract characterization of the space of conformal field theories – and of its close relative, the space of classical ground states of string – might help with the classification of these objects and might also provide a starting point for a characterization of the quantum ground state of string.

This note is a formal and speculative attempt at such a characterization. The space CFT_c of unitary conformal field theories with conformal central charge c is conjectured to be exactly the spectrum $Spec(A_c)$ of a certain commutative $*$-algebra A_c constructed (formally) using only intrinsic structure in a line bundle L_c over the space of all closed Riemann surfaces. Recall that the spectrum of a commutative $*$-algebra is the space of all characters or, equivalently, irreducible representations of the algebra. The algebra can be interpreted as the algebra of functions on its spectrum. In other words, CFT_c is to be constructed by constructing its function algebra A_c from a certain line bundle over the space of all closed Riemann surfaces.

Similarly, the space CGS of all classical ground states of string is constructed (formally and speculatively) by constructing its function algebra A_+ from the space of all closed Riemann surfaces. This construction is purely intrinsic; it does not even refer to the fact that the points of the space stand for Riemann surfaces.

Both constructions are formal and both beg a crucial analytic question. The present goal is only to make a simple formulation; all the hard work is left for later.

The basic underlying suppositions are (1) that the partition function of a conformal field theory for all closed Riemann surfaces determines the conformal field theory uniquely, i.e., that no two conformal field theories have the same partition function, and (2) that any section of a certain line bundle over the space of closed Riemann surfaces satisfying a small number of intrinsic conditions is the partition function of some conformal field theory.

The reason for basing the construction on the *closed* Riemann surfaces is that it avoids introducing a Hilbert space of states as a fundamental object. In string theory, the Hilbert space varies with the classical ground state and might not even make sense in the full quantum theory. It seems attractive to have the Hilbert space be a derivative object in the classical theory.

It should be mentioned that the approach described in the present note has not yet led to any concrete progress on the problem of classifying conformal field theories in general nor on the problem of describing the quantum ground state of string.

The abstract characterization of conformal field theory and classical string theory sketched in this note is essentially a refined version of that given in references 1-3 several years ago. More details on some points can be found there. The idea of seeing the string partition function as an object on the moduli space of Riemann surfaces was independently arrived at by Belavin and Knizhnik[4].

2. THE UNIVERSAL MODULI SPACE

We start by defining the *universal moduli space* of Riemann surfaces, M. It is the space of all smooth, compact, not necessarily connected Riemann surfaces without boundary, none of whose connected components are 2-spheres, plus one extra point, the 2-sphere itself. We will see that M has the structure of a connected, analytic, commutative $*$-semigroup with the 2-sphere $\mathbf{P}$ as its identity element.

It is convenient to define M by way of two larger moduli spaces. Let M_{sm} be the space of all smooth, compact, not necessarily connected Riemann surfaces without boundary. Let $\widetilde{M}$ be the space of all compact, not necessarily connected Riemann surfaces without boundary, which are smooth except for at most a finite number of nodes. The space of surfaces with at least one node is $D = \widetilde{M} - M_{sm}$.

A node in a Riemann surface can be thought of as a circle on the surface which has been pinched down to a single point. Equivalently, a node can be pictured as an infinitely long tube. A more detailed description is given below.

The partition function Z of a conformal field theory is a section of the line bundle $L_c = (EE^*)^{c/24}$ over M_{sm}. Here $E = (\lambda_H)^{12}$ is the holomorphic line bundle over M formed by taking the 12th power of the determinant of the holomorphic differentials on the Riemann surfaces comprising M; E^* is the complex conjugate of E and c is the conformal central charge of the theory. In reference 1 the holomorphic line bundle E is defined in a way more natural to conformal field theory, in terms of families of projective connections on Riemann surfaces.

The partition function satisfies

(I) Z is nonsingular on the surfaces with nodes and thus extends to $\widetilde{M}$,

(II) $Z(\Sigma) = Z(\nu\Sigma)$,

(III) $Z(\mathbf{P}) = 1$,

(IV) $Z(\Sigma^*) = Z(\Sigma)^*$,

(V) $Z(\Sigma_1 \cup \Sigma_2) = Z(\Sigma_1)\, Z(\Sigma_2)$,

where Σ, Σ_1 and Σ_2 are arbitrary Riemann surfaces (in $\widetilde{M}$), Σ^* is the complex conjugate of Σ and $\nu\Sigma$ is the smooth Riemann surface obtained by removing all nodes in Σ (the *normalization* of Σ). These conditions make sense because $(L_c)_{\mathbf{P}} = \mathbf{C}$, the complex numbers; $(L_c)_{\Sigma^*} = (L_c^*)_\Sigma$; L_c extends to $\widetilde{M}$ and $(L_c)_{\Sigma_1 \cup \Sigma_2} = (L_c)_{\Sigma_1} \otimes (L_c)_{\Sigma_2}$.

Condition (I) follows from locality, homogeneity and the fact that the conformal weights of the field theory are nonnegative. This is discussed in more detail below.

Condition (II) follows from locality, homogeneity and the uniqueness of the ground state. This is also discussed below.

Condition (III) follows from the fact that the vacuum state of the conformal field theory has nonzero norm. The partition function can then be normalized to take the value 1 on the 2-sphere.

Condition (IV) follows from *CPT* invariance. In a functional integral formulation of the field theory, *CPT* invariance means that the action is invariant under complex conjugation combined with orientation reversal of the surface.

Condition (V) is a kind of cluster decomposition property. It states the obvious fact that disconnected components of a Riemann surface are decoupled in a conformal field theory.

To explain conditions (I) and (II) we need a detailed description of a node in a Riemann surface. The neighborhood of a node can be parametrized by two local coordinates z_1 and z_2 satisfying a patching equation $(z_1 - x_1)(z_2 - x_2) = 0$. The neigborhood of the node thus consists of the two smooth neighborhoods parametrized by z_1 and z_2 with the point $z_1 = x_1$ in the first neighborhood glued to the point $z_2 = x_2$ in the second neighborhood.

Now restrict the coordinates z_1 and z_2 to annuli $|q| < |z_i - x_i| < 1$ and replace the above patching equation by $(z_1 - x_1)(z_2 - x_2) = q$. For $q \neq 0$ the patching equation is nowhere singular. Then the two coordinates z_1 and z_2 can be replaced by a single coordinate $z = q^{-1/2}(z_1 - x_1) = q^{1/2}(z_2 - x_2)^{-1}$ ranging over the annulus $|q|^{1/2} < |z| < |q|^{-1/2}$. The node has become a tube of length $-\ln|q|$. The limit $q \rightarrow 0$ exhibits the node as an infinitely long tube.

The space D of Riemann surfaces with nodes is generically of complex codimension 1 in $\widetilde{M}$, because the surfaces with node are specified by the one equation $q = 0$. The surfaces with n nodes are specified by n equations of the form $q_i = 0$, with $q_i \rightarrow 0$ describing the closing of the ith node. It is crucial that the closing of multiple nodes is described by the vanishing of *independent* parameters q_i.

Because of the locality and homogeneity of the conformal field theory, the annulus or tube parametrized by q can be represented in the operator formulation of the theory as $q^{L_0}(\bar{q})^{\bar{L}_0}$ acting on the states flowing through the tube (L_0 and $\bar{L}_0$ being the usual Virasoro operators). At $q = 0$ this operator becomes the projection on the ground state(s) of the theory, assuming that L_0 and $\bar{L}_0$ are both nonnegative (which follows from unitarity). The partition function is thus nonsingular in the limit $q \rightarrow 0$. This gives condition (I).

Assuming that the ground state is unique, the node is represented by the projection $|0\rangle\langle 0|$ on the $SL_2 \times SL_2$-invariant ground state $|0\rangle$. The ground state is thus the only state which flows through a node. Picturing the node as an infinitely long tube makes this obvious.

By the operator analysis, the partition function of the Riemann surface with node, i.e., at $q = 0$, is the same as the partition function of the surface obtained by forgetting the identification of the two points $z_1 = x_1$, $z_2 = x_2$, and using the ground state to provide boundary conditions at the two punctures $z_1 = x_1$ and $z_2 = x_2$. The ground state boundary condition at a puncture is equivalent to insertion of the identity quantum field at the puncture, which in turn is equivalent to forgetting the puncture entirely.

The partition function of a Riemann surface Σ with nodes is therefore the same as the partition function of the smooth surface $\nu\Sigma$ obtained by forgetting the patching equations for the nodes, i.e., by removing the nodes and filling in the resulting punctures. This is condition (II).

Now define an equivalence relation on $\widetilde{M}$ by requiring that, for all Riemann surfaces Σ in $\widetilde{M}$,

$$\Sigma \sim \nu\Sigma \sim \Sigma \cup \mathbf{P}\,. \tag{1}$$

The *universal moduli space* is defined to be the quotient $M = \widetilde{M}/\sim$. The line bundle L_c respects the equivalence relation and can be regarded as a line bundle over M.

Under the equivalence relation, every Riemann surface is equivalent to a smooth surface, and every surface except $\mathbf{P}$ is equivalent to a smooth surface none of whose connected components are 2-spheres. Thus, as a point set, M is the space of all smooth, compact, not necessarily connected Riemann surfaces without boundary, none of whose connected components are 2-spheres, plus one extra point, the 2-sphere itself.

It follows from conditions (I)-(III) and (V) that the partition function of a conformal field theory is a section of the line bundle $L_c \to M$.

Henceforth, when we write Σ we mean the equivalence class in M. The equivalence relation respects complex conjugation and disjoint union, so Σ^* and $\Sigma_1 \cup \Sigma_2$ still make sense. In fact, the operation of disjoint union makes M into a commutative $*$-semigroup with product

$$\Sigma_1\Sigma_2 = \Sigma_1 \cup \Sigma_2, \tag{2}$$

identity element

$$1 = \mathbf{P} \tag{3}$$

and conjugation

$$\Sigma \mapsto \Sigma^* . \tag{4}$$

In addition to this algebraic structure, the universal moduli space M is a connected analytic space. This was shown in reference 1 by constructing M from the stable compactifications of the moduli spaces M_g of smooth, connected compact Riemann surfaces of genus g. A more efficient way is to observe* that M is the direct limit as $g \to \infty$ of the Satake compactification M_g^{sat} of the moduli space M_g. The Satake compactification[5] is constructed by embedding M_g in the moduli space of abelian varieties and compactifying that space by algebraic means. The result is that M_g^{sat} is a connected projective algebraic variety for each genus g.

*as was pointed out by S. Bloch and by P. Deligne.

What is added to M_g to obtain M_g^{sat} is precisely the set of Riemann surfaces obtainable from those in M_g by forming and removing nodes and then discarding the genus 0 components (2-spheres) of the resulting surface. In particular, there is a consistent system of natural embeddings $M_{g-1}^{sat} \to M_g^{sat}$. The universal moduli space is the direct limit

$$M = \lim_{g\to\infty} M_g^{sat}\,. \tag{5}$$

M is thus a connected analytic space and, in fact, the direct limit of projective varieties.

The line bundle L_c is real-analytic on M since $E = (\lambda_H)^{12}$ is a well-defined holomorphic line bundle on M. L_c is itself a commutative $*$-semigroup with identity, since $(L_c)_{\Sigma_1\Sigma_2} = (L_c)_{\Sigma_1} \otimes (L_c)_{\Sigma_2}$, $(L_c)_{\mathbf{P}} = \mathbf{C}$ and $(L_c)_{\Sigma^*} = (L_c^*)_{\Sigma}$. The line bundle map $L_c \to M$ respects the semigroup structures, with the multiplication and conjugation laws in L_c being linear in the fibers.

Given that M is a connected topological semigroup with identity, elementary algebraic topology tells us that $\pi_1(M)$ is abelian and thus equal to the first homology group $H_1(M)$. It should be possible to show that $H_1(M) = 0$ and thus that M is simply-connected. This seems a fundamental point to establish, since a non-contractible loop in M would allow global anomalies which could not be eliminated by conditions local on M. Establishing $H_1(M) = 0$ would have the added benefit of leaving the holomorphic line bundles on M classified by $H^2(M)$. It seems possible to show that $H^2(M) = \mathbf{Z}$ (see reference 1), which would then imply that the powers of $E = (\lambda_H)^{12}$ are the only holomorphic line bundles nonsingular on M. The line bundle L_c could then be said to be intrinsic to M.

3. THE SPACE OF CONFORMAL FIELD THEORIES

We have seen that the universal moduli space of Riemann surfaces, M, is a connected, analytic, commutative $*$-semigroup with identity and that the line bundle $L_c \to M$ is a morphism of commutative $*$-semigroups with identity such that the multiplication and conjugation laws in L_c are linear in the fibers.

Conditions (I)-(V) on the partition function Z of a conformal field theory can now be condensed into the condition that $Z : M \to L_c$ should be a section of $\pi : L_c \to M$ which is at the same time a morphism of $*$-semigroups, i.e.,

(1) $\quad \pi \circ Z = id,$

(2) $\quad Z(1) = 1,$

(3) $\quad Z(\Sigma^*) = Z(\Sigma)^*,$

(4) $\quad Z(\Sigma_1\Sigma_2) = Z(\Sigma_1)\, Z(\Sigma_2).$

Conditions (1)-(4) become a formal characterization of the space of conformal field theories, given two suppositions which were made in reference 1:

(S1) A conformal field theory is completely determined by its partition function $Z : M \to L_c$. No two conformal field theories have the same partition function.

(S2) There exist some intrinsic criteria which specify a linear subspace of the local analytic sections of $L_c \to M$ such that *every* global analytic section satisfying Z those local criteria and also satisfying conditions (2)-(4) above is the partition function of some unitary conformal field theory. That is, conditions (1)-(4) plus some appropriate local analyticity condition suffice to permit reconstructing a unique conformal field theory from Z. (The idea that unitarity should follow automatically was not suggested in reference 1. Some support for the idea is given below.)

Given (S1) and (S2), the space CFT_c is exactly the space of analytic (in an as yet unspecified sense) sections of $L_c \to M$ which are morphisms of $*$-semigroups. Not least of the remarkable consequences would be that any such section, restricted to genus 1 surfaces, would have an expansion in powers of q and $\bar{q}$ with nonnegative integer coefficients, since the coefficients would be the multiplicities of representations of the Virasoro algebras.

Neither of these suppositions has been proved (but see reference 1 and below for some arguments in their favor). Supposition (S1) is a reasonable conjecture – it is precisely stated and there are no known counterexamples. (S2) is not a precise conjecture and the essential missing ingredient which must be supplied before it becomes one is the appropriate local analyticity criterion.

It is now a simple formal exercise to translate conditions (1)-(4) into a construction of a commutative $*$-algebra A_c whose spectrum is CFT_c. Let $\Gamma(M, L_c)$ be the linear space of global analytic sections of $L_c \to M$ satisfying the as yet unspecified local analyticity condition. Define A_c to be the dual linear space $\Gamma(M, L_c)^*$ so that $\Gamma(M, L_c) = A_c^*$.

To see that A_c is a commutative $*$-algebra with identity, first construct a dual multiplication on $\Gamma(M, L_c)$. For $s \in \Gamma(M, L_c)$ define $m^*s \in \Gamma(M, L_c) \otimes \Gamma(M, L_c)$ by setting $m^*s(\Sigma_1, \Sigma_2) = s(\Sigma_1\Sigma_2)$. This makes sense because $(L_c)_{\Sigma_1} \otimes (L_c)_{\Sigma_2} = (L_c)_{\Sigma_1\Sigma_2}$. Now define the product of $\alpha_1, \alpha_2 \in A_c$ by $(\alpha_1\alpha_2, s) = (\alpha_1 \otimes \alpha_2, m^*s)$ for all $s \in \Gamma(M, L_c)$. The identity $1 \in A_c$ is given by $(1, s) = s(\mathbf{P})$ for all $s \in \Gamma(M, L_c)$. The conjugation in A_c is given by $(\alpha^*, s) = (\alpha, s^*)^*$.

Since $\Gamma(M, L_c) = A_c^*$, conditions (1)-(4) on the partition function are equivalent to stating that the partition function of a conformal field theory is an element $Z \in A_c^*$ satisfying

(i) $Z(1) = 1$,

(ii) $Z(\alpha^*) = Z(\alpha)^*$,

(iii) $Z(\alpha_1\alpha_2) = Z(\alpha_1)Z(\alpha_2)$.

The elements of A_c^* satisfying (i)-(iii) are the characters of A_c or, equivalently, the irreducible representations or maximal ideals. The space of all irreducible representations is the spectrum $Spec(A_c)$. A_c can be interpreted as the commutative $*$-algebra of functions on its spectrum.

The suppositions (S1) and (S2) now amount to the suggestion that $CFT_c = Spec(A_c)$. This characterization of CFT_c is extremely formal and depends on the imprecise and unproved suppostitions (S1) and (S2). If $\Gamma(M, L_c)$ can be defined precisely, it should then be possible to put a norm or norms on it such that $Spec(A_c)$ becomes a topological space or even a real analytic space. This would be a precise expression of the idea that conformal field theories are close together in CFT_c if their partition functions are close on M.

From what is known of examples, it seems possible that CFT_c is in fact an algebraic variety. It might well be that the solvable conformal field theories are too special. But, for example, the gaussian models, a subset of CFT_c for $c = n$ an integer, form the algebraic variety $O(n,n,\mathbf{Z})\backslash O(n,n,\mathbf{R})/O(n,\mathbf{R}) \times O(n,\mathbf{R})$. It is difficult to see how the present approach would provide such an algebraic structure for $Spec(A_c)$.

This formalism generalizes readily to give the space $SCFT_c$ of superconformal field theories entirely in terms of intrinsic structure in a line bundle over the universal moduli space of super Riemann surfaces[7]. A concrete result in this direction could be of some conventional mathematical interest, since the boundary of $SCFT_c$ consists of the Calabi-Yau spaces of dimension c in the limit of large volume. Reconstructing the superconformal field theories would lead to a construction of the Calabi-Yau metrics. This would be done by considering the conformal fields whose weights approach zero as the field theory approaches the boundary of $SCFT_c$. In the limit, these fields form a commutative associative operator product algebra whose spectrum is the Calabi-Yau space. The conformal weights of these fields approach the eigenvalues of the laplacian on the Calabi-Yau space, from which the metric could be reconstructed, in principle. Reference 6 provides the dictionary, in the large volume limit of a manifold, between the eigenfunctions of the laplacian on the manifold and the fields whose dimensions are the eigenvalues of the laplacian, and between the multiplication of functions on the manifold and the operator products of the corresponding fields.

Comments on unitarity

It might seem surprising to suggest that no additional positivity conditions are needed to ensure unitarity of the reconstructed conformal field theory. There are some obvious positivity conditions which are consequences of unitarity. Suppose Σ to be a doubled surface – a closed surface made by gluing a surface with boundary to its complex conjugate surface. The partition function of a unitary conformal field theory is positive at such a surface Σ. However, it seems impossible to describe the space of doubled surfaces in terms intrinsic to M. Even if this were possible, it is hard to see how such a positivity condition would imply unitarity, at least in as straightforward a way as the reflection positivity condition implies unitarity in euclidean quantum field theory.

Physical considerations suggest that *no* positivity conditions are needed to ensure unitarity. The normalization condition $Z(1) = 1$ and the nonsingularity of Z together imply that a Landau-Ginsburg model exists. That is, the logarithm of the partition function is finite and thus can be expanded in derivatives of order parameters. The reality condition $Z(\Sigma^*) = Z(\Sigma)^*$ implies that the Landau-Ginsburg model will be CPT invariant. By power counting, any such Landau-Ginsburg effective action contains no more than two derivatives of the order parameters and is manifestly unitary.

The available evidence supports this argument. The normalization condition $Z(1) = 1$ and the nonsingularity of Z on M ensure that any reconstructed conformal field theory would have a unique ground state and nonnegative conformal weights. Every known conformal field theory with these two properties is in fact unitary.

Comments on the suppositions

A sketch of a method for substantiating (S1) was given in reference 1. The partition function of a conformal field theory can be expanded in powers of the coordinates q_i and $\bar{q}_i$ which parametrize the opening of nodes. The coefficients in these q-expansions are sums of products of correlation functions of local fields. It should be possible to reconstruct the correlation functions from these coefficients.

The situation is especially simple when the representations of the conformal algebra occur with multiplicity at most 1. The partition function in genus 1 gives the conformal weights. The partition function in genus 2 gives the squares of the 3-point correlation functions and thus determines the operator product coefficients up to signs. It seems plausible that enough information is available at higher genus to fix the signs (up to Z_2 symmetries of the theory). This would determine the theory completely.

A sketch of an argument in favor of supposition (S2) was also given in reference 1. Any section $Z \in \Gamma(M, L_c)$ satisfying conditions (2)-(4) will have q-expansions from whose coefficients correlation functions can be extracted. These will satisfy – by virtue of (2)-(4) and analyticity – the axioms of conformal field theory. The correlation functions of the analytic stress tensor will then be derived by taking derivatives on M.

Comments on $\Gamma(M, L_c)$

If we admit as the sections of $L_c \to M$ all the real analytic sections which can be expressed locally as finite sums of analytic times analytic functions on M, then $Spec(A_c)$ will be the space of so-called *rational* unitary conformal field theories - modulo (S1) and (S2). In the example of the gaussian models, the rational conformal field theories are the rational points in the space of conformal field theories (this was one motivation for the nomenclature). In these special theories it is possible to describe the partition function as a sesquilinear pairing of holomorphic sections of finite rank projectively flat vector bundles over M, as in reference 1. Reference 8 reviews the considerable progress which has been made towards classifying the rational theories.

The rank of the projectively flat vector bundle jumps wildly even when the conformal field theory - the partition function - changes very little. Since our object is to characterize the whole space of conformal field theories, the vector bundle language seems inappropriate. There might be, however, a way to specify $\Gamma(M, L_c)$ based on some analytically precise notion of sesquilinear pairings of sections of infinite rank projectively flat vector bundles. The rational theories would merely be special points at which the infinite rank bundle degenerates to finite rank.

It might seem natural to take as $\Gamma(M, L_c)$ all nonsingular real analytic sections of L_c which have expansions in powers of q and $\bar{q}$. There seems, however, to be a difficulty here. Suppose that Z is the partition function of a conformal field theory with central charge c, such that Z is nowhere zero on M - for example, a partition function defined via a functional integral with positive measure. Any real power Z^γ would be nonsingular and real analytic, would have a q expansion and would satisfy conditions (1)-(4) with central charge γc. But we know that unitarity permits only a discrete set of values of the central charge less than 1. Thus either a simple real analyticity condition for $\Gamma(M, L_c)$ is too weak, or unitarity requires additional conditions beyond (1)-(4).

4. The space of classical string ground states

The classical ground states of string theory differ from conformal field theories in a few respects.

First, the partition function of string theory cannot be normalized to 1 on the 2-sphere. Its value on the 2-sphere is $Z(\mathbf{P}) = \lambda^{-2}$ where λ is the string coupling constant, which is a parameter of the classical ground state. To take the string coupling constant into account it is necessary to define an augmented universal moduli space M_+.

Let $\widetilde{M}_+$ be the compact, not necessarily connected Riemann surfaces without boundary which are smooth except for at most a finite number of nodes and which have no components of genus 0 (no components which are 2-spheres). The empty surface is included in $\widetilde{M}_+$.

For $\Sigma \in \widetilde{M}_+$, let $\nu_+\Sigma$ be the smooth surface obtained by removing the nodes in Σ and discarding all the resulting components of genus 0. Write $\chi(\Sigma)$ for the Euler number of Σ. The difference $[\chi(\nu_+\Sigma) - \chi(\Sigma)]/2$ is equal to the number of nodes removed from Σ minus the number of genus 0 components discarded. It is always nonnegative.

Introduce a new, abstract element $x = x^*$ with Euler number $\chi(x) = -2$. This new element will play the formal role of $\mathbf{P}^{-1}$ in the semigroup. Let $F(x) = \{1, x, x^2, \ldots\}$ be the free commutative $*$-semigroup on x. Define an equivalence relation on $F(x) \times \widetilde{M}_+$ by requiring that

$$x^n \Sigma \sim x^{n+[\chi(\nu_+\Sigma)-\chi(\Sigma)]/2} (\nu_+\Sigma) . \tag{6}$$

The augmented universal moduli space M_+ is defined to be the quotient $F(x) \times \widetilde{M}_+/\sim$.

As before, M_+ is a $*$-semigroup with identity. The product operation is again the disjoint union of Riemann surfaces. But now the identity element is the empty surface, which is equivalent to the singular torus, since removing the node from the singular torus leaves $\mathbf{P}$, which is discarded to give the empty surface. $F(x)$ is a sub-semigroup of M_+. The previously defined universal moduli space M is the quotient $M_+/F(x)$.

M_+ is analytic but it is not connected – in fact it is the infinite symmetric product of the union of all the Satake compactifications M_g^{sat}. The Euler number is a well-defined continuous morphism from $\widetilde{M}_+$ to the additive semigroup of nonpositive even numbers. The connected components of $\widetilde{M}_+$ are the sets of fixed Euler number.

The partition function Z_{str} of a classical string ground state is a section of the line bundle $L_+ \to M_+$, where L_+ is the line bundle of densities or volume elements on M_+. In terms of holomorphic objects, $L = K_{M_+} K^*_{M_+}$, where K_{M_+} is the canonical line bundle of M_+ – the determinant of the holomorphic cotangent bundle. Suppose (q, x_1, x_2) parametrizes a node in a Riemann surface Σ. A local holomorphic section ω of K_{M_+} has a double pole at $q = 0$ of the form

$$\omega(q, x_1, x_2, \Sigma) = q^{-2} dq dx_1 dx_2 \, \omega(\Sigma) + O(q^{-1}) . \tag{7}$$

This is the only definition of the canonical bundle which respects the equivalence relation used to define M_+.

The line bundle $L_+ \to M_+$ is again a morphism of $*$-semigroups. The space CGS of string classical ground states is – modulo the familiar suppositions – the space of all sections $Z_{str} : M_+ \to L_+$ which are $*$-semigroup morphisms.

The coupling constant is given by $Z_{str}(x) = \lambda^2$. The coupling constant is a free parameter of the classical ground state because the Euler number is continuous on M_+. This allows the string partition function to be multiplied by $(\lambda'/\lambda)^{\chi(\Sigma)}$, changing the coupling constant from λ to λ'.

Just as before, the linear space $A_+ = \Gamma(M_+, L_+)^*$ is a $*$-algebra with identity and (conjecturally) $CGS = Spec(A_+)$.

The algebra A_+ is related to the algebra A_{26} by the exact sequence

$$0 \to \mathbf{C}(x) \to A_+ \to A_{26} \to 0\,. \tag{8}$$

This follows from the fact that the central charge of the conformal ghost system is -26.

In order to obtain the analogous abstract characterization of the classical ground states of fermionic string theory, the augmented universal moduli space of ordinary Riemann surfaces should be replaced by the analogous construction for super Riemann surfaces[7].

The string construction is completely intrinsic to the space M_+. Once M_+ is obtained, the fact that the points of M_+ represent Riemann surfaces can be forgotten. Abstracting the string ground state away from the notion of Riemann surface might be desirable, since the interpretation as a theory of strings might make sense only at weak coupling.

In reference 2 it was suggested that the quantum ground state of string might be described in the same language as the classical ground states, after "completing" the universal moduli space, now M_+, to include "infinite genus" surfaces. There is still nothing particularly useful to say about this suggestion, except possibly that the completion of M_+ ought to be connected. Then the Euler number would no longer be continuous and the coupling constant no longer arbitrary.

A more accessible problem might be to construct the perturbative string S-matrix in this abstract approach. There would have to be some way to understand

Wick rotation abstractly (perhaps as analytic continuation in *CGS*). It might also be interesting to try to extend the abstract characterization of the classical ground states described in this note to an analogous characterization of at least the perturbative quantum ground states, to see if the formal simplicity can be maintained.

Acknowledgments

Versions of this note were presented in a variety of forums including the Schloss Ringberg Workshop in March, 1987, the Trieste Spring School in April, 1987, the El Escorial School in June, 1987, the 1987 Cargese Summer School, the Yukawa Symposium in October, 1987, lectures at the Princeton Institute for Advanced Study in 1987-8, the Cargese workshop on Field Theory and Statistical Mechanics in June, 1988, the 1988 Les Houches Summer School, the Nankei University Symposium in September, 1988, and the London Royal Society Workshop on Geometry and Physics in December, 1988. I thank all of the institutions and organizers for their hospitality.

I am grateful to T. Banks, S. Bloch, A. Borel, P. Deligne, D. Kazhdan, E. Martinec, B. Mazur, G. Moore, D. Quillen, G. Segal, S. Shenker, D. Smit and E. Witten, among others, for helpful discussions and suggestions.

REFERENCES

1. D. Friedan and S. H. Shenker, *Nucl. Phys.* B281 (1987) 509.

2. D. Friedan and S. H. Shenker, *Phys. Lett.* 175B (1986) 287.

3. D. Friedan, in *Unification of Fundamental Interactions, Physica Scripta* T15 (1987) 78.

4. A. A. Belavin and V.G. Knizhnik, *Phys. Lett.* 168B (1986) 201.

5. W.L. Baily, *Ann. of Math.* 71 (1960) 303.

6. D. Friedan, *Phys. Rev. Lett.* 45 (1980) 1057, *Ann. Phys.* 163 (1985) 318.

7. M. A. Baranov and A. S. Schwarz, *Pis'ma Zh. Eksp. Teor. Fiz.* 42 (1985) 340 [*JETP Lett.* 42 (1985) 419]; D. Friedan, in *Proceedings of the Workshop on Unified String Theories, Institute for Theoretical Physics, Santa Barbara, July 29– August 16, 1985*, M. Green and D. Gross (eds.), World Scientific (1986); J. Cohn, *Nucl. Phys.* B306 (1988) 239.

8. See the review by G. Moore and N. Seiberg in *Proceedings of the 1989 Banff Summer School,* ed. H.C. Lee and in *Proceedings of the 1989 Trieste Spring School,* and references therein.

www.ingramcontent.com/pod-product-compliance
Lightning Source LLC
Chambersburg PA
CBHW070712220326
41598CB00026B/3694